Biology of Extracellular Matrix

Volume 16

AF812746

Extracellular matrix (ECM) biology, which includes the functional complexities of ECM molecules, is an important area of cell biology. Individual ECM protein components are unique in terms of their structure, composition and function, and each class of ECM macromolecule is designed to interact with other macromolecules to produce the unique physical and signaling properties that support tissue structure and function. ECM ties everything together into a dynamic biomaterial that provides strength and elasticity, interacts with cell-surface receptors, and controls the availability of growth factors. Topics in this series include cellular differentiation, tissue development and tissue remodeling. Each volume provides an in-depth overview of a particular topic, and offers a reliable source of information for post-graduates and researchers alike.

All chapters are systematically reviewed by the series editor and respective volume editor(s).

"Biology of Extracellular Matrix" is published in collaboration with the American Society for Matrix Biology and the International Society for Matrix Biology.

Antonio Rossi • Frank Zaucke
Editors

The Extracellular Matrix in Genetic Skeletal Disorders

Springer

Editors
Antonio Rossi
Department of Molecular Medicine, Unit of Biochemistry
University of Pavia
Pavia, Italy

Frank Zaucke
Department of Trauma Surgery and Orthopedics
University Hospital Frankfurt, Goethe University
Frankfurt am Main, Germany

ISSN 0887-3224 ISSN 2191-1959 (electronic)
Biology of Extracellular Matrix
ISBN 978-3-031-70834-3 ISBN 978-3-031-70835-0 (eBook)
https://doi.org/10.1007/978-3-031-70835-0

This Springer imprint is published by the registered company Springer Nature Switzerland AG
The registered company address is: Gewerbestrasse 11, 6330 Cham, Switzerland

Preface

The extracellular matrix (ECM) of cartilage and bone is composed of different collagen types, a variety of proteoglycans and a large number of glycoproteins. These components form a highly specialized three-dimensional macromolecular network providing tissues not only with structural integrity determining mechanical properties but also a communication hub essential for cartilage and bone function.

Cartilage forms most parts of the skeleton in mammalian embryos before bone formation begins. Bone development is linked to cartilage since several bone elements are formed from a cartilage anlage through endochondral ossification. Cartilage persists through adulthood at the extremities of bones at the articular joints where it serves as a lubricated, wear-resistant, friction-reducing surface designed to distribute load and allow for resilience. Bones support the body and facilitate movements. To allow this, the bone matrix consists of a high degree of mineralization that is required for the formation of a rigid functional structure.

In addition to forming a complex architecture around the cell, ECM molecules mediate cell signaling that is important to regulate tissue development, growth, and homeostasis.

The importance of matrix molecules for normal tissue functions is highlighted by mutations in matrix genes resulting in genetic disorders with diverse clinical phenotypes. The characterization of mutations and their correlation with the pathology is crucial for a better patient prognosis and in order to devise potential therapeutic approaches. Moreover, the study of these disorders using patient biopsies, *in vivo* and *in vitro* models has been instrumental in elucidating the function of many ECM components and the interactions among them.

The ECM is not a static structure but a dynamic network of molecules synthesized and secreted by the cells, in particular during remodeling and repair processes. Recent studies demonstrate that mutations altering ECM protein structure affect also their synthesis and secretion thus impacting cell functions and homeostasis.

This book aims at providing the readers with an overview on how qualitative and quantitative defects in ECM proteins, in their post-translational modifications and intracellular trafficking might affect cartilage and bone integrity and function. Given the wide range of underlying mechanisms causing skeletal disorders, the book has been divided into three sessions. Session I "ECM proteins and skeletal disorders," including Chaps. 1 to 7, describes how structural alterations or defects in post-translational modifications of ECM proteins impact the skeletal system. A comprehensive overview of ECM proteins involved in cartilage and bone disorders is given in Chap. 1 by Maurizi and Teti. Defects in the major fibril-forming collagens, type I and type II collagen, result in bone and cartilage disorders, respectively. Chapter 2 by Gatto et al describes mutations in genes that lead to a reduced quality and/or quantity of collagen type I in skeletal and extraskeletal tissues, while Chap. 4 by Bateman and Lamandé focuses on type II collagenopathies. The Ehlers-Danlos syndromes, described in Chap. 3 by Malfait et al., encompass a heterogeneous group of heritable conditions caused by defects in fibrillar collagens itself or in proteins involved in collagen processing, folding, and cross-linking or in enzymes involved in the biosynthesis of glycosaminoglycans. The synthesis and sulfation of glycosaminoglycans, the oligosaccharide components of proteoglycans, involves several enzymes and transporters whose defects result in altered ECM properties causing different skeletal disorders reviewed in Chaps. 5 and 6 by Uber et al and by Gramegna Tota et al, respectively. Fibrillins are ECM glycoproteins that form supramolecular networks that are also involved in TGF-β signaling. Disorders caused by mutations in fibrillins are described by Morcos and Sengle in Chap. 7.

The ECM acts also as a reservoir of growth factors affecting signal transduction pathways crucial for cartilage and bone which are described in Session II "ECM signaling and skeletal disorders" (Chaps. 8-10). Altered FGF signaling resulting in skeletal disorders is described in Chap. 8 by Fayad and Legeai-Mallet, while the focus is set on Wnt signaling in Chap. 9 by Yorgan and Schinke. The primary cilium is considered a primary antenna for the cells important not only in the reception of extracellular signals but also in mechanotransduction. Mutations in genes encoding for proteins of the primary cilium, described in Chap. 10 by Wann, disrupt several cellular processes including the regulation of the composition, secretion, remodeling, and organization of the ECM.

Finally, Chaps. 11 to 13 in Session III "Intracellular trafficking of ECM proteins and related skeletal disorders," provide relevant examples of how disruption of microtubule-associated molecular motors and of the membrane trafficking machinery can also lead to skeletal dysplasias. Chapter 11 by Brito et al. discusses the microtubule-associated molecular motors and their emerging roles in skeletal dysplasia. Chapter 12 by Ferreira describes altered vesicular trafficking from the Golgi to the ER in Saul-Wilson syndrome and Chap. 13 by Guarino et al. focuses on mutations in proteins involved in membrane trafficking, thus disrupting correct delivery of ECM proteins.

All chapters were written by renowned experts who contributed key discoveries about the molecular basis of skeletal disorders and are involved in the development of therapeutic strategies for the patients. We thank all authors for their valuable contribution to this book and express our gratitude also to Prof. Nikos Karamanos, Series Editor, Dr. Alina Cepraga, Associate Editor, and the Editorial Board Members for their revision and valuable comments in the whole process of preparing the book.

As Editors, we believe that this book will be a valuable resource not only for students, junior and advanced researchers but also clinicians from different disciplines interested in getting a broader view about this growing and exciting field. We really hope you will enjoy reading this new contribution to the book series Biology of Extracellular Matrix.

Pavia, Italy — Antonio Rossi
Frankfurt am Main, Germany — Frank Zaucke

Contents

Part I
ECM Proteins and Skeletal Disorders

Chapter 1
Extracellular Matrix and Skeletal Dysplasias

Antonio Maurizi and Anna Teti

Abstract Skeletal dysplasia is a group of inherited rare diseases affecting bone and cartilage, in some cases along with other tissues, including the connective tissue proper. Several skeletal dysplasias are caused by mutations of genes encoding extracellular matrix (ECM) proteins, which are inherited with autosomal recessive, autosomal dominant or X-linked patterns. In this chapter, we describe the composition, the structure and the function of connective tissue proper, cartilage and bone ECM, identifying mutant genes primarily causing skeletal dysplasias. The chapter is meant to walk the reader from physiology to pathology, explaining the molecular mechanisms governing ECM biology and the genetic modifications involved in skeletal dysplasias caused by ECM alterations.

1.1 Introduction

Extracellular matrix (ECM) is a complex cellular product released in the extracellular microenvironment. Therefore, even if it is distributed between cells, ECM must be considered an integral cellular component that affects not only the microenvironmental functions but also the cellular metabolism. Many tissues, including epithelial, muscle and nervous tissues, are characterised by scarce ECM, while connective tissues are typically constituted by large amount of ECM, with cells distributed several microns apart from each other. Typically, ECM is innervated and vascularised, generally lodging both blood and lymphatic vessels. However, cartilage is refractory to innervation and vasculogenesis, the latter until it mineralises and is replaced by bone tissue (Gentili and Cancedda 2009).

A. Maurizi · A. Teti (✉)
Department of Biotechnological and Applied Clinical Sciences, University of L'Aquila, L'Aquila, Italy
e-mail: antonio.maurizi@univaq.it; annamaria.teti@univaq.it

A. Rossi, F. Zaucke (eds.), *The Extracellular Matrix in Genetic Skeletal Disorders*, Biology of Extracellular Matrix 16,
https://doi.org/10.1007/978-3-031-70835-0_1

ECM is abnormal in many diseases, including cartilage and bone diseases that, all together, are classified as skeletal dysplasias (Unger et al. 2023). In this chapter, we will describe the features of ECM in connective tissue proper, cartilage and bone and will introduce the general concepts of ECM anomalies in skeletal dysplasias.

1.2 ECM General Features

ECM is constituted by ground substance and fibrillar components. Ground substance is formed by a series of proteins, mainly glycoproteins, and by glycosaminoglycans and proteoglycans. They form an amorphous matrix, resulting pale in histological sections because of refractory to conventional dyes due to its high carbohydrate content. In contrast, the ground substance stains well with metachromatic dyes, such as toluidine blue, due to its polyanion enrichment. It is also visible by periodic acid-Schiff, alcian blue and safranin staining, which react with glycoproteins, glycosaminoglycans and proteoglycans (Karamanos et al. 2021).

The ground substance lodges the interstitial fluid leaking out from the capillaries, through which oxygen and nutrients reach the cells, while carbon dioxide and waste products are removed when the fluid returns into the circulation. It also represents a reservoir of many cytokines, growth factors, hormones and other regulators of cell functions, including ectoenzymes. The ground substance is essential for cellular homeostasis and for the innate and acquired immune response (Le and Mayor 2023).

The ground substance fills the space between the fibrillar components of the ECM. These latter represent a kind of mechanical and functional support for cells and ground substance components. Major fibrillar material is represented by collagens. At least 28 different collagen types are known in vertebrates, classified in fibrillar, fibril-associated, network-forming and membrane-associated collagens (Mienaltowski et al. 2021) (Table 1.1).

The *fibrillar collagen* class includes seven collagen types distributed in connective tissue proper, bone and cartilage (Table 1.1) (Holmes et al. 2018). Ninety percent of collagen in the body is represented by fibrillar type I collagen, an acidophilic glycoprotein that stains with conventional acidic dyes, including eosin, fast green and aniline blue (Fig. 1.1).

Type I collagen forms by assembly of triple-helical tropocollagens. In the cell, tropocollagen is protected from auto-assembly by propeptides, which are cleaved after secretion by ecto-pro-peptidases and assemble in fibrils showing typical transverse 67 nm periods. This structure can remain in fibrillar form (1 μm diameter) or fibrils can assemble in fibres (up to 10–12 μm in diameter). The ratio between the ground substance and type I collagen fibres distinguishes the various connective tissue proper variants: loose (or areolar) and dense (or compact). The loose connective tissue is enriched in ground substance and cells and poor in fibres, while the dense tissue is especially enriched in fibres, with lower number of cells, primarily resident fibroblasts, macrophages and mast cells. Typically, fibres are irregularly distributed in all directions in the loose connective tissue, while two variants of

Table 1.1 Collagen classes, types, formulas and tissue distribution in vertebrates

Collagen class	Collagen type	Formula	Tissue distribution
Fibrillar	I	$[\alpha1(I)]_2\alpha2(I)$	Bone, cornea, ligaments, skin, tendons
	II	$[\alpha1(II)]_3$	Cartilage, eye vitreous humour
	III	$[\alpha1(III)]_3$	Blood vessels, skin
	V*	$[\alpha1(V)]_3[\alpha1(V)]_2\alpha2(V)\alpha1(V)\alpha2(V)\alpha3(V)$	Bone, dermis
	XI	$\alpha1(XI)\alpha2(XI)\alpha3(XI)$	Cartilage, intervertebral discs
	XXIV	$[\alpha1(XXIV)]_3$	Bone, cornea
	XXVII	$[\alpha1(XXVII)]_3$	Cartilage
Fibril-associated	VII	$[\alpha1(VII)]_2\alpha2(VII)$	Bladder, dermis
	IX	$\alpha1(IX)\alpha2(IX)\alpha3(IX)$	Cartilage, cornea
	XII	$[\alpha1(XII)]_3$	Dermis, tendons
	XIV	$[\alpha1(XIV)]_3$	Bone, cartilage, dermis
	XVI	$[\alpha1(XVI)]_3$	Dermis, kidney
	XIX	$[\alpha1(XIX)]_3$	Basement membrane
	XX	$[\alpha1(XX)]_3$	Cornea (in chicks)
	XXI	$[\alpha1(XXI)]_3$	Kidney, stomach
	XXII	$[\alpha1(XXII)]_3$	Myotendinous junctions
	XXVI	$[\alpha1(XXVI)]_3$	Gonads
Network-forming	IV*	$[\alpha1(IV)]_2\alpha2(IV)\alpha3(IV)\alpha4(IV)\alpha5(IV)[\alpha5(IV)]_2\alpha6(IV)$	Basement membrane
	VI**	$\alpha1(VI)\alpha2(VI)\alpha3(VI)\alpha1(VI)\alpha2(VI)\alpha4(VI)$	Cartilage, cornea, dermis, muscle
	VIII*	$[\alpha1(VIII)]_3[\alpha2(VIII)]_3[\alpha1(VIII)]_2\alpha2(VIII)$	Brain, heart, kidney, brain
	X	$[\alpha1(X)]_3$	Cartilage
	XXVIII	$[\alpha1(XXVIII)]_3$	Dermis, Ischiatic nerve
Membrane-associated	XIII	$[\alpha1(XIII)]_3$	Capillaries, eyes, dermis
	XVII	$[\alpha1(XVII)]_3$	Hemi desmosomes
	XXIII	$[\alpha1(XXIII)]_3$	Heart, retina
	XXV	$[\alpha1(XXV)]_3$	Brain, heart, testis
	XV	$[\alpha1(XV)]_3$	Capillaries, heart, kidney, testis
	XVIII	$[\alpha1(XVIII)]_3$	Basement membrane, liver

*Three isoforms known; **Two isoforms known

dense connective tissues are known: irregular and regular. In the dense irregular tissue, fibres run in all directions, while in the dense regular tissue, they are ordered in parallel in one direction, such as in tendons, or are ordered in nearly orthogonal layers of parallel fibres, such as in tympanic membrane, cornea and lamellar bone (Siadat and Ruberti 2023).

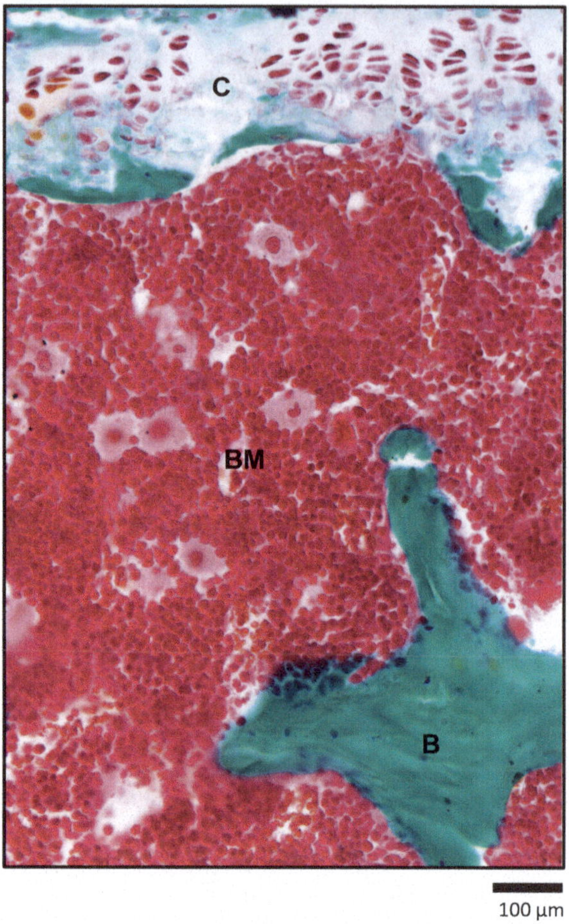

Fig. 1.1 *Masson-Goldner's trichrome staining* showing mouse collagen staining by the acidic dye Fast Green. *C* cartilage; *BM* Bone marrow; *B* Bone

Ten types of collagens are known in the *fibril-associated collagen* class (Table 1.1) (van der Rest et al. 1990). They are distributed in many organs and confer fibrillar collagen their specific organisation. Their structure presents one or two non-helical domains that interrupt the triple-helical structure. Therefore, the molecule is flexible, is not cleaved during its processing and retains the propeptides. Extracellularly, they bind in a periodic manner to fibrillar collagens, guiding fibril spatial organisation and assembly.

Ten *network-forming collagens* are known (Table 1.1) (Knupp and Squire 2003). They are enriched in basement membranes where they support and anchor the cells, representing molecular filters and protecting permeable barriers. Finally, three *membrane-associated collagens* (Table 1.1) (Pihlajaniemi and Rehn 1995) are type II transmembrane proteins with a short N-terminal cytosolic domain, a transmembrane domain with three collagenous sequences interrupted by non-collagenous sequences and an ectodomain that can be shed by furin convertases. The other three members of this group are in the basement membrane, indirectly associated with the plasma membrane through molecular bridges with other ECM components, such as laminin.

The connective tissue ECM also presents reticular and elastic fibres. *Reticular fibres* (Ushiki 2002) belong to the collagen group (type III collagen, Table 1.1) and are ubiquitous in connective tissue proper. However, they are enriched in the so-called reticular tissue, representing the delicate stroma surrounding cells in parenchymal organs, adipose tissue, endomysium and endoneurium. Reticular fibres cannot be detected by staining with conventional dyes, but they are argyrophilic and can be easily observed by silver stain.

Elastic fibres (Ushiki 2002) are formed by a protein called elastin and are ubiquitously distributed in connective tissue proper in small amounts, but when elasticity is necessary for tissue function, they dominate the fibrillar components and the connective tissue is then called elastic tissue. They are made by an amorphic mass of elastin surrounded by fibrillin-1 microfibrils (Halper and Kjaer 2014), which organize elastin to form fibres that interconnect with their branches forming a net-like structure. Elastic fibres are resistant to enzymatic degradation and offer elasticity and resistance to the tissue. Typical connective tissues enriched in elastic fibres are the tunica media of elastic blood vessels, such as the aorta and pulmonary arteries, the vocal cords, the trachea, the bronchial tubes, the derma and various ligaments. Elastic fibres are refractory to conventional histological staining but can be identified by special dyes. These include orcein, which binds to elastic fibres by non-ionic mechanism, and iron–haematoxylin that forms a variety of cationic, anionic and non-ionic bonds with elastin.

The connective tissue ECM is synthesized and released by fibroblasts (Talbott et al. 2022), spindle-shaped cells with elongated and euchromatic nuclei, large amounts of organelles and membrane processes through which they adhere to the ECM components by integrin-mediated interactions. They produce most components of the ECM, and at the end of the ECM formation, their nuclei turn heterochromatic, organelles diminish and cells become quiescent, prone to be reactivated when tissue regeneration is required.

Fibroblasts originate from mesenchymal stem cells (Ding et al. 2011). After differentiation, they can become plastic, and to maintain correct tissue homeostasis, they are able to transdifferentiate into other mesenchymal cell types, including adipocytes, smooth muscle cells, chondrocytes and osteoblasts. They also contribute to ECM remodelling, releasing ectoenzymes and cytokines involved in the regulation of ECM homeostasis and regeneration.

1.3 Cartilage ECM

Cartilage is a specialized connective tissue whose unique ECM composition allows cells to survive in an avascular and aneural microenvironment (Gentili and Cancedda 2009). It forms the rudiment of the axial and appendicular skeleton that will be replaced by bone through the endochondral ossification process (Long and Ornitz 2013). It also forms the diarthrosis articular cartilages; the costal cartilages; the fibrocartilage in the pubis, in the intervertebral discs and in the knees; the larynx cartilages; the tracheal rings; the bronchial rings and plaques; the auricle cartilages; and the nasal cartilages (Umlauf et al. 2010). Finally, it forms the growth plate, which ensures the longitudinal growth of the skeleton, and the endochondral callus to repair fractures. Cartilage has a slow turnover and repair at slow rate relative to other tissues.

Cartilage ECM is formed by the chondrocytes (Archer and Francis-West 2003), deriving from mesenchymal stem cells upon expression of genes encoding chondrogenic factors, including Sox9 transcription factor, type II collagen, aggrecan and cartilage oligomeric matrix protein (COMP). They synthesize both the ground substance and the fibres and lay in cartilage lacunae. They divide to form isogenic groups of cells, producing new matrix for the interstitial growth, or generate from chondrogenic progenitors of the inner perichondrium layer, allowing the apposition of new matrix at the cartilage border.

Cartilage is an avascular tissue because blood vessels would compromise its mechanical properties. This feature is due to the enrichment in anti-angiogenic factors, including chondromodulin (Shukunami and Hiraki 2007) and endostatin (Pufe et al. 2004), which prevent vasculogenesis, making the cartilage ECM a unique extracellular environment in which chondrocytes can survive because of their peculiar cellular metabolism.

Chondrocytes are long-living cells, which are flattened or rounded depending on their peripheral or deep position, respectively (Archer and Francis-West 2003). Flattened cells are in the juxta-perichondrium zone, oriented parallel to the collagen fibrils they secrete, while in the deep zone, chondrocytes are larger and rounded, with a random distribution between irregularly oriented type II collagen fibrils. In the growth plate, they form columns parallel to the long axis of the bone, along the line of the collagen fibrils.

Despite the avascular environment, chondrocytes are very active and synthesize all ECM components. However, deeper chondrocytes have access to low oxygen and contain a few mitochondria, relying on anaerobic metabolism to produce ATP. Here, hypoxia-inducible factor 1 is a major regulator of the chondrocyte hypoxic response and is essential for chondrocyte growth arrest and survival. When, at the end of their lifespan, chondrocytes become hypertrophic and release pro-angiogenic factors, such as vascular endothelial growth factor (VEGF) and Indian hedgehog, they trigger vasculogenesis and subsequent ossification (Stegen et al. 2019).

Chondrocytes synthesize all cartilage ECM components. These include fibres and ground substances. According to the type of fibres, cartilage is classified into three categories: hyaline, elastic and fibrous.

Hyaline cartilage, so-called because of its pale colour and glassy aspect, is the most abundant cartilage variant (Malinin and Malinin 1999) (Fig. 1.1). Fibrillar components are mainly represented by *type II collagen*, which form isolated fibrils that do not assemble in fibres. They are bound to the fibril-associated *type IX collagen*, which also serves as a core protein for a single glycosaminoglycan side chain. Type II collagen is especially abundant in the so-called interterritorial matrix, located far from the chondrocyte lacunae. Another fibrillar component of hyaline cartilage is *type XXVII collagen*, particularly expressed during skeletal development.

Type XIV collagen is another fibril-associated collagen, often expressed in tissues subjected to high mechanical stress, indicating a potential protective role. In fact, hyaline cartilage is found in the larynx, trachea, bronchial tubes, nose, epiphyseal growth plate, sternum and costal ribs, where it shows a resistant surface with minimal friction. It is also found in diarthrosis joints, where it exhibits resistance to compression.

Finally, *type VI and type X collagens* are network-forming collagens, the former stabilising the chondrocyte membrane and the latter expressed in hypertrophic chondrocytes during the process of endochondral ossification.

The other two types of cartilage present additional fibrous elements. The *elastic cartilage* has a yellowish appearance and is found in the larynx and auricles. Its name derives from the high content of *elastic fibres* that provide flexibility and resistance to pressure. *Fibrocartilage* (Pirozzi 2022) is a cartilage type whose resistance is potentiated by *type I collagen* fibres. It can resist to high degrees of tension and compression and is found in the pubic symphysis, in the intervertebral discs and in the menisci. Interestingly, fibrocartilage has no perichondrium.

Besides fibres, cartilage ECM presents an abundant ground substance that masks the fibrillar components especially in hyaline cartilage. Ground substance is mainly formed by GAGs, proteoglycans and glycoproteins, and for this reason, it is pale if stained with conventional dyes, but stains well with metachromatic dyes, periodic acid-Schiff base, alcian blue and safranin red for its high carbohydrate content.

GAGs are linear polysaccharide chains made by disaccharide units (Zhang et al. 2010), in which one monomer of the two is always an amino sugar, either *N*-acetylglucosamine or *N*-acetylgalactosamine, often sulphated. The other monomer is usually a uronic acid, either glucuronic or iduronic acid. Sulphate and carboxyl groups confer GAGs highly negative charges, while sugars allow the classification of GAGs in four main groups. This classification considers the type of sugar linkage and the number and location of sulphates. Group 1 includes *hyaluronic acid*, group 2 *chondroitin sulphate and dermatan sulphate*, group 3 *heparan sulphate* and group 4 *keratan sulphate*. Because of their strong hydrophilicity, GAGs have an extended conformation occupying large volumes, thus forming a gel structure even at low concentrations. This way GAGs fill most of the extracellular space forming a highly hydrated gel. In fact, being highly negatively charged, they attract cations, especially sodium. This creates an osmotic pressure drawing water from the circulation, ending up in tissue turgor that makes the matrix resistant to compression.

Hyaluronic acid (also known as hyaluronan or hyaluronate) (Garg and Hales 2004) is made by non-sulphated disaccharides, not typical of other GAGs, repeated up to 25,000 times. Its molecular weight is over 3–4 million Daltons and, in most cases, it associates non-covalently with proteoglycan core proteins. At variance with the other GAGs that are secreted by exocytosis, hyaluronic acid is synthesized by the membrane-bound hyaluronan synthases (HASs). HASs assemble UDP-α-*N*-acetyl-D-glucosamine and UDP-α-D-glucuronate in dimers to form the nascent hyaluronic acid that is released at the cell surface through the HAS pore, growing extracellularly up to its final molecular weight. Hyaluronic acid binds link protein and aggrecan, forming the so-called aggrecan aggregates, which are very abundant in cartilage.

Hyaluronic acid is the only free GAG found in the cartilage ECM. All other GAGs are associated with core proteins to form proteoglycans. The link of GAGs to the core protein is ensured by a link tetra-saccharide that interacts with a serine radical. This linker serves as the primer for the GAG growth, with one sugar at the time added by specific glycosyl transferases. The sugars, which represent 95% of the proteoglycan weight, form long unbranched chain of around 80 monomers and are then modified by epimerisation or sulphation, this latter increasing the negative charge, thus attracting sodium and water. The major proteoglycan in cartilage has the core protein aggrecan (Nia et al. 2022) encoded by the *ACAN* gene, a mass of about three million Daltons and over 100 GAG chains. Other cartilage proteoglycans include the cell surface syndecans and glypican and the small leucine-rich proteoglycans decorin, biglycan, fibromodulin, lumican and epiphycan.

Besides proteoglycans, cartilage ECM also includes glycoproteins (Table 1.2). These contain 1–60% sugars forming relatively short-branched oligosaccharide chains, thus differing substantially from proteoglycans.

Thrombospondins are a family of glycoproteins with multiple functions, including anti-angiogenic activity. Among the five thrombospondin isoforms, thrombospondin 5 is also called COMP (Alford and Hankenson 2023).

Matrilins (Klatt et al. 2011) are proteoglycan-associated proteins. They form adapter oligomeric proteins, with isoforms 1 and 3 expressed in cartilage where they can form matrilin-3 subunit homodimers or matrilin-1/matrilin-3 heterodimers. They are highly expressed in cartilage and bone and play a role in skeletal development.

WARP is the acronym of von Willebrand factor A-domain-related protein (Fitzgerald 2020). The protein is expressed in portions of ECM, but its function has not been yet fully elucidated. Studies suggested that WARP may represent a fibril-associated collagen with interrupted triple helices that has lost the triple helix. It interacts with perlecan and shares an overlapping expression pattern with type VI collagen in the chondrocyte pericellular matrix.

Chondroadherin (Neame et al. 1999) is a leucine-rich repeat protein promoting cell attachment through binding to integrin $\alpha_2\beta_1$ and syndecans. In cartilage it forms a complex with type II collagen that is considered functionally relevant for the communication between cells and matrix and for the regulation of collagen fibril assembly. It also interacts with the N-terminal and C-terminal non-triple helical domains of type VI collagen.

Asporin, so-called for the presence of a unique aspartate-rich N-terminal sequence, is another leucine-rich repeat protein associated with the cartilage matrix (Fan et al. 2022). It binds type I collagen, competing for this function with decorin

Table 1.2 Cartilage ECM proteins

Protein	Function
Type II collagen	Isolated collagen fibrils abundant in interterritorial matrix
Type IV collagen	Network-forming collagen stabilising the chondrocyte membrane
Type IX collagen	Core protein for a single glycosaminoglycan side chain
Type X collagen	Network-forming collagen involved in endochondral ossification
Type XIV collagen	Fibril-associated collagen with potential mechanical stress protective role
Type XXVII collagen	Skeletal development
Thrombospondin-1	Cell–cell and cell–matrix interactions, antiangiogenic properties
Thrombospondin-2	Collagen fibrillogenesis, chondrogenic differentiation
Thrombospondin-3	Not known
Thrombospondin-4	Promoter of angiogenesis, inhibitor of collagen production
Thrombospondin-5	Collagen fibrillogenesis, growth factor binding
Matrilin-1	Adapter protein, cell attachment, inhibitor of neovascularisation
Matrilin-3	Adapter protein, growth factor binding
WARP	Adapter protein
Chondroadherin	Cell–cell and cell–matrix interactions
Asporin	Matrix protein and growth factor binding
PRELP	Adapter protein, regulator of osteoclastogenesis
UCMA	Control of osteogenic differentiation
CILP	Binding of TGF-β
Fibronectin	Cell–matrix interactions
Nidogens	Not known

and biglycan. It also binds calcium and is a decoy TGFβ1-binding protein that decreases TGFβ1-mediated chondrogenesis.

PRELP stands for proline/arginine-rich end leucine-rich repeat protein, representing a member of the small leucine-rich repeat ECM proteins (Bengtsson et al. 2002). Its function is to anchor the basement membrane to the connective tissue, binding both type I collagen and, in cartilage, type II collagen. PRELP has affinity for heparin and heparan sulphate and mediates cell adhesion to GAGs.

UCMA or "upper zone of growth plate and cartilage matrix-associated" protein is a chondrocyte protein expressed in foetal cartilage and growth plate (Seuffert et al. 2018). It contains glutamic acid residues that undergo vitamin K-dependent γ-carboxylation. Its role is likely to be associated with inhibition of osteogenic differentiation and transition of osteochondrogenic precursors to the osteoblast phenotype.

CILP is the cartilage intermediate layer protein whose high expression is associated with early osteoarthrosis (Neame et al. 1999). The CILP N-terminal domain represents a TGF-β1 and IGF-1 antagonist, playing a role in cartilage scaffolding by suppression of IGF-1-dependent sulphated proteoglycan synthesis and TGFβ-1-mediated induction of cartilage matrix genes.

Fibronectin is a well-known adhesion glycoprotein, recognised by integrins, that binds cells to ECM components, including collagen, fibrin and heparan sulphate proteoglycans (Mosher 1984). While soluble fibronectin is a circulating liver product, insoluble fibronectin is a member of the ECM, which forms a dimer involved in cell adhesion, growth, migration and differentiation.

Nidogens, also known as entactins, are sulphated glycoproteins present in ECM in two isoforms: nidogen-1 and nidogen-2 (Töpfer 2023). They stabilize the basement membranes and are involved in embryo organogenesis and development, especially of the limb. They are largely expressed in articular cartilage and increased in osteoarthritis.

Bone ECM
Bone is a peculiar tissue that mineralises physiologically (Fig. 1.2). Therefore, its ECM has two components: the organic matrix, containing an assortment of

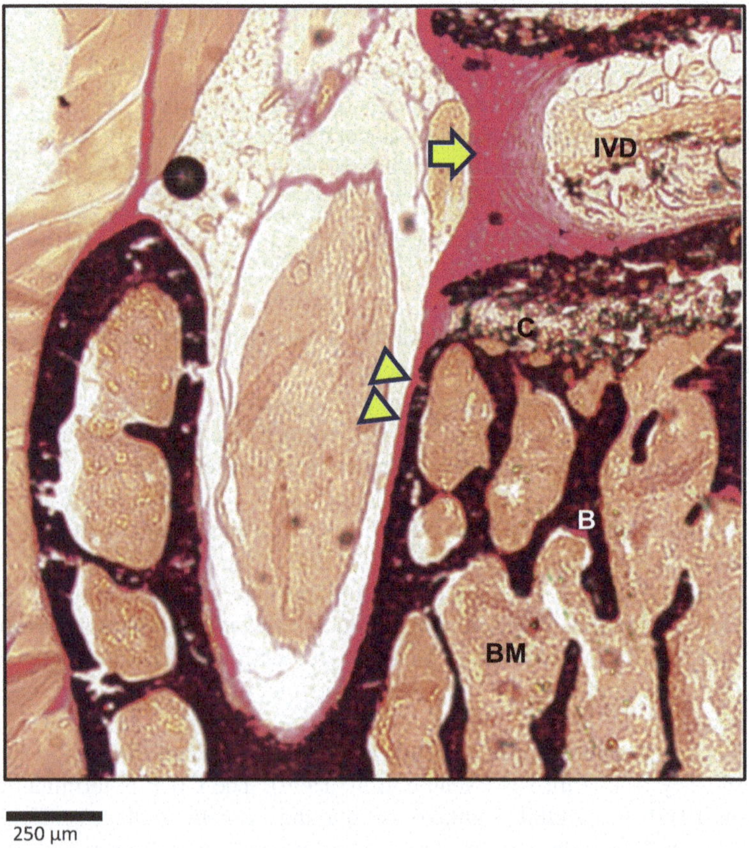

Fig. 1.2 *Undecalcified mouse vertebra* stained with the Von Kossa-Gieson's method. B: Trabecular bone (black) lined by non-mineralised osteoid (pink); *C* Cartilage; *IVD* Intervertebral disk. Arrow: IVD annulus; Arrowheads; periosteum

Table 1.3 Bone ECM proteins

Protein	Function
Type I collagen	Scaffold for bone cells and mechanical support of the tissue
Types III and V collagen	Regulate fibre diameter and type I collagen fibrillogenesis
Biglycan	Promotes collagen fibrillogenesis and bone formation
Decorin	Promotes collagen fibrillogenesis and bone formation
Keratocan	Promotes mineral deposition rates
Asporin	Promotes collagen mineralisation
Osteocalcin	Regulates calcium metabolism, indicates bone formation
Matrix Gla protein (MGP)	Inhibits bone formation and mineralisation
Periostin	Regulates collagen fibrillogenesis, maintains bone strength
Osteonectin	Promotes bone formation and mineralisation, regulates collagen fibrillogenesis, maintains biomechanical properties
Thrombospondins	Promotes bone formation, regulates collagen fibrillogenesis
R-spondins	Promotes Wnt/b-catenin signalling, regulates bone development
Bone sialoprotein	Promotes bone formation and mineralisation
Osteopontin	Promotes bone formation and mineralisation, regulates bone remodelling
Dentin matrix acidic phosphoprotein 1 (DMP1)	Regulates phosphate metabolism, promotes bone mineralisation
Matrix extracellular phosphoglycoprotein (MEPE)	Regulates phosphate metabolism, promotes bone mineralisation

glycoproteins (Table 1.3), including some shared with cartilage, and the inorganic matrix, largely composed of mineral crystals, with small amounts of amorphous minerals.

Collagens have been largely described in the previous section. In bone, type I collagen represents 90–95% of the bone ECM. It is the structural protein that provides a scaffold for bone cells and supports the tissue mechanically (Fig. 1.3). Type III and V collagens are barely represented and act as regulators of type I collagen fibrillogenesis.

Biglycan, *decorin*, *keratocan* and *asporin* are small leucine-rich proteoglycans or SLRPs (Hua and Jiang 2021), which contribute to many functions of the bone, including cell proliferation, osteogenesis, mineral deposition and bone remodelling. They also regulate collagen fibrillogenesis.

Osteocalcin, matrix Gla protein (MPG) and *periostin* are members of the γ-carboxyglutamic acid-containing class of proteins, with a modified glutamic acid (Gla) formed because of vitamin K-dependent post-translational modification (Cancela et al. 2012). Osteocalcin is expressed primarily by active osteoblasts, contains three Gla residues and binds calcium, mediating its association with hydroxyapatite. It is also a hormone that is thought to regulate glucose and energy metabolism. MPG is released by osteoblasts, osteocytes and chondrocytes, negatively regulating mineralisation. Periostin is a ligand for $\alpha V/\beta 3$ and $\alpha V/\beta 5$ integrins supporting cell adhesion.

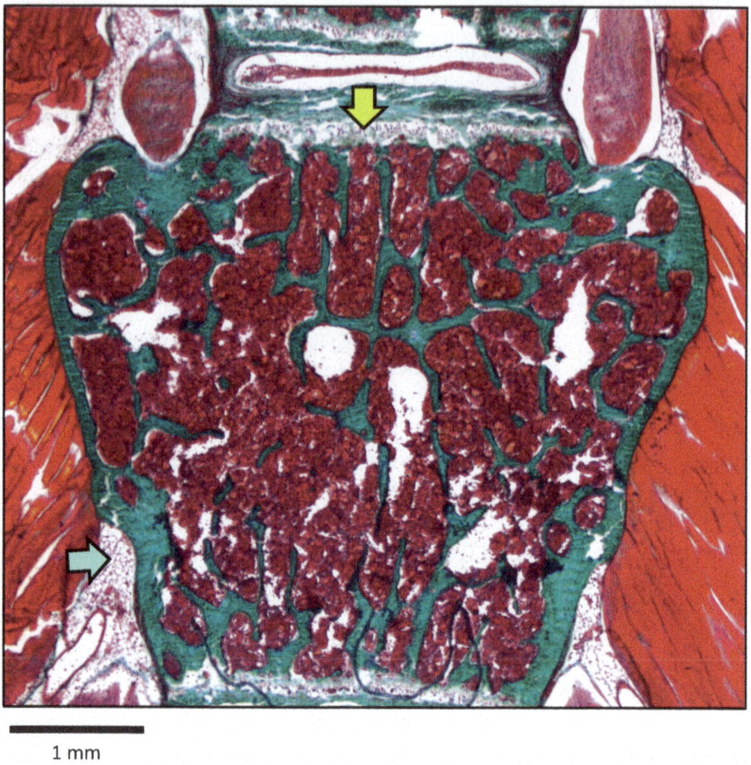

Fig. 1.3 *Masson-Goldner's trichrome staining* of a mouse vertebra showing collagen stained in green, bone marrow in dark red and skeletal muscle in bright red; yellow arrow: cartilage; azure arrow: adipose tissue

The nature and structure of the bone ECM vary during the development (Salhotra et al. 2020). The first bone ECM secreted and mineralised by osteoblasts is poorly organised and mechanically incompetent. It is called primary or woven bone and is made by irregularly oriented type I collagen fibrils that represent around 95% of the organic matrix, intermingled with non-collagenous glycoproteins shared with other connective tissues, riddled by numerous lenticular lacunae hosting mature bone cells, named osteocytes (Webster et al. 2013). Osteoblasts are instead located at the bone surface in a cuboidal epithelioid stratus, lining newly formed not yet mineralised ECM, called osteoid (Fig. 1.2), which will later mineralise through the so-called matrix vesicles released by osteoblasts (Maciel et al. 2013). Mineralisation follows the organic matrix deposition and is induced by the removal of the ECM mineralisation inhibitor pyrophosphate through transporter-mediated internalisation and alkaline phosphatase-induced breakdown in two inorganic phosphates. Phosphate and calcium are then combined in hydroxyapatite crystals, deposited initially within the matrix vesicles, and then growing outside along the type I collagen fibrils. A similar process occurs in mineralising cartilage during the process of endochondral ossification, preceded by vasculogenesis, chondrocyte hypertrophy and release of matrix vesicles (Hasegawa et al. 2022).

Primary bone is formed in foetus, but after birth, it must be replaced by secondary or lamellar bone, which has a solid and well-organised structure suitable to resist the mechanical load that increases during growth (IsoJima and Sims 2021). Here, collagen fibrils are deposited by osteoblasts in layers in which mature cells remain entrapped to form osteocytes. Each layer is again formed by 95% type I collagen and 5% non-collagenous glycoproteins, in which collagen fibrils are oriented in parallel forming a lamella, with alternation of right-handed and left-handed lamellae, each accompanied by a layer of osteocyte lacunae. This structure is observed both in the thin bone spiculae that intrude the medullary canal in the form of bone trabeculae (spongy or trabecular bone) and in the thick peripheral bone that, for its dense appearance, is called compact bone. Compact bone differs from the trabecular bone because, due to its thickening, it requires the penetration of vessels and nerves through the transverse Volkmann canals. These run from the outer and inner surface into the compact bone, converging in longitudinal Havers canals, which distribute vessels and nerves in any part of this thick bone wall. In contrast, trabecular bone is thin enough to be nourished and oxygenated by diffusion from the medullary vessels.

Of note, compact bone is organised in cylindrical longitudinal structures, called osteons, formed around each Havers canal (Chang and Liu 2022). Osteons are made by concentring bone lamellae, alternating with layers of osteocyte lacunae connected each other and with the surface by thin canaliculi in which osteocyte processes penetrate and connect by gap junctions. After bone remodelling, incomplete residual osteon parts remain in the interstitial areas between newly formed osteons, in which osteocytes stay connected with the osteon osteocytes. Furthermore, some layers of circumferential lamellae run at the inner (endosteal) and the outer (periosteal) surface, the former covered by a single layer of bone cells and the latter by a double connective tissue layer, called periosteum. The external periosteum layer is enriched in dense connective tissue and cells as well as in blood vessels penetrating the Volkmann canals. It also presents thick transverse type I collagen fibres, called Sharpey fibres, that remain entrapped in the periosteal circumferential lamellae, anchoring the periosteum to the bone. The inner layer is instead a loose connective tissue, enriched in osteoprogenitor cells that differentiate in osteoblasts and deposit and mineralise new periosteal bone during accrual.

Because of its mineralized nature, bone EMC requires a specialised mechanism of matrix renewal, called bone remodelling (Raggatt and Partridge 2010). This consists of a phase of bone resorption by monocyte-derived multinucleated cells, called osteoclasts, in which the bone matrix is first demineralised and then digested, coupled in the same site with a subsequent deposition and mineralisation of new bone matrix by osteoblasts. These latter remain entrapped in the newly formed bone as osteocytes, which participate in calcium and phosphate homeostasis both by perilacunar remodelling and FGF23-mediated hormonal regulation of calcium/phosphate excretion by the kidney. During accrual, bone resorption and bone formation may also be uncoupled, occurring in different sites, and contributing to bone growth and shape modelling. At the end of the deposition phase, excess osteoblasts die by apoptosis, while the remaining osteoblasts reduce their cytoplasm and become quiescent lining cells that cover the bone surface until a new resorption or formation phase starts.

1.4 Skeletal Dysplasias

Skeletal dysplasias are pathological conditions with heterogeneous genetic aetiologies. They are rare diseases that affect bone and cartilage development, often inducing neurological dysfunctions (Unger et al. 2023). Subjects affected by skeletal dysplasias show developmental abnormalities in the skull and axial and long bones (Table 1.4). Altogether they represent over 500 conditions affecting 2.4 in 10,000 born live (Unger et al. 2023). Frequent symptoms associated with skeletal dysplasias consist in dwarfism, cervical spine instability, kyphosis and spinal stenosis, genu varum or valgum, joint stiffness and pain and club foot. When the stature is short, the head may be large with prominent forehead, with disproportionate long trunk and short legs. Teeth and bones can be fragile, while the palate can be cleft. Patients may present with hearing and vision impairment, cervical medullary compression and hydrocephalus, which strongly compromise neural functions.

Many skeletal dysplasias feature altered ECM components and organisation, affecting connective tissue proper, bone and cartilage. Anomalies are very heterogeneous, depending on the underlying altered molecular pathways and, in bone, associated with compromised tissue mineralisation. In this section, the major skeletal dysplasias due to ECM structural proteins will be discussed (Table 1.4).

Skeletal dysplasias can be grouped according to the ECM component anomalies involved in its pathogenesis (Table 1.4). Major conditions are affected by aggrecan, fibrillin, fibronectin, matrilin, proteoglycans and collagens, with diverse and severe consequences that affect not only bone and cartilage but also the connective tissue proper.

1.4.1 Aggrecan, Perlecan and Biglycan Diseases

Aggrecan is a chondroitin sulphate proteoglycan that is involved in familial osteochondritis dissecans, spondyloepimetaphyseal dysplasia aggrecan type and spondyloepiphyseal dysplasia Kimberley type.

Familial osteochondritis dissecans is an autosomal dominant disease characterised by areas of bone lesions caused by cartilage detachment (Gibson and Briggs 2016). It affects joints in which both cartilage and the underlying bone are damaged. Lesions can affect several joints, including knees, elbows, hips, and ankles. Patients suffer from pain and joint swelling and have short stature and osteoarthritis that causes pain and joint stiffness.

Table 1.4 Skeletal dysplasias caused by mutations of ECM structural proteins

Syndrome	OMIM number	Mutated protein	Phenotype
Proteoglycan core proteins			
Familial osteochondritis dissecans	165800	Aggrecan	Areas of bone lesions caused by cartilage detachment
Spondyloepimetaphyseal dysplasia aggrecan type	612813	Aggrecan	Severe dwarfism, short fingers and toes, facial dysmorphism
Spondyloepiphyseal dysplasia, Kimberley type	608361	Aggrecan	Short stature and early development of osteoarthritis, particularly in knees, ankles and hips
Schwartz–Jampel syndrome	255800	Perlecan	Dwarfism, chondrodysplasia, permanent myotonia, eyelid anomalies
Spondyloepimetaphyseal dysplasia biglycan type	300106	Biglycan	Dwarfism, brachydactyly, osteoarthritis
Fibrillin			
Acromelic dysplasia	102370	Fibrillin-1	Short stature, pseudomuscular build, stiff joints and tight skin
Marfan syndrome	154700	Fibrillin-1	Tall and slender individuals, arachnodactyly, loose joints, arm span exceeding body height
Marfanoid congenital contractural arachnodactyly	121050	Fibrillin-2	Arachnodactyly, flexion contractures of multiple joints, kyphoscoliosis
Fibronectin			
Spondylometaphyseal dysplasia with "corner fractures"	184255	Fibronectin	Irregularly shaped metaphyses with "corner fractures", appearance, coxa vara, ovoid vertebral bodies and severe scoliosis
Matrilin			
Multiple epiphyseal dysplasia matrilin type	607078	Matrilin-3	Joint pain of hips and knees, early-onset arthritis, waddling walk
Spondyloepimetaphyseal dysplasia matrilin type	608728	Matrilin-3	Dwarfism; bowing of the lower limbs; short, wide and stocky long bones with severe epiphyseal and metaphyseal changes; lumbar lordosis; hypoplastic iliac bones; flat ovoid vertebral bodies
Cartilage oligomeric matrix protein			
Pseudoachondroplasia	177170	Cartilage oligomeric matrix protein	Dwarfism; waddling walk; joint pain; osteoarthritis; hyperextensibility in hands, knees and ankles; limited motion at the elbows and hips; valgus, varus or windswept deformity; scoliosis; lordosis

(continued)

Table 1.4 (continued)

Syndrome	OMIM number	Mutated protein	Phenotype
Collagens			
Caffey disease	114000	Collagen I α-1 chain	Hyperostosis of the jawbone, scapulae, clavicles and long bone diaphyses
Osteogenesis imperfecta types 1–4	1: 166200 2: 166210 3: 259420 4: 166220	Collagen I α-1 chain	Bone fragility and spontaneous fractures, blue sclerae, short stature, scoliosis, contractures, hearing loss, respiratory problems, dentinogenesis imperfecta
Osteogenesis imperfecta types 1–4		Collagen I α-2 chain	
Osteogenesis imperfecta types 2–4		Cartilage-associated protein	
Osteogenesis imperfecta type 8	610915	Leprecan	
Achondrogenesis type 2	200610	Collagen II α-1 chain	Dwarfism, narrow chest with short ribs, underdeveloped lungs, lack of ossification in the spine and pelvis, prominent forehead, small jawbone and cleft palate
Hypochondrogenesis	200610	Collagen II α-1 chain	Dwarfism, abnormal ossification of the spine and pelvis
Kniest dysplasia	156550	Collagen II α-1 chain	Dwarfism, large joints with pain and movement restriction, arthritis, kyphoscoliosis, platyspondyly, dumbbell-shaped long bones, long and knobby fingers, club foot
Spondyloepiphyseal dysplasia congenita	183900	Collagen II α-1 chain	Dwarfism, kyphoscoliosis, lordosis and neck vertebral instability with risk of spinal cord damage
Stickler dysplasia type 1	108300	Collagen II α-1 chain	Middle face underdeveloped bones, Pierre Robin sequence (cleft palate, glossoptosis, micrognathia), dyspnoea
Multiple epiphyseal dysplasia	226900	Collagen IX α-1/2/3 chains	Joint pain of hips and knees, early-onset arthritis, waddling walk
Schmid type metaphyseal chondrodysplasia	156500	Collagen X α-1 chain	Dwarfism, bowing of the legs, metaphyseal irregularities with fraying and splaying, more severe at the knee
Fibrochondrogenesis	228520	Collagen XI α-1/2 chain	Narrow chest impairing lung development. Generally lethal shortly after birth for respiratory failure, rarely lethal into childhood
Marshall syndrome	154780	Collagen XI α-1/2 chain	Hearing loss, large eyes, double-jointedness, early-onset osteoarthritis, micrognathia, cleft palate

(continued)

Table 1.4 (continued)

Syndrome	OMIM number	Mutated protein	Phenotype
Otospondylomega-epiphyseal dysplasia Deafness	215150	Collagen XI α-2 chain	Dwarfism, platyspondyly, back and joint pain, limited joint movement, arthritis, severe high-frequency hearing loss, protruding eyes, flattened bridge of the nose, upturned nose with large, rounded tip, micrognathia, cleft palate
Stickler dysplasia type 2	604841	Collagen XI α-1 chain	Middle face underdeveloped bones, Pierre Robin sequence (cleft palate, glossoptosis, micrognathia), dyspnoea, eye abnormalities, high chance of hearing loss
Weissenbacher–Zweymüller syndrome	215150	Collagen XI α-2 chain	Dwarfism, dumbbell-shaped upper arm bones, vertebral deformities, hearing wide-set, protruding eyes, small and upturned nose with a flat bridge, micrognathia, cleft palate
Steel syndrome	615155	Collagen XXVII α-1 chain	Dwarfism, bilateral congenital hip and radial head dislocation, carpal coalition, scoliosis, pes cavus, atlantoaxial subluxation, broad forehead and nasal bridge, hypertelorism, mild midface hypoplasia, hearing loss

Spondyloepimetaphyseal dysplasia aggrecan type is an autosomal recessive disease (Gibson and Briggs 2016). Patients show severe dwarfism, macrocephaly, midface hypoplasia, almost absent nasal cartilage, prognathism, slightly low-set and posteriorly rotated ears, short neck, barrel chest and mild lumbar lordosis. They are also affected by rhizomelia and mesomelia, brachydactyly, horizontal nails and telescoping interphalangeal joints. Long bones show irregular epiphyses and expanded metaphyses, while platyspondyly and multiple cervical–vertebral clefts affect the spine.

Spondyloepiphyseal dysplasia Kimberley type is an autosomal dominant disease (Gibson and Briggs 2016). Patients show proportionate short stature, stocky habitus, osteoarthropathy of the weight-bearing joints, prominently irregular vertebral endplates and sclerotic vertebral bodies.

Schwartz–Jampel syndrome is a disease due to autosomal recessive mutations of the gene encoding perlecan, characterised by myotonia and chondrodysplasia that appear at birth or in early childhood (Stum et al. 2005). Continued muscle contraction impairs various movements, including eating, sitting and walking. Patients show "mask-like" facial expression, blepharophimosis and pursed lips very typical of this syndrome. They may also show myopia and blepharospasm. Patients have short stature, the long bones can be bowed, the pectus carinatum and the spine curvature altered. Platyspondyly and joint contractures could severely limit movements.

Spondyloepimetaphyseal dysplasia biglycan type is a X-linked syndrome due to mutations of the gene encoding biglycan. The disease is characterised by dwarfism, brachydactyly and osteoarthritis (Cho et al. 2016).

1.4.2 Fibrillin Diseases

Mutations can occur in both fibrillin-1 (acromelic dysplasia and Marfan syndrome) and fibrillin-2 (marfanoid congenital contractural arachnodactyly) (see Chap. 7).

Fibrillin-1 mutations may result in distinct disorders (Peeters et al. 2022). *Acromelic dysplasias* represent a group of skeletal dysplasias induced by autosomal dominant mutations confined to specific domains of the fibrillin-1 gene. Other forms are due to mutations of ADAMTS proteins L2, 10, and 17, latent binding proteins 2 and 3 and SMAD4 transcription factor. These dysplasias share dwarfism, pseudomuscular build, stiff joints and tight skin. Patients may also show distinctive round face, sharply defined eyebrows, small mouth, thick lips, prominent philtrum, narrow palpebral fissures and bulbous nose. At birth the body length is unremarkable, but postnatal growth is severely impaired resulting in very short stature.

Marfan syndrome is also due to fibrillin-1 autosomal dominant mutations (Coelho and Almeida 2020). These are significantly more than in the acromelic dysplasias and can occur in any of the 56 gene domains. Opposite to acromelic dysplasias, patients affected by the Marfan syndrome are tall and slender, with long fingers and arachnodactyly, loose joints and arm length that exceed the body length. They also have long and narrow face, crowded teeth, scoliosis or kyphosis, stretch marks and pectus excavatum or carinatum. Spontaneous pneumothorax can be a complication, along with dura ectasia that may cause pain in various regions. Several patients present with myopia, cataract or glaucoma. Symptoms may appear any time in infancy and adulthood and in severe cases the disease can be lethal.

Heterozygous mutations in fibrillin-2 occur in *marfanoid congenital contractural arachnodactyly* (Callewaert 2001), which presents with arachnodactyly; flexion contractures of elbows, knees, hips, ankles and fingers; progressive kyphoscoliosis; marfanoid slender build; dolichostenomelia; pectus deformity; muscular hypoplasia; highly arched palate; and wrinkled ears. In severe cases, patients suffer from cardiovascular and gastrointestinal anomalies.

1.4.3 Fibronectin Diseases

Spondylometaphyseal dysplasia with "corner fractures" is caused by fibronectin heterozygous mutations (England et al. 2020). Patients show flake-like, triangular or curvilinear ossification centres at the edges of irregular metaphyses that simulate fractures. This anomaly becomes prominent in older children but disappears after growth. The syndrome also shows scoliosis and, in some cases, coxa vara.

1.4.4 Matrilin Diseases

Matrilin-3 mutations are involved in multiple epiphyseal dysplasia matrilin type and spondyloepimetaphyseal dysplasia matrilin type.

Multiple epiphyseal dysplasia matrilin type is an autosomal dominant disease characterised by pain in the hips and/or the knees after exercise and fatigue after long walking (Anthony et al. 2015). They may also present with waddling gait. At the end of growth, patient's stature is normal or moderately shorter. However, the limbs remain relatively short compared to the trunk and pain can appear with the joint deformity progression. This may lead to early-onset osteoarthritis especially in the weight-bearing joints.

Spondyloepimetaphyseal dysplasia matrilin type is a member of the spondyloepimetaphyseal dysplasia family, which is caused by mutations in a multitude of genes, one of which encodes matrilin-3 (Dennis et al. 2021). It has an autosomal recessive inheritance and features wide metaphyses with lateral spurs and small and under-ossified ischia, which result in dwarfism and hypoplastic iliac bones.

Pseudoachondroplasia is caused by autosomal dominant mutations in the gene encoding COMP (Briggs and Wright 2004). COMP mutations prevent secretion of the protein that remains entrapped in the cells. Patients show dwarfism. Stature is normal at birth, but growth rate decreases within the second year of age. Consistent with the dwarfism, arms and legs are short, and patient's walk is waddling. Patients also suffer from joint pain associated with osteoarthritis; hyperextensibility of hands, knees and ankles; and limitation in elbow and hip movements. Some subjects have valgus or varus legs, sometimes with asymmetry of the two legs (windswept deformity). Scoliosis and lordosis are other features, while facial features and head size are normal.

1.4.5 Type I Collagen Diseases

Given that collagens are the most common proteins in the body, their mutations induce a plethora of different diseases according to the collagen chain affected.

Caffey disease is caused by autosomal dominant mutations of the gene encoding the collagen I α-1 chain. It is also called cortical hyperostosis due to the excess of bone formation. Patients show anomalies in the jawbone, scapulae, clavicles and arm and leg long bone diaphysis. Thickening of the bones is significant, and sometimes adjacent bones can fuse. Swelling joints and muscles can also affect patients, with pain, redness, fever and irritability.

A large group of collagen I α-1 or α-2 chain mutations features *osteogenesis imperfecta types 1 to 4* and have an autosomal dominant pattern or inheritance (see Chap. 2). This is a brittle bone disease that can be characterised by alterations of other connective tissues, including sclerae and tooth. The less severe form is type 1, characterised by few fractures lifelong. Type 2 is the most severe. Fractures are frequent and can occur also before birth, inducing perinatal lethality. Severity of type 3 and type 4 falls between the extremes of type 1 and 2. Specific variants in type 1 collagen genes results also in rare forms of Ehlers–Danlos syndrome (see Chap. 3).

Osteogenesis imperfecta types 2 to 4 and 8 can also be due to mutations of other genes, including the gene encoding the cartilage-associated protein (type 2 and 4), important for collagen prolyl 3-hydroxylation and chaperone function in the endoplasmic reticulum, and leprecan (type 8), an enzyme involved in proline hydroxylase activity, relevant for collagen synthesis (Rossi et al. 2019).

1.4.6 Type II Collagen Diseases

Mutations of collagen II α-1 chain are involved in various syndromes, including achondrogenesis type 2, hypochondrogenesis, kniest dysplasia, spondyloepiphyseal dysplasia congenita and Stickler dysplasia type 1 (see Chap. 4).

Achondrogenesis type 2 is an autosomal dominant disorder, also called Langer–Saldino type (Handa et al. 2021). Patients show short arms and legs, narrow chest with short ribs and underdeveloped lungs. They lack ossification in the spine and pelvis and have prominent forehead, small chin and, in some cases, cleft palate. They also show enlarged abdomen and *hydrops fetalis* due to excess of body fluid gained before birth.

Hypochondrogenesis is a lethal syndrome with autosomal dominant inheritance, in which patients show small body, short limbs and abnormal ossification in the spine and pelvis. Affected newborns die immediately or shortly after birth from respiratory failure (Gregersen and Savarirayan 2019).

Kniest dysplasia is another autosomal dominant disorder featuring short trunk, arms and legs, abnormally large joints, pain and movement limitations associated with arthritis, kyphoscoliosis, platyspondyly, arm and leg dumbbell-shaped bones, long and knobby fingers and club foot (56). Some subjects show round flat face, bulging and wide-set eyes. Some infants present with cleft palate and respiratory insufficiency. Severe myopia and retinal detachment can be associated with blindness, while deafness can result from recurrent ear infections.

Spondyloepiphyseal dysplasia congenita is also an autosomal dominant syndrome that induces disorders of bone growth. Patients present with dwarfism. As indicated by the name, the condition affects especially the spine and the long bone epiphyses and is apparent already at birth. Other symptoms include kyphoscoliosis and lordosis and spinal cord damage caused by the vertebral instability. They also show platyspondyly, coxa vara, club foot and board barrel-shaped chest that cause breath insufficiency. Arthritis and joint stiffness are further complications that appear early in life. Facial deformities include flattened cheek bones and cleft palate. Finally, patients may display severe myopia and deafness (Nenna et al. 2019).

Stickler dysplasia type 1 is again an autosomal dominant condition (Soh et al. 2022). Approximately 80–90% of patients classified as type I are affected by mutations of type II collagen α-1 chain, while the remaining cases display mutations of collagen XI α-2 chain. The syndrome presents with symptoms affecting face, eyes, ears and joints which vary among individuals. Face appears flat due to the underdevelopment of middle face bones, including the cheek bones and nose bridge bones. They also frequently show the so-called "Pierre Robin" sequence that include cleft palate, glossoptosis and micrognathia. This combination of anomalies may induce breath insufficiency. Severe myopia, abnormal vitreous, glaucoma, cataracts and retinal detachment can be severe complications that may lead to blindness. Patients may also suffer from deafness that can be either sensorineural or conductive. Joints are also affected, which may be loose and hypermobile in young age and becoming stiff in older age. These joints also suffer from arthritis. Scoliosis, kyphosis or platyspondyly can also be seen, inducing severe back pain.

1.4.7 Type IX Collagen Diseases

Type IX collagen anomalies induce *multiple epiphyseal dysplasia*, a disease inherited by both autosomal recessive and autosomal dominant patterns (Anthony et al. 2015). It affects cartilage and long bones during development. Both inheritance patterns induce a relatively mild syndrome, which includes hip and knee joint pain, early-onset arthritis and waddling walking. Infrequently, mild short stature can be observed. The recessive forms present also with malformations of hands, feet and knees, scoliosis, club foot, cleft palate, clinodactyly, ear swelling and double-layered patella.

1.4.8 Type X Collagen Diseases

Type X collagen anomalies cause the *Schmid type metaphyseal chondrodysplasia*, which is inherited in an autosomal dominant manner. The syndrome features dwarfism, apparent by the second year of age, characterised by short limbs, genu varum and waddling walk. Facial appearance and size are normal, while bones show metaphyseal irregularities, widened growth plates, coxa vara, anterior cupping, sclerosis and spaying of the ribs. Metacarpal bones and proximal phalanges may also be short, while vertebras may infrequently show platyspondyly and irregularities, which improve with age (Richmond and Savarirayan 2019).

1.4.9 *Type XI Collagen Diseases*

Mutations in type XI collagen may induce otospondylomega-epiphyseal dysplasia deafness, Stickler dysplasia type 2 and Weissenbacher–Zweymüller syndrome.

While *Stickler dysplasia type 2* and *Weissenbacher–Zweymüller syndrome* are very similar between them and to Stickler dysplasia type 1 (see section on type II collagen diseases) (Gregersen and Savarirayan 2019), *otospondylomega-epiphyseal dysplasia deafness* has specific features (Mutlu et al. 2023). It has an autosomal recessive pattern of inheritance and affects ears, spine and long bone epiphyses. Subjects are affected by dwarfism, enlarged joints and platyspondyly, associated with joint and back pain, stiffness and arthritis. The syndrome induces deafness, exophthalmos and nose flattened bridge and rounded tip. Micrognathia and cleft palate are also frequent.

1.4.10 *Type XXVII Collagen Diseases*

Mutations in type XXVII collagen induce the *Steel syndrome* (Girisha et al. 2022), which is not yet fully characterised and is thought to be inherited in an autosomal recessive manner. Patients show dwarfism, face dysmorphism, dislocated hips and radial heads and absence of ossification centres, fusion of carpal bones, scoliosis and anomalies of the cervical vertebras, externally rotated feet, mild skin syndactyly, short mid phalanges and sensorineural deafness.

1.5 Conclusions and Future Perspectives

In this chapter, we have described the ECM features and its structural alterations in genetic cartilage and bone diseases. Given the complexity and the multifaceted aspects of ECM, related diseases are manyfold and affect several organs in the body. Novel sequencing techniques and bioinformatic technologies, especially after the full description of the Human Genome, has largely improved the molecular diagnosis of diseases affecting the ECM. Management of patients is also expected to improve in the next future. Nevertheless, we must be aware that this is still a pioneer field, and much effort will be necessary at least to reduce the burden of ECM mutation-induced genetic alterations and give patients a hope for a better life.

Acknowledgements We are indebted to Rita Di Massimo for editing the manuscript.

Compliance with ethical standard Authors declare no conflict of interest. The chapter reports only data described in the literature. No original experiments on humans or laboratory animals were performed. Histological images derive from archive material.

References

Alford AI, Hankenson KD (2023) Thrombospondins modulate cell function and tissue structure in the skeleton. Semin Cell Dev Biol 7:S1084–9521(23)00138–6. https://doi.org/10.1016/j.semcdb.2023.06.011

Anthony S, Munk R, Skakun W, Masini M (2015) Multiple epiphyseal dysplasia. J Am Acad Orthop Surg 23:164–172. https://doi.org/10.5435/jaaos-d-13-00173

Archer CW, Francis-West P (2003) The chondrocyte. Int J Biochem Cell Biol 35:401–404. https://doi.org/10.1016/s1357-2725(02)00301-1

Bengtsson E, Mörgelin M, Sasaki T, Timpl R, Heinegård D, Aspberg S (2002) The leucine-rich repeat protein PRELP binds perlecan and collagens and may function as a basement membrane anchor. J Biol Chem 277:15061–15068. https://doi.org/10.1074/jbc.m108285200

Briggs MD, Wright MJ (2004) Pseudoachondroplasia [updated 2018 Aug 16]. In: Adam MP, Feldman J, Mirzaa GM, Pagon RA, Wallace SE, LJH B, Gripp KW, Amemiya A (eds) GeneReviews®. University of Washington, Seattle (WA), pp 1993–2023

Callewaert B (2001) Congenital Contractural Arachnodactyly [updated 2022 Jul 14]. In: Adam MP, Feldman J, Mirzaa GM, Pagon RA, Wallace SE, LJH B, Gripp KW, Amemiya A (eds) GeneReviews®. University of Washington, Seattle (WA), pp 1993–2023

Cancela ML, Conceição N, Laizé V (2012) Gla-rich protein, a new player in tissue calcification? Adv Nutr 3:174–181. https://doi.org/10.3945/an.111.001685

Chang B, Liu X (2022) Osteon: structure, turnover, and regeneration. Tissue Eng Part B Rev 28:261–278. https://doi.org/10.1089/ten.teb.2020.0322

Cho SY, Bae JS, Kim NKD (2016) BGN mutations in X-linked spondyloepimetaphyseal dysplasia. Am J Hum Genet 98:1243–1248. https://doi.org/10.1016/j.ajhg.2016.04.004

Coelho SG, Almeida AG (2020) Marfan syndrome revisited: from genetics to the clinic. Rev Port Cardiol (Engl Ed) 39:215–226. https://doi.org/10.1016/j.repc.2019.09.008

Dennis EP, Greenhalgh-Maychell PL, Briggs MD (2021) Multiple epiphyseal dysplasia and related disorders: molecular genetics, disease mechanisms, and therapeutic avenues. Dev Dyn 250:345–359. https://doi.org/10.1002/dvdy.221

Ding DC, Shyu WC, Lin SZ (2011) Mesenchymal stem cells. Cell Transplant 20:5–14. https://doi.org/10.3727/096368910x

England J, McFarquhar A, Campeau PM (2020) Spondylometaphyseal dysplasia, corner fracture type. In: Adam MP, Feldman J, Mirzaa GM, Pagon RA, Wallace SE, Bean LJH, Gripp KW, Amemiya A (eds) GeneReviews®. University of Washington, Seattle, Seattle (WA), pp 1993–2023

Fan R, Yan X, Zhang W (2022) Relationship between asporin and extracellular matrix behavior: a literature review. Medicine (Baltimore) 101:e32490. https://doi.org/10.1097/md.0000000000032490

Fitzgerald J (2020) WARP: a unique extracellular matrix component of cartilage, muscle, and endothelial cell basement membranes. Anat Rec (Hoboken) 303:1619–1623. https://doi.org/10.1002/ar.24087

Garg HG, Hales CA (2004) Chemistry and biology of hyaluronan. Elsevier Ltd

Gentili C, Cancedda R (2009) Cartilage and bone extracellular matrix. Curr Pharm Des 15:1334–1348. https://doi.org/10.2174/138161209787846739

Gibson BG, Briggs MD (2016) The aggrecanopathies: an evolving phenotypic spectrum of human genetic skeletal diseases. Orphanet J Rare Dis 11:86. https://doi.org/10.1186/s13023-016-0459-2

Girisha KM, Jacob P, SriLakshmi Bhavani G (2022) Steel syndrome: report of three patients, including monozygotic twins and review of clinical and mutation profiles. Eur J Med Genet 65:104521. https://doi.org/10.1016/j.ejmg.2022.104521

Gregersen PA, Savarirayan R (2019) Type II collagen disorders overview. In: Adam MP, Feldman J, Mirzaa GM, Pagon RA, Wallace SE, LJH B, Gripp KW, Amemiya A (eds) GeneReviews®. University of Washington, Seattle (WA), pp 1993–2023

Halper J, Kjaer M (2014) Basic components of connective tissues and extracellular matrix: elastin, fibrillin, fibulins, fibrinogen, fibronectin, laminin, tenascins and thrombospondins. Adv Exp Med Biol 802:31–47. https://doi.org/10.1007/978-94-007-7893-1_3

Handa A, Grigelioniene G, Nishimura G (2021) Radiologic features of type II and type XI collagenopathies. Radiographics 41:192–209. https://doi.org/10.1148/rg.2021200075

Hasegawa T, Hongo H, Yamamoto T, Abe M, Yoshino H, Haraguchi-Kitakamae M, Ishizu H, Shimizu T, Iwasaki N, Amizuka N (2022) Matrix vesicle-mediated mineralization and osteocytic regulation of bone mineralization. Int J Mol Sci 23:9941. https://doi.org/10.3390/ijms23179941

Holmes DF, Lu Y, Starborg T, Kadler KE (2018) Collagen fibril assembly and function. Curr Top Dev Biol 130:107–142. https://doi.org/10.1016/bs.ctdb.2018.02.004

Hua R, Jiang JX (2021) Small leucine-rich proteoglycans in physiological and biomechanical function of bone. Matrix Biol Plus 11:100063. https://doi.org/10.1016/j.mbplus.2021.100063

Isojima T, Sims NA (2021) Cortical bone development, maintenance and porosity: genetic alterations in humans and mice influencing chondrocytes, osteoclasts, osteoblasts and osteocytes. Cell Mol Life Sci 78:5755–5773. https://doi.org/10.1007/s00018-021-03884-w

Karamanos NK, Theocharis AD, Piperigkou Z, Manou D, Passi A, Skandalis SS, Vynios DH, Orian-Rousseau V, Ricard-Blum S, Schmelzer CEH, Duca L, Durbeej M, Afratis NA, Troeberg L, Franchi M, Masola V, Onisto M (2021) A guide to the composition and functions of the extracellular matrix. FEBS J 288:6850–6912. https://doi.org/10.1111/febs.15776

Klatt AR, Becker AK, Neacsu CD, Paulsson M, Wagener R (2011) The matrilins: modulators of extracellular matrix assembly. Int J Biochem Cell Biol 43:320–330. https://doi.org/10.1016/j.biocel.2010.12.010

Knupp C, Squire JM (2003) Molecular packing in network-forming collagens. Sci World J 3:558–577. https://doi.org/10.1100/tsw.2003.40

Le HA, Mayor R (2023) Cell-matrix and cell-cell interaction mechanics in guiding migration. Biochem Soc Trans 51:1733–1745. https://doi.org/10.1042/bst20230211

Long F, Ornitz DM (2013) Development of the endochondral skeleton. Cold Spring Harb Perspect Biol 5:a008334. https://doi.org/10.1101/cshperspect.a008334

Maciel GBM, Maciel RM, Danesi CC (2013) Bone cells and their role in physiological remodeling. Mol Biol Rep 50:2857–2863. https://doi.org/10.1007/s11033-022-08190-7

Malinin GI, Malinin TI (1999) Microscopic and histochemical manifestations of hyaline cartilage dynamics. Prog Histochem Cytochem 34:163–242. https://doi.org/10.1016/s0079-6336(99)80006-0

Mienaltowski MJ, Gonzales NL, Beall JM, Pechanec MY (2021) Basic structure, physiology, and biochemistry of connective tissues and extracellular matrix collagens. Adv Exp Med Biol 1348:5–43. https://doi.org/10.1007/978-3-030-80614-9_2

Mosher DF (1984) Physiology of fibronectin. Annu Rev Med 35:561–575. https://doi.org/10.1146/annurev.me.35.020184.003021

Mutlu H, Elçioğlu N, Kiliç E (2023) Autosomal recessive otospondylo-mega-epiphyseal dysplasia: comprehensive clinical review of a pediatric cohort. Clin Dysmorphol 32:151–155. https://doi.org/10.1097/mcd.0000000000000467

Neame PJ, Tapp H, Azizan A (1999) Noncollagenous, non-proteoglycan macromolecules of cartilage. Cell Mol Life Sci 55:1327–1340. https://doi.org/10.1007/s000180050373

Nenna R, Turchetti A, Mastrogiorgio G, Midulla F (2019) *COL2A1* gene mutations: mechanisms of spondyloepiphyseal dysplasia congenita. Appl Clin Genet 12:235–238. https://doi.org/10.2147/tacg.s197205

Nia HT, Ortiz C, Grodzinsky A (2022) Aggrecan: approaches to study biophysical and biomechanical properties. In: Balagurunathan K, Nakato H, Saijoh DU (eds) Glycosaminoglycans methods and protocols, pp 209–226

Peeters S, De Kinderen P, Meester JAN, Verstraeten A, Loeys BL (2022) The fibrillinopathies: new insights with focus on the paradigm of opposing phenotypes for both FBN1 and FBN2. Hum Mutat 43:815–831. https://doi.org/10.1002/humu.24383

Pirozzi KM (2022) Histophysiology of fibrocartilage. Clin Podiatr Med Surg 39:363–370. https://doi.org/10.1016/j.cpm.2022.02.002

Pufe T, Petersen WJ, Miosge N, Goldring MB, Mentlein R, Varoga DJ, Tillmann BN (2004) Endostatin/collagen XVIII–an inhibitor of angiogenesis–is expressed in cartilage and fibrocartilage. Matrix Biol 23:267–276. https://doi.org/10.1016/j.matbio.2004.06.003

Raggatt LJ, Partridge NC (2010) Cellular and molecular mechanisms of bone remodeling. J Biol Chem 285:25103–25108. https://doi.org/10.1074/jbc.r109.041087

Richmond CM, Savarirayan R (2019) Schmid metaphyseal chondrodysplasia. In: Adam MP, Feldman J, Mirzaa GM, Pagon RA, Wallace SE, LJH B, Gripp KW, Amemiya A (eds) GeneReviews®. University of Washington, Seattle, Seattle (WA), pp 1993–2023

Rossi V, Lee B, Marom R (2019) Osteogenesis imperfecta: advancements in genetics and treatment. Curr Opin Pediatr 31:708–715. https://doi.org/10.1097/mop.0000000000000813

Salhotra A, Shah HN, Levi B, Longaker MT (2020) Mechanisms of bone development and repair. Nat Rev Mol Cell Biol 21:696–711. https://doi.org/10.1038/s41580-020-00279-w

Seuffert F, Weidner D, Baum W, Schett G, Stock M (2018) Upper zone of growth plate and cartilage matrix associated protein protects cartilage during inflammatory arthritis. Arthritis Res Ther 20:88. https://doi.org/10.1186/s13075-018-1583-2

Shukunami C, Hiraki Y (2007) Chondromodulin-I and tenomodulin: the negative control of angiogenesis in connective tissue. Curr Pharm Des 13:2101–2112. https://doi.org/10.2174/138161207781039751

Siadat SM, Ruberti JW (2023) Mechanochemistry of collagen. Acta Biomater 163:50–62. https://doi.org/10.1016/j.actbio.2023.01.025

Soh Z, Richards AJ, McNinch A, Alexander A, Martin H, Snead MP (2022) Dominant Stickler syndrome. Genes 13:1089. https://doi.org/10.3390/genes13061089

Stegen S, Laperre K, Eelen G, Rinaldi G, Fraisl P, Torrekens S, Van Looveren R, Loopmans S, Bultynck G, Vinckier S, Meersman MPH, Rai J, Weis MA, Eyre DR, Ghesquière B, Fendt S-M, Carmeliet P, Carmeliet G (2019) HIF-1α metabolically controls collagen synthesis and modification in chondrocytes. Nature 565:511–515. https://doi.org/10.1038/s41586-019-0874-3

Stum M, Davoine CS, Fontaine B, Nicole S (2005) Schwartz-Jampel syndrome and perlecan deficiency. Acta Myol 24:89–92

Pihlajaniemi T, Rehn M (1995) Two new collagen subgroups: membrane-associated collagens and types XV and XVII. Prog Nucleic Acid Res Mol Biol 50:225–262. https://doi.org/10.1016/s0079-6603(08)60816-8

Talbott HE, Mascharak S, Griffin M, Wan DC, Longaker MT (2022) Wound healing, fibroblast heterogeneity, and fibrosis. Cell Stem Cell 29:1161–1180. https://doi.org/10.1016/j.stem.2022.07.006

Töpfer U (2023) Basement membrane dynamics and mechanics in tissue morphogenesis. Biol Open 12:bio059980. https://doi.org/10.1242/bio.059980

Umlauf D, Frank S, Pap T, Bertrand J (2010) Cartilage biology, pathology, and repair. Cell Mol Life Sci 67:4197–4211. https://doi.org/10.1007/s00018-010-0498-0

Unger S, Ferreira CR, Mortier GR, Ali H, Bertola DR, Calder A, Cohn DH, Cormier-Daire V, Girisha KM, Hall C, Krakow D, Makitie O, Mundlos NG, Robertson SP, Savarirayan R, Sillence D, Simon MS, Sutton VR, Warman ML, Superti-Furga A (2023) Nosology of genetic skeletal disorders: 2023 revision. Am J Med Genet A 191:1164–1209. https://doi.org/10.1002/ajmg.a.63132

Ushiki T (2002) Collagen fibers, reticular fibers and elastic fibers. A comprehensive understanding from a morphological viewpoint. Arch Histol Cytol 65:109–126. https://doi.org/10.1679/aohc.65.109

van der Rest M, Dublet B, Champliaud MF (1990) Fibril-associated collagens. Biomaterials 11:28–31

Webster DJ, Schneider P, Dallas SL, Müller R (2013) Studying osteocytes within their environment. Bone 54:285–295. https://doi.org/10.1016/j.bone.2013.01.004

Zhang F, Zhang Z, Linhardt RJ (2010) Glycosaminoglycans. In: Cummings RD, Pierce JM (eds) Handbook of Glycomics chapter 3. Academic Press, pp 59–80

Chapter 2
Osteogenesis Imperfecta and Related Diseases: Collagen I Alterations in Skeletal and Extraskeletal Tissues

Giulio Gatto, Wendy Pérez Franco, Heike-Katharina Hoyer-Kuhn, Oliver Semler, Antonella Forlino, Roberta Besio, and Frank Zaucke

Abstract In this chapter, we review the current knowledge about the complex intracellular biosynthesis, secretion, and extracellular maturation of collagen I. We focus on the specific molecular roles of the numerous proteins involved in these processes and discuss how mutations in the encoding genes lead to alterations in the collagen I quality and/or quantity. We then provide an overview of the important role of collagen I in bone and in extraskeletal tissues. We use osteogenesis imperfecta as a prototype for collagen I-related diseases and address how changes in collagen I quality and/or quantity affect the function of these tissues by providing specific examples based on observations made in patients and various animal models. This chapter shall hence provide a deeper understanding not only of the molecular basis of collagen I biochemistry but also of collagen I-related disorders and their diverse skeletal and extraskeletal manifestations.

G. Gatto · F. Zaucke (✉)
Department of Trauma Surgery and Orthopedics, University Hospital Frankfurt, Goethe University, Frankfurt am Main, Germany
e-mail: frank.zaucke@ukffm.de

W. Pérez Franco · A. Forlino · R. Besio (✉)
Department of Molecular Medicine, Biochemistry Unit, University of Pavia, Pavia, Italy
e-mail: roberta.besio@unipv.it

H.-K. Hoyer-Kuhn · O. Semler
Department of Pediatrics and Adolescent Medicine, Faculty of Medicine and University Hospital Cologne, University of Cologne, Cologne, Germany

A. Rossi, F. Zaucke (eds.), *The Extracellular Matrix in Genetic Skeletal Disorders*, Biology of Extracellular Matrix 16,
https://doi.org/10.1007/978-3-031-70835-0_2

2.1 Introduction

Collagen I is the most abundant fibrillar collagen in vertebrates and a major component of the extracellular matrix (ECM) in several tissues. It is synthesized as a procollagen precursor consisting of two proα1(I) and one proα2(I) chains assembled into a trimeric protein composed of an amino-terminal globular domain (N-propeptide) linked through a short linear N-telopeptide to a triple helical region characterized by 338 Gly-Xaa-Yaa repeats connected to a carboxyl-terminal globular domain (C-propeptide) by a C-telopeptide. After secretion, the proteolytic removal of the N- and C- propeptides allows the assembly of collagen I molecules into fibrils providing the tissue with tensile strength (Besio et al. 2019; Ishikawa et al. 2015, 2017; Ishikawa and Bächinger 2013).

Collagen I is the main constituent of the bone ECM, representing over 90% of its organic components where it has also a peculiar role acting as scaffold for mineralization (Gelse et al. 2003). Its negatively charged amino acids and carbonyl groups favor the nucleation sites for the initial formation of hydroxyapatite-like minerals by binding calcium ions. As a consequence, alterations in its quality and/or quantity might lead to bone fragility. Collagen I is present in the skin and tendon as well as in the connective tissues of many other organs. It is expressed in all major structures of lungs, including the airways, the interstitium of the lung parenchyma, and the basement membranes of epithelial and endothelial cell (Laurent 1986). In the skeletal muscle, collagen I represents the major structural protein of the ECM where it accounts for 1–10% of muscle mass dry weight and it plays an important role in muscle fiber force transmission, maintenance, and repair (Gillies and Lieber 2011). Collagen I is also the main component of the myocardium, cardiac valves, and aortic wall, and its altered synthesis can affect their biomechanical properties (De Vlaming et al. 2012). Indeed, type I collagen was isolated from the cornea, sclera, uveal tissues (Schmut 1978), and gut (Ramadan et al. 2022), where its function, besides the structural support, is still poorly investigated.

In this chapter, we will (I) review what is known about the complex intracellular biosynthesis and extracellular processing of collagen I, (II) provide an overview of the role of collagen I in different tissues, and (III) address how alterations in the collagen I quality and/or quantity might affect the function of these tissues. This chapter shall provide a deep understanding of the molecular basis of collagen-related disorders and their skeletal and extraskeletal manifestations using osteogenesis imperfecta (OI) as a prototype for collagen I-related diseases.

2.2 From Intracellular Synthesis to Extracellular Matrix Assembly

Upon nuclear transcription of *COL1A1* and *COL1A2* genes, their mRNA translation occurs in the endoplasmic reticulum (ER) where the proα1(I) and proα2(I) chains are independently synthesized and assembled with a 2:1 stoichiometry at a specific sequence located in their C-terminal propeptide (Fig. 2.1). Covalent disulfide bonds are formed between cysteine residues to covalently anchor the proα chains allowing their folding from the carboxyl- to the amino-terminal end. The presence of glycine residues at every third position in the triple helical domain is necessary for the zipper-like folding of the three chains, since only the small hydrogen atom side

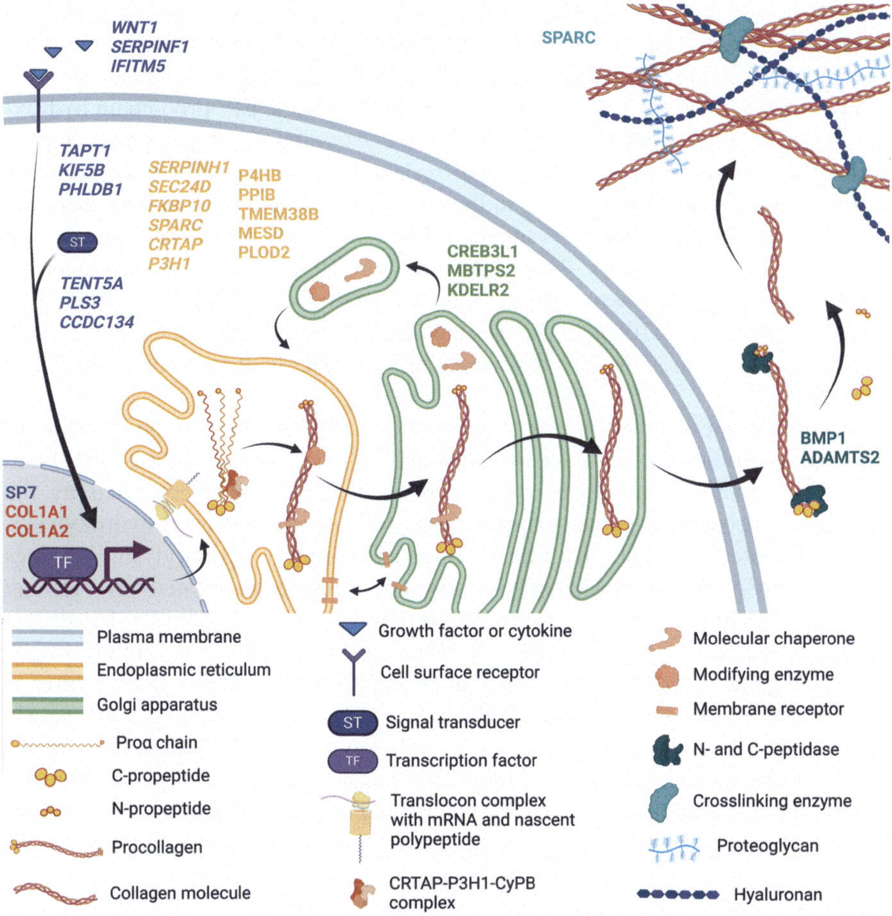

Fig. 2.1 Genes encoding for proteins involved in synthesis, secretion, and extracellular assembly of collagen I whose mutations cause OI or OI-like diseases. The genes are assigned to the cellular compartments where the encoded proteins belong to

chain of this amino acid can fit inside the triple helix without causing distortion (Engel and Bächinger 2005; Brinckmann 2005).

Prior to procollagen chain assembly and folding, C-propeptide interacts with the ER resident proteins protein disulfide isomerase (PDI) and immunoglobulin heavy-chain-binding protein (BIP) (Duran et al. 2017; Chessler and Byers 1993). Procollagen interactions with these chaperones facilitate the retention of unassembled chains within the ER, directly preventing the secretion of procollagen until the correct native structure is reached. After completion of the stoichiometric chain selection through the C-propeptides, following *cis* to *trans* prolyl peptide bonds isomerization, triple helix formation proceeds from the carboxyl-terminal to the amino-terminal end (Bachinger et al. 1978). Thus, the *cis-trans* isomerization of prolyl peptide bonds catalyzed by peptidyl prolyl cis-trans isomerases (PPIase) represents the rate-limiting step in triple helix folding (Bächinger 1987; Vranka et al. 2004; Marini et al. 2007). Collagen folding also requires the specialized, substrate-specific chaperone heat shock protein 47 (HSP47) that preferentially binds to the folded chains and prevents local triple helical unfolding and intracellular aggregation (Ishida and Nagata 2011; Ishikawa et al. 2008). In the ER, proα chain hydroxylations and glycosylations occur, which are essential for the formation of a functional protein. Briefly, prolyl 4-hydroxylase hydroxylates in C4 almost every proline residue at the Yaa position in the collagenic triplet Gly-Xaa-Yaa, ensuring interchain hydrogen bonds relevant for folding and stability of the triple helical domain (Rappu et al. 2019). Prolyl 3-hydroxylase (P3H1), in a complex with the helper protein cartilage-associated protein (CRTAP) and the PPIase cyclophilin B (CyPB), hydroxylates 100% of the $\alpha 1$ chains at position Pro986 in C3 and 70% of the $\alpha 2$ chain at position Pro707, allowing intermolecular hydrogen bonds (Ishikawa et al. 2009).

Specific lysine residues in the triple helical and N-/C-telopeptide domains of unfolded procollagen chains are hydroxylated by lysyl 1-hydroxylase (LH1) and LH2, respectively. Some hydroxylysine residues are then glycosylated by the enzymes hydroxylysyl-galactosyltransferase (GT) and galactosyl-hydroxylysyl-glucosyltransferase (GGT) that transfer a galactose unit to hydroxylysine residues and a glucose unit to galactosyl-hydroxylysine residues, respectively (Hennet 2019). The O-linked glycosylation of the modified Lys residues in the ER differs from conventional O-glycosylation of Ser/Thr occurring in the Golgi. Interestingly, a collagen glycosyltransferase activity has been found associated with the hydroxylase activity in the enzyme lysyl 3-hydroxylase (LH3), likely playing a still poorly understood role in collagen synthesis (Sricholpech et al. 2012; Scietti et al. 2018). Despite these findings, the precise molecular loci and the extent/type of glycosylation in cross-linked and non-cross-linked residues are still not fully characterized (Hanson and Eyre 1996). The intracellular collagen biosynthesis is further complicated by the assembly of the posttranslational modifying enzymes in different complexes, whose existence and role have only recently been identified. Prolyl 3-hydroxylase 3 (P3H3) and LH1 assemble with synaptonemal complex 65 (Sc65) and, likely, CyPB, regulating triple helical Lys87 and Lys930 hydroxylation (Gruenwald et al. 2014; Heard et al. 2016), whereas the LH2, HSP47, FKBP65, and

BIP complex regulates type I procollagen telopeptide lysyl hydroxylation (Duran et al. 2017). HSP47 is also bridging procollagen molecules with the ER protein transmembrane protein transport and Golgi organization 1 (TANGO1), a key player in the formation of the COPII vesicles/tunnels used by procollagen to leave the ER and move to the Golgi intermediate compartment (ERGIC) (Bunel et al. 2024). The procollagen transport to post-Golgi is mediated by tubular saccular carriers originated by the detachment of large regions of the trans Golgi, and bundles of procollagen are then released from the *trans*-face of the Golgi to form secretory vacuoles, as shown by ultrastructural studies in human fibroblasts (Leblond 1989; Marchi and Leblond 1984; Trelstad and Hayashi 1979).

Following procollagen secretion, the globular N- and C-terminal propeptides are removed in the pericellular space by the specific zinc-proteases ADAM metallopeptidase domain 2 (ADAMT2) and bone morphogenetic protein 1 (BMP-1), respectively. This step leads to decreased protein solubility, increasing the spontaneous self-assembly of collagen molecules into highly organized supermolecular fibrillar structures (Zhang et al. 2005). A model for connective tissue growth has been proposed in tendons in which single collagen fibrils are produced in cellular recesses and brought together to form collagen fibril bundles that are assembled into higher-order structures, such as tendon fascicles (Graham et al. 2000). Collagen fibril growth may then occur at the base of the fibril in the fibripositor by the addition of individual collagen molecules to the ends of the fibrils or by end-to-end fusion with nascent short fibrils (Holmes et al. 1992, 1998; Kadler 2017). The collagen fibrils are the result of the lateral assembly of many collagen units in a staggered manner, stabilized by cross-links between lysine and hydroxylysine residues. Covalent intermolecular cross-linking of collagen is initiated by the conversion of Lys/Hyl residues of the non-helical collagen C- and N-telopeptides to the respective aldehydes Lysald/Hylald, catalyzed by lysyl oxidase. The aldehyde can then form an iminium bond to a juxtaposed amino group of Lys/Hyl in the helical domain of the staggered adjacent molecule. In skeletal tissues, these divalent cross-links can then mature into trivalent cross-links, pyridinoline (Pyr), deoxypyridinoline (d-Pyr), and pyrrole (Prl), tying two or three collagen molecules (Eyre and Weis 2013). The extracellular assembly of collagen molecules (~300 nm in length) in a quarter staggered fashion creates the so-called "gap" and "overlap" zones (D-period). Arrays of collagen maintain precise registration of their gap and overlap regions, so that channels are created through the array.

2.3 Effect on Collagen I Quantity and Quality in OI and OI-Like Disorders

Mutations in the collagen I α chains and in proteins involved in its synthesis (Fig. 2.1), as well as defects affecting differentiation and/or activity of the cells producing high amount of this protein, are associated with connective tissue disorders most impairing skeleton structure and function as a consequence of quantitative or qualitative defects of collagen I (Jovanovic et al. 2022).

Osteogenesis imperfecta (OI) is the most common among the rare heritable bone disorders with an incidence of 1:10–15,000 and represents a prototype for collagen I-related diseases. It is a family of disorders characterized mainly by growth delay, skeletal deformity, and bone fragility, although due to the widespread distribution of collagen I in the body, several extraskeletal features, such as muscle weakness and cardiovascular and lung impairment, frequently present in aged people, have been described starting in young age and worsening with time. The dominant "classical" OI forms, caused by mutations in collagen I genes (*COL1A1* and *COL1A2*), are the most common types accounting for 85% of the OI cases (Garibaldi et al. 2022; Marini et al. 2017). In 1979, Sillence classified OI on the basis of clinical, radiographic, and genetic features in four types from types I to IV, including patients with mild (OI type I), moderate (OI type IV), severe (OI type III), and lethal (OI type II) phenotype (Sillence et al. 1979). From 2006 the discovery of mutations in different causative genes (Jovanovic et al. 2022) required a revision of the original classification, and sequential numbers were assigned to the newly discovered defects opening indeed a still running discussion in the field regarding the best way to classify the disease. In the presence of a wide range of clinical overlapping features, mutations affecting specific molecular pathways paved the way for grouping the different forms of OI based on the affected biochemical pathways (Forlino et al. 2011). Such classification provides help for better and personalized therapy; nevertheless, it groups patients with similar phenotypic severity in different OI categories. The discussion on classification is beyond the purpose of this chapter and will not be further addressed. In this chapter, the numerical OI type reference, as reported in OMIM, is used (Table 2.1). The causative genes responsible for OI and OI-like disorders and their effects on quantity and quality of collagen I are grouped based on the type of transmission (autosomal dominant, autosomal recessive, or X-linked) and listed in alphabetical order (Table 2.1). The biochemical role of the encoded proteins is also indicated as well as the models used to dissect the effect of the mutation on collagen I.

Traditionally quantitative defects have been associated to *COL1A1* haploinsufficiency responsible for synthesis of half of the amount of otherwise normally structured collagen I (Willing et al. 1994). Nowadays, thanks to deep in vitro and animal model investigation, it became more and more clear that matrix insufficiency is a common feature for all classical dominant OI forms in which the abnormal collagen is only partially secreted and the rest is intracellularly retained, causing cellular stress and impairing cellular homeostasis (Garibaldi et al. 2021, 2022; Besio et al. 2018; Gorrell et al. 2022).

Among the non-collagen I genes causing OI, the ones exclusively associated with a quantitative defect are either transcription factors (*CREB3L1* (Symoens et al. 2013), *SP7* (Lapunzina et al. 2010)), a ligand modulating nuclear translocation of the transcription factors co-activator β-catenin (*WNT1*) (Laine et al. 2013; Keupp et al. 2013; Pyott et al. 2013) (see Chap. 9), protease responsible for transcription factor activation (*MBPTS2*) (Lindert et al. 2016), chaperone modulating

Table 2.1 OI-causing genes and the effects of the encoded mutant proteins on collagen quantity and quality

Gene	Protein	Role	OI type/other diseases/ MIM	Mutation effects on collagen	References
Autosomal-dominant inheritance					
COL1A1	α1(I)	Collagen α-1(I) chain	I (166200) II (166210) III (166220) IV (259420)	**Quantitative** (human: human samples. Dog**:** *COL1A1*$^{G208A/+}$; *COL1A1* $^{G382A/+}$. Mouse**:** *Col1a1*$^{Mov13/Mov13}$; *Col1a1*$^{Mov13/+}$; *Col1a1* minigene; *Col1a1*$^{Jrt/+}$; *Col1a1*$^{\Delta 365/+}$; *Col1a1*$^{Aga2/+}$; *Col1a1*$^{Aga2/Aga2}$; *Col1a1*$^{seal/+}$. Zebrafish**:** *col1a1a*$^{+/-}$; *col1a1b*$^{+/-}$; *col1a1a*$^{+/-}$ *col1a1b*$^{+/-}$) **Qualitative** (human: human samples. Mouse**:** *Col1a1*$^{Brtl/+}$; *Col1a1*$^{Brtl/Brtl}$; *Col1a1*$^{Aga2/+}$; *Col1a1*$^{Aga2/Aga2}$. Zebrafish**:** *col1a1*$^{chi/+}$; *col1a1b*$^{dmh29/+}$; *col1a1a*$^{dmh13/+}$; *col1a1a*$^{dmh14/+}$; *col1a1a* $^{med/med}$)	(Asharani et al. 2012; Fisher et al. 2003; Forlino and Marini 2016; Gistelinck et al. 2018; Bonadio et al. 1990; Stacey et al. 1988; Pereira et al. 1993; Forlino et al. 1999; Lisse et al. 2008; Chen et al. 2014; Tabeta et al. 2017; Campbell et al. 2000; Garibaldi et al. 2022)
COL1A2	α2(II)	Collagen α-2(I) chain	I (166200) II (166210) III (166220) IV (259420)	**Quantitative** (Mouse**:** *Col1a2*$^{Amish/+}$. Zebrafish**:** *col1a2*$^{-/-}$) **Qualitative** (human: human samples. Dog**:** *COL1A2*$^{-/-}$. Mouse**:** *Col1a2*$^{oim/oim}$. Zebrafish**:** *col1a2*$^{dmh15/+}$)	(Gistelinck et al. 2018; Daley et al. 2010; Phillips et al. 2000; Campbell et al. 2001; Ju et al. 2020; Dimitra et al. 2020; Cundy et al. 2018)

(continued)

Table 2.1 (continued)

Gene	Protein	Role	OI type/other diseases/ MIM	Mutation effects on collagen	References
IFITM5	BRIL	Endochondral bone ossification and mineralization regulator	V (610967)	**Quantitative** (human: human samples)	(Reich et al. 2015)
KIF5B	Kinesin-1 heavy chain	Kinesin-1 subunit Transport of secretory vesicles, mitochondria, and RNA. Regulation of late cytokinesis in chondrocytes		**Quantitative** (human: human samples. Zebrafish: $kif5Bb^{-/-}$, $kif5Ba^{-/-}Kif5Bb^{-/-}$)	(Marom et al. 2023; Santos-Ledo et al. 2017; Gan et al. 2019)
P4HB	PDI	Protein disulfide isomerase, collagen modification regulator	Cole-Carpenter syndrome (112240)		(Rauch et al. 2015)
Autosomal recessive inheritance					
ADAMTS2	ADAM metallopeptidase domain 2	N-propeptide cleavage prior to fibril assembly	Ehlers-Danlos syndrome type VIIC (225410)	**Qualitative** (human: human samples)	(De Coster et al. 2007)
BMP1	BMP1	C-propeptide cleavage prior to fibril assembly	XIII (614856)	**Qualitative** (human: human samples. Mouse: $Bmp1^{-/-}$. Zebrafish: $bmp1^{frf/frf}$)	(Muir et al. 2014; Asharani et al. 2012)
CCDC134	CCDC134	Inhibition of MAPK phosphorylation, regulation of osteogenesis, and bone development	XXII (619795)	**Quantitative** (human: human samples)	(Dubail et al. 2020)
CREB3L1	OASIS	*COL1A1* promoter activation, osteoblast bone matrix protein secretion	XVI (616229)	**Quantitative** (human: human samples. Mouse: $Creb3l1^{-/-}$)	(Sekiya et al. 2010; Murakami et al. 2009; Lindahl et al. 2018)

CRTAP	CRTAP	Intracellular PTMs (α1(I) Pro986 and α2(I)Pro707 3-hydroxylation)	VII (610682)	**Qualitative** (human: human samples. Mouse: *Crtap*$^{-/-}$. Zebrafish: Crtap$^{-/-}$.) **Quantitative** (human: human samples. Mouse: *Crtap*$^{-/-}$)	(Tonelli et al. 2020; Morello et al. 2006; Kuptanon et al. 2022; Udupa et al. 2023)
FKBP10	FKBP10	Procollagen chaperone	XI (610968) Bruck syndrome type 1 (607063)	**Quantitative** (human: human samples. Mouse: *Fkbp10*$^{-/-}$) **Qualitative** (human: human samples. Mouse: *Fkbp10*$^{-/-}$, *ScxCre Fkbp10*$^{fl/fl}$)	(Lietman et al. 2014; Lim et al. 2021; Lietman et al. 2017; Tran et al. 2022; Bank et al. 1999; Li et al. 2019)
KDELR2	KDELR2	Recycling of ER-resident proteins from cis-Golgi to the ER through COPI retrograde transport	XXI (619131)	**Quantitative and qualitative** (human: human samples)	(van Dijk et al. 2020)
MESD	MESD	Embryogenesis regulator, LRP5, LRP6, and collagen chaperone	XX (618644)	**Quantitative** (human: human samples)	(Ghosh et al. 2023)
P3H1	P3H1	Intracellular PTMs (α1(I) Pro986 and α2(I)Pro707 3-hydroxylation)	VIII (610915)	**Qualitative** (human: human samples. Mouse: *P3h1*$^{-/-}$; *P3h1*$^{H662A/H662A}$. Zebrafish: *p3h1*$^{-/-}$.) **Quantitative** (human: human samples. Mouse: *P3h1*$^{-/-}$)	(Tonelli et al. 2020; Vranka et al. 2010; Homan et al. 2014; Kantaputra et al. 2021; Bala and Bala 2021)
PLOD2	PLOD2	Intracellular PTMs (telopeptides hydroxylation)	Bruck syndrome type 2 (609220)	**Qualitative** (human: human samples; zebrafish: *plod2*$^{sa1768/sa1768}$) **Quantitative** (zebrafish: *plod2*$^{sa1768/sa1768}$)	(Gistelinck et al. 2021; Gistelinck et al. 2016)

(continued)

Table 2.1 (continued)

Gene	Protein	Role	OI type/other diseases/ MIM	Mutation effects on collagen	References
PPIB	Cyclophilin B	Intracellular PTMs (α1(I) Pro986 and α2(I)Pro707 3-hydroxylation)	IX (259440)	**Quantitative** (human: human samples. Mouse: *Ppib*$^{-/-}$) **Qualitative** (human: human samples. Mouse: *Ppib*$^{-/-}$)	(Choi et al. 2009; Cabral et al. 2014; Pyott et al. 2011; Van Dijk et al. 2009)
SEC24D	SEC24D	ER protein traffic regulator	Cole-Carpenter syndrome type 2 (616294)	**Quantitative** (human: human samples. Zebrafish: *sec24d*$^{Bulldog/Bulldog}$)	(Garbes et al. 2015; Sarmah et al. 2010; Baines et al. 2013)
SERPINF1	PEDF	Anti-angiogenic factor, bone homoeostasis, and osteoid mineralization regulator	VI (613982)	**Quantitative** (humans: human samples. Mouse: Serpinf1$^{-/-}$)	(Bogan et al. 2013; Al-Jallad et al. 2015)
SERPINH1	HSP47	Procollagen chaperone	X (613848)	**Quantitative and qualitative** (humans: human samples. Mouse: Serpinh1$^{-/-}$. Dog: *SerpinH1*$^{Dachshund/ Dachshund}$)	(Lindert et al. 2015; Nagai et al. 2000; Lindert et al. 2016)
SP7	Osterix	Regulation of osteoblast differentiation, maturation, and activity	XII (613849)	**Quantitative** (humans: human samples. Mouse: *Sp7*$^{-/-}$; *Col1a1-CreERT2 Sp7*$^{flox/flox}$)	(Kague et al. 2016; Nakashima et al. 2002; Baek et al. 2010; Lui et al. 2022; Fiscaletti et al. 2018)
SPARC	Osteonectin	Procollagen assembly, cross-linking, mineralization, and osteoblast/osteoclast differentiation regulator	XVII (616507)	**Qualitative** (mouse: *Sparc*$^{tm1cam/tm1cam}$; human: human samples)	(Boskey et al. 2003; Mendoza-Londono et al. 2015)
TAPT1	TAPT1	Regulation of primary cilia formation, role in cartilage and bone development	–	**Quantitative and qualitative** (human: human samples; zebrafish: *tapt1a*$^{-/-}$/ *tapt1b*$^{-/-}$)	(Symoens et al. 2015; Etich et al. 2023)

TENT5A	TENT5A	Nucleotidyltransferase fold protein TGF-β signal transduction regulation	XVIII (617952)	**Quantitative and qualitative** (mouse: *Tent5a*$^{-/-}$)	(Gewartowska et al. 2021)
TMEM38B	TRIC-B	ER-cation channel formation, cell differentiation	XIV (615066)	**Quantitative and qualitative** (human: human samples and immortalized cell line hFOB. Mouse: *Tric-B*$^{-/-}$ and *Runx2-Cre;Tmem38b*$^{fl/fl}$. Zebrafish: *tmem38b*$^{-/-}$, *tmem38b*$^{\Delta 120-7/\Delta 120-7}$)	(Zhao et al. 2016; Forlino and Marini 2016; Tonelli et al. 2023; Leoni et al. 2021; Cabral et al. 2016; Besio et al. 2023; Webb et al. 2017)
WNT1	WNT1	Osteoblast lineage cell regulation, bone formation	XV (615220)	**Quantitative** (mouse: *Wnt1*$^{sw/sw}$)	(Joeng et al. 2014)
PHLDB1	PHLDB1	Insulin-dependent Akt phosphorylation	XXIII (620639)	No quantitative nor qualitative defect reported	(Tuysuz et al. 2023)
X-linked inheritance					
MBTPS2	MBTPS2; S2P	Regulation of osteoblast Development and ER Stress response	XIX (200294)	**Quantitative**: Human: human samples	(Schlombs et al. 2003; Lindert et al. 2016)
PLS3	PLS3	Regulation of bone remodeling, mechanosensitivity, osteoblastic mineralization, and bone resorption	X-linked osteoporosis (300910)	No quantitative nor qualitative defects reported	(Zhong et al. 2023; Van Dijk et al. 2013)

transcription factors intracellular trafficking (*MESD*) (Moosa et al. 2019), protein implicated in collagen I expression through ERK1/2 pathway (*CCDC134*) (Dubail et al. 2020), protein implicated in collagen secretion (*SEC24D*) (Moosa et al. 2016), or protein involved in intracellular transport (*KIF5B*) (Moosa et al. 2016; Marom et al. 2023). The two exceptions are the mutations in *IFITM5* encoding bone-restricted interferon-induced transmembrane (BRIL) (Cho et al. 2012) and *SERPINF1* encoding epithelium-derived factor (PEDF) (Homan et al. 2011; Becker et al. 2011).

In addition to OI types I–IV caused by mutations in collagen I encoding genes, also OI type V caused by mutations in *IFITM5* shows an autosomal dominant inheritance pattern. OI type V is mainly caused by a single point mutation in the 5′-UTR of *IFITM5* (c. − 14C > T), resulting in the addition of five amino acid residues (MALEP) to the N-terminus of BRIL, a transmembrane protein expressed in osteoblasts. Studies utilizing primary patient cells revealed a decreased collagen I expression, secretion, and ECM incorporation (Reich et al. 2015). The loss of function of the well-known antiangiogenic factor PEDF (encoded by *SERPINF1*) caused a reduced collagen quantity in cultured osteoblasts, but not in patient-derived fibroblasts where it showed normal modification, processing, and secretion (Al-Jallad et al. 2015). Interestingly, an additional *IFITM5* mutation found in a patient with atypical OI type VI has provided evidence that BRIL and PEDF are connected in a novel intracellular signaling pathway that regulates bone mineralization (Hedjazi et al. 2022; Guillén-Navarro et al. 2014). Recently, dominant inheritance has also been associated with OI forms caused by mutations in KIF5B (Marom et al. 2023) and *P4HB* (Rauch et al. 2015).

Causative genes encoding proteins involved in collagen I posttranslational modification (prolyl-3-hydroxylase complex, namely, CRTAP (Morello et al. 2006), P3H1 (Cabral et al. 2007), PPIB (Van Dijk et al. 2009), and PLOD2 (Ha-Vinh et al. 2004)), proteins with chaperone function ((FKBP10) (Alanay et al. 2010), SERPINH1 (Drögemüller et al. 2009; Christiansen et al. 2010), SPARC (Mendoza-Londono et al. 2015)), proteins involved in the retrograde transport of chaperone from Golgi to ER ((SEC24D) (Moosa et al. 2016), KDEL2 (van Dijk et al. 2020)), and protein involved in maintaining Golgi morphology (TAPT1) (Symoens et al. 2015) share qualitative and quantitative effects on collagen I. Impairment of calcium flux due to mutations in the endoplasmic reticulum trimeric intracellular potassium channel type B (TRICB) also affects quantity and quality likely due to the necessity of a fine tune calcium intracellular concentration for collagen expression, folding, posttranslational modification, and secretion (Shaheen et al. 2012). TENT5A, a cytoplasmic poly(A) polymerase expressed in osteoblasts and osteocytes, polyadenylates mRNAs encoding *COL1A1* and *COL1A2* as well as other proteins involved in bone formation. Mutations in this gene drastically decrease the synthesis of collagen impacting both its quantity and quality, as revealed by the aberrant structure of collagen fibers (Gewartowska et al. 2021).

Not surprisingly, mutations in bone morphogenic protein 1 (*BMP1*), the extracellular C-proteinase that removes the C-propeptide of procollagen, exclusively cause qualitative defects. The extracellular presence of pC-collagen molecules strongly affects the structure of fibrils with irregular shape and larger diameter, likely providing an increased number of sites for mineral nucleation in the bone matrix (Martínez-Glez et al. 2012; Hoyer-Kuhn et al. 2013).

No direct link on quantitative or qualitative collagen I defects has so far been described for *PL3* mutations (Van Dijk et al. 2013). This actin-binding protein seems to play crucial role in osteocytes, osteoblasts, and osteoclasts related to its function in regulating mechanotransduction, calcium regulation, vesicle trafficking, cell differentiation, and mineralization (Apperley et al. 2023).

2.4 Collagen I Containing Tissues and How OI-Causing Mutations Affect the Tissue Function

2.4.1 Bone

Bone includes living cells, osteocytes, osteoblasts, and osteoclasts, embedded in a mineralized matrix that consists of an organic and an inorganic phase, providing strength and stiffness and ensuring energy absorption, respectively. The organic portion of bone tissue is mainly represented by collagens, together with lower amounts of non-collagenous proteins (NCPs). Collagen I represents over 90% of total bone proteins, and it provides the scaffold for crystal deposition and contributes to tissue organization and hence to bone tissue mechanical properties. Noteworthy, low amounts of collagen V and XI are also part of the collagenous matrix functioning as regulators of fiber size (Gorski 2011; Alcorta-Sevillano et al. 2020).

NCPs mainly include proteoglycans, small integrin binding ligands (SIBLINGs), glycoproteins, and carboxylated proteins (Boskey 1989). These molecules are able to interact with collagen fibers, mediating and regulating their assembly and, generally, influencing bone development and maintenance. Among the SIBLINGs synthesized by osteoblasts and osteocytes, osteopontin is involved in bone remodeling, being able to inhibit calcification and hence required for bone resorption, while dentin matrix acidic phosphoprotein 1 (DMP-1), depending on its phosphorylation status, acts as a hydroxyapatite nucleator (non-phosphorylated form) or as a hydroxyapatite crystal growth inhibitor (phosphorylated form) (Gorski 2011).

The bone inorganic phase contains a combination of calcium and phosphorus salts predominantly hydroxyapatite ($Ca_{10}(PO_4)_6(OH)_2$). Calcium and phosphate ions are then actively secreted within matrix vesicles, and hydroxyapatite crystal are released onto the bone surface where initial formation of mineral crystals takes place (Blair et al. 2017). Collagen I fibrils at the level of the trivalent cross-links constitute the protein scaffold upon which osteoblasts deposit mineral, triggering mineralization. The maintenance of a net-negative charge close to the C-terminal end of the collagen molecules is thought to promote infiltration of amorphous calcium phosphate into the 40-nm-long gap zone (Nudelman et al. 2010; Landis and Jacquet 2013). Clusters of charged amino acids, both in the gap and overlap regions, form nucleation sites and control the conversion of amorphous calcium phosphate into apatite crystals. Once nucleated in the gap region, the hydroxyapatite crystals grow parallel to each other and to the collagen molecules associated with the crystals and serve as template for further propagation of the mineralization process of the osteoid, a process that takes several days and that is still not completely understood (Murshed 2018). Mineralization is inhibited by the presence of pyrophosphate, while it is promoted by its hydrolysis mediated by the osteoblast enzyme alkaline phosphatase.

2.4.2 Bone and OI

Bone is the tissue most affected in OI and thus the most investigated (Fig. 2.2). The main feature of OI bone is fragility, which usually results in fractures. Low bone mass and abnormalities in bone material properties characterize most of the OI types and are responsible for the bone fragility. The OI bone shows a reduced collagen amount, but also high number of nonenzymatic cross-links and disordered alignment of collagen fibrils (Carriero et al. 2014; Fratzl et al. 1996; Glorieux et al. 2000). Histomorphometric studies have shown a decrease in cortical width and trabecular bone volume in the classical OI types I, II, III, and IV, as well as in recessive

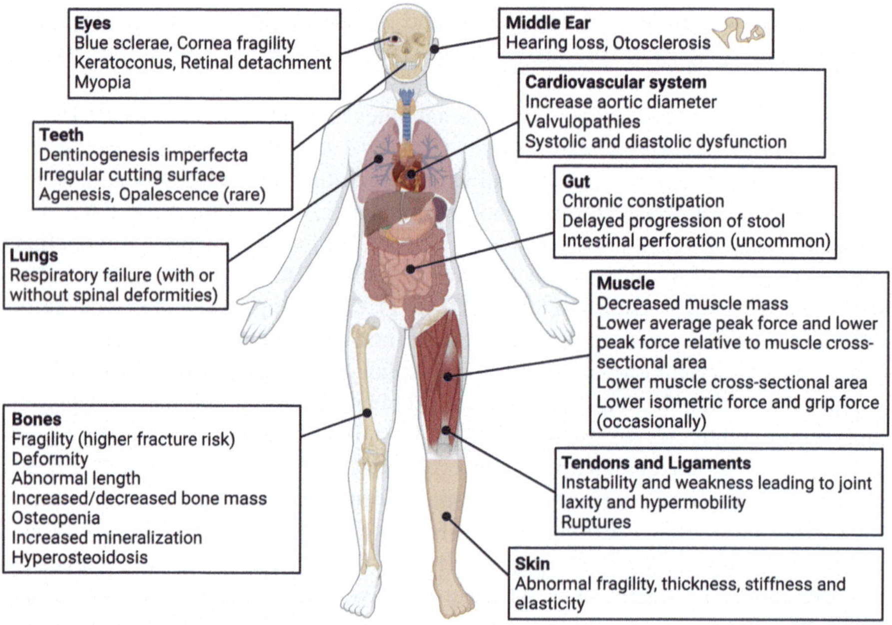

Fig. 2.2 Tissues affected by OI and main clinical phenotypes. Created with BioRender.com

type VII and type VIII. Interestingly, matrix production at the single cell level is markedly reduced, although bone turnover is increased due to the presence of a higher number of osteoblasts and osteoclasts. *TMEM38B* mutant bone (OI type XIV) has instead peculiar features since reduced osteoblast and osteoclast numbers and decreased bone resorption were reported (Webb et al. 2017).

A peculiar general property of OI bone is an abnormal mineral distribution, with an increased phosphate to amide ratio, indicating hypermineralization. Consistently, biopsies from children with a mild phenotype and either qualitative or quantitative *COL1A1* or *COL1A2* mutations or with a more severe phenotype as in type III and in recessive type VII and VIII revealed a similar increase in mean calcium content of the bone matrix (Roschger et al. 2008; Fratzl-Zelman et al. 2016; Weber et al. 2006; Fratzl-Zelman et al. 2010). In samples from children and murine models with mild osteogenesis imperfecta, small-angle X-ray scattering revealed that the thickness of bone mineral particles is similar or reduced compared to controls (Vanleene et al. 2012; Fratzl-Zelman et al. 2014). The combination of increased matrix mineral content and similar particle size implies that mineral particles are more tightly packed in the ECM, leaving less space for water and making the bone stiffer.

OI types V and VI constitute an exception to this rule. The presence of unmineralized osteoid and a fish-scale pattern under polarized light characterizes OI type VI bone deficient for PEDF (Glorieux et al. 2002; Homan et al. 2011), while mutations in the gene that encodes BRIL (OI type V) can either induce increased bone formation (gain-of-function effect if the protein is elongated), which leads to a mesh-like lamellation pattern under polarized light, hyperplastic callus and calcification of the interosseous membrane of the forearm, or impaired mineralization (mutations in the coding regions) (Glorieux et al. 2000; Semler et al. 2012; Farber et al. 2014). Normal matrix mineralization was reported in OI type XIV (Webb et al. 2017), while the effects of the very rare X-linked *PLS3* mutations on human bone mineral deposition is controversial since variable levels were reported in different individuals (Laine et al. 2015; Fahiminiya et al. 2014; Kämpe et al. 2017).

Resistance to fracture, not determinable in OI individuals, was extensively analyzed in OI murine models. Mechanical testing showed that the OI bone is generally less stiff, bends with less force, and needs a reduced loading to reach a point of non-recoverable deformity, leading to fracture (Carriero et al. 2014; Kozloff et al. 2004). The reduced resistance to fracture and the inability to dissipate energy efficiently are likely explained by the abnormalities of the mineralized matrix affecting structures that normally hinder crack propagation (Wagermaier et al. 2015).

2.4.3 Skin

In the dermal interstitial matrix, the major collagen fibrils are heterotypic and contain collagens I, III, and V. Collagen I is again the most abundant type, collagen III constitutes around one-third of the fibril content, while collagen V is only a minor component (Nyström and Bruckner-Tuderman 2019). The formation of fibrils occurs spontaneously and is largely regulated by the main components of the collagen fibrils themselves with collagen V controlling the initiation of this process (Nyström and Bruckner-Tuderman 2019). In addition, the complex fibril formation is regulated by many other fibril-interacting proteins such as small leucine-rich proteoglycans (SLRPs) and other so-called matricellular proteins, e.g., SPARC (Gubbiotti et al. 2016; Rentz et al. 2007). Mutations in genes encoding these proteins cause many different types of osteochondrodysplasia (Unger et al. 2023), and interestingly, mutations in *SPARC* have been associated with OI (Durkin et al. 2022; Mendoza-Londono et al. 2015). Depending on the location within the dermis, collagen fibrils have a variable diameter. The so-called papillary dermis contains mostly thinner fibrils, while the reticular matrix is characterized by thicker fibrils (Marcos-Garcés et al. 2014). The mechanical stability and functionality of the tissue are determined not only by fibril diameter but also by cross-links between individual collagen molecules. Cross-linking is mediated by lysyl oxidases between lysine and hydroxylysine residues in telopeptides of different collagen monomers (Eyre et al. 1984). Therefore, cross-linking can only be achieved if lysine residues were properly hydroxylated in the ER by lysyl hydroxylases, a reaction that is modulated by a set of enzyme and chaperone complexes in which mutations have also been linked to OI (Eyre and Weis 2013).

2.4.4 Skin and OI

Due to their easy accessibility, primary skin fibroblasts are often isolated from patients to perform in vitro analysis at the cellular and molecular level. In many cases, collagen I synthesis, posttranslational modification, secretion, and fibril formation were studied providing important insight into underlying disease mechanisms (Van Dijk et al. 2012). However, data from primary skin fibroblasts might not always be transferable to other collagen I-expressing cells, in particular osteoblasts, as expression levels, enzyme repertoire, cross-linking types, etc. might differ considerably (Sarafova et al. 1998). Therefore, results from cell culture experiments have to be interpreted carefully and conclusions should be drawn with caution.

Concerning the skin phenotype in patients, there are apparently contradictory statements in the literature. Some studies show that the skin of individuals with OI is more stiff and less elastic than the normal skin (Hansen and Jemec 2002), while others describe patients with thin and fragile, translucent, stretchy, and hyperelastic skin that is highly prone to tearing and bruising. Even patients with normal skin tensile strength and ratios of collagen I/III within the normal range were reported (Oxlund et al. 1985). These observations again indicate that the phenotype might strongly depend on the specific mutation. In a skin-focused study of 16 OI patients, in-depth ultrastructural and histological analysis in four patients provides novel insights into genotype-phenotype correlation, and common and distinguishing features between the different subtypes of OI were identified (Balasubramanian et al. 2016). Using electron microscopy, the authors consistently observed variable diameters of collagen fibrils and the presence of so-called collagen flowers when looking at cross sections of individual fibrils. Data from patients with mutations in non-collagen genes are scarce, and in most cases, the number of patients is too low to draw a valid conclusion.

Importantly, a skin phenotype has also been described in several animal species and models. An early study described general skin fragility in a newborn New Zealand Romney lamb with a lethal OI caused by a defect in collagen production (Arthur et al. 1992). However, many cat and dog species with OI have been described without any obvious skin phenotype, again suggesting that the skin phenotype might strongly depend on the specific mutation. In the meantime, more than ten different zebrafish lines to study the effects of OI-causing mutations have been generated (Marí-Beffa et al. 2021). However, the skin has been investigated only in depth in a model in which the α2(I) chain was completely absent. Compared to wild-type fish, the skin of *col1a2*-deficient fish was more fragile and easily damaged. The dermis in zebrafish is composed of two layers, and histological analysis revealed that adult skin has only half the thickness in mutant fish (Gistelinck et al. 2018). The skin has been characterized also in more detail in several mouse lines. The $Col1a1^{+/Mov13}$ mouse, carrying a single copy of the Moloney murine leukemia virus stably integrated into *Col1a1* gene, showed significantly lower *Col1a1* expression in the skin tissue (Bonadio et al. 1990; Claeys et al. 2023). A different murine model carrying a *Col1a1 trans*gene with the Gly859Cys substitution resulted in a strong reduction of the total collagen I content that was proportional to the level of mutant gene expression (Stacey et al. 1988). Significant qualitative as well as quantitative differences in protein expression profiles of skin samples were also identified by mass spectrometry $Col1a1^{Brtl/+}$ mice (Bianchi et al. 2012), carrying a α1(I)-Gly349Cys substitution. A very detailed analysis of skin properties has also been performed in the $Col1a2^{oim/+}$ mouse model, with a defect in the α2(I) that impairs its ability to assemble in heterotrimeric molecules (Canuto et al. 2012). By using high-resolution magnetic resonance imaging and Fourier transform infrared imaging spectroscopy (FT-IRIS), this study gave important insight into alterations at the molecular level. In mutant $Col1a2^{oim/oim}$ mice, the appearance of a collagen-depleted lower dermal layer was observed. The collagen network was more highly hydrated and FT-IRIS-defined measure of collagen integrity was 30% lower in the mutant mice (Canuto et al. 2012). In $Crtap^{-/-}$ and $Ppib^{-/-}$ mice, altered mechanical properties were reported. The peak load and stiffness of the skin in $Crtap^{-/-}$ mice were 60% lower than in wild-type animals (Baldridge et al. 2010), and the skin in $Ppib^{-/-}$ animals was characterized by a decreased stiffness and an increased laxity (Choi et al. 2009). Also, transgenic mice with inactive alleles for procollagen N-proteinase (ADAMTS-2) developed fragile skin (Li et al. 2001).

2.4.5 Cardiovascular System

In the heart, collagen I constitutes approximately 75–85% of the total collagen in the myocardium. It is deposited specifically in the interstitial space between cardiomyocytes, and it is also present in the valves and blood vessels (Del Monte-Nieto et al. 2020). Collagen I self-assembly is facilitated by fibronectin, integrins, and collagen V, resulting in the creation of fibers from 50 to 150 nm of diameter with a highly organized structure primarily aligned parallel to the long axis of cardiac muscle fibers (Medugorac and Jacob 1983).

The tensile strength of the heart relies on a precise organization of connective tissue layers, which allows electrical conduction, myocardial contraction, and relaxation (MacKenna et al. 1997), known as the epimysium, perimysium, and endomysium. The epimysium is the more external layer, enveloping the entire heart and anchoring it within the pericardial cavity. It consists of dense irregular connective tissue, primarily collagen fibers, which confer strength and protection to the heart (Janicki and Brower 2002). Moving inward, the perimysium surrounds bundles of cardiac muscle fibers, forming a middle layer. It plays a crucial role in distributing blood vessels and nerves, and it is one of the main contributors to the passive stiffness of the myocardium (MacKenna et al. 1997). The collagen fibers within the perimysium, predominantly composed of collagen I and elastin, are oriented initially in a wavy conformation along the length of the muscle fibers, and they become straighter as left ventricular pressure increases during diastole (Mackenna et al. 1996). Finally, the endomysium constitutes the inner layer, surrounding and interconnecting individual cardiac muscle cells. This layer consists of loose connective tissue, containing bundles of collagen fibril that bridge cells and fill the ECM (Robinson et al. 1983; MacKenna et al. 1997; Janicki and Brower 2002). Collagen I in the endomysium contributes to the elasticity of cardiomyocytes and ensures the coordinated movement of muscle fibers during the cardiac cycle (MacKenna et al. 1997). The presence of collagen III and in minor amount of elastin contributes to the formation of fine fibrils and guarantees elasticity to the heart ECM. The different ratios of collagens I and III influence the mechanical properties of the heart, with higher ratios of collagen III in the epimysium and decreasing concentrations in the perimysium. Collagens also serve as a scaffold for non-cardiomyocytes, such as microvascular endothelial cells and fibroblasts (Medugorac and Jacob 1983).

The cardiac valve leaflets are composed of three distinct layers: the ventricularis, spongiosa, and fibrosa. The fibrosa layer, situated nearest to the outflow, is rich in fibrillar collagens I, III, and V, contributing to the stiffness and ensuring the integrity of the valve leaflets (Wiltz et al. 2013). Collagen I is the dominant collagen type (74%) and is mainly restricted to the fibrosa layer, while collagen III is more widely expressed (24%) and the rest is collagen type V (Millington-Sanders et al. 1998). Fibrous cords that connect the valve leaflets to the papillary muscles in the ventricles exhibit a pattern of collagen fibrils forming bundles of planar waves that confer elasticity (Millington-Sanders et al. 1998). Additionally, a subendothelial basement membrane layer is present, composed of collagens I and IV, as well as laminin.

Various other minor components of the ECM, including vitronectin, fibronectin, osteonectin, and periostin, are also expressed throughout the leaflets (Latif et al. 2005).

Blood vessels consist of three main layers: the tunica intima, which is the inner layer composed of endothelial cells; the tunica media, a middle layer primarily consisting of smooth muscle cells and elastic fibers; and the outer layer, the tunica externa (adventitia), composed of connective tissue. Collagen I is primarily localized in the tunica media, surrounding the smooth muscle cells, and in the tunica adventitia, where it provides mechanical strength, maintains vessel integrity, and regulates vascular tone and endothelial cell function. The tunica adventitia consists of a loose ECM that includes collagen fibers, collagen-binding proteoglycans (such as biglycan and decorin), and hyaluronic acid (Del Monte-Nieto et al. 2020). Arteries are richer in elastin than in collagens I and III which provides the elasticity needed for their specific function, and in the case of capillaries, they lack a distinct tunica media and instead have scattered pericytes within the basement membrane (Wittig and Szulcek 2021).

2.4.6 Cardiovascular System and OI

Changes in the amounts or in the organization of collagen fibers in the ECM of the myocardium result in increased stiffness that ultimately compromises cardiac function (Nikolov and Popovski 2022). Not surprisingly, adults diagnosed with OI face a higher risk of cardiovascular diseases compared to the general population, with an increase in cumulative incidence for conditions such as mitral valve regurgitation, aortic valve regurgitation, atrial fibrillation, and heart failure (Folkestad et al. 2016; Verdonk et al. 2024; Radunovic et al. 2011). Although most studies show that increased aortic diameter and valvulopathies are the prevalent issues, there is evidence of diminished systolic and diastolic function in the left ventricle and, although in minor extent, in the right ventricle among individuals with OI types I, III, and IV (Chougui et al. 2020; Radunovic and Steine 2015).

While the relationship between specific mutations and cardiovascular impairments in OI remains an active area of investigation, data from children with OI highlight the prominent negative effects of *COL1A1* mutations. Patients with quantitative or qualitative changes in collagen I had larger diameters in the main pulmonary artery, increased left atrial and left ventricular end-diastolic dimension, and lower left ventricular ejection fraction than controls, whereas children with mutations in α2(I) or with mutations in non-collagen genes showed no significant differences (Zhao et al. 2022). Interestingly, only mild mitral, tricuspid, and/or pulmonic valve regurgitation has been reported in children diagnosed with moderate to severe OI, whereas adult OI patients showed a severe phenotype, suggesting a worsening in this condition with age (Pinheiro et al. 2020; Zhao et al. 2022).

COL1A1 mutations have been reported impacting on the vascular system. OI individuals with α1(I) mutations suffer from capillary fragility, reduced platelet retention, decreased levels of factor VIII-related antigen, impaired collagen-induced platelet aggregation, and deficiency in plasminogen activator inhibitor-1. These phenotypes cause susceptibility to bleeding, hyperfibrinolysis, and instability in clot formation (Chughtai et al. 2022).

Cardiovascular phenotype in OI murine models replicated some key outcomes observed in humans. *Col1a1*$^{Aga2/+}$ mice, carrying a point mutation into intron 50 resulting in a 16-base expansion of the transcript and showing a severe phenotype, exhibited enlarged septa, right ventricular hypertrophy, compromised blood vessel integrity, increased left ventricular end-systolic dimension, and lower ejection fraction than controls. In cardiac fibroblasts from *Col1a1*$^{Aga2/+}$ with a perinatal lethal phenotype, ECM was impaired due to lower *Col1a1* mRNA expression (Thiele et al. 2012). The *Col1a2*$^{oim/oim}$ mice showed normal systolic function despite thicker septal and posterior walls. These mice presented reduced fibril diameter and quantity of myocardial collagen with a significant reduction in fiber and chamber stiffness and an increase in collagen cross-linking (Weis et al. 2000). Another mouse model with a knock-out mutation in the *Col1a2* gene showed affected heart ECM and abnormal activation of cardiac fibroblasts inducing a fibrotic response that led to an impaired heart function at 9 months (Bowers et al. 2023). Valve disease has also been observed in mice. Aortic valves of *Col1a2*$^{oim/oim}$ presented distal thickening of the leaflets and increased proteoglycan deposition compared to control mice. Interestingly, the affectation was described when mice were 5 months old, and it increased with age, with more severe thickening at 9 months old (Cheek et al. 2012).

Additionally, also among patients with recessive forms of OI, individuals carrying mutation in *TMEM38B* (type XIV OI) causing loss of function of TRIC-B protein showed cardiac abnormalities such as atrial septal defect and ventricular septal defect, heart failure, tricuspid regurgitation, asymmetric septal hypertrophy of the left ventricle, and mild aortic root dilation (Webb et al. 2017). Accordingly, double knockout mutations in TRIC-A and TRIC-B caused embryonic lethality in mice due to cardiac arrest, and *Tric-a*$^{-/-}$ and *Tric-b*$^{+/-}$ mice are susceptible to stress-induced heart failure (Zhou et al. 2014).

2.4.7 Muscle

Collagens form a scaffold, together with proteoglycans, and elastin, that supports the muscle fibers, blood vessels, and nerves, maintaining the overall shape and organization of the muscle, distributing the mechanical stress evenly across the tissue, and preventing excessive deformation during contraction. Thus, they play an important role in transmitting muscle force to tendons and bones, allowing movement and locomotion. As in the heart, collagens form a network of intramuscular connective tissue organized in endomysium, perimysium, and epimysium. Collagen I assembles in parallel fibers providing tensile strength and rigidity to the muscle, whereas collagen III forms a loose meshwork of fibers that confers elasticity to endo- and perimysium (Kovanen 2002). Collagen V, another fibril-forming collagen, associates with collagens I and III and may form a core for nucleation of collagen fibrils in perimysium and endomysium (Kovanen 2002).

Other collagens, such as IV and VI, are involved in the formation and stabilization of the basement membrane surrounding muscle fibers. The endomysium interfaces with the myofiber sarcolemma at a specialized basement membrane, which consists primarily of collagen IV and laminin, the latter serving as ligand for two sarcolemma receptors—the dystrophin-associated glycoprotein complex and the α7β1 integrin (Sanes 1982; Grounds et al. 2005). Of relevance, collagen IV and laminin concentrations differ based on muscle fiber type, with slow twitch fibers showing higher concentrations of collagen IV and lower concentrations of laminin (Kovanen et al. 1988).

Collagens have unique interactions with proteoglycans that maintain the structure and organization of the muscle matrix, and indeed, decorin, the major proteoglycan in the perimysium, is a main regulator of collagen type I fibrillogenesis in skeletal muscle. Biglycan, which competes with decorin for the same binding site on collagen, is likely to play a role in both muscle formation and regeneration (Grounds et al. 2005; Kovanen et al. 1988).

2.4.8 Muscle and OI

A variable decrease of muscle mass is found both in adults and children with OI, correlating with OI type (Tauer et al. 2019). Reduced mobility, due to the bone deformity and fragility, can indirectly cause muscle wasting; nevertheless, also the presence of mutant collagen I can directly affect the muscle (Veilleux et al. 2014, 2017). Abnormalities in OI muscle were first found in the *Col1a2*$^{oim/oim}$ model of severe OI. In *Col1a2*$^{oim/oim}$ less fibrillar collagen and reduced peak tetanic force relative to muscle cross-sectional area were reported, but with a normal muscle fiber size and distribution (Gentry et al. 2010; Jeong et al. 2018).

In humans, impaired muscle force and, to a lesser extent, muscle power are associated even with mild OI forms (Takken et al. 2004; Veilleux et al. 2014). Lower isometric force and grip force were found in mild OI I together with a smaller muscle size and normal muscle density. OI individuals showed lower average peak force and lower peak force relative to muscle cross-sectional area in hopping tests. Despite these significant differences, many children with OI type I had normal muscle function, supporting the hypothesis that quantitative collagen defects in the muscle are less harmful than qualitative ones.

Interestingly, children with type I OI who experienced muscle force deficits were active as their healthy peers, suggesting that reduced physical activity is not the leading cause of muscle weakness. Proprioceptive deficit might also contribute to explain the muscle phenotype in OI since many tissues that are involved in postural control, such as skin, tendons, ligaments, and muscles, are rich in collagen I (Engelbert et al. 2004).

The low functional mobility due to bone deformities makes it difficult to perform muscle function tests in individuals with moderate to severe OI (OI types III and IV), where only muscle size at the forearm can be generally successfully measured. In a large study of OI children with OI types III and IV, muscle cross-sectional area adjusted for forearm length was significantly lower in patients with OI types III, but not in patients with OI type IV (Palomo et al. 2016). Also, grip force was found unaltered in individuals with OI type IV when compared to height-matched controls (Veilleux et al. 2017), suggesting that the involvement in OI varies with the disease-causing *COL1A1/COL1A2* mutation and severity (Gremminger and Phillips 2021). However, in a recent mechanographic study on 27 children and adolescents with OI type IV, in the presence of normal muscle mass and function at the upper extremities, OI individuals revealed weakness in the lower extremities, suggesting a muscle impairment also in this OI type (Veilleux et al. 2017).

The presence of altered collagen in muscle ECM and of an altered signal transduction and mechanotransduction between muscle and bone seem to be major contributors of OI muscle weakness. The middle part of many muscles is in direct contact with a periosteal bone surface allowing a direct interaction through paracrine and endocrine factors between the two tissues. For instance, the release of the cytokine TGF-β, which is abundant in OI bone, can contribute to explain the decreased muscle mass (Klein 2022).

$Col1a1^{Jrt/+}$, a combined model of severe type IV OI and Ehlers-Danlos syndrome (EDS) due to a *Col1a1* gene mutation leading to skipping of exon 9 and subsequent 18-amino acid deletion in the triple helical domain, exhibited skeletal muscle weakness (Boraschi-Diaz et al. 2017; Abdelaziz et al. 2015) with a downregulation of genes involved in muscle contraction (Moffatt et al. 2021). In contrast, no muscle defect was found in the murine model with a milder form of OI, the $Col1a2^{Amish/+}$ mouse, carrying a Gly substitution in the collagen I α2 chain. $Col1a2^{Amish/+}$ showed normal locomotor activity, muscle mass, fiber size, and force production in functional tests, suggesting that the muscle abnormality in the OI mouse model depends on the specific collagen mutation (Jeong et al. 2015). Severe motor impairments and reduced grip strength were also found in mice without collagen mutations but showing collagen qualitative defects associated with mutations in the CRTAP protein (Grol et al. 2021).

Furthermore, evidence from the $Col1a1^{Jrt/+}$ mouse suggests that metabolic perturbations exist, consisting of a hypermetabolic state with increased whole-body oxygen consumption and energy expenditures, potentially contributing also to the pathophysiology of the muscle phenotype (Boraschi-Diaz et al. 2017). A recent study in the homozygous $Col1a2^{oim/oim}$ mouse demonstrated altered energy metabolism supported by skeletal muscle mitochondrial dysfunction, as evidenced by drastically reduced gastrocnemius mitochondrial respiration rates (Gremminger et al. 2019). Unfortunately, limited studies evaluating energy metabolism in OI patients are available, and the presence of a hypermetabolic state in OI individuals was only described in a study from 1972 (Cropp and Myers 1972). Since skeletal muscles account for roughly 20% of the basal metabolic rate, an effect on energy metabolism in OI is likely to occur (Zurlo et al. 1990; Heymsfield et al. 2019).

The involvement of the skeletal muscle in OI is a relatively recent finding, and understanding the mechanisms underneath its dysfunction is crucial to identify novel therapeutic ways to improve the muscle phenotype and to benefit in bone strength through the muscle-bone crosstalk.

2.4.9 Tendons and Ligaments

In general, the organization of the tendon and ligament ECM is rather comparable, with a slightly lesser degree of fibril organization in the latter. Collagen I is the major protein component in both structures, accounting for 65–80% of the tissue dry weight. The total collagen amount can vary slightly in distinct anatomical localizations. Approximately 95% of the total collagen in tendons is organized in collagen fibrils in a quarter-staggered orientation. Collagen I can also polymerize together with collagen III into heterotypic fibrils (Siadat et al. 2021). The diameter of fibrils can vary, and fibrils are eventually bundled together to form collagen fibers that determine tensile strength and thus the mechanical properties of the tissue. Collagens V and XI can also be found in tendons and have been shown to exert a regulatory effect on the diameter of collagen I-containing fibrils (Wenstrup et al. 2011). Minor amounts of collagens XII and XIV were also detected, and a role in the establishment of tendon structure and function was suggested (Izu et al. 2021). Together with elastin and fibrillins (only 1–2% of the dry weight) forming elastic fibers, collagen I is embedded in the tendinous ground substance that consists of different proteoglycans, GAGs, structural glycoproteins, and various other small molecules (Kannus 2000). The best characterized and functionally most relevant non-collagenous components of the tendon ECM belong to the family of small leucine-rich proteoglycans (SLRPs). Compelling evidence for their role in regulating the collagen fibril diameter is provided by mouse models that are deficient in individual SLRPs. Decorin and biglycan are found in the interfibrillar and interfascicular matrix and do not only share a binding site for collagen I (Schönherr et al. 1995) but also play a specific role in collagen fibril assembly at different postnatal developmental stages (Zamboulis et al. 2020). Biglycan promotes fibril diameter growth, while decorin controls the lateral fusion of the fibrils (Zhang et al. 2005). Fibromodulin and lumican also interact with collagen I via a shared binding site, and their lack in mouse lines results in an abnormal collagen fibril structure. Lumican-deficient animals show an increase in fibrils with larger diameter, while in fibromodulin-deficient mice, an increase in smaller diameter fibrils was observed (Chen and Birk 2013). Among the glycoproteins, members of the thrombospondin family have been described as collagen-interacting proteins being expressed in tendons and ligaments. In particular, the lack of thrombospondin-4 and thrombospondin-5 resulted in an altered collagen fibril phenotype that might affect the mechanical properties of tendons and ligaments (Schulz et al. 2016; Frolova et al. 2014).

2.4.10 Tendons and Ligaments and OI

Many individual case reports describe ruptures of Achilles, patellar, quadriceps, and palmar flexor tendon. In addition, avulsion fractures associated with the patellar and triceps tendon have been described. Ligament injuries are less frequently reported, but there are also individual cases describing anterior cruciate ligament and medial collateral ligament tears (Puvanesarajah and Sponseller 2014). Ligament instability and weakness is a major reason for joint hypermobility that has been reported in two-thirds of patients with OI (McKiernan 2005). Another study assessing prevalence of hypermobility in adults with OI reported hypermobility in 44% of analyzed individuals with a positive association with the severity of OI (Wekre et al. 2010). Alterations in tendons and ligaments are partially responsible for the development of a variety of pathologies such as scoliosis, spondylolisthesis, ankle laxity, and flatfoot (Puvanesarajah and Sponseller 2014). In questionnaires, OI patients also reported an increased prevalence of osteoarthritis (OA) that started at an earlier age compared to healthy controls (Andersen et al. 2022).

Tendons have been investigated in several OI mouse models. In the tail tendon fascicle of 8-week-old *Col1a1*$^{\mathrm{Mov13/Mov13}}$ mice, similar amounts of total collagen and no alterations of tissue mechanics were reported (Derwin et al. 2001; Derwin and Soslowsky 1999). Also in *Col1a1*$^{Jrt/+}$ mice, the amount of collagen was unchanged; however, the thickness and cross-sectional area of Achilles tendons were reduced leading to a diminished stiffness and peak and yield force (Sinkam et al. 2023). Recently, it has been reported that in young (5 weeks old) *Col1a2*$^{oim/oim}$ mice, tendons were less rigid and less resistant than those of wild-type animals. In addition, the elastic modulus was significantly lower in *Col1a2*$^{oim/oim}$ tendons. In older animals (14 weeks old), the cross-sectional area of flexor digitorum superficialis (FDS) tendons was reduced and the collagen organization was altered. Mechanical testing revealed that the stress relaxation of *Col1a2*$^{oim/oim}$ tendons was significantly lower. However, these studies also demonstrated that not all tendons in this model are affected to the same extent (Chretien et al. 2022, 2023). In *Col1a1*$^{\mathrm{Brtl/+}}$ mice, the tendon structure was analyzed using X-ray diffraction indicating that the Gly349Cys substitution appeared to disrupt tendon crystallinity, in particular in 1-month-old animals (Kuznetsova et al. 2004). Alterations in nanoscale morphology and mechanics were confirmed in *Col1a1*$^{\mathrm{Brtl/+}}$ tendons by atomic force microscopy (Kemp et al. 2012).

Tendons have also been analyzed in some models of nonclassical OI. Mice with a tendon-specific deletion of *Fkbp10* showed reduced lysyl hydroxylation and, as a consequence, a reduced collagen cross-linking in tendons. These alterations induced a site-specific fibrosis, inflammation, and ectopic chondrogenesis resulting in joint deformities (Lim et al. 2021). Lysyl hydroxylation was also affected in cyclophilin B-deficient mice, but in a site-specific manner. Lysyl hydroxylation was decreased at helical sites, but increased in the telopeptides. These alterations led to tendon collagen with severe size and organizational abnormalities (Terajima et al. 2016). Prolyl 3-hydroxylase 1 (P3H1)-deficient mice displayed an abnormal tendon morphology with acute ultrastructural defects in overall shape and diameter (Vranka et al. 2010). In mutant mice, larger fibrils seemed to branch into numerous smaller fibrils, and axial twists were observed in longitudinal sections using electron microscopy. In addition, there was an increase in the number of fibrils with smaller diameter. At the molecular level, collagen molecules were overmodified and the secretion of collagen I in fibroblasts was delayed. P3H1 and CypB form a stable complex with CRTAP and also *Crtap*-deficient mice displayed a tendon phenotype. Mutant Achilles and patellar tendons were thinner with increased collagen cross-links and reduced fibril size. Gene expression analysis indicated an excessive matrix remodeling and inflammation correlating with a tendon pathology accompanied by severe motor impairments and reduced grip strength (Grol et al. 2021). Interestingly, a Dachshund puppy with a mutation in *SERPINH1* showed a wobbly gait due to ligament hyperlaxity and unstable joints, especially the carpal joints (Drögemüller et al. 2009).

2.4.11 Teeth

Collagen I represents the major protein (90%) of intertubular dentin, being assembled in 20-nm-thick fibrils together with trace amounts of collagens III and V and non-collagenous proteins (NCPs) (Goldberg et al. 2011). This dense network of collagen fibrils and NCPs provides an organized and rigid scaffold, required for the deposition of hydroxyapatite crystals (i.e., mineralization) and hence for correct dentin development and maintenance. Being composed of dentin tubules surrounded by peritubular and intertubular dentin, dentin represents one of the major components of teeth, together with enamel, cementum, and pulp. The enamel is the extremely hard outer covering of the tooth crown, that is, the part of the tooth which is outside the gingiva (Gil-Bona and Bidlack 2020). Cementum is the thin coating surrounding the tooth root, connecting it to the periodontal ligament and hence the socket alveolar bone (Foster 2017). Dental pulp is the soft, innermost part of each tooth, containing blood vessels, nerves, and odontoblasts intercalated inside connective tissue and protected by dentin (Yu and Abbott 2007). Collagen I has a crucial role in the correct development, maintenance, and functionality of the whole tooth. It is actively secreted into the predentin by odontoblast processes, residing inside the dentin tubules. Therefore, collagen fibrils can assemble forming a rigid matrix, becoming mature intertubular dentin upon mineralization. Dentin NCPs include SLRPs such as decorin, lumican, fibromodulin, and osteoadherin. These proteoglycans interact with collagen fibrils, regulating their diameter. In addition, dentin contains members of the SIBLING (small integrin-binding ligand, N-linked glycoproteins) family such as bone sialoprotein (BSP) and DMP-1promoting collagen fibrillogenesis and hydroxyapatite deposition. Growth factors (e.g., FGF2, TGFβ1, BMPs), enzymes (e.g., phosphatases, collagenases, and ADAMs), and calcium-binding proteins (e.g., calmodulin) have also been detected in the intertubular dentin matrix (Goldberg et al. 2011).

The periodontal ligament (PDL) connects the cementum to the alveolar bone, protecting teeth from mechanical stress and infections. Its elastic properties are conferred by the abundance of collagens I and III, secreted by PDL cells together with fibronectin and proteoglycans (e.g., decorin and biglycan). PDL collagen fibers are directly connected to the cellular and the acellular cementum which mainly consists of collagens I and III (Nakamura et al. 2021; Häkkinen et al. 1993).

2.4.12 Teeth and OI

Since collagen is the major component of dentin, abnormalities in its assembly and maturation due to genetic mutations can result in dentinogenesis imperfecta (DI). This pathology causes dentin structural defects both in primary and permanent dentitions, abnormal tooth coloring, and thinner enamel layer. This condition can lead to enamel fractures upon mastication and subsequent dentin exposure and damage. *COL1A1/2*-OI (types I, III, and IV) can cause variable degrees of DI in primary dentition, improving in permanent teeth, that can lead to impaction, malocclusion, and anomalous tooth eruption in patients (Forlino et al. 2011; Garrocho-Rangel et al. 2019). Even mutations in non-collagenous genes can result in DI. For example, individuals with missense mutations in *SERPINH1* (OI type X) can display DI. Interestingly, also $SerpinH1^{Dachshund/\ Dachshund}$ dogs with a missense mutation on the same gene suffer from DI (Drögemüller et al. 2009). Mutations in *SEC24D*, the gene encoding a component of the COP-II vesicles responsible for Cole-Carpenter syndrome type 2, can also cause DI (Forlino et al. 2011; Garrocho-Rangel et al. 2019; Garbes et al. 2015).

2.4.13 Sclera and Cornea

The sclera, the hard white opaque outer covering of the eyeball, and the cornea, the transparent dome-shaped structure present on the front of the eye, are components necessary to accurately focus light onto the retina allowing vision. They together constitute the external coat of the eyeball. The main function of the sclera is supporting the retina and protecting the other mechanical vulnerable internal structures of the eye against trauma and breakage, whereas the cornea acts as a camera lens, helping to focus light coming into the eye onto the retina (Boote et al. 2020; Espana and Birk 2020).

Their structure and functions are highly dependent on the organization of their ECM mainly made by fibrillar and non-fibrillar collagens interacting with different types of extracellular matrix proteins. Of the four layers of the sclera (from out to in: episclera, stroma, lamina fusca, and endothelium), the major and most distinct one is the stroma, a collagen-rich ECM immersed in a hydrated interfibrillar matrix of proteoglycans and glycoproteins. About 95% of the collagens consist of collagen type I, the rest being represented by collagens III, V, and VI. The fibrils are heterotypic, including a core of fibrillar collagen types I and III decorated outside by collagen V and interacting with collagen VI molecules distributed in the inter-bundle space (Marshall et al. 1993). The role of collagens V and VI is likely to regulate fibril assembly and diameter (Izu et al. 2011; Linsenmayer et al. 1993; Wenstrup et al. 2004). Present in the interfibrillar space, the proteoglycans regulate fibril size and organization. The main proteoglycan is aggrecan, even if its role is still unclear, but also the SLRPs decorin and biglycan are present (Rada et al. 2000; Rada et al. 1997). Thanks to their negative charge, largely depending on the presence of sulfate, the proteoglycans retain water favoring the formation of an incompressive gel-like structure necessary to favor the load transfer between the embedded collagen fibrils. An amorphous elastic fiber core surrounded by a well-aligned fibrillin-rich microfibril network is also detectable even if it represents a limited proportion of the tissue. Nevertheless, elastic fibers have a relevant role in allowing slight eye deformation following internal and external pressure (Marshall 1995).

Of the three recognizable layers in the cornea, from out to in the cornea epithelial layer, the stroma and the endothelium layer, the main layer is the stroma (Espana and Birk 2020). In the cornea stroma from the anterior to the posterior side, it is possible to recognize Bowman's layer, present in humans but not in all mammals, the cornea stromal proper, and Descemet's membrane. The main components of Bowman's layer are thin collagen I fibrils (18–22 μm) highly decorated by collagen type V homo- and heterotrimers (Gordon et al. 1994; Linsenmayer et al. 1993; Birk et al. 1986). The collagen fibrils here do not have lamellar organization. Collagens V and IV are anchoring Bowman's layer to the stromal proper that guarantees the cornea's transparency and shape. The stroma proper is constituted by fibrillar collagens I, III, and V, fibril-associated collagens with interrupted triple helices (FACITS, namely, XII and XIV), network-forming collagen (types IV, VI, and VIII), and SLRPs such as decorin, biglycan, lumican, keratocan, and fibromodulin (Quantock and Young 2008). The collagen type I fibrils present a rigid packing highly ordered and organized in lamellae with orthogonal orientation (Hassell and Birk 2010). Collagen V modulates the diameter of collagen I fibrils by regulating the number of nucleation sites (Komai and Ushiki 1991). Collagens XII and XIV are located on the collagen I fibril surface in the interfibrillar space, and their expression largely depends on mechanical forces. Collagen XII stabilizes the hierarchical organization of the lamella, and collagen XIV is involved in stroma compaction (Sun et al. 2020). The networking-forming collagens IV and VI play a role in anchoring the ECM to basement membrane and cells, respectively. Collagen VII is necessary to anchor the epithelial basement membrane to Bowman's layer, while collagen VIII is mainly present in Descemet's layer. The presence of proteoglycans ensures tissue organization, cell growth, and proper growth factor reservoir, as well as water storage and collagen fibrillogenesis. Among glycoproteins, fibronectin is the most abundant stimulating, especially after injury, collagen I and III deposition. A series of matricellular proteins, mainly thrombospondin-1 and thrombospondin-2, SPARC, and osteopontin, mediate cellular/ECM interactions (Espana and Birk 2020).

2.4.14 Sclera and Cornea and OI

A careful literature search of ocular defects in OI individuals points out that, although most information is available for the classical dominant OI forms caused by collagen I gene mutations, almost every component of the eye is affected. Not surprisingly, the most affected structures are sclera and cornea mainly composed of collagen I whose structure, synthesis, secretion and extracellular processing, assembly, and/or quantity are compromised in all OI forms (Treurniet et al. 2022).

One of the most frequent described extraskeletal features in OI, the presence of bluish-gray sclera discoloration, is caused by reduced thickness of both the sclera and central cornea often found in OI individuals. The thinning of these structures, due to reduced quantity and quality of collagen I, allows the underlying darker choroid vasculature to be seen. The lower collagen amount and altered corneal structure have been linked to keratoconus predisposition in OI, although a direct correlation is missing (Correia Barão et al. 2023). Ocular rigidity has also been reported in OI in which reduced corneal diameter and axial length cause reduced eye size. Axial length reduction and presence of keratoconus seem to be associated also with myopia that, among the refractive errors, is the most common in OI (Kaiser-Kupfer et al. 1981).

Few cases of retinal detachment have also been described as well as open angle glaucoma and a case of posterior staphyloma of the left eye in a 40-year-old woman with OI type I and blue sclerae. Cornea and sclera thinning increase the susceptibility to ruptures and laceration of the cornea after minor trauma or hemorrhages after small falls as well as surgery complications that may occur and compromise eye function in OI (Natarajan et al. 2003; Jonisch and Deramo 2011; Polat and Ulucan 2015; Oh et al. 2016; Campagna et al. 2018). Among incidental findings in OI, strabismus (Andalib and Gharabaghi 2009), absence of Bowman's layer (Kasner et al. 1993; Akira Kobayashi et al. 2014), detachment of Descemet's layer (Gorovoy et al. 2012; Polat and Ulucan 2015), corneal epithelial edema (Nwosu et al. 2005) and retinal artery occlusion (Bradish and Flowers 1987), macula (Benzimra et al. 2012; Sendul et al. 2016), and retinoblastoma (Haik Jr. 1984) have been described.

2.4.15 Muscle

In the ECM of lungs, collagen I is extensively distributed and acts as a key structural protein. Collagen fibers surround the alveoli, bronchioles, blood vessels, and airway walls (Karakioulaki et al. 2020). The interstitium of the lung parenchyma, bronchi, and alveolar walls are predominantly composed of collagens I and III, and the ratio between them determines the stiffness and the ability to resist mechanical forces during breathing, therefore influencing the overall biomechanical characteristics of the lungs (Sobin et al. 1988; Onursal et al. 2021; Karakioulaki et al. 2020). In the interstitial space of the lung tissue, collagen I creates a continuum of fibers networks. One network is spanning from the central airways to the alveolar ducts, another one extends centrally from the visceral pleura, and a third one is connecting these two (Suki and Bates 2008; Toshima et al. 2004). Collagen fibers are more condensed in the alveolar mouths, and they are divided into smaller fibers in the alveolar septa with varying thicknesses than range from several hundred nanometers to more than a micron (Toshima et al. 2004).

Moreover, in the airway cartilage, 95% of collagen is collagen II which supports chondrocyte synthesis of ECM, and collagen IV is predominantly found in the basement membrane of the alveolar epithelium (Karakioulaki et al. 2020; Liu et al. 2021).

Elastic fibers, composed mainly of elastin and microfibrils (fibrillin and fibulin), and collagen fibers are interconnected, and both play a vital role in providing mechanical strength to the lung tissue (Suki and Bates 2008; Toshima et al. 2004). Elastic fibers facilitate the necessary elasticity for the lungs' expansion and contraction during respiration at normal breathing volumes, and collagen fibers become more involved as lungs reach their total capacity. This is because collagen fibers adopt a wavy pattern in collapsed lungs, whereas during lung's inflation, they adopt a straight configuration that allows alveoli to expand (Toshima et al. 2004; Onursal et al. 2021).

2.4.16 Muscle and OI

In the lungs, imbalances in collagen synthesis and degradation can lead to pathological conditions (Liu et al. 2021). Given the impaired synthesis and release of collagen I in individuals with OI, it is not surprising that an increase prevalence of pulmonary complications is observed across subtypes. Patients with severe OI can exhibit lung abnormalities, and respiratory failure is one of the leading causes of death in these individuals (Khan et al. 2020).

Although scoliosis and rib fractures are often identified as primary factors, it is common that patients exhibit restrictive lung function even with mild spinal curvature or even complete absence of scoliosis. This implies a direct impact of collagen mutations on lung physiology, a phenomenon that has been explored in numerous studies (Khan et al. 2020; Gerdhem 2020; Radunovic et al. 2012; Keuning et al. 2022). A prospective observational study with adult OI patients reported no correlation between the magnitude of the scoliosis curve and the deficient pulmonary function and that patients with OI type III had lower forced vital capacity (FVC) and total lung capacity than other types of OI (Khan et al. 2020). An examination of pediatric patients with collagen structural mutations without scoliosis revealed a notable decrease in tidal lung capacity (TLC) and FVC throughout childhood, although progressing more gradually than in children with scoliosis. Decline in FVC, TLC, and lung volume with age was significantly greater for severe type III than the moderate type IV OI patients (Thiele et al. 2012).

Additional observations include thickening of the bronchial walls in individuals exhibiting both normal and restrictive physiology. While further research is required to delve into this matter, the authors propose that this condition is inherent to individuals with OI and may potentially play a role in pulmonary dysfunction (Khan et al. 2020). Moreover, a study performed by Arponen et al. (2018) revealed that 52% of adults diagnosed with OI had obstructive sleep apnea (Arponen et al. 2018).

Beyond the skeletal affectations, mouse models of OI also exhibit alterations in lung structure. In *Col1a1*$^{Aga2/Aga2}$ mice with a severe phenotype, the lungs exhibited alveolar bleeding, infiltration of polymorphonuclear neutrophils, and alveolar macrophages. Gas exchange is also compromised with results showing a 44% reduction in arterial pO2, a 40% increase in pCO2, and a 61% decrease in O_2 saturation, indicating hypoxic conditions. Pulmonary fibroblasts isolated from these mice showed a decline in *Col1a1* transcript expression, causing changes in the tissue that ultimately could lead to mortality (Thiele et al. 2012). The *Col1a1*$^{Jrt/+}$ mouse model, representing severe dominant OI, exhibited an increased average size of alveolar airspaces (Baglole et al. 2018). Interestingly, it seems that there is a correlation between the severity of lung dysfunction with the collagen deficiency in the ECM and that different mutations can result in a similar phenotype. As reported by Dimori et al. (2023), *Col1a2*$^{oim/oim}$ mice also showed alveolar simplification with enlarged acinar airspace and reduced alveolar surface that caused changes in mechanical parameters, while no significant differences were seen in heterozygous *Col1a2*$^{oim/+}$ or *Col1a2*$^{Amish/+}$ compared to controls (Dimori et al. 2023).

There is evidence suggesting that mutations in collagen-related genes could also cause respiratory impairment. In humans, biallelic *CRTAP* mutations resulted in severe to lethal OI, and newborns frequently display life-threatening respiratory distress (Dimori et al. 2020). *Crtap*$^{-/-}$ mice, lacking a functional prolyl 3-hydroxylation complex in the ER, also showed enlargement in the alveolar airway space in lungs and increased TGF-β activity (Grafe et al. 2014).

2.4.17 Gut

The intestinal ECM represents a complex network of proteins that not only does provide a structural scaffold for the tissue but also confers tensile strength and prevents overstretching of the organ. The intestinal epithelium is constituted of a single layer of epithelial cells lining on ECM mainly composed of interstitial matrix, providing the 3D tissue structure and by basement membrane (BM) that supports the epithelial linen, forming a barrier between the epithelium and the stroma (Theocharis et al. 2016). Collagens I and III, elastin, and glycoproteins, such as tenascin-C and fibronectin, are the major components of interstitial matrix, while collagen IV, laminin 111 and 211, nidogen, and perlecan are the main constituents of BMs (Perreault et al. 1998; Theocharis et al. 2016). In colonic specimens, both the hydroxyproline content and hydroxyproline per unit α-aminonitrogen are increased compared to values found in the ileum, reflecting both a higher absolute amount of collagen and relative amount of collagen compared with other proteins. From a structural point of view, gut collagen fibers are organized in a cross-cross pattern with 30 degrees angle in longitudinal direction (Orberg et al. 1983; Komuro 1988). Of relevance, the gut ECM can undergo peculiar compartmentalization, creating specialized 3D invagination in both the small and large intestines, known as the intestinal stem cell niche that acts in physiological epithelial tissue regeneration (Pompili et al. 2021).

2.4.18 Gut and OI

Gut function was reported to be affected in OI patients. Bowel symptoms, mainly chronic constipation, have been reported in several OI patients and, in the majority of the cases, associated with pelvic deformity, severe acetabular protrusion, and reduced mobility (Lee et al. 1995; Martins et al. 2020). Nevertheless, it is still not clear if this phenotype is directly caused by the collagen impairment in the bowel tissue or it is just a secondary effect of pelvic deformity or of wheelchair ambulation (Basel and Steiner 2009).

Of relevance, ECM controls the specification of stem cells that will renew the epithelial lining and crypts in the gut (Ramadan et al. 2022), and, in particular, collagen I was shown to specifically mediate mechanotransduction that controls epithelial cell fate (K obayashi et al. 2022). In a single case of OI type III diagnosed with Hirschsprung's disease, the progression of stool through the intestine was delayed as a consequence of improper ganglion cell development (Stockwell and Wallace 2022). A case of spontaneous intestinal perforation was reported characterized by absent musculature with no evidence of ischemia, as revealed by histological analyses (Wheatley et al. 2010).

2.5 Aging and OI

Generally, as individuals age, the synthesis and degradation of collagen I are impaired resulting in abnormal levels of collagens fibers (Sherratt 2009). Additionally, in aging, collagen I undergoes a series of posttranslational modifications, namely, deamidation and oxidation, enzymatic cross-linking reactions, and an accumulation in advanced glycation end-products (AGEs) formed non-enzymatically through molecular reactions with sugars (Birch 2018). Altogether, these alterations are associated with the overall changes in remodeling and turnover of the ECM, to an impaired matrix function and a decline in tissue elasticity and strength with age (Boskey and Coleman 2010).

In the different tissues, these changes manifest differently due to tissue-specific collagen requirements and functions. The expression and synthesis of collagen I decline in bone with age, and the process of osteoblastic bone formation decreases, while osteoclastic bone resorption increases, contributing to decreased bone strength, increased brittleness, and a higher susceptibility to fractures in older individuals (Kanis 2002; Bailey et al. 1999). In the teeth, collagen fibers become less aligned and more bundled with age which changes the properties of the ECM (Weerakoon et al. 2022). The skin undergoes structural changes such as impaired collagen synthesis that, consequently, reduces collagen density and mechanical tension (Varani et al. 2006). There is an increase in the fibril diameter due to an increase in collagen content and increase in collagen AGEs, altering the biomechanical properties and functions in the cornea (Daxer et al. 1998). Human male and female colons showed increased total collagen content with age, potentially affecting its function and causing increased incidence of lower gastrointestinal disorders (Baidoo et al. 2022). In the aged heart, collagen content and collagen cross-linking are increased in the left ventricles (Thomas et al. 1992; Meschiari et al. 2017), leading to fibrosis, stiffening of the ECM, and higher risk of cardiovascular diseases (Gazoti Debessa et al. 2001; Bailey et al. 1998). Increased collagen AGEs occur also in aged muscle where, although collagen concentration is unchanged, collagen is heavily cross-linked by glycation, stiffening the matrix and causing it to be resistant to enzymatic turnover leading to reduced muscle strength and function (Haus et al. 2007). Finally, lungs' fibroblasts produce more collagen and less elastin with age, which alongside with increased collagen cross-linking results in an increase in pulmonary stiffness (Liu et al. 2021).

The aging-related implications specific to OI in bone and in the other tissues remain unknown, but it is expected that they also involve changes in the content and cross-linking of collagen I, altering matrix composition and impacting cell functions.

In the *Col1a1* minigene mouse, with a mutated type I procollagen gene, age-related effects were studied in bone. In this mouse, the collagen content was lower than in controls at 0.5 and 1.5 months, equal at 6 months, and lower again at 24 months (Pereira et al. 1995). The age-related decline may also stem from the degradation of collagen I, marked by the loss of enzymatic cross-linking and the accumulation of AGEs, as revealed by studies linking the accumulations of AGEs with osteoclast-induced bone resorption and the osteoclastic differentiation process (Viguet-Carrin et al. 2010). Indeed, the analyses of enzymatic and nonenzymatic collagen cross-links in the OI $Col1a2^{oim/oim}$ model revealed a decrease in the first and an increase in the latter, explaining the increased bone stiffness and the decreased mechanical strength (Carriero et al. 2014).

Compromised collagen production and secretion may exacerbate the impact of aging on OI individuals. Premature aging is a condition characterized by symptoms typically associated with aging occurring at an earlier age than usual. A literature review reveals that the adult OI patient population suffers from age-related diseases such as cardiorespiratory diseases, vision impairment, increased risk of fractures, joint problems, and hearing loss (Marom et al. 2020). According to Folkestad et al. (2016), OI individuals are also at greater risk of early death, with an average life span approximately 7–10 years lower than the general population, although very few are known about the mechanism underneath this difference (Folkestad et al.

2016). Recent evidence in a murine model of OI type IX, characterized by the absence of the CypB protein and displaying early onset of aging-related features, showed an activation of the *P21*-dependent pathway, closely associated with cellular senescence (Zhang et al. 2022). Higher expression of *p21* mRNA was also found in adult *Crtap*$^{-/-}$ mice calvaria bone (Grafe et al. 2014). Furthermore, it is interesting to mention that OI patients with mutations in *TAPT1* suffer from a progeroid syndrome with wrinkled and thin skin as well as with premature depigmentation (Nabavizadeh et al. 2023).

While there is evidence suggesting a link between OI and premature aging, the exact mechanisms remain unknown. OI is a complex disease, characterized by mutations affecting the synthesis and structure of collagen I in a tissue-specific manner, and it is probable that several mechanisms beyond accelerated senescence may be at play. Further research to delve deeper into these aspects is needed, continuing the search for novel therapeutic strategies to benefit individuals with OI.

2.6 Conclusion and Future Perspectives

The crucial role of collagen I as one of the major components in many tissues of the body is supported by a highly complex and tightly regulated biosynthesis and assembly into functional fibers. The organization and composition of collagen I fibers in the ECM are very heterogeneous and are fully adapted to fulfill the functional needs of each tissue. Although structural and mechanical aspects of collagen I have been extensively studied in most tissues, its implication in processes such as biosensing and biosignaling is less understood. Further research on this subject will be key to provide a full vision about the biology of the ECM and ECM-related diseases.

A wide spectrum of mutations in many genes affects the synthesis of collagen I as well as its incorporation into a functional ECM. Alterations in collagen I quality and quantity can result in OI, a disease that is mainly characterized by brittle bones. However, as collagen I is a major component of almost all ECMs in the body, OI patients and animal models face specific defects also in many other tissues. The exact molecular mechanisms might also be tissue-specific and need to be investigated. In addition, changes in collagen quantity and quality also occur during the biological process of aging. Therefore, it is not surprising that OI patients show premature features of aging-associated frailty. This aspect is a relatively new area of research that might lead to the identification of novel targets and create new treatment options.

Hopefully, this chapter will provide researchers, medical professionals, and patients with a deeper understanding on collagen I biochemistry and the molecular and physiological basis of collagen-related diseases increasing the awareness of OI as a multi-organ disease and thus stimulating them to deeply investigate the extraskeletal disease manifestations.

Acknowledgments This chapter was supported by MUR "Dipartimenti di Eccellenza 2023-2027," by the European Community's Horizon-MSCA-DN-01-01 under grant agreement n. 101072766 (CHANGE), by the German Research Foundation (DFG) under Grant FOR 2722 (Project-ID 407168728, ZA 561/3-1 (FZ) and Project-ID 384170921, SE2373/1-2 (OS)), and by the "OI-Stiftung Baden-Württemberg."

Conflict of Interest No conflict of interest to declare.

References

Abdelaziz DM, Abdullah S, Magnussen C, Ribeiro-da-Silva A, Komarova SV, Rauch F, Stone LS (2015) Behavioral signs of pain and functional impairment in a mouse model of osteogenesis imperfecta. Bone 81:400–406. https://doi.org/10.1016/j.bone.2015.08.001

Akira Kobayashi A, Higashide T, Yokogawa H, Yamazaki N, Masaki T, Sugiyama K (2014) In vivo laser confocal microscopy findings of a cornea with osteogenesis imperfecta. Clin Ophthalmol 429. https://doi.org/10.2147/opth.s58087

Al-Jallad H, Palomo T, Roughley P, Glorieux FH, McKee MD, Moffatt P, Rauch F (2015) The effect of SERPINF1 in-frame mutations in osteogenesis imperfecta type VI. Bone 76:115–120. https://doi.org/10.1016/j.bone.2015.04.008

Alanay Y, Avaygan H, Camacho N, Utine GE, Boduroglu K, Aktas D, Alikasifoglu M, Tuncbilek E, Orhan D, Bakar FT, Zabel B, Superti-Furga A, Bruckner-Tuderman L, Curry CJR, Pyott S, Byers PH, Eyre DR, Baldridge D, Lee B, Merrill AE, Davis EC, Cohn DH, Akarsu N, Krakow D (2010) Mutations in the Gene Encoding the RER Protein FKBP65 Cause Autosomal-Recessive Osteogenesis Imperfecta. Am J Hum Genet 86:551–559. https://doi.org/10.1016/j.ajhg.2010.02.022

Alcorta-Sevillano N, Macías I, Infante A, Rodríguez CI (2020) Deciphering the relevance of bone ECM signaling. Cells 9. https://doi.org/10.3390/cells9122630

Andalib D, Gharabaghi D (2009) Unilateral congenital third nerve palsy associated with osteogenesis imperfecta, Iranian. J Ophthalmol 21:62–64

Andersen JD, Folkestad L, Hald JD, Harsløf T, Langdahl BL, Abrahamsen B (2022) Osteoarthritis in osteogenesis imperfecta: a nationwide register-based cohort study. Bone 154:116222. https://doi.org/10.1016/j.bone.2021.116222

Apperley LJ, Albaba S, Dharmaraj P, Balasubramanian M (2023) PLS3 whole gene deletion as a cause of X-linked osteoporosis: clinical report with review of published PLS3 literature. Clin Dysmorphol 32:43–47. https://doi.org/10.1097/mcd.0000000000000442

Arponen H, Bachour A, Bäck L, Valta H, Mäkitie A, Waltimo-Sirén J, Mäkitie O (2018) Is sleep apnea underdiagnosed in adult patients with osteogenesis imperfecta? -a single-center cross-sectional study. Orphanet J Rare Dis 13:231. https://doi.org/10.1186/s13023-018-0971-7

Arthur DG, Thompson KG, Swarbrick P (1992) Lethal osteogenesis imperfecta and skin fragility in newborn New Zealand Romney lambs. N Z Vet J 40:112–116. https://doi.org/10.1080/00480169.1992.35712

Asharani PV, Keupp K, Semler O, Wang W, Li Y, Thiele H, Yigit G, Pohl E, Becker J, Frommolt P, Sonntag C, Altmüller J, Zimmermann K, Greenspan DS, Akarsu NA, Netzer C, Schönau E, Wirth R, Hammerschmidt M, Nürnberg P, Wollnik B, Carney TJ (2012) Attenuated BMP1 function compromises osteogenesis, leading to bone fragility in humans and zebrafish. Am J Hum Genet 90:661–674. https://doi.org/10.1016/j.ajhg.2012.02.026

Bachinger HP, Engel J, Bruckner P, Timpl R (1978) The role of cis-trans isomerization of peptide bonds in the coil triple helix conversion of collagen. Eur J Biochem 90:605–613. https://doi.org/10.1111/j.1432-1033.1978.tb12641.x

Baek WY, de Crombrugghe B, Kim JE (2010) Postnatally induced inactivation of osterix in osteoblasts results in the reduction of bone formation and maintenance. Bone 46:920–928. https://doi.org/10.1016/j.bone.2009.12.007

Baglole CJ, Liang F, Traboulsi H, Rico de Souza A, Giordano C, Tauer JT, Rauch F, Petrof BJ (2018) Pulmonary and diaphragmatic pathology in collagen type I α1 mutant mice with osteogenesis imperfecta. Pediatr Res 83:1165–1171. https://doi.org/10.1038/pr.2018.36

Baidoo N, Crawley E, Knowles CH, Sanger GJ, Belai A (2022) Total collagen content and distribution is increased in human colon during advancing age. PLoS One 17:e0269689. https://doi.org/10.1371/journal.pone.0269689

Bailey AJ, Paul RG, Knott L (1998) Mechanisms of maturation and ageing of collagen. Mech Ageing Dev 106:1–56. https://doi.org/10.1016/s0047-6374(98)00119-5

Bailey AJ, Sims TJ, Ebbesen EN, Mansell JP, Thomsen JS, Mosekilde L (1999) Age-related changes in the biochemical properties of human cancellous bone collagen: relationship to bone strength. Calcif Tissue Int 65:203–210. https://doi.org/10.1007/s002239900683

Baines AC, Adams EJ, Zhang B, Ginsburg D (2013) Disruption of the Sec24d gene results in early embryonic lethality in the mouse. PLoS One 8:e61114. https://doi.org/10.1371/journal.pone.0061114

Bala MM, Bala KA (2021) Severe cases of osteogenesis imperfecta type VIII due to a homozygous mutation in P3H1 (LEPRE1) and review of the literature. Adv Clin Exp Med 30:1233–1238. https://doi.org/10.17219/acem/141367

Balasubramanian M, Sobey GJ, Wagner BE, Peres LC, Bowen J, Bexon J, Javaid MK, Arundel P, Bishop NJ (2016) Osteogenesis imperfecta: ultrastructural and histological findings on examination of skin revealing novel insights into genotype-phenotype correlation. Ultrastruct Pathol 40:71–76. https://doi.org/10.3109/01913123.2016.1140253

Baldridge D, Lennington J, Weis M, Homan EP, Jiang M-M, Munivez E, Keene DR, Hogue WR, Pyott S, Byers PH, Krakow D, Cohn DH, Eyre DR, Lee B, Morello R (2010) Generalized connective tissue disease in Crtap−/− mouse. PLoS One 5:e10560. https://doi.org/10.1371/journal.pone.0010560

Bank RA, Robins SP, Wijmenga C, Breslau-Siderius LJ, Bardoel AF, van der Sluijs HA, Pruijs HE, TeKoppele JM (1999) Defective collagen crosslinking in bone, but not in ligament or cartilage, in Bruck syndrome: indications for a bone-specific telopeptide lysyl hydroxylase on chromosome 17. Proc Natl Acad Sci USA 96:1054–1058. https://doi.org/10.1073/pnas.96.3.1054

Basel D, Steiner RD (2009) Osteogenesis imperfecta: recent findings shed new light on this once well-understood condition. Genet Med 11:375–385. https://doi.org/10.1097/GIM.0b013e3181a1ff7b

Becker J, Semler O, Gilissen C, Li Y, Bolz HJ, Giunta C, Bergmann C, Rohrbach M, Koerber F, Zimmermann K, de Vries P, Wirth B, Schoenau E, Wollnik B, Veltman JA, Hoischen A, Netzer C (2011) Exome sequencing identifies truncating mutations in human SERPINF1 in autosomal-recessive osteogenesis imperfecta. Am J Hum Genet 88:362–371. https://doi.org/10.1016/j.ajhg.2011.01.015

Benzimra JD, Kapoor B, Edmeades N, Burger A, Simcock PR (2012) Surgical repair of bilateral full thickness macular holes in a patient with blue sclera secondary to osteogenesis imperfecta. Eye 26:1023–1024. https://doi.org/10.1038/eye.2012.63

Besio R, Chow CW, Tonelli F, Marini JC, Forlino A (2019) Bone biology: insights from osteogenesis imperfecta and related rare fragility syndromes. FEBS J 286:3033–3056. https://doi.org/10.1111/febs.14963

Besio R, Iula G, Garibaldi N, Cipolla L, Sabbioneda S, Biggiogera M, Marini JC, Rossi A, Forlino A (2018) 4-PBA ameliorates cellular homeostasis in fibroblasts from osteogenesis imperfecta patients by enhancing autophagy and stimulating protein secretion. Biochim Biophys Acta Mol basis Dis 1864:1642–1652. https://doi.org/10.1016/j.bbadis.2018.02.002

Besio R, Contento BM, Garibaldi N, Filibian M, Sonntag S, Shmerling D, Tonelli F, Biggiogera M, Brini M, Salmaso A, Jovanovic M, Marini JC, Rossi A, Forlino A (2023) CaMKII inhibition due to TRIC-B loss-of-function dysregulates SMAD signaling in osteogenesis imperfecta. Matrix Biol 120:43–59. https://doi.org/10.1016/j.matbio.2023.05.002

Bianchi L, Gagliardi A, Gioia R, Besio R, Tani C, Landi C, Cipriano M, Gimigliano A, Rossi A, Marini JC, Forlino A, Bini L (2012) Differential response to intracellular stress in the skin from osteogenesis imperfecta Brtl mice with lethal and non lethal phenotype: a proteomic approach. J Proteome 75:4717–4733. https://doi.org/10.1016/j.jprot.2012.01.038

Birch HL (2018) Extracellular matrix and ageing. Subcell Biochem 90:169–190. https://doi.org/10.1007/978-981-13-2835-0_7

Birk DE, Fitch JM, Linsenmayer TF (1986) Organization of collagen types I and V in the embryonic chicken cornea. Invest Ophthalmol Vis Sci 27:1470–1477

Blair HC, Larrouture QC, Li Y, Lin H, Beer-Stoltz D, Liu L, Tuan RS, Robinson LJ, Schlesinger PH, Nelson DJ (2017) Osteoblast Differentiation and Bone Matrix Formation *In Vivo* and *In Vitro*. Tissue Eng Part B Rev 23:268–280. https://doi.org/10.1089/ten.teb.2016.0454

Bogan R, Riddle RC, Li Z, Kumar S, Nandal A, Faugere MC, Boskey A, Crawford SE, Clemens TL (2013) A mouse model for human osteogenesis imperfecta type VI. J Bone Miner Res 28:1531–1536. https://doi.org/10.1002/jbmr.1892

Bonadio J, Saunders TL, Tsai E, Goldstein SA, Morris-Wiman J, Brinkley L, Dolan DF, Altschuler RA, Hawkins JE, Bateman JF (1990) Transgenic mouse model of the mild dominant form of osteogenesis imperfecta. Proc Natl Acad Sci 87:7145–7149. https://doi.org/10.1073/pnas.87.18.7145

Boote C, Sigal IA, Grytz R, Hua Y, Nguyen TD, Girard MJA (2020) Scleral structure and biomechanics. Prog Retin Eye Res 74:100773. https://doi.org/10.1016/j.preteyeres.2019.100773

Boraschi-Diaz I, Tauer JT, El-Rifai O, Guillemette D, Lefebvre G, Rauch F, Ferron M, Komarova SV (2017) Metabolic phenotype in the mouse model of osteogenesis imperfecta. J Endocrinol 234:279–289. https://doi.org/10.1530/joe-17-0335

Boskey AL (1989) Noncollagenous matrix proteins and their role in mineralization. Bone Miner 6:111–123. https://doi.org/10.1016/0169-6009(89)90044-5

Boskey AL, Coleman R (2010) Aging and bone. J Dent Res 89:1333–1348. https://doi.org/10.1177/0022034510377791

Boskey AL, Moore DJ, Amling M, Canalis E, Delany AM (2003) Infrared analysis of the mineral and matrix in bones of osteonectin-null mice and their wildtype controls. J Bone Miner Res 18:1005–1011. https://doi.org/10.1359/jbmr.2003.18.6.1005

Bowers SLK, Meng Q, Kuwabara Y, Huo J, Minerath R, York AJ, Sargent MA, Prasad V, Saviola AJ, Galindo DC, Hansen KC, Vagnozzi RJ, Yutzey KE, Molkentin JD (2023) Col1a2-deleted mice have defective type I collagen and secondary reactive cardiac fibrosis with altered hypertrophic dynamics. Cells 12:2174. https://doi.org/10.3390/cells12172174

Bradish CF, Flowers M (1987) Central retinal artery occlusion in association with osteogenesis imperfecta. Spine (Phila Pa 1976) 12:193–194. https://doi.org/10.1097/00007632-198703000-00018

Brinckmann J (2005) Collagens at a Glance. In: Brinckmann J, Notbohm H, Müller PK (eds) Collagen: Primer in Structure, Processing and Assembly. Springer, pp 1–6. https://doi.org/10.1007/b103817

Bunel L, Pincet L, Malhotra V, Raote I, Pincet F (2024) A model for collagen secretion by intercompartmental continuities. Proc Natl Acad Sci U S A 121:e2310404120. https://doi.org/10.1073/pnas.2310404120

Bächinger HP (1987) The influence of peptidyl-prolyl cis-trans isomerase on the in vitro folding of type III collagen. J Biol Chem 262:17144–17148

Cabral WA, Chang W, Barnes AM, Weis M, Scott MA, Leikin S, Makareeva E, Kuznetsova NV, Rosenbaum KN, Tifft CJ, Bulas DI, Kozma C, Smith PA, Eyre DR, Marini JC (2007) Prolyl 3-hydroxylase 1 deficiency causes a recessive metabolic bone disorder resembling lethal/severe osteogenesis imperfecta. Nat Genet 39:359–365. https://doi.org/10.1038/ng1968

Cabral WA, Perdivara I, Weis M, Terajima M, Blissett AR, Chang W, Perosky JE, Makareeva EN, Mertz EL, Leikin S, Tomer KB, Kozloff KM, Eyre DR, Yamauchi M, Marini JC (2014) Abnormal type I collagen post-translational modification and crosslinking in a cyclophilin B KO mouse model of recessive osteogenesis imperfecta. PLoS Genet 10:e1004465. https://doi.org/10.1371/journal.pgen.1004465

Cabral WA, Ishikawa M, Garten M, Makareeva EN, Sargent BM, Weis M, Barnes AM, Webb EA, Shaw NJ, Ala-Kokko L, Lacbawan FL, Högler W, Leikin S, Blank PS, Zimmerberg J, Eyre DR, Yamada Y, Marini JC (2016) Absence of the ER Cation Channel TMEM38B/TRIC-B disrupts intracellular calcium homeostasis and dysregulates collagen synthesis in recessive osteogenesis imperfecta. PLoS Genet 12:e1006156. https://doi.org/10.1371/journal.pgen.1006156

Campagna G, Al-Mohtaseb Z, Khandelwal S, Chang E (2018) Sequential traumatic corneal open globe rupture in a patient with osteogenesis imperfecta type I. Am J Ophthalmol Case Rep 11:35–36. https://doi.org/10.1016/j.ajoc.2018.05.001

Campbell BG, Wootton JA, MacLeod JN, Minor RR (2000) Sequence of normal canine COL1A1 cDNA and identification of a heterozygous alpha1(I) collagen Gly208Ala mutation in a severe case of canine osteogenesis imperfecta. Arch Biochem Biophys 384:37–46. https://doi.org/10.1006/abbi.2000.2099

Campbell BG, Wootton JA, Macleod JN, Minor RR (2001) Canine COL1A2 mutation resulting in C-terminal truncation of pro-alpha2(I) and severe osteogenesis imperfecta. J Bone Miner Res. 16(6):1147–53. https://doi.org/10.1359/jbmr.2001.16.6.1147

Canuto HC, Fishbein KW, Huang A, Doty SB, Herbert RA, Peckham J, Pleshko N, Spencer RG (2012) Characterization of skin abnormalities in a mouse model of osteogenesis imperfecta using high resolution magnetic resonance imaging and Fourier transform infrared imaging spectroscopy. NMR Biomed 25:169–176. https://doi.org/10.1002/nbm.1732

Carriero A, Zimmermann EA, Paluszny A, Tang SY, Bale H, Busse B, Alliston T, Kazakia G, Ritchie RO, Shefelbine SJ (2014) How tough is brittle bone? Investigating osteogenesis imperfecta in mouse bone. J Bone Miner Res 29:1392–1401. https://doi.org/10.1002/jbmr.2172

Cheek JD, Wirrig EE, Alfieri CM, James JF, Yutzey KE (2012) Differential activation of valvulogenic, chondrogenic, and osteogenic pathways in mouse models of myxomatous and calcific aortic valve disease. J Mol Cell Cardiol 52:689–700. https://doi.org/10.1016/j.yjmcc.2011.12.013

Chen F, Guo R, Itoh S, Moreno L, Rosenthal E, Zappitelli T, Zirngibl RA, Flenniken A, Cole W, Grynpas M, Osborne LR, Vogel W, Adamson L, Rossant J, Aubin JE (2014) First mouse model for combined osteogenesis imperfecta and Ehlers-Danlos syndrome. J Bone Miner Res 29:1412–1423. https://doi.org/10.1002/jbmr.2177

Chen S, Birk DE (2013) The regulatory roles of small leucine-rich proteoglycans in extracellular matrix assembly. FEBS J 280:2120–2137. https://doi.org/10.1111/febs.12136

Chessler SD, Byers PH (1993) BiP binds type I procollagen pro alpha chains with mutations in the carboxyl-terminal propeptide synthesized by cells from patients with osteogenesis imperfecta. J Biol Chem 268:18226–18233

Cho T-J, Lee K-E, Lee S-K, Song J et al (2012) A single recurrent mutation in the 5′-UTR of IFITM5 causes osteogenesis imperfecta type V. Am J Hum Genet 91:343–348. https://doi.org/10.1016/j.ajhg.2012.06.005

Choi JW, Sutor SL, Lindquist L, Evans GL, Madden BJ, Bergen HR 3rd, Hefferan TE, Yaszemski MJ, Bram RJ (2009) Severe osteogenesis imperfecta in cyclophilin B-deficient mice. PLoS Genet 5:e1000750. https://doi.org/10.1371/journal.pgen.1000750

Chougui K, Addab S, Palomo T, Morin SN, Veilleux LN, Bernstein M, Thorstad K, Hamdy R, Tsimicalis A (2020) Clinical manifestations of osteogenesis imperfecta in adulthood: an integrative review of quantitative studies and case reports. Am J Med Genet A 182:842–865. https://doi.org/10.1002/ajmg.a.61497

Chretien A, Couchot M, Mabilleau G, Behets C (2022) Biomechanical, microstructural and material properties of tendon and bone in the Young Oim mice model of osteogenesis imperfecta. Int J Mol Sci 23. https://doi.org/10.3390/ijms23179928

Chretien A, Mabilleau G, Lebacq J, Docquier PL, Behets C (2023) Beneficial effects of Zoledronic acid on tendons of the osteogenesis imperfecta mouse (Oim). Pharmaceuticals (Basel) 16:832. https://doi.org/10.3390/ph16060832

Christiansen HE, Schwarze U, Pyott SM, Alswaid A, Al Balwi M, Alrasheed S, Pepin MG, Weis MA, Eyre DR, Byers PH (2010) Homozygosity for a missense mutation in SERPINH1, which encodes the collagen chaperone protein HSP47, results in severe recessive osteogenesis imperfecta. Am J Hum Genet 86:389–398. https://doi.org/10.1016/j.ajhg.2010.01.034

Chughtai A, Dey R, Chattopadhyay S (2022) Arterial vascular access complication in osteogenesis imperfecta. J Vasc Access 23:822–824. https://doi.org/10.1177/11297298211009989

Claeys L, Zhytnik L, Wisse LE, Van Essen HW, Eekhoff EMW, Pals G, Bravenboer N, Micha D (2023) Exploration of the skeletal phenotype of the Col1a1+/Mov13 mouse model for haploinsufficient osteogenesis imperfecta type 1. Front Endocrinol 14. https://doi.org/10.3389/fendo.2023.1145125

Correia Barão R, Santos M, Marques RE, Quintas AM, Guerra P (2023) Keratoconus tomographic indices in osteogenesis imperfecta. Graefes Arch Clin Exp Ophthalmol 261:2585–2592. https://doi.org/10.1007/s00417-023-06059-4

Cropp GJ, Myers DN (1972) Physiological evidence of hypermetabolism in osteogenesis imperfecta. Pediatrics 49:375–391

Cundy T, Dray M, Delahunt J, Hald JD, Langdahl B, Li C, Szybowska M, Mohammed S, Duncan EL, Mcinerney-Leo AM, Wheeler PG, Roschger P, Klaushofer K, Rai J, Weis M, Eyre D, Schwarze U, Byers PH (2018) Mutations that Alter the Carboxy-terminal-Propeptide cleavage site of the chains of type I procollagen are associated With a unique osteogenesis imperfecta phenotype. J Bone Miner Res 33:1260–1271. https://doi.org/10.1002/jbmr.3424

Daley E, Streeten EA, Sorkin JD, Kuznetsova N, Shapses SA, Carleton SM, Shuldiner AR, Marini JC, Phillips CL, Goldstein SA, Leikin S, McBride DJ Jr (2010) Variable bone fragility associated with an Amish COL1A2 variant and a knock-in mouse model. J Bone Miner Res 25:247–261. https://doi.org/10.1359/jbmr.090720

Daxer A, Misof K, Grabner B, Ettl A, Fratzl P (1998) Collagen fibrils in the human corneal stroma: structure and aging. Invest Ophthalmol Vis Sci 39:644–648

De Coster PJ, Cornelissen M, De Paepe A, Martens LC, Vral A (2007) Abnormal dentin structure in two novel gene mutations [COL1A1, Arg134Cys] and [ADAMTS2, Trp795-to-ter] causing rare type I collagen disorders. Arch Oral Biol 52:101–109. https://doi.org/10.1016/j.archoralbio.2006.08.007

De Vlaming A, Sauls K, Hajdu Z, Visconti RP, Mehesz AN, Levine RA, Slaugenhaupt SA, Hagège A, Chester AH, Markwald RR, Norris RA (2012) Atrioventricular valve development: new perspectives on an old theme. Differentiation 84:103–116. https://doi.org/10.1016/j.diff.2012.04.001

Del Monte-Nieto G, Fischer JW, Gorski DJ, Harvey RP, Kovacic JC (2020) Basic biology of extracellular matrix in the cardiovascular system, part 1/4: JACC focus seminar. J Am Coll Cardiol 75:2169–2188. https://doi.org/10.1016/j.jacc.2020.03.024

Derwin KA, Soslowsky LJ (1999) A quantitative investigation of structure-function relationships in a tendon fascicle model. J Biomech Eng 121:598–604. https://doi.org/10.1115/1.2800859

Derwin KA, Soslowsky LJ, Kimura JH, Plaas AH (2001) Proteoglycans and glycosaminoglycan fine structure in the mouse tail tendon fascicle. J Orthop Res 19:269–277. https://doi.org/10.1016/s0736-0266(00)00032-2

Dimitra M, Gerard P, Theo HS, Samaneh G (2020) An in vitro model to evaluate the properties of matrices produced by fibroblasts from osteogenesis imperfecta and Ehlers-Danlos syndrome patients. Biochem Biophys Res Commun 521:310–317. https://doi.org/10.1016/j.bbrc.2019.09.081

Dimori M, Fett J, Leikin S, Otsuru S, Thostenson JD, Carroll JL, Morello R (2023) Distinct type I collagen alterations cause intrinsic lung and respiratory defects of variable severity in mouse models of osteogenesis imperfecta. J Physiol 601:355–379. https://doi.org/10.1113/jp283452

Dimori M, Heard-Lipsmeyer ME, Byrum SD, Mackintosh SG, Kurten RC, Carroll JL, Morello R (2020) Respiratory defects in the CrtapKO mouse model of osteogenesis imperfecta. Am J Physiol Lung Cell Mol Physiol 318:L592–l605. https://doi.org/10.1152/ajplung.00313.2019

Drögemüller C, Becker D, Brunner A, Haase B, Kircher P, Seeliger F, Fehr M, Baumann U, Lindblad-Toh K, Leeb T (2009) A missense mutation in the SERPINH1 gene in dachshunds with osteogenesis imperfecta. PLoS Genet 5:e1000579. https://doi.org/10.1371/journal.pgen.1000579

Dubail J, Brunelle P, Baujat G, Huber C, Doyard M, Michot C, Chavassieux P, Khairouni A, Topouchian V, Monnot S, Koumakis E, Cormier-Daire V (2020) Homozygous loss-of-function mutations in CCDC134 are responsible for a severe form of osteogenesis imperfecta. J Bone Miner Res 35:1470–1480. https://doi.org/10.1002/jbmr.4011

Duran I, Martin JH, Weis MA, Krejci P, Konik P, Li B, Alanay Y, Lietman C, Lee B, Eyre D, Cohn DH, Krakow D (2017) A chaperone complex formed by HSP47, FKBP65, and BiP modulates Telopeptide Lysyl hydroxylation of type I procollagen. J Bone Miner Res 32:1309–1319. https://doi.org/10.1002/jbmr.3095

Durkin A, Devile C, Arundel P, Bull M, Walsh J, Bishop NJ, Hupin E, Parekh S, Nadarajah R, Offiah AC, Calder A, Brock J, Baker D, Balasubramanian M (2022) Expanding the phenotype of *SPARC*-related osteogenesis imperfecta: clinical findings in two patients with pathogenic variants in *SPARC* and literature review. J Med Genet 59:810–816. https://doi.org/10.1136/jmedgenet-2021-107942

Engel J, Bächinger HP (2005) Structure, stability and folding of the collagen triple helix. In: Brinckmann J, Notbohm H, Müller PK (eds) Collagen: primer in structure, processing and assembly. Springer Berlin Heidelberg, Berlin, Heidelberg, pp 7–33. https://doi.org/10.1007/b103818

Engelbert RH, Uiterwaal CS, Gerver WJ, van der Net JJ, Pruijs HE, Helders PJ (2004) Osteogenesis imperfecta in childhood: impairment and disability. A prospective study with 4-year follow-up. Arch Phys Med Rehabil 85:772–778. https://doi.org/10.1016/j.apmr.2003.08.085

Espana EM, Birk DE (2020) Composition, structure and function of the corneal stroma. Exp Eye Res 198:108137. https://doi.org/10.1016/j.exer.2020.108137

Etich J, Semler O, Stevenson NL, Stephan A, Besio R, Garibaldi N, Reintjes N, Dafinger C, Liebau MC, Baumann U, Mörgelin M, Forlino A, Stephens DJ, Netzer C, Zaucke F, Rehberg M (2023) TAPT1-at the crossroads of extracellular matrix and signaling in Osteogenesis imperfecta. EMBO Mol Med, 7, © 2023 The Authors. Published under the terms of the CC BY 4.0 license, England, p e17528. https://doi.org/10.15252/emmm.202317528

Eyre DR, Weis MA (2013) Bone collagen: new clues to its mineralization mechanism from recessive osteogenesis imperfecta. Calcif Tissue Int 93:338–347. https://doi.org/10.1007/s00223-013-9723-9

Eyre DR, Paz MA, Gallop PM (1984) Cross-linking in collagen and elastin. Annu Rev Biochem 53:717–748. https://doi.org/10.1146/annurev.bi.53.070184.003441

Fahiminiya S, Majewski J, Al-Jallad H, Moffatt P, Mort J, Glorieux FH, Roschger P, Klaushofer K, Rauch F (2014) Osteoporosis caused by mutations in *PLS3*: clinical and bone tissue characteristics. J Bone Miner Res 29:1805–1814. https://doi.org/10.1002/jbmr.2208

Farber CR, Reich A, Barnes AM, Becerra P, Rauch F, Cabral WA, Bae A, Quinlan A, Glorieux FH, Clemens TL, Marini JC (2014) A novel *IFITM5* mutation in severe atypical osteogenesis imperfecta type VI impairs osteoblast production of pigment epithelium-derived Factor. J Bone Miner Res 29:1402–1411. https://doi.org/10.1002/jbmr.2173

Fiscaletti M, Biggin A, Bennetts B, Wong K, Briody J, Pacey V, Birman C, Munns CF (2018) Novel variant in Sp7/Osx associated with recessive osteogenesis imperfecta with bone fragility and hearing impairment. Bone 110:66–75. https://doi.org/10.1016/j.bone.2018.01.031

Fisher S, Jagadeeswaran P, Halpern ME (2003) Radiographic analysis of zebrafish skeletal defects. Dev Biol 264:64–76. https://doi.org/10.1016/s0012-1606(03)00399-3

Folkestad L, Hald JD, Gram J, Langdahl BL, Hermann AP, Diederichsen AC, Abrahamsen B, Brixen K (2016) Cardiovascular disease in patients with osteogenesis imperfecta—a nationwide, register-based cohort study. Int J Cardiol 225:250–257. https://doi.org/10.1016/j.ijcard.2016.09.107

Forlino A, Marini JC (2016) Osteogenesis imperfecta. Lancet 387:1657–1671. https://doi.org/10.1016/s0140-6736(15)00728-x

Forlino A, Cabral WA, Barnes AM, Marini JC (2011) New perspectives on osteogenesis imperfecta. Nat Rev Endocrinol 7:540–557. https://doi.org/10.1038/nrendo.2011.81

Forlino A, Porter FD, Lee EJ, Westphal H, Marini JC (1999) Use of the Cre/lox recombination system to develop a non-lethal knock-in murine model for osteogenesis imperfecta with an alpha1(I) G349C substitution. Variability in phenotype in BrtlIV mice. J Biol Chem 274:37923–37931. https://doi.org/10.1074/jbc.274.53.37923

Foster BL (2017) On the discovery of cementum. J Periodontal Res 52:666–685. https://doi.org/10.1111/jre.12444

Fratzl P, Paris O, Klaushofer K, Landis WJ (1996) Bone mineralization in an osteogenesis imperfecta mouse model studied by small-angle x-ray scattering. J Clin Invest 97:396–402. https://doi.org/10.1172/jci118428

Fratzl-Zelman N, Morello R, Lee B, Rauch F, Glorieux FH, Misof BM, Klaushofer K, Roschger P (2010) CRTAP deficiency leads to abnormally high bone matrix mineralization in a murine model and in children with osteogenesis imperfecta type VII. Bone 46:820–826. https://doi.org/10.1016/j.bone.2009.10.037

Fratzl-Zelman N, Schmidt I, Roschger P, Glorieux FH, Klaushofer K, Fratzl P, Rauch F, Wagermaier W (2014) Mineral particle size in children with osteogenesis imperfecta type I is not increased independently of specific collagen mutations. Bone 60:122–128. https://doi.org/10.1016/j.bone.2013.11.023

Fratzl-Zelman N, Barnes AM, Weis M, Carter E, Hefferan TE, Perino G, Chang W, Smith PA, Roschger P, Klaushofer K, Glorieux FH, Eyre DR, Raggio C, Rauch F, Marini JC (2016) Non-lethal type VIII osteogenesis imperfecta has elevated bone matrix mineralization. J Clin Endocrinol Metabol 101:3516–3525. https://doi.org/10.1210/jc.2016-1334

Frolova EG, Drazba J, Krukovets I, Kostenko V, Blech L, Harry C, Vasanji A, Drumm C, Sul P, Jenniskens GJ, Plow EF, Stenina-Adognravi O (2014) Control of organization and function of muscle and tendon by thrombospondin-4. Matrix Biol 37:35–48. https://doi.org/10.1016/j.matbio.2014.02.003

Gan H, Xue W, Gao Y, Zhu G, Chan D, Cheah KSE, Huang J (2019) KIF5B modulates central spindle organization in late-stage cytokinesis in chondrocytes. Cell Biosci 9:85. https://doi.org/10.1186/s13578-019-0344-5

Garbes L, Kim K, Rieß A, Hoyer-Kuhn H, Beleggia F, Bevot A, Kim J, Netzer C et al (2015) Mutations in SEC24D, encoding a component of the COPII machinery, cause a syndromic form of osteogenesis imperfecta. Am J Hum Genet 96:432–439. https://doi.org/10.1016/j.ajhg.2015.01.002

Garibaldi N, Besio R, Dalgleish R, Villani S, Barnes AM, Marini JC, Forlino A (2022) Dissecting the phenotypic variability of osteogenesis imperfecta. Dis Model Mech 15. https://doi.org/10.1242/dmm.049398

Garibaldi N, Contento BM, Babini G, Morini J, Siciliani S, Biggiogera M, Raspanti M, Marini JC, Rossi A, Forlino A, Besio R (2021) Targeting cellular stress in vitro improves osteoblast homeostasis, matrix collagen content and mineralization in two murine models of osteogenesis imperfecta. Matrix Biol 98:1–20. https://doi.org/10.1016/j.matbio.2021.03.001

Garrocho-Rangel A, Dávila-Zapata I, Martínez-Rider R, Ruiz-Rodríguez S, Pozos-Guillén A (2019) Dentinogenesis Imperfecta Type II in Children: a scoping review. J Clin Pediatr Dent 43:147–154. https://doi.org/10.17796/1053-4625-43.3.1

Gazoti Debessa CR, Mesiano Maifrino LB, Rodrigues de Souza R (2001) Age related changes of the collagen network of the human heart. Mech Ageing Dev 122:1049–1058. https://doi.org/10.1016/s0047-6374(01)00238-x

Gelse K, Pöschl E, Aigner T (2003) Collagens—structure, function, and biosynthesis. Adv Drug Deliv Rev 55:1531–1546. https://doi.org/10.1016/j.addr.2003.08.002

Gentry BA, Ferreira JA, Mccambridge AJ, Brown M, Phillips CL (2010) Skeletal muscle weakness in osteogeneis imperfecta mice. Matrix Biol 29:638–644. https://doi.org/10.1016/j.matbio.2010.06.006

Gerdhem P (2020) CORR insights®: cardiopulmonary status in adults with osteogenesis imperfecta: intrinsic lung disease May contribute more than scoliosis. Clin Orthop Relat Res 478:2844–2845. https://doi.org/10.1097/corr.0000000000001469

Gewartowska O, Aranaz-Novaliches G, Krawczyk PS, Mroczek S, Kusio-Kobiałka M, Tarkowski B, Spoutil F, Benada O, Kofroňová O, Szwedziak P, Cysewski D, Gruchota J, Szpila M, Chlebowski A, Sedlacek R, Prochazka J, Dziembowski A (2021) Cytoplasmic polyadenylation by TENT5A is required for proper bone formation. Cell Rep 35:109015. https://doi.org/10.1016/j.celrep.2021.109015

Ghosh DK, Udupa P, Shrikondawar AN, Bhavani GS, Shah H, Ranjan A, Girisha KM (2023) Mutant MESD links cellular stress to type I collagen aggregation in osteogenesis imperfecta type XX. Matrix Biol 115:81–106. https://doi.org/10.1016/j.matbio.2022.12.001

Gil-Bona A, Bidlack FB (2020) Tooth enamel and its dynamic protein matrix. Int J Mol Sci 21:4458. https://doi.org/10.3390/ijms21124458

Gillies AR, Lieber RL (2011) Structure and function of the skeletal muscle extracellular matrix. Muscle Nerve 44:318–331. https://doi.org/10.1002/mus.22094

Gistelinck C, Weis M, Rai J, Schwarze U, Niyazov D, Song KM, Byers PH, Eyre DR (2021) Abnormal bone collagen cross-linking in osteogenesis imperfecta/Bruck syndrome caused by compound heterozygous <scp><i>PLOD2</i></scp> mutations. JBMR Plus 5. https://doi.org/10.1002/jbm4.10454

Gistelinck C, Witten PE, Huysseune A, Symoens S, Malfait F, Larionova D, Simoens P, Dierick M, Van Hoorebeke L, De Paepe A, Kwon RY, Weis M, Eyre DR, Willaert A, Coucke PJ (2016) Loss of type I collagen Telopeptide Lysyl hydroxylation causes musculoskeletal abnormalities in a zebrafish model of Bruck syndrome. J Bone Miner Res 31:1930–1942. https://doi.org/10.1002/jbmr.2977

Gistelinck C, Kwon RY, Malfait F, Symoens S, Harris MP, Henke K, Hawkins MB, Fisher S, Sips P, Guillemyn B, Bek JW, Vermassen P, De Saffel H, Witten PE, Weis M, De Paepe A, Eyre DR, Willaert A, Coucke PJ (2018) Zebrafish type I collagen mutants faithfully recapitulate human type I collagenopathies. Proc Natl Acad Sci USA 115:E8037–e8046. https://doi.org/10.1073/pnas.1722200115

Glorieux FH, Ward LM, Rauch F, Lalic L, Roughley PJ, Travers R (2002) Osteogenesis imperfecta type VI: a form of brittle bone disease with a mineralization defect. J Bone Miner Res 17:30–38. https://doi.org/10.1359/jbmr.2002.17.1.30

Glorieux FH, Travers R, Taylor A, Bowen JR, Rauch F, Norman M, Parfitt AM (2000) Normative data for iliac bone histomorphometry in growing children. Bone 26:103–109. https://doi.org/10.1016/s8756-3282(99)00257-4

Goldberg M, Kulkarni AB, Young M, Boskey A (2011) Dentin: structure, composition and mineralization. Front Biosci (Elite Ed) 3:711. https://doi.org/10.2741/e281

Gordon MK, Foley JW, Birk DE, Fitch JM, Linsenmayer TF (1994) Type V collagen and Bowman's membrane. Quantitation of mRNA in corneal epithelium and stroma. J Biol Chem 269:24959–24966

Gorovoy MS, Gorovoy IR, Ullman S, Gorovoy JB (2012) Descemet stripping automated endothelial keratoplasty for spontaneous descemet membrane detachment in a patient with osteogenesis imperfecta. Cornea 31:832–835. https://doi.org/10.1097/ICO.0b013e318225418b

Gorrell L, Makareeva E, Omari S, Otsuru S, Leikin S (2022) ER, mitochondria, and ISR regulation by mt-HSP70 and ATF5 upon procollagen Misfolding in osteoblasts. Adv Sci 9:2201273. https://doi.org/10.1002/advs.202201273

Gorski JP (2011) Biomineralization of bone: a fresh view of the roles of non-collagenous proteins. Front Biosci 16:2598. https://doi.org/10.2741/3875

Grafe I, Yang T, Alexander S, Homan EP, Lietman C, Jiang MM, Bertin T, Munivez E, Chen Y, Dawson B, Ishikawa Y, Weis MA, Sampath TK, Ambrose C, Eyre D, Bächinger HP, Lee B (2014) Excessive transforming growth factor-β signaling is a common mechanism in osteogenesis imperfecta. Nat Med 20:670–675. https://doi.org/10.1038/nm.3544

Graham HK, Holmes DF, Watson RB, Kadler KE (2000) Identification of collagen fibril fusion during vertebrate tendon morphogenesis. The process relies on unipolar fibrils and is regulated by collagen-proteoglycan interaction. J Mol Biol 295:891–902. https://doi.org/10.1006/jmbi.1999.3384

Gremminger VL, Phillips CL (2021) Impact of intrinsic muscle weakness on muscle–bone crosstalk in osteogenesis imperfecta. Int J Mol Sci 22:4963. https://doi.org/10.3390/ijms22094963

Gremminger VL, Jeong Y, Cunningham RP, Meers GM, Rector RS, Phillips CL (2019) Compromised exercise capacity and mitochondrial dysfunction in the osteogenesis imperfecta murine (<i>oim</i>) mouse model. J Bone Miner Res 34:1646–1659. https://doi.org/10.1002/jbmr.3732

Grol MW, Haelterman NA, Lim J, Munivez EM, Archer M, Hudson DM, Tufa SF, Keene DR, Lei K, Park D, Kuzawa CD, Ambrose CG, Eyre DR, Lee BH (2021) Tendon and motor phenotypes in the Crtap(−/−) mouse model of recessive osteogenesis imperfecta. Elife 10. https://doi.org/10.7554/eLife.63488

Grounds MD, Sorokin L, White J (2005) Strength at the extracellular matrix-muscle interface. Scand J Med Sci Sports 15:381–391. https://doi.org/10.1111/j.1600-0838.2005.00467.x

Gruenwald K, Castagnola P, Besio R, Dimori M, Chen Y, Akel NS, Swain FL, Skinner RA, Eyre DR, Gaddy D, Suva LJ, Morello R (2014) Sc65 is a novel endoplasmic reticulum protein that regulates bone mass homeostasis. J Bone Miner Res 29:666–675. https://doi.org/10.1002/jbmr.2075

Gubbiotti MA, Vallet SD, Ricard-Blum S, Iozzo RV (2016) Decorin interacting network: a comprehensive analysis of decorin-binding partners and their versatile functions. Matrix Biol 55:7–21. https://doi.org/10.1016/j.matbio.2016.09.009

Guillén-Navarro E, Ballesta-Martínez MJ, Valencia M, Bueno AM, Martinez-Glez V, López-González V, Burnyte B, Utkus A, Lapunzina P, Ruiz-Perez VL (2014) Two mutations in <i>IFITM5</i> causing distinct forms of osteogenesis imperfecta. Am J Med Genet A 164:1136–1142. https://doi.org/10.1002/ajmg.a.36409

Haik GM Jr (1984) Occurrence of unilateral retinoblastoma in three generations of a Louisiana family. J La State Med Soc 136:41–44

Hansen B, Jemec GB (2002) The mechanical properties of skin in osteogenesis imperfecta. Arch Dermatol 138:909–911. https://doi.org/10.1001/archderm.138.7.909

Hanson DA, Eyre DR (1996) Molecular site specificity of Pyridinoline and pyrrole cross-links in type I collagen of human bone. J Biol Chem 271:26508–26516. https://doi.org/10.1074/jbc.271.43.26508

Hassell JR, Birk DE (2010) The molecular basis of corneal transparency. Exp Eye Res 91:326–335. https://doi.org/10.1016/j.exer.2010.06.021

Haus JM, Carrithers JA, Trappe SW, Trappe TA (2007) Collagen, cross-linking, and advanced glycation end products in aging human skeletal muscle. J Appl Physiol 103:2068–2076. https://doi.org/10.1152/japplphysiol.00670.2007

Ha-Vinh R, Alanay Y, Bank RA, Campos-Xavier AB, Zankl A, Superti-Furga A, Bonafé L (2004) Phenotypic and molecular characterization of Bruck syndrome (osteogenesis imperfecta with contractures of the large joints) caused by a recessive mutation in <i>PLOD2</i>. Am J Med Genet A 131A:115–120. https://doi.org/10.1002/ajmg.a.30231

Heard ME, Besio R, Weis M, Rai J, Hudson DM, Dimori M, Zimmerman SM, Kamykowski JA, Hogue WR, Swain FL, Burdine MS, Mackintosh SG, Tackett AJ, Suva LJ, Eyre DR, Morello R (2016) Sc65-null mice provide evidence for a novel endoplasmic reticulum complex regulating collagen Lysyl hydroxylation. PLoS Genet 12:e1006002. https://doi.org/10.1371/journal.pgen.1006002

Hedjazi G, Guterman-Ram G, Blouin S, Schemenz V, Wagermaier W, Fratzl P, Hartmann MA, Zwerina J, Fratzl-Zelman N, Marini JC (2022) Alterations of bone material properties in growing Ifitm5/BRIL p.S42 knock-in mice, a new model for atypical type VI osteogenesis imperfecta. Bone 162:116451. https://doi.org/10.1016/j.bone.2022.116451

Hennet T (2019) Collagen glycosylation. Curr Opin Struct Biol 56:131–138. https://doi.org/10.1016/j.sbi.2019.01.015
Heymsfield SB, Thomas DM, Bosy-Westphal A, Müller MJ (2019) The anatomy of resting energy expenditure: body composition mechanisms. Eur J Clin Nutr 73:166–171. https://doi.org/10.1038/s41430-018-0319-3
Holmes DF, Graham HK, Kadler KE (1998) Collagen fibrils forming in developing tendon show an early and abrupt limitation in diameter at the growing tips. J Mol Biol 283:1049–1058. https://doi.org/10.1006/jmbi.1998.2153
Holmes DF, Chapman JA, Prockop DJ, Kadler KE (1992) Growing tips of type I collagen fibrils formed in vitro are near-paraboloidal in shape, implying a reciprocal relationship between accretion and diameter. Proc Natl Acad Sci 89:9855–9859. https://doi.org/10.1073/pnas.89.20.9855
Homan EP, Lietman C, Grafe I, Lennington J, Morello R, Napierala D, Jiang MM, Munivez EM, Dawson B, Bertin TK, Chen Y, Lua R, Lichtarge O, Hicks J, Weis MA, Eyre D, Lee BH (2014) Differential effects of collagen prolyl 3-hydroxylation on skeletal tissues. PLoS Genet 10:e1004121. https://doi.org/10.1371/journal.pgen.1004121
Homan EP, Rauch F, Grafe I, Lietman C, Doll JA, Dawson B, Bertin T, Napierala D, Morello R, Gibbs R, White L, Miki R, Cohn DH, Crawford S, Travers R, Glorieux FH, Lee B (2011) Mutations in SERPINF1 cause osteogenesis imperfecta type VI. J Bone Miner Res 26:2798–2803. https://doi.org/10.1002/jbmr.487
Hoyer-Kuhn H, Semler O, Schoenau E, Roschger P, Klaushofer K, Rauch F (2013) Hyperosteoidosis and hypermineralization in the same bone: bone tissue analyses in a boy with a homozygous BMP1 mutation. Calcif Tissue Int 93:565–570. https://doi.org/10.1007/s00223-013-9799-2
Häkkinen L, Oksala O, Salo T, Rahemtulla F, Larjava H (1993) Immunohistochemical localization of proteoglycans in human periodontium. J Histochem Cytochem 41:1689–1699. https://doi.org/10.1177/41.11.8409375
Ishida Y, Nagata K (2011) Hsp47 as a collagen-specific molecular chaperone. Methods Enzymol 499:167–182. https://doi.org/10.1016/b978-0-12-386471-0.00009-2
Ishikawa Y, Bächinger HP (2013) A molecular ensemble in the rER for procollagen maturation. Biochim Biophys Acta 1833:2479–2491. https://doi.org/10.1016/j.bbamcr.2013.04.008
Ishikawa Y, Boudko S, Bächinger HP (2015) Ziploc-ing the structure: triple helix formation is coordinated by rough endoplasmic reticulum resident PPIases. Biochim Biophys Acta 1850:1983–1993. https://doi.org/10.1016/j.bbagen.2014.12.024
Ishikawa Y, Mizuno K, Bächinger HP (2017) Ziploc-ing the structure 2.0: endoplasmic reticulum-resident peptidyl prolyl isomerases show different activities toward hydroxyproline. J Biol Chem 292:9273–9282. https://doi.org/10.1074/jbc.m116.772657
Ishikawa Y, Vranka J, Wirz J, Nagata K, Bächinger HP (2008) The rough endoplasmic reticulum-resident FK506-binding protein FKBP65 is a molecular chaperone that interacts with collagens. J Biol Chem 283:31584–31590. https://doi.org/10.1074/jbc.m802535200
Ishikawa Y, Wirz J, Vranka JA, Nagata K, Bächinger HP (2009) Biochemical characterization of the prolyl 3-hydroxylase 1·cartilage-associated protein·Cyclophilin B complex. J Biol Chem 284:17641–17647. https://doi.org/10.1074/jbc.m109.007070
Izu Y, Adams SM, Connizzo BK, Beason DP, Soslowsky LJ, Koch M, Birk DE (2021) Collagen XII mediated cellular and extracellular mechanisms regulate establishment of tendon structure and function. Matrix Biol 95:52–67. https://doi.org/10.1016/j.matbio.2020.10.004
Izu Y, Ansorge HL, Zhang G, Soslowsky LJ, Bonaldo P, Chu M-L, Birk DE (2011) Dysfunctional tendon collagen fibrillogenesis in collagen VI null mice. Matrix Biol 30:53–61. https://doi.org/10.1016/j.matbio.2010.10.001
Janicki JS, Brower GL (2002) The role of myocardial fibrillar collagen in ventricular remodeling and function. J Card Fail 8:S319–S325. https://doi.org/10.1054/jcaf.2002.129260
Jeong Y, Daghlas SA, Kahveci AS, Salamango D, Gentry BA, Brown M, Rector RS, Pearsall RS, Phillips CL (2018) Soluble activin receptor type IIB decoy receptor differentially impacts

murine osteogenesis imperfecta muscle function. Muscle Nerve 57:294–304. https://doi.org/10.1002/mus.25706

Jeong Y, Carleton SM, Gentry BA, Yao X, Ferreira JA, Salamango DJ, Weis M, Oestreich AK, Williams AM, Mccray MG, Eyre DR, Brown M, Wang Y, Phillips CL (2015) Hindlimb skeletal muscle function and skeletal quality and strength in +/<i>G610C</i> mice With and without weight-bearing exercise. J Bone Miner Res 30:1874–1886. https://doi.org/10.1002/jbmr.2518

Joeng KS, Lee YC, Jiang MM, Bertin TK, Chen Y, Abraham AM, Ding H, Bi X, Ambrose CG, Lee BH (2014) The swaying mouse as a model of osteogenesis imperfecta caused by WNT1 mutations. Hum Mol Genet 23:4035–4042. https://doi.org/10.1093/hmg/ddu117

Jonisch J, Deramo VA (2011) Sutureless 25-gauge vitrectomy for retinal detachment repair in a patient with osteogenesis imperfecta. Retin Cases Brief Rep 5:67–69. https://doi.org/10.1097/ICB.0b013e3181babe6c

Jovanovic M, Guterman-Ram G, Marini JC (2022) Osteogenesis imperfecta: mechanisms and signaling pathways connecting classical and rare OI types. Endocr Rev 43:61–90. https://doi.org/10.1210/endrev/bnab017

Ju M, Bai X, Zhang T, Lin Y, Yang L, Zhou H, Chang X, Guan S, Ren X, Li K, Wang Y, Li G (2020) Mutation spectrum of COL1A1/COL1A2 screening by high-resolution melting analysis of Chinese patients with osteogenesis imperfecta. J Bone Miner Metab 38:188–197. https://doi.org/10.1007/s00774-019-01039-3

Kadler KE (2017) Fell Muir Lecture: Collagen fibril formation <i>in vitro</i> and <i>in vivo</i>. Int J Exp Pathol 98:4–16. https://doi.org/10.1111/iep.12224

Kague E, Roy P, Asselin G, Hu G, Simonet J, Stanley A, Albertson C, Fisher S (2016) Osterix/Sp7 limits cranial bone initiation sites and is required for formation of sutures. Dev Biol 413:160–172. https://doi.org/10.1016/j.ydbio.2016.03.011

Kaiser-Kupfer MI, McCain L, Shapiro JR, Podgor MJ, Kupfer C, Rowe D (1981) Low ocular rigidity in patients with osteogenesis imperfecta. Invest Ophthalmol Vis Sci 20:807–809

Kanis JA (2002) Diagnosis of osteoporosis and assessment of fracture risk. Lancet 359:1929–1936. https://doi.org/10.1016/s0140-6736(02)08761-5

Kannus P (2000) Structure of the tendon connective tissue. Scand J Med Sci Sports 10:312–320. https://doi.org/10.1034/j.1600-0838.2000.010006312.x

Kantaputra PN, Dejkhamron P, Intachai W, Ngamphiw C, Ketudat Cairns JR, Kawasaki K, Ohazama A, Olsen B, Tongsima S, Angkurawaranon S (2021) A novel P3H1 mutation is associated with osteogenesis imperfecta type VIII and dental anomalies. Oral Surg Oral Med Oral Pathol Oral Radiol 132:e198–e207. https://doi.org/10.1016/j.oooo.2021.01.023

Karakioulaki M, Papakonstantinou E, Stolz D (2020) Extracellular matrix remodelling in COPD. Eur Respir Rev 29. https://doi.org/10.1183/16000617.0124-2019

Kasner L, Mietz H, Green WR (1993) Agenesis of Bowman's layer. A histopathological study of four cases. Cornea 12:163–170. https://doi.org/10.1097/00003226-199303000-00013

Kemp AD, Harding CC, Cabral WA, Marini JC, Wallace JM (2012) Effects of tissue hydration on nanoscale structural morphology and mechanics of individual type I collagen fibrils in the Brtl mouse model of osteogenesis imperfecta. J Struct Biol 180:428–438. https://doi.org/10.1016/j.jsb.2012.09.012

Keuning MC, Leeuwerke SJG, van Dijk PR, Harsevoort AGJ, Grotjohan HP, Franken AAM, Janus GJM (2022) Prevalence of scoliosis and impaired pulmonary function in patients with type III osteogenesis imperfecta. Eur Spine J 31:2295–2300. https://doi.org/10.1007/s00586-022-07260-5

Keupp K, Beleggia F, Kayserili H, Barnes M, Amling M, Kornak U, Marini J, Wollnik B et al (2013) Mutations in WNT1 cause different forms of bone fragility. Am J Hum Genet 92:565–574. https://doi.org/10.1016/j.ajhg.2013.02.010

Khan SI, Yonko EA, Carter EM, Dyer D, Sandhaus RA, Raggio CL (2020) Cardiopulmonary status in adults with osteogenesis imperfecta: intrinsic lung disease May contribute more than scoliosis. Clin Orthop Relat Res 478:2833–2843. https://doi.org/10.1097/corr.0000000000001400

Klein GL (2022) Transforming growth Factor-Beta in skeletal muscle wasting. Int J Mol Sci 23:1167. https://doi.org/10.3390/ijms23031167

Kobayashi S, Ogasawara N, Watanabe S, Yoneyama Y, Kirino S, Hiraguri Y, Inoue M, Nagata S, Okamoto-Uchida Y, Kofuji S, Shimizu H, Ito G, Mizutani T, Yamauchi S, Kinugasa Y, Kano Y, Nemoto Y, Watanabe M, Tsuchiya K, Nishina H, Okamoto R, Yui S (2022) Collagen type I-mediated mechanotransduction controls epithelial cell fate conversion during intestinal inflammation. Inflamm Regen 42:49. https://doi.org/10.1186/s41232-022-00237-3

Komai Y, Ushiki T (1991) The three-dimensional organization of collagen fibrils in the human cornea and sclera. Invest Ophthalmol Vis Sci 32:2244–2258

Komuro T (1988) The lattice arrangement of the collagen fibres in the submucosa of the rat small intestine: scanning electron microscopy. Cell Tissue Res 251:117–121. https://doi.org/10.1007/bf00215455

Kovanen V (2002) Intramuscular extracellular matrix: complex environment of muscle cells. Exerc Sport Sci Rev 30:20–25. https://doi.org/10.1097/00003677-200201000-00005

Kovanen V, Suominen H, Risteli J, Risteli L (1988) Type IV collagen and laminin in slow and fast skeletal muscle in rats—effects of age and life-time endurance training. Coll Relat Res 8:145–153. https://doi.org/10.1016/s0174-173x(88)80026-8

Kozloff KM, Carden A, Bergwitz C, Forlino A, Uveges TE, Morris MD, Marini JC, Goldstein SA (2004) Brittle IV mouse model for osteogenesis imperfecta IV demonstrates Postpubertal adaptations to improve whole bone strength. J Bone Miner Res 19:614–622. https://doi.org/10.1359/jbmr.040111

Kuptanon C, Thamkunanon V, Srichomthong C, Theerapanon T, Suphapeetiporn K, Porntaveetus T, Shotelersuk V (2022) Novel BMP1, CRTAP, and SERPINF1 variants causing autosomal recessive osteogenesis imperfecta. Clin Genet 102:242–243. https://doi.org/10.1111/cge.14172

Kuznetsova NV, Forlino A, Cabral WA, Marini JC, Leikin S (2004) Structure, stability and interactions of type I collagen with GLY349-CYS substitution in alpha 1(I) chain in a murine osteogenesis imperfecta model. Matrix Biol 23:101–112. https://doi.org/10.1016/j.matbio.2004.03.002

Kämpe AJ, Costantini A, Levy-Shraga Y, Zeitlin L, Roschger P, Taylan F, Lindstrand A, Paschalis EP, Gamsjaeger S, Raas-Rothschild A, Hövel M, Jiao H, Klaushofer K, Grasemann C, Mäkitie O (2017) <i>PLS3</i> deletions Lead to severe spinal osteoporosis and disturbed bone matrix mineralization. J Bone Miner Res 32:2394–2404. https://doi.org/10.1002/jbmr.3233

Laine CM, Wessman M, Toiviainen-Salo S, Kaunisto MA, Mäyränpää MK, Laine T, Pekkinen M, Kröger H, Välimäki V-V, Välimäki MJ, Lehesjoki A-E, Mäkitie O (2015) A novel splice mutation in <i>PLS3</i> causes X-linked early onset low-turnover osteoporosis. J Bone Miner Res 30:510–518. https://doi.org/10.1002/jbmr.2355

Laine CM, Joeng KS, Campeau PM, Kiviranta R, Tarkkonen K, Grover M, Lu JT, Pekkinen M, Wessman M, Heino TJ, Nieminen-Pihala V, Aronen M, Laine T, Kröger H, Cole WG, Lehesjoki A-E, Nevarez L, Krakow D, Curry CJR, Cohn DH, Gibbs RA, Lee BH, Mäkitie O (2013) <i>WNT1</i> mutations in early-onset osteoporosis and osteogenesis imperfecta. N Engl J Med 368:1809–1816. https://doi.org/10.1056/nejmoa1215458

Landis WJ, Jacquet R (2013) Association of Calcium and Phosphate Ions with collagen in the mineralization of vertebrate tissues. Calcif Tissue Int 93:329–337. https://doi.org/10.1007/s00223-013-9725-7

Lapunzina P, Aglan M, Temtamy S, Caparrós-Martín JA, Valencia M, Letón R, Martínez-Glez V, Elhossini R, Amr K, Vilaboa N, Ruiz-Perez VL (2010) Identification of a frameshift mutation in Osterix in a patient with recessive osteogenesis imperfecta. Am J Hum Genet 87:110–114. https://doi.org/10.1016/j.ajhg.2010.05.016

Latif N, Sarathchandra P, Taylor PM, Antoniw J, Yacoub MH (2005) Localization and pattern of expression of extracellular matrix components in human heart valves. J Heart Valve Dis 14:218–227

Laurent GJ (1986) Lung collagen: more than scaffolding. Thorax 41:418–428. https://doi.org/10.1136/thx.41.6.418

Leblond CP (1989) Synthesis and secretion of collagen by cells of connective tissue, bone, and dentin. Anat Rec 224:123–138. https://doi.org/10.1002/ar.1092240204

Lee JH, Gamble JG, Moore RE, Rinsky LA (1995) Gastrointestinal problems in patients who have type-III osteogenesis imperfecta. J Bone Joint Surg Am 77:1352–1356. https://doi.org/10.2106/00004623-199509000-00010

Leoni L, Tonelli F, Besio R, Gioia R, Moccia F, Rossi A, Forlino A (2021) Knocking out TMEM38B in human foetal osteoblasts hFOB 1.19 by CRISPR/Cas9: a model for recessive OI type XIV. PLoS One 16:e0257254. https://doi.org/10.1371/journal.pone.0257254

Li L, Mao B, Li S, Xiao J, Wang H, Zhang J, Ren X, Wang Y, Wu Y, Cao Y, Lu C, Gao J, You Y, Zhao F, Geng X, Xiao Y, Jiang C, Ye Y, Yang T, Zhao X, Zhang X (2019) Genotypic and phenotypic characterization of Chinese patients with osteogenesis imperfecta. Hum Mutat 40:588–600. https://doi.org/10.1002/humu.23718

Li S-W, Arita M, Fertala A, Bao Y, Kopen GC, Långsjö TK, Hyttinen MM, Helminen HJ, Prockop DJ (2001) Transgenic mice with inactive alleles for procollagen N-proteinase (ADAMTS-2) develop fragile skin and male sterility. Biochem J 355:271. https://doi.org/10.1042/0264-6021:3550271

Lietman CD, Rajagopal A, Homan EP, Munivez E, Jiang MM, Bertin TK, Chen Y, Hicks J, Weis M, Eyre D, Lee B, Krakow D (2014) Connective tissue alterations in Fkbp10−/− mice. Hum Mol Genet 23:4822–4831. https://doi.org/10.1093/hmg/ddu197

Lietman CD, Lim J, Grafe I, Chen Y, Ding H, Bi X, Ambrose CG, Fratzl-Zelman N, Roschger P, Klaushofer K, Wagermaier W, Schmidt I, Fratzl P, Rai J, Weis M, Eyre D, Keene DR, Krakow D, Lee BH (2017) Fkbp10 deletion in osteoblasts leads to qualitative defects in bone. J Bone Miner Res 32:1354–1367. https://doi.org/10.1002/jbmr.3108

Lim J, Lietman C, Grol MW, Castellon A, Dawson B, Adeyeye M, Rai J, Weis M, Keene DR, Schweitzer R, Park D, Eyre DR, Krakow D, Lee BH (2021) Localized chondro-ossification underlies joint dysfunction and motor deficits in the Fkbp10 mouse model of osteogenesis imperfecta. Proc Natl Acad Sci U S A 118. https://doi.org/10.1073/pnas.2100690118

Lindahl K, Åström E, Dragomir A, Symoens S, Coucke P, Larsson S, Paschalis E, Roschger P, Gamsjaeger S, Klaushofer K, Fratzl-Zelman N, Kindmark A (2018) Homozygosity for CREB3L1 premature stop codon in first case of recessive osteogenesis imperfecta associated with OASIS-deficiency to survive infancy. Bone 114:268–277. https://doi.org/10.1016/j.bone.2018.06.019

Lindert U, Weis MA, Rai J, Seeliger F, Hausser I, Leeb T, Eyre D, Rohrbach M, Giunta C (2015) Molecular consequences of the SERPINH1/HSP47 mutation in the dachshund natural model of osteogenesis imperfecta. J Biol Chem 290:17679–17689. https://doi.org/10.1074/jbc.M115.661025

Lindert U, Cabral WA, Ausavarat S, Tongkobpetch S, Ludin K, Barnes AM, Yeetong P, Weis M, Krabichler B, Srichomthong C, Makareeva EN, Janecke AR, Leikin S, Röthlisberger B, Rohrbach M, Kennerknecht I, Eyre DR, Suphapeetiporn K, Giunta C, Marini JC, Shotelersuk V (2016) MBTPS2 mutations cause defective regulated intramembrane proteolysis in X-linked osteogenesis imperfecta. Nat Commun 7:11920. https://doi.org/10.1038/ncomms11920

Linsenmayer TF, Gibney E, Igoe F, Gordon MK, Fitch JM, Fessler LI, Birk DE (1993) Type V collagen: molecular structure and fibrillar organization of the chicken alpha 1(V) NH2-terminal domain, a putative regulator of corneal fibrillogenesis. J Cell Biol 121:1181–1189. https://doi.org/10.1083/jcb.121.5.1181

Lisse TS, Thiele F, Fuchs H, Hans W, Przemeck GK, Abe K, Rathkolb B, Quintanilla-Martinez L, Hoelzlwimmer G, Helfrich M, Wolf E, Ralston SH, Hrabé de Angelis M (2008) ER stress-mediated apoptosis in a new mouse model of osteogenesis imperfecta. PLoS Genet 4:e7. https://doi.org/10.1371/journal.pgen.0040007

Liu L, Stephens B, Bergman M, May A, Chiang T (2021) Role of collagen in airway mechanics. Bioengineering 8:13. https://doi.org/10.3390/bioengineering8010013

Lui JC, Raimann A, Hojo H, Dong L, Roschger P, Kikani B, Wintergerst U, Fratzl-Zelman N, Jee YH, Haeusler G, Baron J (2022) A neomorphic variant in SP7 alters sequence specific-

ity and causes a high-turnover bone disorder. Nat Commun 13. https://doi.org/10.1038/s41467-022-28318-4
Mackenna DA, Omens JH, Covell JW (1996) Left ventricular perimysial collagen fibers uncoil rather than stretch during diastolic filling. Basic Res Cardiol 91:111–122. https://doi.org/10.1007/bf00799683
MacKenna DA, Vaplon SM, McCulloch AD (1997) Microstructural model of perimysial collagen fibers for resting myocardial mechanics during ventricular filling. Am J Phys 273:H1576–H1586. https://doi.org/10.1152/ajpheart.1997.273.3.H1576
Marchi F, Leblond CP (1984) Radioautographic characterization of successive compartments along the rough endoplasmic reticulum-Golgi pathway of collagen precursors in foot pad fibroblasts of [3H]proline-injected rats. J Cell Biol 98:1705–1709. https://doi.org/10.1083/jcb.98.5.1705
Marcos-Garcés V, Molina Aguilar P, Bea Serrano C, García Bustos V, Benavent Seguí J, Ferrández Izquierdo A, Ruiz-Saurí A (2014) Age-related dermal collagen changes during development, maturation and ageing—a morphometric and comparative study. J Anat 225:98–108. https://doi.org/10.1111/joa.12186
Marini JC, Cabral WA, Barnes AM, Chang W (2007) Components of the collagen prolyl 3-hydroxylation complex are crucial for Normal bone development. Cell Cycle 6:1675–1681. https://doi.org/10.4161/cc.6.14.4474
Marini JC, Forlino A, Bächinger HP, Bishop NJ, Byers PH, Paepe A, Fassier F, Fratzl-Zelman N, Kozloff KM, Krakow D, Montpetit K, Semler O (2017) Osteogenesis imperfecta. Nat Rev Dis Primers 3:17052. https://doi.org/10.1038/nrdp.2017.52
Marom R, Rabenhorst BM, Morello R (2020) Osteogenesis imperfecta: an update on clinical features and therapies. Eur J Endocrinol 183:R95–r106. https://doi.org/10.1530/eje-20-0299
Marom R, Zhang B, Washington ME, Song I-W, Burrage LC, Rossi VC, Berrier AS, Lindsey A, Lesinski J, Nonet ML, Chen J, Baldridge D, Silverman GA, Sutton VR, Rosenfeld JA, Tran AA, Hicks MJ, Murdock DR, Dai H, Weis M, Jhangiani SN, Muzny DM, Gibbs RA, Caswell R, Pottinger C, Cilliers D, Stals K, Eyre D, Krakow D, Schedl T, Pak SC, Lee BH (2023) Dominant negative variants in KIF5B cause osteogenesis imperfecta via down regulation of mTOR signaling. PLoS Genet 19:e1011005. https://doi.org/10.1371/journal.pgen.1011005
Marshall GE (1995) Human scleral elastic system: an immunoelectron microscopic study. Br J Ophthalmol 79:57–64. https://doi.org/10.1136/bjo.79.1.57
Marshall GE, Konstas AG, Lee WR (1993) Collagens in the aged human macular sclera. Curr Eye Res 12:143–153. https://doi.org/10.3109/02713689308999482
Martins G, Siedlikowski M, Coelho AKS, Rauch F, Tsimicalis A (2020) Bladder and bowel symptoms experienced by children with osteogenesis imperfecta. J Pediatr (Rio J) 96:472. https://doi.org/10.1016/j.jped.2018.12.008
Martínez-Glez V, Valencia M, Caparrós-Martín JA, Aglan M, Temtamy S, Tenorio J, Pulido V, Lindert U, Rohrbach M, Eyre D, Giunta C, Lapunzina P, Ruiz-Perez VL (2012) Identification of a mutation causing deficient BMP1/mTLD proteolytic activity in autosomal recessive osteogenesis imperfecta. Hum Mutat 33:343–350. https://doi.org/10.1002/humu.21647
Marí-Beffa M, Mesa-Román AB, Duran I (2021) Zebrafish models for human skeletal disorders. Front Genet 12. https://doi.org/10.3389/fgene.2021.675331
McKiernan FE (2005) Musculoskeletal manifestations of mild osteogenesis imperfecta in the adult. Osteoporos Int 16:1698–1702. https://doi.org/10.1007/s00198-005-1905-5
Medugorac I, Jacob R (1983) Characterisation of left ventricular collagen in the rat. Cardiovasc Res 17:15–21. https://doi.org/10.1093/cvr/17.1.15
Mendoza-Londono R, Fahiminiya S, Majewski J, Tétreault M, Nadaf J, Kannu P, Sochett E, Howard A, Stimec J, Dupuis L, Roschger P, Klaushofer K, Palomo T, Ouellet J, Al-Jallad H, Mort JS, Moffatt P, Boudko S, Bächinger HP, Rauch F (2015) Recessive osteogenesis imperfecta caused by missense mutations in SPARC. Am J Hum Genet 96:979–985. https://doi.org/10.1016/j.ajhg.2015.04.021
Meschiari CA, Ero OK, Pan H, Finkel T, Lindsey ML (2017) The impact of aging on cardiac extracellular matrix. GeroScience 39:7–18. https://doi.org/10.1007/s11357-017-9959-9

Millington-Sanders C, Meir A, Lawrence L, Stolinski C (1998) Structure of chordae tendineae in the left ventricle of the human heart. J Anat 192:573–581. https://doi.org/10.1046/j.1469-7580.1998.19240573.x

Moffatt P, Boraschi-Diaz I, Bardai G, Rauch F (2021) Muscle transcriptome in mouse models of osteogenesis imperfecta. Bone 148:115940. https://doi.org/10.1016/j.bone.2021.115940

Moosa S, Chung BHY, Tung JYL, Altmüller J, Thiele H, Nürnberg P, Netzer C, Nishimura G, Wollnik B (2016) Mutations in <i><scp>SEC24D</scp></i> cause autosomal recessive osteogenesis imperfecta. Clin Genet 89:517–519. https://doi.org/10.1111/cge.12678

Moosa S, Yamamoto GL, Garbes L, Keupp K, Beleza-Meireles A, Moreno CA, Valadares ER, De Sousa SB, Maia S, Saraiva J, Honjo RS, Kim CA, Cabral De Menezes H, Lausch E, Lorini PV, Lamounier A, Carniero TCB, Giunta C, Rohrbach M, Janner M, Semler O, Beleggia F, Li Y, Yigit G, Reintjes N, Altmüller J, Nürnberg P, Cavalcanti DP, Zabel B, Warman ML, Bertola DR, Wollnik B, Netzer C (2019) Autosomal-recessive mutations in mesd cause osteogenesis imperfecta. Am J Hum Genet 105:836–843. https://doi.org/10.1016/j.ajhg.2019.08.008

Morello R, Bertin TK, Chen Y, Hicks J, Tonachini L, Monticone M, Castagnola P, Rauch F, Glorieux FH, Vranka J, Bächinger HP, Pace JM, Schwarze U, Byers PH, Weis M, Fernandes RJ, Eyre DR, Yao Z, Boyce BF, Lee B (2006) CRTAP is required for prolyl 3- hydroxylation and mutations cause recessive osteogenesis imperfecta. Cell 127:291–304. https://doi.org/10.1016/j.cell.2006.08.039

Muir AM, Ren Y, Butz DH, Davis NA, Blank RD, Birk DE, Lee SJ, Rowe D, Feng JQ, Greenspan DS (2014) Induced ablation of Bmp1 and Tll1 produces osteogenesis imperfecta in mice. Hum Mol Genet 23:3085–3101. https://doi.org/10.1093/hmg/ddu013

Murakami T, Saito A, Hino S, Kondo S, Kanemoto S, Chihara K, Sekiya H, Tsumagari K, Ochiai K, Yoshinaga K, Saitoh M, Nishimura R, Yoneda T, Kou I, Furuichi T, Ikegawa S, Ikawa M, Okabe M, Wanaka A, Imaizumi K (2009) Signalling mediated by the endoplasmic reticulum stress transducer OASIS is involved in bone formation. Nat Cell Biol 11:1205–1211. https://doi.org/10.1038/ncb1963

Murshed M (2018) Mechanism of bone mineralization. Cold Spring Harb Perspect Med 8:a031229. https://doi.org/10.1101/cshperspect.a031229

Nabavizadeh N, Bressin A, Shboul M, Moreno Traspas R, Chia PH, Bonnard C, Szenker-Ravi E, Sarıbaş B, Beillard E, Altunoglu U, Hojati Z, Drutman S, Freier S, El-Khateeb M, Fathallah R, Casanova JL, Soror W, Arafat A, Escande-Beillard N, Mayer A, Reversade B (2023) A progeroid syndrome caused by a deep intronic variant in TAPT1 is revealed by RNA/SI-NET sequencing. EMBO Mol Med 15:e16478. https://doi.org/10.15252/emmm.202216478

Nagai N, Hosokawa M, Itohara S, Adachi E, Matsushita T, Hosokawa N, Nagata K (2000) Embryonic lethality of molecular chaperone Hsp47 knockout mice is associated with defects in collagen biosynthesis. J Cell Biol 150:1499–1506. https://doi.org/10.1083/jcb.150.6.1499

Nakamura T, Yamashita M, Ikegami K, Suzuki M, Yanagita M, Kitagaki J, Kitamura M, Murakami S (2021) Autophagy facilitates type I collagen synthesis in periodontal ligament cells. Sci Rep 11:1291. https://doi.org/10.1038/s41598-020-80275-4

Nakashima K, Zhou X, Kunkel G, Zhang Z, Deng JM, Behringer RR, de Crombrugghe B (2002) The novel zinc finger-containing transcription factor osterix is required for osteoblast differentiation and bone formation. Cell 108:17–29. https://doi.org/10.1016/s0092-8674(01)00622-5

Natarajan R, Shah GY, Rao SK, Padamanabhan P (2003) Penetrating keratoplasty as a globe-saving procedure in fragile cornea. Cornea 22:164–165. https://doi.org/10.1097/00003226-200303000-00017

Nikolov A, Popovski N (2022) Extracellular matrix in heart disease: focus on circulating collagen type I and III derived peptides as biomarkers of myocardial fibrosis and their potential in the prognosis of heart failure: a concise review. Meta 12:297. https://doi.org/10.3390/metabo12040297

Nudelman F, Pieterse K, George A, Bomans PHH, Friedrich H, Brylka LJ, Hilbers PAJ, De With G, Sommerdijk NAJM (2010) The role of collagen in bone apatite formation in the presence of hydroxyapatite nucleation inhibitors. Nat Mater 9:1004–1009. https://doi.org/10.1038/nmat2875

Nwosu BU, Raygada M, Tsilou ET, Rennert OM, Stratakis CA (2005) Rieger's anomaly and other ocular abnormalities in association with osteogenesis imperfecta and a COL1A1 mutation. Ophthalmic Genet 26:135–138. https://doi.org/10.1080/13816810500228993

Nyström A, Bruckner-Tuderman L (2019) Matrix molecules and skin biology. Semin Cell Dev Biol 89:136–146. https://doi.org/10.1016/j.semcdb.2018.07.025

Oh EK, Choi HJ, Oh JY, Kim MK, Wee WR (2016) Sequential traumatic and spontaneous corneal rupture in patient with osteogenesis imperfecta. Can J Ophthalmol 51:e81–e84. https://doi.org/10.1016/j.jcjo.2016.01.001

Onursal C, Dick E, Angelidis I, Schiller HB, Staab-Weijnitz CA (2021) Collagen biosynthesis, processing, and maturation in lung ageing. Front Med (Lausanne) 8:593874. https://doi.org/10.3389/fmed.2021.593874

Orberg J, Baer E, Hiltner A (1983) Organization of collagen fibers in the intestine. Connect Tissue Res 11:285–297. https://doi.org/10.3109/03008208309004861

Oxlund H, Pedersen U, Danielsen CC, Oxlund I, Elbrøond O (1985) Reduced strength of skin in osteogenesis imperfecta. Eur J Clin Investig 15:408–411. https://doi.org/10.1111/j.1365-2362.1985.tb00293.x

Palomo T, Glorieux FH, Schoenau E, Rauch F (2016) Body composition in children and adolescents with osteogenesis imperfecta. J Pediatr 169:232–237. https://doi.org/10.1016/j.jpeds.2015.10.058

Pereira R, Khillan JS, Helminen HJ, Hume EL, Prockop DJ (1993) Transgenic mice expressing a partially deleted gene for type I procollagen (COL1A1). A breeding line with a phenotype of spontaneous fractures and decreased bone collagen and mineral. J Clin Invest 91:709–716. https://doi.org/10.1172/jci116252

Pereira RF, Hume EL, Halford KW, Prockop DJ (1995) Bone fragility in transgenic mice expressing a mutated gene for type I procollagen (COL1A1) parallels the age-dependent phenotype of human osteogenesis imperfecta. J Bone Miner Res 10:1837–1843. https://doi.org/10.1002/jbmr.5650101202

Perreault N, Herring-Gillam FE, Desloges N, Bélanger I, Pageot LP, Beaulieu JF (1998) Epithelial vs mesenchymal contribution to the extracellular matrix in the human intestine. Biochem Biophys Res Commun 248:121–126. https://doi.org/10.1006/bbrc.1998.8919

Phillips CL, Bradley DA, Schlotzhauer CL, Bergfeld M, Libreros-Minotta C, Gawenis LR, Morris JS, Clarke LL, Hillman LS (2000) Oim mice exhibit altered femur and incisor mineral composition and decreased bone mineral density. Bone 27:219–226. https://doi.org/10.1016/s8756-3282(00)00311-2

Pinheiro BS, Barrios PM, Souza LT, Félix TM (2020) Echocardiographic study in children with osteogenesis imperfecta. Cardiol Young 30:1490–1495. https://doi.org/10.1017/s1047951120002474

Polat N, Ulucan PB (2015) Nontraumatic Descemet membrane detachment with tear in osteogenesis imperfecta. Ophthalmol Therapy 4:59–63. https://doi.org/10.1007/s40123-015-0034-2

Pompili S, Latella G, Gaudio E, Sferra R, Vetuschi A (2021) The charming world of the extracellular matrix: a dynamic and protective network of the Intestinal Wall. Front Med (Lausanne) 8:610189. https://doi.org/10.3389/fmed.2021.610189

Puvanesarajah V, Sponseller PD (2014) Chapter 37–Muscle, tendon and ligament in osteogenesis imperfecta. In: Shapiro JR, Byers PH, Glorieux FH, Sponseller PD (eds) Osteogenesis imperfecta. Academic Press, San Diego, pp 349–352. https://doi.org/10.1016/B978-0-12-397165-4.00037-X

Pyott M et al (2013) WNT1 mutations in families affected by moderately severe and progressive recessive osteogenesis imperfecta. Am J Hum Genet 92:590–597. https://doi.org/10.1016/j.ajhg.2013.02.009

Pyott SM, Schwarze U, Christiansen HE, Pepin MG, Leistritz DF, Dineen R, Harris C, Burton BK, Angle B, Kim K, Sussman MD, Weis M, Eyre DR, Russell DW, Mccarthy KJ, Steiner RD, Byers PH (2011) Mutations in PPIB (cyclophilin B) delay type I procollagen chain association and result in perinatal lethal to moderate osteogenesis imperfecta phenotypes. Hum Mol Genet 20:1595–1609. https://doi.org/10.1093/hmg/ddr037

Quantock AJ, Young RD (2008) Development of the corneal stroma, and the collagen–proteoglycan associations that help define its structure and function. Dev Dyn 237:2607–2621. https://doi.org/10.1002/dvdy.21579

Rada JA, Achen VR, Perry CA, Fox PW (1997) Proteoglycans in the human sclera. Evidence for the presence of aggrecan. Invest Ophthalmol Vis Sci 38:1740–1751

Rada JA, Achen VR, Penugonda S, Schmidt RW, Mount BA (2000) Proteoglycan composition in the human sclera during growth and aging. Invest Ophthalmol Vis Sci 41:1639–1648

Radunovic Z, Steine K (2015) Prevalence of cardiovascular disease and cardiac symptoms: left and right ventricular function in adults With osteogenesis imperfecta. Can J Cardiol 31:1386–1392. https://doi.org/10.1016/j.cjca.2015.04.016

Radunovic Z, Wekre LL, Steine K (2012) Right ventricular and pulmonary arterial dimensions in adults with osteogenesis imperfecta. Am J Cardiol 109:1807–1813. https://doi.org/10.1016/j.amjcard.2012.01.402

Radunovic Z, Wekre LL, Diep LM, Steine K (2011) Cardiovascular abnormalities in adults with osteogenesis imperfecta. Am Heart J 161:523–529. https://doi.org/10.1016/j.ahj.2010.11.006

Ramadan R, Wouters VM, van Neerven SM, de Groot NE, Garcia TM, Muncan V, Franklin OD, Battle M, Carlson KS, Leach J, Sansom OJ, Boulard O, Chamaillard M, Vermeulen L, Medema JP, Huels DJ (2022) The extracellular matrix controls stem cell specification and crypt morphology in the developing and adult mouse gut. Biol Open 11. https://doi.org/10.1242/bio.059544

Rappu P, Salo M, Antti Myllyharju J, Heino J (2019) Role of prolyl hydroxylation in the molecular interactions of collagens. Essays Biochem 63:325–335. https://doi.org/10.1042/ebc20180053

Rauch F, Fahiminiya S, Majewski J, Carrot-Zhang J, Boudko S, Glorieux F, Mort S, John Bächinger H-P, Moffatt P (2015) Cole-carpenter syndrome is caused by a heterozygous missense mutation in P4HB. Am J Hum Genet 96:425–431. https://doi.org/10.1016/j.ajhg.2014.12.027

Reich A, Bae AS, Barnes AM, Cabral WA, Hinek A, Stimec J, Hill SC, Chitayat D, Marini JC (2015) Type V OI primary osteoblasts display increased mineralization despite decreased COL1A1 expression. J Clin Endocrinol Metab 100:E325–E332. https://doi.org/10.1210/jc.2014-3082

Rentz TJ, Poobalarahi F, Bornstein P, Sage EH, Bradshaw AD (2007) SPARC regulates processing of procollagen I and collagen Fibrillogenesis in dermal fibroblasts. J Biol Chem 282:22062–22071. https://doi.org/10.1074/jbc.m700167200

Robinson TF, Cohen-Gould L, Factor SM (1983) Skeletal framework of mammalian heart muscle. Arrangement of inter- and pericellular connective tissue structures. Lab Investig 49:482–498

Roschger P, Fratzl-Zelman N, Misof BM, Glorieux FH, Klaushofer K, Rauch F (2008) Evidence that abnormal high bone mineralization in growing children with osteogenesis imperfecta is not associated with specific collagen mutations. Calcif Tissue Int 82:263–270. https://doi.org/10.1007/s00223-008-9113-x

Sanes JR (1982) Laminin, fibronectin, and collagen in synaptic and extrasynaptic portions of muscle fiber basement membrane. J Cell Biol 93:442–451. https://doi.org/10.1083/jcb.93.2.442

Santos-Ledo A, Garcia-Macia M, Campbell PD, Gronska M, Marlow FL (2017) Kinesin-1 promotes chondrocyte maintenance during skeletal morphogenesis. PLoS Genet 13:e1006918. https://doi.org/10.1371/journal.pgen.1006918

Sarafova AP, Choi H, Forlino A, Gajko A, Cabral WA, Tosi L, Reing CM, Marini JC (1998) Three novel type I collagen mutations in osteogenesis imperfecta type IV probands are associated with discrepancies between electrophoretic migration of osteoblast and fibroblast collagen. Hum Mutat 11:395–403. https://doi.org/10.1002/(sici)1098-1004(1998)11:5<395::aid-humu7>3.0.co;2-4

Sarmah S, Barrallo-Gimeno A, Melville DB, Topczewski J, Solnica-Krezel L, Knapik EW (2010) Sec24D-dependent transport of extracellular matrix proteins is required for zebrafish skeletal morphogenesis. PLoS One 5:e10367. https://doi.org/10.1371/journal.pone.0010367

Schlombs K, Wagner T, Scheel J (2003) Site-1 protease is required for cartilage development in zebrafish. Proc Natl Acad Sci USA 100:14024–14029. https://doi.org/10.1073/pnas.2331794100

Schmut O (1978) The organization of tissues of the eye by different collagen types. Albrecht Von Graefes Arch Klin Exp Ophthalmol 207:189–199. https://doi.org/10.1007/bf00411053

Schulz JN, Nüchel J, Niehoff A, Bloch W, Schönborn K, Hayashi S, Kamper M, Brinckmann J, Plomann M, Paulsson M, Krieg T, Zaucke F, Eckes B (2016) COMP-assisted collagen secretion—a novel intracellular function required for fibrosis. J Cell Sci 129:706–716. https://doi.org/10.1242/jcs.180216

Schönherr E, Hausser H, Beavan L, Kresse H (1995) Decorin-type I collagen interaction. Presence of separate core protein-binding domains. J Biol Chem 270:8877–8883. https://doi.org/10.1074/jbc.270.15.8877

Scietti L, Chiapparino A, De Giorgi F, Fumagalli M, Khoriauli L, Nergadze S, Basu S, Olieric V, Cucca L, Banushi B, Profumo A, Giulotto E, Gissen P, Forneris F (2018) Molecular architecture of the multifunctional collagen lysyl hydroxylase and glycosyltransferase LH3. Nat Commun 9. https://doi.org/10.1038/s41467-018-05631-5

Sekiya H, Murakami T, Saito A, Hino S, Tsumagari K, Ochiai K, Imaizumi K (2010) Effects of the bisphosphonate risedronate on osteopenia in OASIS-deficient mice. J Bone Miner Metab 28:384–394. https://doi.org/10.1007/s00774-009-0142-y

Semler O, Garbes L, Keupp K, Swan D, Zimmermann K, Becker J, Iden S, Wirth B, Eysel P, Koerber F, Schoenau E, Bohlander K (2012) Stefan, Wollnik, B., and Netzer, C.: a mutation in the 5′-UTR of IFITM5 creates an in-frame start codon and causes autosomal-dominant osteogenesis imperfecta type V with hyperplastic callus. Am J Hum Genet 91:349–357. https://doi.org/10.1016/j.ajhg.2012.06.011

Sendul SY, Atilgan CU, Tiryaki S, Guven D (2016) Bilateral papilledema in a child with osteogenesis imperfecta. Eye Vis (Lond) 3:25. https://doi.org/10.1186/s40662-016-0056-4

Shaheen R, Alazami AM, Alshammari MJ, Faqeih E, Alhashmi N, Mousa N, Alsinani A, Ansari S, Alzahrani F, Al-Owain M, Alzayed ZS, Alkuraya FS (2012) Study of autosomal recessive osteogenesis imperfecta in Arabia reveals a novel locus defined by TMEM38B mutation. J Med Genet 49:630–635. https://doi.org/10.1136/jmedgenet-2012-101142

Sherratt MJ (2009) Tissue elasticity and the ageing elastic fibre. Age 31:305–325. https://doi.org/10.1007/s11357-009-9103-6

Siadat SM, Zamboulis DE, Thorpe CT, Ruberti JW, Connizzo BK (2021) Tendon extracellular matrix assembly, maintenance and dysregulation throughout life. Adv Exp Med Biol 1348:45–103. https://doi.org/10.1007/978-3-030-80614-9_3

Sillence DO, Rimoin DL, Danks DM (1979) Clinical variability in osteogenesis imperfecta-variable expressivity or genetic heterogeneity. Birth Defects Orig Artic Ser 15:113–129

Sinkam L, Boraschi-Diaz I, Svensson RB, Kjaer M, Komarova SV, Bergeron R, Rauch F, Veilleux LN (2023) Tendon properties in a mouse model of severe osteogenesis imperfecta. Connect Tissue Res 64:285–293. https://doi.org/10.1080/03008207.2022.2161376

Sobin SS, Fung YC, Tremer HM (1988) Collagen and elastin fibers in human pulmonary alveolar walls. J Appl Physiol (1985) 64:1659–1675. https://doi.org/10.1152/jappl.1988.64.4.1659

Sricholpech M, Perdivara I, Yokoyama M, Nagaoka H, Terajima M, Tomer KB, Yamauchi M (2012) Lysyl hydroxylase 3-mediated Glucosylation in type I collagen. J Biol Chem 287:22998–23009. https://doi.org/10.1074/jbc.m112.343954

Stacey A, Bateman J, Choi T, Mascara T, Cole W, Jaenisch R (1988) Perinatal lethal osteogenesis imperfecta in transgenic mice bearing an engineered mutant pro-alpha 1(I) collagen gene. Nature 332:131–136. https://doi.org/10.1038/332131a0

Stockwell E, Wallace M (2022) Obstructive constipation in two patients With severe osteogenesis imperfecta and acetabular Protrusio. J Am Acad Orthop Surg Glob Res Rev 6. https://doi.org/10.5435/JAAOSGlobal-D-21-00194

Suki B, Bates JH (2008) Extracellular matrix mechanics in lung parenchymal diseases. Respir Physiol Neurobiol 163:33–43. https://doi.org/10.1016/j.resp.2008.03.015

Sun M, Zafrullah N, Devaux F, Hemmavanh C, Adams S, Ziebarth NM, Koch M, Birk DE, Espana EM (2020) Collagen XII is a regulator of corneal stroma structure and function. Invest Ophthalmol Vis Sci 61:61. https://doi.org/10.1167/iovs.61.5.61

Symoens S, Malfait F, D'Hondt S, Callewaert B, Dheedene A, Steyaert W, Bächinger HP, De Paepe A, Kayserili H, Coucke PJ (2013) Deficiency for the ER-stress transducer OASIS causes severe recessive osteogenesis imperfecta in humans. Orphanet J Rare Dis 8:154. https://doi.org/10.1186/1750-1172-8-154

Symoens S, Barnes M, Aileen Gistelinck C, Malfait F, Guillemyn B, Steyaert W, Syx D, D'Hondt S, Biervliet M, Backer D, Julie Witten P, Eckhard Leikin S, Makareeva E, Gillessen-Kaesbach G, Huysseune A, Vleminckx K, Willaert A, Paepe D, Anne Marini C, Coucke J (2015) Genetic defects in TAPT1 disrupt ciliogenesis and cause a complex lethal osteochondrodysplasia. Am J Hum Genet 97:521–534. https://doi.org/10.1016/j.ajhg.2015.08.009

Tabeta K, Du X, Arimatsu K, Yokoji M, Takahashi N, Amizuka N, Hasegawa T, Crozat K, Maekawa T, Miyauchi S, Matsuda Y, Ida T, Kaku M, Hoebe K, Ohno K, Yoshie H, Yamazaki K, Moresco EMY, Beutler B (2017) An ENU-induced splice site mutation of mouse Col1a1 causing recessive osteogenesis imperfecta and revealing a novel splicing rescue. Sci Rep 7:11717. https://doi.org/10.1038/s41598-017-10343-9

Takken T, Terlingen HC, Helders PJ, Pruijs H, Van der Ent CK, Engelbert RH (2004) Cardiopulmonary fitness and muscle strength in patients with osteogenesis imperfecta type I. J Pediatr 145:813–818. https://doi.org/10.1016/j.jpeds.2004.08.003

Tauer JT, Robinson ME, Rauch F (2019) Osteogenesis imperfecta: new perspectives from clinical and translational research. JBMR Plus 3:e10174. https://doi.org/10.1002/jbm4.10174

Terajima M, Taga Y, Chen Y, Cabral WA, Hou-Fu G, Srisawasdi S, Nagasawa M, Sumida N, Hattori S, Kurie JM, Marini JC, Yamauchi M (2016) Cyclophilin-B modulates collagen cross-linking by differentially affecting lysine hydroxylation in the helical and Telopeptidyl domains of tendon type I collagen. J Biol Chem 291:9501–9512. https://doi.org/10.1074/jbc.M115.699470

Theocharis AD, Skandalis SS, Gialeli C, Karamanos NK (2016) Extracellular matrix structure. Adv Drug Deliv Rev 97:4–27. https://doi.org/10.1016/j.addr.2015.11.001

Thiele F, Cohrs CM, Flor A, Lisse TS, Przemeck GK, Horsch M, Schrewe A, Gailus-Durner V, Ivandic B, Katus HA, Wurst W, Reisenberg C, Chaney H, Fuchs H, Hans W, Beckers J, Marini JC, Hrabé de Angelis M (2012) Cardiopulmonary dysfunction in the osteogenesis imperfecta mouse model Aga2 and human patients are caused by bone-independent mechanisms. Hum Mol Genet 21:3535–3545. https://doi.org/10.1093/hmg/dds183

Thomas DP, McCormick RJ, Zimmerman SD, Vadlamudi RK, Gosselin LE (1992) Aging- and training-induced alterations in collagen characteristics of rat left ventricle and papillary muscle. Am J Phys 263:H778–H783. https://doi.org/10.1152/ajpheart.1992.263.3.H778

Tonelli F, Leoni L, Daponte V, Gioia R, Cotti S, Fiedler IAK, Larianova D, Willaert A, Coucke PJ, Villani S, Busse B, Besio R, Rossi A, Witten PE, Forlino A (2023) Zebrafish Tric-b is required for skeletal development and bone cells differentiation. Front Endocrinol (Lausanne) 14:1002914. https://doi.org/10.3389/fendo.2023.1002914

Tonelli F, Cotti S, Leoni L, Besio R, Gioia R, Marchese L, Giorgetti S, Villani S, Gistelinck C, Wagener R, Kobbe B, Fiedler IAK, Larionova D, Busse B, Eyre D, Rossi A, Witten PE, Forlino A (2020) Crtap and p3h1 knock out zebrafish support defective collagen chaperoning as the cause of their osteogenesis imperfecta phenotype. Matrix Biol 90:40–60. https://doi.org/10.1016/j.matbio.2020.03.004

Toshima M, Ohtani Y, Ohtani O (2004) Three-dimensional architecture of elastin and collagen fiber networks in the human and rat lung. Arch Histol Cytol 67:31–40. https://doi.org/10.1679/aohc.67.31

Tran CT, Smet ME, Forsey J, Zankl A, Nayyar R (2022) Bruck syndrome: beyond the obvious. Fetal Diagn Ther 49:479–485. https://doi.org/10.1159/000527594

Trelstad RL, Hayashi K (1979) Tendon collagen fibrillogenesis: intracellular subassemblies and cell surface changes associated with fibril growth. Dev Biol 71:228–242. https://doi.org/10.1016/0012-1606(79)90166-0

Treurniet S, Burger P, Ghyczy EAE, Verbraak FD, Curro-Tafili KR, Micha D, Bravenboer N, Ralston SH, De Vries R, Moll AC, Eekhoff EMW (2022) Ocular characteristics and complications in patients with osteogenesis imperfecta: a systematic review. Acta Ophthalmologica 100. https://doi.org/10.1111/aos.14882

Tuysuz B, Alkaya DU, Geyik F, Alaylıoğlu M, Kasap B, Kurugoğlu S, Akman YE, Vural M, Bilguvar k (2023) Biallelic frameshift variants in PHLDB1 cause mild-type osteogenesis imperfecta with regressive spondylometaphyseal changes. J Med Genet. 60(8):819–826. https://doi.org/10.1136/jmg-2022-108763. Epub 2022 Dec 21

Udupa P, Shrikondawar AN, Nayak SS, Shah H, Ranjan A, Girisha KM, Bhavani GS, Ghosh DK (2023) Deep intronic mutation in CRTAP results in unstable isoforms of the protein to induce type I collagen aggregation in a lethal type of osteogenesis imperfecta type VII. Biochim Biophys Acta Mol basis Dis 1869:166741. https://doi.org/10.1016/j.bbadis.2023.166741

Unger S, Ferreira CR, Mortier GR, Ali H, Bertola DR, Calder A, Cohn DH, Cormier-Daire V, Girisha KM, Hall C, Krakow D, Makitie O, Mundlos S, Nishimura G, Robertson SP, Savarirayan R, Sillence D, Simon M, Sutton VR, Warman ML, Superti-Furga A (2023) Nosology of genetic skeletal disorders: 2023 revision. Am J Med Genet A 191:1164–1209. https://doi.org/10.1002/ajmg.a.63132

Van Dijk FS, Byers PH, Dalgleish R, Malfait F, Maugeri A, Rohrbach M, Symoens S, Sistermans EA, Pals G (2012) EMQN best practice guidelines for the laboratory diagnosis of osteogenesis imperfecta. Eur J Hum Genet 20:11–19. https://doi.org/10.1038/ejhg.2011.141

Van Dijk FS, Nesbitt IM, Zwikstra EH, Nikkels PGJ, Piersma SR, Fratantoni SA, Jimenez CR, Huizer M, Morsman AC, Cobben JM, Van Roij MHH, Elting MW, Verbeke JIML, Wijnaendts LCD, Shaw NJ, Högler W, Mckeown C, Sistermans EA, Dalton A, Meijers-Heijboer H, Pals G (2009) PPIB mutations cause severe osteogenesis imperfecta. Am J Hum Genet 85:521–527. https://doi.org/10.1016/j.ajhg.2009.09.001

Van Dijk FS, Zillikens MC, Micha D, Riessland M, Marcelis CLM, De Die-Smulders CE, Milbradt J, Franken AA, Harsevoort AJ, Lichtenbelt KD, Pruijs HE, Rubio-Gozalbo ME, Zwertbroek R, Moutaouakil Y, Egthuijsen J, Hammerschmidt M, Bijman R, Semeins CM, Bakker AD, Everts V, Klein-Nulend J, Campos-Obando N, Hofman A, Te Meerman GJ, Verkerk AJMH, Uitterlinden AG, Maugeri A, Sistermans EA, Waisfisz Q, Meijers-Heijboer H, Wirth B, Simon MEH, Pals G (2013) <i>PLS3</i> mutations in X-Linked osteoporosis with fractures. N Engl J Med 369:1529–1536. https://doi.org/10.1056/nejmoa1308223

van Dijk FS, Semler O, Etich J, Köhler A, Jimenez-Estrada JA, Bravenboer N, Claeys L, Riesebos E, Gegic S, Piersma SR, Jimenez CR, Waisfisz Q, Flores CL, Nevado J, Harsevoort AJ, Janus GJM, Franken AAM, van der Sar AM, Meijers-Heijboer H, Heath KE, Lapunzina P, Nikkels PGJ, Santen GWE, Nüchel J, Plomann M, Wagener R, Rehberg M, Hoyer-Kuhn H, Eekhoff EMW, Pals G, Mörgelin M, Newstead S, Wilson BT, Ruiz-Perez VL, Maugeri A, Netzer C, Zaucke F, Micha D (2020) Interaction between KDELR2 and HSP47 as a key determinant in osteogenesis imperfecta caused by Bi-allelic variants in KDELR2. Am J Hum Genet 107:989–999. https://doi.org/10.1016/j.ajhg.2020.09.009

Vanleene M, Porter A, Guillot PV, Boyde A, Oyen M, Shefelbine S (2012) Ultra-structural defects cause low bone matrix stiffness despite high mineralization in osteogenesis imperfecta mice. Bone 50:1317–1323. https://doi.org/10.1016/j.bone.2012.03.007

Varani J, Dame MK, Rittie L, Fligiel SEG, Kang S, Fisher GJ, Voorhees JJ (2006) Decreased collagen production in chronologically aged skin. Am J Pathol 168:1861–1868. https://doi.org/10.2353/ajpath.2006.051302

Veilleux L-N, Darsaklis VB, Montpetit K, Glorieux FH, Rauch F (2017) Muscle function in osteogenesis imperfecta type IV. Calcif Tissue Int 101:362–370. https://doi.org/10.1007/s00223-017-0287-y

Veilleux L-N, Lemay M, Pouliot-Laforte A, Cheung MS, Glorieux FH, Rauch F (2014) Muscle anatomy and dynamic muscle function in osteogenesis imperfecta type I. J Clin Endocrinol Metabol 99:E356–E362. https://doi.org/10.1210/jc.2013-3209

Verdonk SJE, Storoni S, Micha D, Van Den Aardweg JG, Versacci P, Celli L, De Vries R, Zhytnik L, Kamp O, Bugiani M, Eekhoff EMW (2024) Is osteogenesis imperfecta associated with

cardiovascular abnormalities? A systematic review of the literature. Calcif Tissue Int 114:210. https://doi.org/10.1007/s00223-023-01171-3
Viguet-Carrin S, Follet H, Gineyts E, Roux JP, Munoz F, Chapurlat R, Delmas PD, Bouxsein ML (2010) Association between collagen cross-links and trabecular microarchitecture properties of human vertebral bone. Bone 46:342–347. https://doi.org/10.1016/j.bone.2009.10.001
Vranka JA, Sakai LY, Bächinger HP (2004) Prolyl 3-hydroxylase 1, enzyme characterization and identification of a novel family of enzymes. J Biol Chem 279:23615–23621. https://doi.org/10.1074/jbc.m312807200
Vranka JA, Pokidysheva E, Hayashi L, Zientek K, Mizuno K, Ishikawa Y, Maddox K, Tufa S, Keene DR, Klein R, Bächinger HP (2010) Prolyl 3-hydroxylase 1 null mice display abnormalities in fibrillar collagen-rich tissues such as tendons, skin, and bones. J Biol Chem 285:17253–17262. https://doi.org/10.1074/jbc.M110.102228
Wagermaier W, Klaushofer K, Fratzl P (2015) Fragility of bone material controlled by internal interfaces. Calcif Tissue Int 97:201–212. https://doi.org/10.1007/s00223-015-9978-4
Webb EA, Balasubramanian M, Fratzl-Zelman N, Cabral WA, Titheradge H, Alsaedi A, Saraff V, Vogt J, Cole T, Stewart S, Crabtree NJ, Sargent BM, Gamsjaeger S, Paschalis EP, Roschger P, Klaushofer K, Shaw NJ, Marini JC, Högler W (2017) Phenotypic Spectrum in osteogenesis imperfecta due to mutations in TMEM38B: unraveling a complex cellular defect. J Clin Endocrinol Metab 102:2019–2028. https://doi.org/10.1210/jc.2016-3766
Weber M, Roschger P, Fratzl-Zelman N, Schöberl T, Rauch F, Glorieux FH, Fratzl P, Klaushofer K (2006) Pamidronate does not adversely affect bone intrinsic material properties in children with osteogenesis imperfecta. Bone 39:616–622. https://doi.org/10.1016/j.bone.2006.02.071
Weerakoon AT, Condon N, Cox TR, Sexton C, Cooper C, Meyers IA, Thomson D, Ford PJ, Roy S, Symons AL (2022) Dynamic dentin: a quantitative microscopic assessment of age and spatial changes to matrix architecture, peritubular dentin, and collagens types I and III. J Struct Biol 214:107899. https://doi.org/10.1016/j.jsb.2022.107899
Weis SM, Emery JL, Becker KD, Mcbride DJ, Omens JH, Mcculloch AD (2000) Myocardial mechanics and collagen structure in the osteogenesis imperfecta murine oim. Circ Res 87:663–669. https://doi.org/10.1161/01.res.87.8.663
Wekre LL, Frøslie KF, Haugen L, Falch JA (2010) A population-based study of demographical variables and ability to perform activities of daily living in adults with osteogenesis imperfecta. Disabil Rehabil 32:579–587. https://doi.org/10.3109/09638280903204690
Wenstrup RJ, Florer JB, Brunskill EW, Bell SM, Chervoneva I, Birk DE (2004) Type V collagen controls the initiation of collagen fibril assembly. J Biol Chem 279:53331–53337. https://doi.org/10.1074/jbc.m409622200
Wenstrup RJ, Smith SM, Florer JB, Zhang G, Beason DP, Seegmiller RE, Soslowsky LJ, Birk DE (2011) Regulation of collagen fibril nucleation and initial fibril assembly involves coordinate interactions with collagens V and XI in developing tendon. J Biol Chem 286:20455–20465. https://doi.org/10.1074/jbc.M111.223693
Wheatley K, Heng EL, Sheppard M, Schneider H, Moat N, Cordingley J, Kaul S (2010) A case of spontaneous intestinal perforation in osteogenesis imperfecta. J Clin Med Res 2:198–200. https://doi.org/10.4021/jocmr369w
Willing MC, Deschenes SP, Scott DA, Byers PH, Slayton RL, Pitts SH, Arikat H, Roberts EJ (1994) Osteogenesis imperfecta type I: molecular heterogeneity for COL1A1 null alleles of type I collagen. Am J Hum Genet 55:638–647
Wiltz D, Alexander CR, Blancas A, Zhang X, Jane K (2013) Extracellular matrix organization, structure, and function. In: Calcific aortic valve disease. InTech. https://doi.org/10.5772/52842
Wittig C, Szulcek R (2021) Extracellular matrix protein ratios in the human heart and vessels: how to distinguish pathological from physiological changes? Front Physiol 12. https://doi.org/10.3389/fphys.2021.708656
Yu C, Abbott P (2007) An overview of the dental pulp: its functions and responses to injury. Aust Dent J 52:S4–S6. https://doi.org/10.1111/j.1834-7819.2007.tb00525.x

Zamboulis DE, Thorpe CT, Ashraf Kharaz Y, Birch HL, Screen HR, Clegg PD (2020) Postnatal mechanical loading drives adaptation of tissues primarily through modulation of the non-collagenous matrix. Elife 9. https://doi.org/10.7554/eLife.58075
Zhang G, Young BB, Ezura Y, Favata M, Soslowsky LJ, Chakravarti S, Birk DE (2005) Development of tendon structure and function: regulation of collagen fibrillogenesis. J Musculoskelet Neuronal Interact 5:5–21
Zhang Y, Pignolo RJ, Bram RJ (2022) Accelerated aging in Cyclophilin B–deficient mice downstream of <scp>p21-Cip1</scp>/Waf1. JBMR Plus 6. https://doi.org/10.1002/jbm4.10674
Zhao C, Ichimura A, Qian N, Iida T, Yamazaki D, Noma N, Asagiri M, Yamamoto K, Komazaki S, Sato C, Aoyama F, Sawaguchi A, Kakizawa S, Nishi M, Takeshima H (2016) Mice lacking the intracellular cation channel TRIC-B have compromised collagen production and impaired bone mineralization. Sci Signal 9:ra49. https://doi.org/10.1126/scisignal.aad9055
Zhao D, Liu Y, Liu J, Hu J, Zhang Q, Wang O, Jiang Y, Xia W, Xing X, Li M (2022) Cardiovascular abnormalities and its correlation with genotypes of children with osteogenesis imperfecta. Front Endocrinol (Lausanne) 13. https://doi.org/10.3389/fendo.2022.1004946
Zhong W, Pathak JL, Liang Y, Zhytnik L, Pals G, Eekhoff EMW, Bravenboer N, Micha D (2023) The intricate mechanism of PLS3 in bone homeostasis and disease. Front Endocrinol 14. https://doi.org/10.3389/fendo.2023.1168306
Zhou X, Lin P, Yamazaki D, Park KH, Komazaki S, Chen SRW, Takeshima H, Ma J (2014) Trimeric intracellular cation channels and sarcoplasmic/endoplasmic reticulum calcium homeostasis. Circ Res 114:706–716. https://doi.org/10.1161/circresaha.114.301816
Zurlo F, Larson K, Bogardus C, Ravussin E (1990) Skeletal muscle metabolism is a major determinant of resting energy expenditure. J Clin Invest 86:1423–1427. https://doi.org/10.1172/jci114857

Chapter 3
Bone Involvement in the Ehlers-Danlos Syndromes

Fransiska Malfait, Delfien Syx, and Yoshihiro Ishikawa

Abstract The Ehlers-Danlos syndromes (EDS) encompass a heterogeneous group of heritable conditions that share several characteristics such as soft and hyperextensible skin, abnormal wound healing, easy bruising, and (generalized) joint hypermobility. The currently used 2017 International EDS classification recognizes 13 EDS types, caused by genetic defects in 20 different genes. Defects in different fibrillar collagens, which are distributed throughout the connective tissues of the skin, ligaments, tendons, blood vessels, and bone, lie at the molecular basis of many EDS types. Interestingly, fragility of the bone is not traditionally considered a hallmark of EDS, despite the importance of many of the causative molecules in bone development and/or homeostasis. Only a few studies have directly addressed fracture rate, bone quality, and/or density in EDS patients. Reduced bone mass density, abnormal bone structure, and/or a tendency to low-impact fractures has been reported in several EDS patients across different types, but these studies are usually confined to case studies or small case series and/or do not provide detailed information on the specific EDS type. This chapter provides a brief overview of the structure and biosynthesis of fibrillar collagens and focuses on the clinical and molecular characteristics of different types of genetically defined types of EDS, with a special emphasis on bone phenotypes reported for these conditions.

F. Malfait (✉) · D. Syx
Center for Medical Genetics, Ghent University Hospital, Department for Biomolecular Medicine, Ghent University, Ghent, Belgium
e-mail: Fransiska.Malfait@Ugent.be

Y. Ishikawa
Department of Ophthalmology, University of California San Francisco, School of Medicine, San Francisco, CA, USA

A. Rossi, F. Zaucke (eds.), *The Extracellular Matrix in Genetic Skeletal Disorders*, Biology of Extracellular Matrix 16,
https://doi.org/10.1007/978-3-031-70835-0_3

3.1 Introduction

The Ehlers-Danlos syndromes (EDS) encompass a heterogeneous group of heritable conditions that share several characteristics such as soft and hyperextensible skin, abnormal wound healing, easy bruising, and (generalized) joint hypermobility. Musculoskeletal and ligamentous complications are frequently observed and include congenital and/or progressive spinal ((kypho-)scoliosis) and pectoral deformities (excavatum, carinatum), congenital hip dislocations, congenital club feet and other congenital or progressive contractures, and pes planus. Additional clinical features include fragility of soft tissues leading to recurrent soft tissue herniations and organ prolapses, fragility of the wall of blood vessels and hollow organs, and involvement of the ocular system (e.g., keratoconus, severe myopia, strabismus) (Malfait et al. 2020; Steinmann and Superti-Furga 2002). The severity of these symptoms and the extent of organ involvement vary widely across affected individuals, and this clinical heterogeneity is mirrored by an underlying genetic heterogeneity. The currently used 2017 International EDS classification recognizes 13 EDS types, caused by genetic defects in 20 different genes (Malfait et al. 2017) (Table 3.1).

In general, EDS is considered to occur in about 1 in 5000 individuals with no ethnic predisposition (Steinmann and Superti-Furga 2002), but well-designed epidemiological studies are lacking. Classical and vascular EDS are the most common genetically defined types of EDS, with incidences of 1:20,000 and 1:50,000–1:200,000, respectively (Bowen et al. 2017; Byers et al. 2017). For the other EDS types for which the underlying genetic cause is known, there are no incidence estimates, but the numbers of reported individuals or families are generally low (<150 individuals) (Brady et al. 2017). For all these types, incidence rates are similar in males and females. Patients with milder forms frequently do not seek medical attention, and as such, incidence rates are likely an underestimation. The pathogenic cause of the hypermobile type of EDS (hEDS) remains unresolved. No clear genetic defects have been identified to date, and the diagnosis relies on clinical criteria and exclusion of other (genetic) conditions that cause joint hypermobility (Malfait et al. 2017; Tinkle et al. 2017). It is generally considered to be the most common type of EDS; however, it forms a phenotypic continuum with the hypermobility spectrum disorders (HSD), a group of conditions characterized by symptomatic joint hypermobility and a range of additional extra-articular signs and symptoms including (but not limited to) fatigue, chronic pain, headaches, and functional gastrointestinal disorders (Castori et al. 2017). The incidence of HSD is also unknown, but it likely is a common disorder. Diagnoses of hEDS and HSD are more commonly made in females than in males, but whether this is due to an increased incidence or more severe manifestations is unknown.

Defects in different fibrillar collagens, which are distributed throughout the connective tissues of the skin, ligaments, tendons, blood vessels, and bone, lie at the molecular basis of many EDS types (Malfait et al. 2021). The key collagens directly involved in these diseases are mainly type III and type V collagen and to a lesser

Table 3.1 The 13 EDS types ranked according to the underlying pathomechanism with indication of the causal genes and proteins. IP: inheritance pattern; AD: autosomal dominant; AR: autosomal recessive

EDS type (abbreviation)	Gene	Protein	IP	Phenotype MIM
Defects in collagen structure and processing				
Classical (cEDS)	*COL5A1* *COL5A2* (*COL1A1* p.(Arg312Cys))	Pro-α1-chain of type V collagen Pro-α2-chain of type V collagen Pro-α1-chain of type I collagen	AD AD AD	130000 130010 /
Vascular (vEDS)	*COL3A1*	Pro-α1-chain of type III collagen	AD	130050
Arthrochalasia (aEDS)	*COL1A1* *COL1A2*	Pro-α1-chain of type I collagen Pro-α2-chain of type I collagen	AD AD	130060 617821
Dermatosparaxis (dEDS)	*ADAMTS2*	A disintegrin and metalloproteinase with thrombospondin motifs 2 (ADAMTS-2)	AR	225410
Cardiac-valvular (cvEDS)	*COL1A2*	Pro-α2-chain of type I collagen	AR	225320
Defects in collagen folding and cross-linking				
Kyphoscoliotic (kEDS)	*PLOD1* *FKBP14*	Lysyl hydroxylase 1 (LH1) FK506 binding protein, 22 kDa (FKBP22)	AR AR	225400 614557
Defects in extracellular matrix bridging molecules				
Classical-like (clEDS)	*TNXB*	Tenascin X (TNX)	AR	606408
Myopathic (mEDS)	*COL12A1*	Pro-α1-chain of type XII collagen	AD/AR	616471
Defects in intracellular processes				
Brittle cornea syndrome (BCS)	*ZNF469* *PRDM5*	Zinc finger protein 469 (ZNF469) PR/SET domain 5 (PRDM5)	AR AR	229200 614170
Spondylodysplastic (spEDS)	*SLC39A13*	Zrt/Irt-like protein 13 (ZIP13)	AR	612350
Defects in glycosaminoglycan biosynthesis				
Musculocontractural (mcEDS)	*CHST14* *DSE*	Dermatan 4-*O*-sulfotranferase-1 (D4ST1) Dermatan sulfate epimerase-1 (DS-epi1)	AR AR	601776 615539
Spondylodysplastic (spEDS)	*B4GALT7* *B3GALT6*	Galactosyltransferase I (ß4GalT7) Galactosyltransferase II (ß3GalT6)	AR AR	130070 615349
Defects in the complement pathway				
Periodontal (pEDS)	*C1S* *C1R*	Complement C1s (C1s) Complement C1r (C1r)	AD AD	617174 130080

(continued)

Table 3.1 (continued)

EDS type (abbreviation)	Gene	Protein	IP	Phenotype MIM
Molecularly unsolved				
Hypermobile (hEDS)	*?*	Unknown	?	130020
Novel EDS type *(identified after the 2017 classification)*				
Classical-like II (clEDS II)	*AEBP1*	Adipocyte enhancer-binding protein 1 (AEBP1)	AR	618000

extent type I collagen, due to (likely) pathogenic variants in *COL3A1*, *COL5A1*, *COL5A2*, *COL1A1*, and *COL1A2*. Other EDS genes encode proteins involved in collagen processing, folding, and cross-linking (*ADAMTS2, PLOD1*, and *FKBP14*), enzymes involved in the biosynthesis of glycosaminoglycans (*B4GALT7, B3GALT6, CHST14*, and *DSE*), and a few other genes coding for extracellular matrix proteins (e.g., *TNXB, AEBP1, COL12A1, PRDM5*, and *ZNF469*) [2]. Despite this genetic heterogeneity, a common histological pattern, seen across all EDS types but to a variable extent, is a disrupted biosynthesis and/or supramolecular organization of the collagen fibrils in different affected tissues such as the dermis, tendons, ligaments, and vascular wall (Malfait et al. 2020; Malfait et al. 2017).

Interestingly, fragility of the bone is not traditionally considered a hallmark of EDS, despite the importance of many of the causative molecules in bone development and/or homeostasis. Only a few studies have directly addressed fracture rate, bone quality, and/or density in EDS patients. Reduced bone mass density, abnormal bone structure, and/or a tendency to low-impact fractures has been reported in several EDS patients across different types, but these studies are usually confined to case studies or small case series and/or do not provide detailed information on the distribution of EDS types or have analyzed several EDS types as a single group. In general, low mineral density and fractures have mostly been reported in some rare (mostly autosomal recessive) types of EDS, such as the dermatosparaxis and kyphoscoliotic, spondylodysplastic, and classical-like EDS, and in brittle cornea syndrome, where they may occur at birth, during infancy and childhood, and/or in adulthood. For the more frequently occurring autosomal dominant types (classical, vascular) and the hypermobile type, mildly reduced bone density has been reported in some adults, without a clear increase in fracture rate.

This chapter provides a brief overview of the structure and biosynthesis of fibrillar collagens to help guide the reader from the nature of the molecular defect to the clinical presentation of the EDS, which are considered paradigm collagen-related disorders. Next it will focus on the clinical and molecular characteristics of different types of genetically defined types of EDS, with a special focus on bone phenotypes observed in these conditions.

3.2 Collagens I, III, and V and Bone

Collagen, a protein with a simple triple helix structure, is one of the most abundant proteins in mammals (Tarnutzer et al. 2023). Fibrillar collagens are trimeric molecules that consist of three identical ("homotrimer") or genetically distinct ("heterotrimer") polypeptide chains, designated as α-chains. These form typical triple helix structures characterized by a repeating Gly-Xaa-Yaa triple repeat. Fibrillar collagens are formed as precursor molecules, called procollagen chains or proα-chains. The amino (N-) and carboxy (C-) terminal ends of the procollagen chains are typically cleaved to form mature collagen molecules by specific enzymes, thereby initiating collagen fibril formation. These collagen fibrils are further stabilized by the formation of intermolecular cross-links.

The main compound of the organic bone matrix is collagen type I. It constitutes about 85–90% of the total bone protein content and is secreted by osteoblasts. It plays a major role in bone structure and mineralization by forming a tissue scaffold, but also provides a substrate for cell anchorage and regulates bioavailability of growth factors and cytokines (Ricard-Blum 2011). It is a heterotrimer consisting of two α(I) and one α2(I) chains, encoded by the *COL1A1* and *COL1A2* genes, respectively. Most pathogenic variants in these two genes result in osteogenesis imperfecta, the quintessential brittle bone disorder, but a smaller number of specific recurrent alterations give rise to rare forms of EDS.

In contrast, collagen type V comprises only 2–5% of the total collagen content in most tissues including bone (Birk 2001). Type V collagen is present mainly as a heterotrimer of two α1(V) and one α2(V) chains (encoded by *COL5A1* and *COL5A2*, respectively). It forms the initial scaffold onto which type I collagen molecules assemble to form heterotypic fibrils (fibrils consisting of type I and type V collagen) and as such is a major regulator of collagen fibrillogenesis and plays a critical role during the early process of fibril nucleation (Birk 2001). Its deficiency in classical EDS results in an abnormal architecture of these heterotypic collagen fibers, which are required for adequate bone mineralization (Bowen et al. 2017). Therefore, type V collagen alterations may also be hypothesized to result in abnormally mineralized bone and osteoporosis.

Type III collagen is a homotrimer of three α1(III) chains (encoded by *COL3A1*). Type III collagen also co-assembles with type I collagen to form heterotypic fibrils, but it is most abundant in extensible tissues such as blood vessels, skin, and lung. A small amount of collagen type III is found in the heterotypic collagen fibrils of the bone (Keene et al. 1987, 1991; Reddi et al. 1977). Type III collagen appears in mesenchymal condensations preceding cartilage and bone formation and is required for growth acceleration of osteoblasts, suggesting a critical role in skeletal development (Maehata et al. 2007; Silver et al. 1981). It is possibly also involved in preserving osteogenic potential of mesenchymal stem cells (Chen et al. 2007).

The collagen proteins are synthesized in the endoplasmic reticulum (ER) and then move through the Golgi apparatus before being transported to the extracellular matrix (ECM) (Fig. 3.1). Despite their abundance and simple structure, the

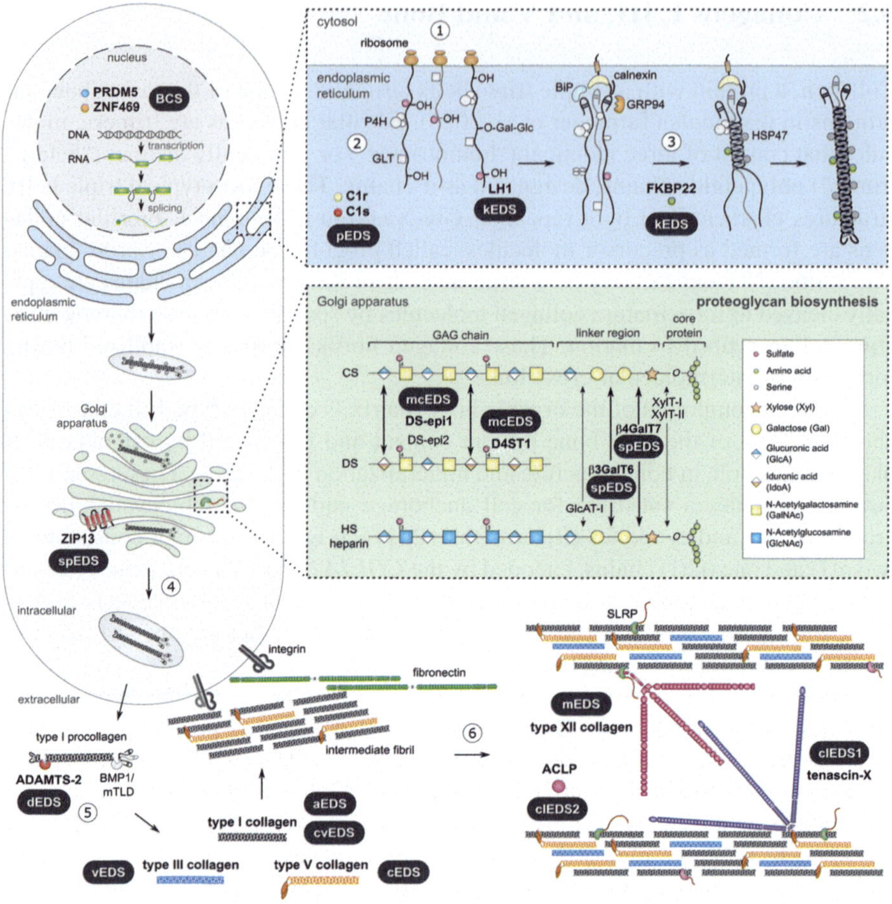

Fig. 3.1 Schematic representation of the biosynthetic pathway of fibrillar collagens and proteoglycans and visualization of the molecules defective in the different EDS types. Molecules implicated in EDS pathogenesis are depicted in bold and the corresponding EDS type is indicated with a black background. (1) Biosynthesis of fibrillar collagen starts with transcription and translation of pro-α-chains. (2) In the endoplasmic reticulum (ER), nascent pro-α-chains are co- and post-translationally modified by several enzymes, including proline and lysine hydroxylases (LH) and galactosyltransferases (GLT). (3) Triple helix formation is initiated by the association of three pro-α-chains at their C-terminal propeptides and propagates toward the N-terminus in a zipperlike fashion. During this process, several molecular chaperones assist with the correct folding. (4) Trimeric procollagen molecules aggregate laterally and are transported in secretory vesicles toward the extracellular environment. (5) The removal of the N- and C-propeptides, by specific enzymes (e.g., ADAMTS-2 and BMP-1/mTLD for type I procollagen), results in the formation of mature collagen molecules. (6) Subsequently, these collagens assemble into highly ordered striated fibrils, a process that is assisted by several proteins in a tissue-specific manner. Fibril assembly is mediated at the plasma membrane by fibronectin and integrins that serve as *organizers*. At the cell surface, some collagens function as *nucleators* (e.g., type V collagen) and initiate immature fibril assembly. Intermediate fibrils are deposited into the ECM and stabilized by interactions with *regulators* such as the small leucine-rich proteoglycan (SLRP) decorin, the glycoprotein tenascin X, and type XII collagen. These molecules regulate the rate of assembly, size, and structure of the collagen fibrils (together with other molecules). Subsequent fibril growth occurs through linear and lateral fusion of intermediate collagen fibrils that are then stabilized by intra- and intermolecular cross-links

synthesis of collagen involves a complex process necessitating the intricate coordination of over 20 enzymes and molecular chaperones, known as the molecular ensemble (Ishikawa and Bachinger 2013). The complex biosynthesis of collagen in the ER is divided into three stages: translation and post-translation modification, triple helix formation, and transport. In the extracellular space, the secreted collagens undergo processing and form collagen fibrils, a process assisted by ECM proteins during maturation. Initially, the synthesis of all types of collagens was thought to be similar, based on studies centered on collagen type I. However, the reality of this "most abundant protein's biosynthesis" remains a partially unexplored area. There are still many unresolved questions, such as tissue preference and collagen type-specific biosynthetic pathways.

1) *Translation and post-translation modifications (PTMs)*—Newly synthesized collagen polypeptides undergo post-translation modification before forming a triple helix structure. In the collagen-like sequence Gly-Xaa-Yaa (where Xaa/Yaa can be any amino acid), when Yaa is proline or lysine, these amino acids are hydroxylated to form 4-hydroxyproline (4Hyp) and hydroxylysine (Hyl), respectively. 4Hyp is hydroxylated by procollagen prolyl 4-hydroxylase (P4H) and Hyl by lysyl hydroxylase (LH). If Yaa is 4Hyp, Xaa's proline is modified to 3-hydroxyproline (3Hyp) by prolyl 3-hydroxylase (P3H) (Gorres and Raines 2010; Salo and Myllyharju 2021; Yamauchi and Shiiba 2008). The frequency of Hyl and 3Hyp modifications varies with organ type, collagen type, and age (Montgomery et al. 2018; Taga et al. 2016). Procollagen hydroxylase enzymes (P4H, P3H, LH) commonly exist in three isoforms each (P4HA1–3, P3H1–3, and LH1–3). Most hydroxylases participate in collagen biosynthesis not individually, but rather in complexes with ER chaperones and other enzymes (Ishikawa and Bachinger 2013; Ito and Nagata 2021). P4Hs form a 1:1 strong-binding tetrameric complex with the oxidoreductase protein disulfide isomerase (PDI). P4Hs have low solubility and lack an ER retention signal, so they bind to PDI to resolve these issues (Koski et al. 2017; Murthy et al. 2022). P3H1, P3H3, LH1, and LH2 are also suggested to work in complexes. While P3H1 forms a strong-binding complex with cartilage-associated protein (CRTAP) and cyclophilin B (CypB) similar to the P4H/PDI tetramer (Ishikawa et al. 2009; Vranka et al. 2004), the other enzymes likely form functional complexes with transient or less tight interaction (Gjaltema et al. 2016) (Duran et al. 2017; Heard et al. 2016). In the context of EDS, LH1 forms a complex with P3H3, CypB, and SC65, which is significantly associated with lysine modification in cross-link formation sites (Hudson et al. 2017; Ishikawa et al. 2021; Terajima et al. 2019; Terajima et al. 2016). LH1, traditionally thought to modify lysines in collagen's triple helix structure, has shown collagen type preference for lysine modification—the lysine residues in collagen type I are predominantly hydroxylated by LH1 (Ishikawa et al. 2022; Ishikawa et al. 2021; Terajima et al. 2022) and that in collagen type V should be modified by LH2 and/or LH3 (Ishikawa et al. 2021). Recent research indicates that the activity of P4HA1 and P4HA2 varies with the Xaa position of amino acid adjacent to the Yaa position of hydroxylated proline

in the Gly-Xaa-Yaa triplets (Salo et al. 2023; Wilhelm et al. 2023). We speculate that a similar rule, substrate prediction based on collagen amino acid sequences, could exist for P3Hs and LHs. After the primary modification of lysine to hydroxylysine, collagen undergoes two successive and coordinated O-glycosylation steps as secondary modifications (Hennet 2019; Yamauchi et al. 2019a; Yamauchi et al. 2019b). In the first step, hydroxylysines are modified to galactosyl hydroxylysine (GHL) by collagen β (1-O) galactosyl transferases, namely, GLT25D1 and D2. Subsequently, GHL is further modified by the addition of glucose to form glucosyl-galactosyl-hydroxylysine (GGHL). This conversion is carried out by the LH3, which has recently been identified as responsible for most of GHL glycosylation in vitro (Mattoteia et al. 2023). The sequential modifications—from lysine to hydroxylysine, then to GHL, and finally to GGHL—are crucially orchestrated by the enzymatic actions of LHs and GLT25Ds during the biosynthesis of collagen.

Collectively, the temporal and spatial coordination is crucial in these collagen PTMs. As described above, these enzymes only modify single strands before triple helix formation. Once the triple helix is formed, the enzymes cannot access the modification sites unless the triple helix unravels. Especially, modifications at the carboxyl terminus are time-sensitive due to the folding characteristics of collagens, which starts at this end (more details are in the next section). Conversely, the amino terminus can be exposed to collagen-modifying enzymes right from the start of translation until the completion of triple helix formation. In addition to accessibility of collagen-modifying enzymes, the amount of collagen PTMs is precisely tailored to collagens depending on collagen types and tissues. Thus, collagen PTMs are likely among the most diverse but specific events in the ER.

2) *Formation of the triple helix structure and transport*—Collagen's triple helix formation begins with the completion of a single collagen chain translation with precise PTMs. In heterotrimers, such as collagen types I and V, this process involves the correct assembly of two α1 chains and one α2 chain (Boudko et al. 2012). The assembly starts from the carboxyl terminal propeptides and progresses like a zipper toward the amino terminal end (Frank et al. 2003) (Bachinger et al. 1980). The amino acids in the triple helix structure have a high ratio of proline, with all prolines in the trans state. The rate-limiting step in forming the triple helix is the transition time of proline from cis to trans state (Qiu et al. 2023). To optimize this formation, ER-resident peptidyl-prolyl cis/trans isomerases (PPIases) are involved. There are seven PPIases in the ER, belonging to two families (Hendershot et al. 2023; Ishikawa et al. 2015; Ishikawa et al. 2017b): six from the FK506 binding protein (FKBP) family (FKBP13, FKBP19, FKBP22, FKBP23, FKBP60, FKBP65), characterized by their FKBP domain that binds to the immunosuppressant drug FK506, and cyclophilin B (CypB) from the cyclophilin family, the most abundant PPIase in the ER (Gidalevitz et al. 2013), known to accelerate the formation of collagen's triple helix structure. CypB exists both in complexes and as an independent molecule within the ER (Morello et al. 2006). In vitro experiments confirm that CypB alone can promote triple

helix formation (Ishikawa et al. 2012, 2017b). CypB, along with P3H1 and CRTAP, forms a complex, and deficiencies of these proteins slow down the triple helix formation rate (Cabral et al. 2007a; Morello et al. 2006), leading to excessive O-glycosylation and giving rise to osteogenesis imperfecta.

In the context of EDS, FKBP22 uniquely binds to some collagen types (III, IV, VI, X) but not others (I, II, V, XI) (Colman et al. 2022b; Ishikawa and Bachinger 2014; Ishikawa et al. 2020), potentially explaining the phenotypic spectrum of FKBP22 mutations. In the ER, collagen molecules, once formed into a triple helix, may bundle and induce fibril formation, potentially causing ER stress. To prevent this, collagen chaperone HSP47 binds to the collagen molecules, inhibiting their aggregation (Ito and Nagata 2017, 2019). In addition to HSP47, FKBP65 is likely involved in type I collagen quality control (Ishikawa et al. 2008, 2017a), while FKBP22 is associated with type III collagen quality control (Ishikawa and Bachinger 2014). Although the entire mechanism of FKBP22 binding to specific collagen types is not fully characterized, FKBP22 likely recognizes specific positively charged motifs in the collagen triple helix generated by its chain registration (Ishikawa et al. 2023).

Correctly folded proteins in the ER are transported to the extracellular space via the Golgi apparatus, facilitated by coat complex type II (COPII) vesicles (Gomez-Navarro and Miller 2016; Raote et al. 2023). The large size of ECM proteins, particularly fibrillar collagens such as collagen type I, which take a linear form of about 300 nm, cannot fit into the usual COPII size of 60–90 nm (Hutchings and Zanetti 2019; McCaughey and Stephens 2019). This led to the hypothesis of special COPII vesicles for collagen. The discovery of TANGO1 (Transport ANd Golgi Organization 1), a membrane protein, was pivotal in understanding how collagens are guided into these special transport vesicles (Saito et al. 2009) (see Chap. 13). TANGO1 forms a ring around the ER exit site (Raote et al. 2017, 2018), synchronizing ER-lumen extended arm and cytoplasmic domains to expand COPII size and accommodate collagens (Bunel et al. 2024; Ma and Goldberg 2016). Initially, it was thought that TANGO1's SH3 domain recognized collagen (Saito et al. 2009), but it has been shown that HSP47 acts as an anchor between collagen and the SH3 domain (Ishikawa et al. 2016). This suggests that only triple helical collagens are allowed on the secretory pathway since HSP47 interacts with triple helical collagens but not unfolded collagens (Ono et al. 2012; Widmer et al. 2012). This represents a crucial biological event in the ER for quality control.

3) *Maturation*—The processing of procollagen propeptides is key for fibril formation, involving A Disintegrin And Metalloproteinase with ThromboSpondin (ADAMTS) type 1 motif enzymes for N-propeptide cleavage (Bekhouche and Colige 2015; Le Goff et al. 2006) and bone morphogenetic protein (BMP) 1/tolloid-like enzymes for C-propeptide removal (Hopkins et al. 2007). Additionally, meprin α and β enzymes also process collagen types I and III propeptides (Prox et al. 2015). The release of propeptides reduces collagen solubility, triggering self-assembly into fibrils. These fibrils, which are strengthened through cross-linking, form flexible yet strong macromolecular structures.

Collagen fibrils further assemble into striated fibers via linear and lateral fusion, aided by the ECM proteins such as organizers (fibronectin and integrins) (Kadler et al. 2008; Musiime et al. 2021), nucleators (collagens type V and XI) (Birk et al. 1990; Sun et al. 2020; Wenstrup et al. 2011), regulators (small leucine-rich proteoglycans, tenascin-X) (Bristow et al. 2005; Kalamajski and Oldberg 2010; Lopez and Bonassar 2022)), and modulators (the FACIT—Fibril Associated Collagens with Interrupted Triple helices collagens) (Izu and Birk 2023; Ricard-Blum 2011; Shaw and Olsen 1991). This process varies across tissues, forming structures like tendons or the cornea's hexagonal fiber array. Collagen fibers, generally protease-resistant, are broken down by specific matrix metalloproteinases (MMPs) during connective tissue degradation (Amar et al. 2017; Manka et al. 2019), resulting in gelatin-like structures that are further degraded by other proteinases.

In summary, each stage of collagen biosynthesis is intricately regulated, with spatiotemporal dynamics being crucial. This complex process reflects the diversity and specificity of collagen molecular ensembles in the ER.

Proteoglycan biosynthesis starts with the synthesis of a core protein that is subsequently modified by several enzymes in the Golgi apparatus. First, a common tetrasaccharide linker region is formed by the stepwise addition of a xylose (Xyl) residue to a serine residue of the core protein catalyzed by xylosyltransferase-I and -II (XylT-I/-II), two galactose (Gal) residues by galactosyltransferase-I (GalT-I or β4GalT7) and galactosyltransferase-II (GalT-II or β3GalT6), and a glucuronic acid (GlcA) catalyzed by glucuronosyltransferase-I (GlcAT-I). The alternating addition of either N-acetylglucosamine (GlcNAc) or N-acetylgalactosamine (GalNAc) and GlcA to the nascent glycosaminoglycan (GAG) chain results in the formation of heparan sulfate (HS) proteoglycans and chondroitin sulfate (CS)/dermatan sulfate (DS) proteoglycans. The GAG chains are further modified by epimerization and sulfation reactions. Epimerization of GlcA toward iduronic acid (IdoA) catalyzed by DS epimerase-1 and epimerase-2 (DS-epi1 and DS-epi2) is required for the formation of DS. Subsequently, dermatan 4-*O*-sulfotransferase 1 (D4ST1) catalyzes 4-*O*-sulfation of GalNAc, thereby preventing back-epimerization of the adjacent IdoA. Figure adapted from Van Damme et al. (2022)), used under the Creative Commons License (CC BY 4.0)

3.3 Bone Phenotypes in EDS Types Caused by Defects in Collagen Primary Structure and Collagen Processing

3.3.1 Classical EDS and Type V Collagen

The classical type of EDS (cEDS) is caused by heterozygous variants in the genes coding for type V collagen (*COL5A1, COL5A2*). These defects mainly affect the skin, tendon, and ligaments, and the major hallmarks of this condition include skin hyperextensibility and fragility, slow wound healing with formation of large atrophic scars, easy bruising, generalized joint hypermobility with a propensity to (sub-) luxations, and other signs of soft connective tissue fragility (Bowen et al. 2017, 2020; Colman et al. 2021).

Only a few studies have addressed bone quality in cEDS patients. Mazziotti et al. evaluated bone mineral density (BMD) at the lumbar spine, total hip and femoral neck by dual-energy X-ray absorptiometry, and vertebral morphometry (by quantitative morphometry) in 52 adult cEDS/hEDS patients (12 cEDS patients) (age range between 21 and 71 years) and compared them to age- and sex-matched control subjects. Interestingly, abnormal vertebral morphology was reported to be significantly more prevalent in the 52 EDS patients (38.5% vs 5.1%; $p < 0.001$) without significant differences in BMD at either skeletal site. The prevalence of abnormal vertebral morphology was not significantly different between cEDS and hEDS. BMD was not significantly different between EDS patients with or without abnormal vertebral morphology at either site. Moreover, abnormal vertebral morphology was associated with more severe back pain, suggesting a potential impact on quality of life (Mazziotti et al. 2016). Eller-Vainicher subsequently evaluated bone metabolism, BMD, bone quality (measured by trabecular bone score, TBS), and vertebral morphology in a group of 50 eugonadal adult EDS patients with either cEDS or hEDS. They found a reduced TBS and an increased prevalence of abnormal vertebral morphology in EDS patients (38% vs 8%; $p < 0.05$). The patients with abnormal vertebral morphology had a comparable BMD as those with a normal vertebral morphology, but they had a lower TBS. The authors concluded that bone quality, rather than bone density, is the most important determinant of bone strength in these EDS patients (Eller-Vainicher et al. 2016). In both studies, the data were reported for the EDS group as a whole (cEDS and hEDS), and no detailed data were provided for the cEDS group alone. The authors of both studies concluded that vertebral fractures are more common in EDS, but it should be noted that the abnormal vertebral morphology could also result from dysplastic changes that occurred during bone development (Basalom and Rauch 2020).

In a recent retrospective study, Ritelli et al. reported osteopenia (T-scores between −1.0 and −2.4) in 11 out of 21 adults with cEDS (35.5%) (age range 29–58 years), whereas it was absent in 10 examined young cEDS individuals. Osteoporosis (T-scores <−2.5) was diagnosed in 5/21 adults (16.1%) (Ritelli et al. 2020). There is one report of successful denosumab therapy in a 59-year-old cEDS male with osteoporosis (Yasukawa et al. 2021).

Wenstrup and colleagues generated a homozygous *Col5a1* knockout (*Col5a1*$^{-/-}$) mouse model that showed lethality around embryonic day 10 due to cardiovascular failure. Ultrastructural evaluation with transmission electron microscopy (TEM) of these embryos revealed virtual absence of predermal mesenchymal fibril formation, revealing the critical role of type V collagen in fibril formation and nucleation (Wenstrup et al. 2004). Haploinsufficient *Col5a1*$^{+/-}$ were viable but they had a phenotype strongly reminiscent of human cEDS. Fractures were not reported in these mice (Wenstrup et al. 2006). TEM analysis of the dermis and of tendon showed abnormal collagen fibril density and architecture; ultrastructural studies of the collagen fibrils in bone of these mice were not reported.

3.3.2 Vascular EDS and Type III Collagen

The vascular type of EDS (vEDS) is caused by heterozygous variants in the gene coding for type III collagen (*COL3A1*). The phenotype of vEDS is predominated by fragility of the vascular, gastrointestinal, and urogenital tissues, with risk of arterial dissections and ruptures and nontraumatic rupture of the gastrointestinal and uterine wall (the latter mainly during pregnancy). Patients usually have a thin and fragile skin that bruises easily and heals slowly. vEDS patients may also display some skeletal abnormalities including a distinctive facial appearance, pectoral and spinal deformities, and congenital club foot. No systematic studies on bone densitometry, vertebral morphology, and/or fracture rate are available for vEDS (Byers et al. 2017; Malfait et al. 2017); osteoporosis was reported in a few vEDS patients (Yen et al. 2006).

Successful denosumab therapy was reported in only one 42-year-old male with vEDS and osteoporosis (Yasukawa et al. 2021).

The role of type III collagen on bone development and homeostasis was studied in a homozygous *Col3a1* knockout mouse model (*Col3a1*^{tm1Jae}) (Liu et al. 1997), as well as in heterozygous haploinsufficient mice (*Col3a1*$^{tm1Jae/J}$). Most of the homozygous mice died within the first 48 h after birth of unknown etiology, and the small number of surviving mice were also shown to have a reduced life span, mostly due to vascular rupture. TEM of several tissues of these mice showed abnormal collagen fibril architecture, hinting at a critical role for type III collagen in collagen fibrillogenesis; unfortunately, however, TEM of bone was not performed. The haploinsufficient mice were phenotypically normal with a normal lifespan, but showed subclinical late-onset histological lesions in the aorta. In order to gain insight in the role of type III collagen on bone development and homeostasis, Volk et al. studied the skeletal and craniofacial phenotype of young (6–8 weeks) *Col3a1*^{tm1Jae}, and young and middle-aged (>1 year) *Col3a1*$^{tm1Jae/J}$ mice using microCT scan, and compared them to wild-type mice. They did not find significant craniofacial abnormalities, but the trabecular bone of the distal femur showed an abnormal geometry and density by 6–8 weeks of age (in both the knockout and the haploinsufficient model), and this was sustained in the older (haploinsufficient) mice. These findings suggest that collagen III is important in establishing early accumulation of bone. In contrast, no effect was seen on cortical bone geometry. Since cortical bone forms by a process resembling intramembranous bone formation, this (together with the normal craniofacial structure) suggests that intramembranous bone is less severely affected than bones that are formed through endochondral ossification. An osteogenesis assay on cultures of mesenchymal progenitors of *Col3a1*^{tm1Jae} mice compared to controls revealed a reduced capacity to undergo mineralization and a reduction in osteogenic markers such as collagen I, osteocalcin, and bone sialoprotein (Volk et al. 2014). In a follow-up study, Miedel et al. studied long bone fracture healing after bilateral tibial fracture in young *Col3a1*$^{tm1Jae/J}$ mice. Their studies showed impaired remodeling during fracture healing, without an impact on regaining mechanical function in these young mice (Miedel et al. 2015). Together, these studies suggest a role for type III collagen in trabecular bone development through its effect on osteoblast differentiation, as well as a role in fracture repair.

3.3.3 EDS Types Caused by Impaired Procollagen I N-Propeptide Cleavage

Whereas defects in type I collagen are generally associated with osteogenesis imperfecta, some specific abnormalities give rise to rare forms of EDS. This is particularly true for defects that hamper the cleavage of the amino (N-)terminal propeptide of type I procollagen, i.e., the arthrochalasia and dermatosparaxis types of EDS, and the more recently described collagen I-related overlap disorder (C1ROD).

The arthrochalasia type of EDS (aEDS) (previously arthrochalasia multiplex congenita, EDS type VIIA and B) is clinically hallmarked by congenital bilateral hip dislocation, generalized joint hypermobility with recurrent (sub)luxations, and skin hyperextensibility and fragility. Foot deformities (pes equinovarus, planus or valgus, and hallux valgus) and spinal deformities are frequently observed. Fractures were reported in a small number of affected individuals (Brady et al. 2017). Radiologically mild osteopenia and Wormian bones, suggesting bone fragility, were also observed in a few individuals. In addition, some aEDS patients were noted to have blue sclerae and (rarely) signs of dentinogenesis imperfecta.

aEDS is typically caused by acceptor or donor site variants adjacent to exon 6 in either *COL1A1* or *COL1A2*, leading to deletion of part or all of this exon (Byers et al. 1997). These amino acid sequences contain the procollagen N-proteinase cleavage site, the telopeptide cross-link lysine residue, a cleavage site for proteinases such as pepsin and α-chymotrypsin, and the first triplet of the major triple helical Gly-Xaa-Yaa domain. Individuals harboring a splice site alteration that leads to use of a cryptic acceptor site, which results in loss of only the N-proteinase cleavage site, but not the cross-linking site, display a full aEDS phenotype, suggesting that persistence of the N-propeptide, rather than loss of the cross-linking site, lies at the basis of the aEDS phenotype (Chiodo et al. 1992). Sodium dodecyl sulfate-polyacrylamide gel electrophoresis (SDS-PAGE) of pepsin-digested collagens produced by fibroblasts of EDS VIIA and VIIB patients shows, in addition to normal α1(I) and α2(I) chains, the presence of mutant pNα1(I)- and pNα2(I) chains, respectively, a feature that was for a long time used as a diagnostic marker. Incorporation of these pN-collagen molecules into the mature collagen fibrils results in fibrils with irregular contours and smaller diameters, as observed by TEM of the dermis of affected individuals (Byers et al. 1997). Since type I collagen consists of two α1(I) and one α2(I) chains, 75% of collagen I molecules due to a *COL1A1* mutation will contain one or two mutant pNα1(I) chains, whereas in case of a *COL1A2* mutation, only half of the collagen I molecules are affected. This difference in stoichiometry suggests that *COL1A1*-associated aEDS (EDS VIIA) might be more severe than *COL1A2*-associated aEDS (EDS VIIB). At the ultrastructural level, a more severe effect on collagen fibrillogenesis is indeed observed for *COL1A1* defects. Also at the clinical level, patients with *COL1A1* mutations seem to be at the more severe end of the spectrum, especially in terms of musculoskeletal features (Klaassens et al. 2012; Liu et al. 2020).

The dermatosparaxis type of EDS (dEDS) is caused by deficient activity of the enzyme that cleaves off the N-terminal propeptides from the proα1(I) and proα2(I) chains, ADAMTS2, due to biallelic mutations in its gene, *ADAMTS2*. In contrast to aEDS, this affects all three proα(I) chains. SDS-PAGE of dermal procollagen I of affected individuals indeed shows accumulation of bands representing both pNα1(I)- and pNα2(I) chains, and ultrastructural studies of the dermis show loosely packed, thin, and twisted ribbonlike collagen fibrils with a typical "hieroglyphic" aspect on TEM cross sections (Van Damme et al. 2016).

To date, only 15 dEDS have been reported. They display a severe phenotype mainly characterized by extreme skin fragility with spontaneous tearing of the skin and easy bruisability, typical craniofacial features, and generalized tissue fragility. Most dEDS patients had a normal height, weight, and orofacial circumference at birth, but postnatal growth retardation was noted in all patients, often in association with non-rhizomelic shortening of the limbs and short, plump hands and feet with stubby fingers and toes. Four of the 15 patients had a history of fractures, including (congenital) skull fractures in three. Several other skeletal abnormalities were reported, including delayed ossification of the cranial vault, Wormian bones, delayed bone age, and persistence of woven bone in the ribs of one patient. Osteopenia was reported in only two patients (Colige et al. 1997). Targeted inactivation of adamts2$^{-/-}$ in mice was shown to lead to a phenotype with strong resemblances to human dEDS. These mice were grossly normal at birth but developed thin and fragile skin after 1–2 months. A large fraction of the N-propeptides of dermal type I procollagen and cartilaginous type II procollagen were cleaved, suggesting that these tissues contain one or more additional enzymes that slowly process the proteins. TEM did not reveal any defects in the morphology of dermal collagen fibrils in newborn mice, but at 2 months of age, severe dermal ultrastructural abnormalities were seen similar to those observed in human dEDS patients. No defects in the morphology of the collagen fibrils of articular cartilage or growth plate were observed. In addition, there were no differences in subchondral plate thickness, bone volume fraction, or birefringence values obtained from the subchondral bone (Li et al. 2001).

The surprising clinical differences observed between aEDS (mainly congenital bilateral hip dislocation and severe joint hypermobility) and dEDS (mainly excessive skin fragility) are not completely understood. Several factors may contribute to these phenotypic outcomes. First, in contrast to aEDS (where alterations in either the proα1(I) or the proα2(I) chain are heterozygous and still produce some normal molecules), variants in *ADAMTS2* render the enzyme nonfunctional, meaning that both the proα1(I) and proα2(I) chains are not cleaved into mature chains, unless there is some residual enzyme function and/or some cleavage by other enzymes. Second, a few other enzymes, mainly ADAMTS3 and ADAMTS14, have been shown to display procollagen N-proteinase activity (Bekhouche and Colige 2015; Colige et al. 2002; Fernandes et al. 2001; Le Goff et al. 2007). Studies in dermatosparactic animals revealed that the proportion of pNα(I) chains to fully processed α(I) chains was much higher in the skin than in other tissues, such as tendon, further underscored by TEM analyses revealing much more severe collagen fibril abnormalities in the skin than in the tendon (Cassidy et al. 1980). This suggests that in non-dermal tissues, a larger proportion of procollagen I might be cleaved by other procollagen N-proteinases. Third, ADAMTS2 also possesses N-endopeptidase activity for types II, III, and V procollagen and perhaps other substrates (Bekhouche and Colige 2015; Colige et al. 2005). It is possible that deficient ADAMTS2 activity toward other substrates adds to the severity of the disorder, such as the extreme skin fragility and laxity (type V collagen) and the propensity for bruising and bleeding (type III collagen) (Van Damme et al. 2016).

Besides these well-defined aEDS and dEDS phenotypes, a few studies have reported on patients with either *COL1A1* or *COL1A2* variants (exon skipping variants or glycine substitutions) who show features of both a mild OI phenotype (blue sclerae, a few low impact fractures) and of EDS (severe joint hypermobility, skin, vascular, and other soft tissue fragility) (Cabral et al. 2005; Malfait et al. 2013b). This condition is currently coined as "collagen I related overlap phenotype" (C1ROD) (Morlino et al. 2020; Takeda et al. 2022). Many of these variants reside within the 85 most N-terminal amino acid residues of the triple helical domain of either the α1(I) or the α2(I) of type I collagen. This region acts as a stabilizing anchor for the N-terminal end of the type I collagen triple helix, and defects in this region lead to a conformational change of the N-proteinase cleavage site. As a result, the N-propeptides are not properly cleaved. Incorporation of these inefficiently cleaved collagen molecules into the mature collagen fibrils leads to ultrastructural changes of the collagen fibrils, at least in the skin (Valli et al. 2012). It should be noted that mutations within this region can also lead to a "true" OI phenotype (without overt features of EDS). Why some of these variants lead to OI and others to C1ROD remains currently unclear (Takeda et al. 2022).

3.3.4 EDS Types Caused by Other Rare Defects in Type I Collagen

Defects in type I collagen have been associated with two other forms of EDS. The first is a very rare autosomal recessive condition referred to as cardiac-valvular EDS. It is caused by biallelic loss-of-function variants in *COL1A2* which render unstable *COL1A2* mRNA molecules that undergo nonsense-mediated mRNA decay (Malfait et al. 2006; Schwarze et al. 2004). This results in the total absence of α2(I) collagen chains and the production of $[\alpha1(I)]_3$ homotrimers. Affected individuals have typical EDS skin and joint features, but they are particularly at risk for developing severe cardiac valvular disease. Some were reported to have osteopenia without increased fracture rate (Brady et al. 2017).

Interestingly a biallelic loss-of-function *COL1A2* variant that leads to stable mRNA, but unstable proteins, was associated with a moderately severe form of osteogenesis imperfecta (Nicholls et al. 1984). This observation suggests that production of unstable mutant protein triggers an unfolded protein response, which contributes to a more severe OI phenotype (Deak et al. 1983) (Pihlajaniemi et al. 1984). *COL1A2* mRNA instability and complete absence of proα2(I) chains on the other hand appear to lead to a phenotype that affects the soft connective tissues rather than the bone. The exact pathogenic mechanisms by which α1(I) homotrimers affect ECM structure and homeostasis are unknown.

The second comprises a group of arginine to cysteine substitutions in the triple helix domain of the proα1(I) chain. One of these, p.(Arg312Cys), is a recurrent variant that has been identified in 35 individuals from 12 distinct families and that is associated with a classical EDS phenotype with a particular propensity to vascular dissection (Colman et al. 2022a; Malfait et al. 2007). These patients do not seem to have an increased risk for low-impact fractures; osteopenia was reported in a few individuals (Colman et al. 2022a). Other proa1(I) Arg-to-Cys substitutions (positions 1036 and 1066) were reported in families with an EDS/OI overlap phenotype, without signs of vascular fragility (Cabral et al. 2007b; Lund et al. 2008), or, in contrast (positions 574 and 1093), in patients with a propensity for arterial rupture without other bone or connective tissue fragility (Malfait et al. 2007). The molecular basis for these particular phenotypic outcomes is unclear.

3.4 EDS Types Caused by Defects in Collagen Folding and Cross-Linking

Kyphoscoliotic EDS is a rare autosomal recessive form of EDS that was first linked to deficiency of lysyl hydroxylase 1 (LH1) (EDS VIA), encoded by *PLOD1*. The typical features of EDS VIA are neonatal muscle hypotonia, progressive kyphoscoliosis, abnormal scarring and easy bruising of the skin, increased risk of fatal arterial ruptures, and in some cases, ocular fragility. Patients do not seem to have an increased fracture rate, but in about 20% of reported patients, osteopenia was noted (Brady et al. 2017; Giunta et al. 2018).

LH1 deficiency results in underhydroxylation of triple helical lysyl residues and underglycosylation of hydroxylysyl residues in collagens (Rohrbach et al. 2011). As a result of underhydroxylation and underglycosylation, the collagen α-chains display a faster electrophoretic mobility in SDS-PAGE. In addition, patients with LH1 deficiency have a high urinary excretion of LP, resulting in a high LP/HP ratio (range 2–9, compared to a normal ratio of about 0.2, as measured by high-performance liquid chromatography), a finding that is highly sensitive and specific for EDS VIA (Rohrbach et al. 2011). The mechanical instability of the tissues that is observed in EDS VIA patients is attributed to the impaired cross-link formation; however, the role of other factors, e.g., the underglycosylation, remains not fully understood.

Biochemical data on human patients with LH1 deficiency are limited almost entirely to the skin, where the hydroxylysine content was shown to be maximal 5% of normal, whereas its content in the other tissue samples that were analyzed was few tendon and cartilage samples of about 50 and 90%, respectively (Steinmann and Superti-Furga 2002). The analysis of different collagen types extracted from various tissues from an EDS VIA patient gave similar results: the extent of lysyl hydroxylation of collagen I extracted from the skin was 0%, from the bone 17%, from the tendon 36%, and from the lung 73%. Collagen III extracted from the skin and lung contained 0 and 50%, respectively, of the normal hydroxylysine content, and collagens II, IV, and V from the cartilage, kidney, and bone, respectively, were almost normally hydroxylated (Ihme et al. 1984). Targeted inactivation of the *Plod1* gene in mice resulted in a phenotype reminiscent of human EDS VIA with muscle hypotonia and soft and flaccid tissues; about 15% of the mice died of aortic rupture (Takaluoma et al. 2007). The amounts of total hydroxylysine residues in all *Plod1*-null mouse tissues were decreased, but the magnitudes of these decreases also varied markedly between tissues, the amount of hydroxylysine in the *Plod1*$^{-/-}$ lung and femur being as high as 86% and 75% of that in the wild type, respectively, while the corresponding values in the skin and tail tendon were only 22% and 24%. These data suggest that the other two LH isoenzymes hydroxylate lysine residues effectively in the triple helical region, the large differences in the extent of the hydroxylysine deficiency probably being due to differences in the levels of the three isoenzymes in various (Takaluoma et al. 2007).

Mutations in *FKBP14,* coding for the ER-resident mixed function protein FKBP22, were identified in patients that show a phenotype that overlaps with EDS VIA but with normal urinary LP/HP ratio (currently classified as kyphoscoliotic EDS_*FKBP14*) (Baumann et al. 2012; Colman et al. 2022b). A small percentage of currently reported kEDS_*FKBP14* patients had a history of fractures.

A common mechanism that bridges kEDS due to LH1 deficiency and kEDS due to FKBP22 defects remains elusive. Intriguingly, alterations in *PLOD2* (encoding LH2) and *FKBP10* (encoding FKBP65, another rER-resident PPIase) result in overlapping OI phenotypes and could be explained by the role of FKBP65 in LH2 dimerization, essential for LH2 function (Gjaltema et al. 2016). To date no similar model has emerged for LH1 and FKBP22.

3.5 Bone Phenotypes in EDS Caused by Defects in *SLC39A13, ZNF469*, and *PRDM5*

ZIP13 is a homodimeric transmembrane Zrt/irt-like protein 13 (ZIP13) protein and a member of the SLC39A/ZIP family that regulates the influx of Zn^{2+} into the cytosol (Giunta et al. 2008). Biallelic variants in *SLC39A13*, which encodes ZIP13, give rise to a very rare EDS phenotype that is classified as spondylodysplastic EDS_*SLC39A13*. To date only about 14 individuals have been identified (Agrawal et al. 2023; Dusanic et al. 2018; Giunta et al. 2008). Their phenotype is characterized by hypotonia, ophthalmological abnormalities (e.g., thin, bluish sclerae, corneal deformities), dental abnormalities, and progressive growth retardation leading to proportionate short stature. Radiological findings include mild to moderate platyspondyly, mild to moderate osteopenia of the spine, small ileum, flat proximal femoral epiphyses, and short, wide femoral necks. The defects in ZIP13 lead to generalized underhydroxylation of collagen lysyl and prolyl residues and to abnormal cross-linking of collagen in the ECM (Giunta et al. 2008). The molecular pathways leading to these abnormal cross-links remain not well understood. Several pathophysiological mechanisms were suggested, including Zn^{2+} overload in the ER and competition with Fe^{++} for binding to lysyl hydroxylase and prolyl 4-hydroxylase (Giunta et al. 2008), trapping of Zn^{2+} in cytosolic vesicular stores ('zincosomes') leading to the reduced availability of Zn^{2+} in the ER and other cellular components and induction of ER stress (Jeong et al. 2012), and alterations in the activation of BMP/TGFβ signaling via regulation of the intracellular localization of Smad proteins in connective tissue-forming cells (Fukada et al. 2008). Recent findings in *Drosophila melanogaster* show that the fruit fly homologue of human ZIP13 is an Fe^{2+} exporter on the ER/Golgi membrane. Its absence might therefore result in Fe^{2+}-depletion in the ER/Golgi compartment, which could lead to underhydroxylation of lysyl and prolyl residues in collagen (Xiao et al. 2014).

Brittle cornea syndrome (BCS) is a rare autosomal recessive condition, the hallmarks of which are thin and fragile corneas that are at increased risk for spontaneous perforation, keratoglobus and keratoconus, and often also hearing loss (Burkitt Wright et al. 2013; Cameron 1993; Ticho et al. 1980). BCS was initially considered to be a form of EDS, because of a subset of patients presented signs of generalized connective tissue fragility, including joint hypermobility, muscle hypotonia, kyphoscoliosis, and mild EDS-like skin abnormalities (Cameron 1993; Macsai et al. 2000; Ogur et al. 1994). At the time, these patients were also labeled as EDS VIB. As such, BCS is included in the EDS classification.

To date, biallelic mutations have been identified in two genes, *ZNF469* and *PRDM5*, which lead to BCS1 and BCS2, the phenotypes of which are indistinguishable. In a few families with either BCS1 or BCS2, increased fracture rate has been reported, and in a recent literature review on BCS, signs of bone fragility were found in about 15% of affected individuals (Dhooge et al. 2021). In a few of them, the bone phenotype was investigated in more detail. Christensen et al. described two adult siblings with BCS1 who had osteopenia without fractures (Christensen et al.

2010). In contrast, Rolvien et al. found no reduced bone mass or increased fracture rate in two siblings with BCS1 (Rolvien et al. 2020). Cundy et al. reported on two siblings with BCS who both sustained more than ten fractures before the age of 15 years. When investigated as adults, they had osteopenia on bone densitometry. A bone biopsy from one of the siblings showed reduced cortical porosity. Both parents, who were heterozygous carriers, also had an increased fracture rate, but had normal bone density (Cundy et al. 2021). Together, these observations suggest that BCS may have a bone fragility phenotype. In view of the combined presence of thin, blue sclerae, hearing loss, and bone fragility, BCS should be considered as a differential diagnosis of osteogenesis imperfecta (Dhooge et al. 2021).

Galli et al. performed a detailed characterization of the cellular and molecular functions of Prdm5 (PR domain zinc finger protein 5) during murine bone formation in vivo and found that Prdm5 is expressed in skeletal structures during development and that its loss impacts the ossification process, leading to a decrease in bone mineral density. They showed that PRDM5 is a regulator of the expression of multiple collagen- and small leucine-rich proteoglycan-encoding, especially decorin, genes (Galli et al. 2012). More specifically, PRDM5 appears to regulate collagen transcription and fibrillogenesis by binding collagen genes and maintaining RNA polymerase II activity. Little is known about the exact function of ZNF469, but based on a limited homology of approximately 30% to the helical domains of certain collagens (α1(I), α2(I), α1(IV) collagen), it was suggested that ZNF469 could act as (transcriptional) regulator in the synthesis or assembly of collagen fibrils (Abu et al. 2008), though this has not been shown in experimental studies. Transcript analyses of fibroblasts from patients with *ZNF469* or *PRDM5* mutations showed dysregulation of several genes involved in the development and maintenance of the ECM, such as downregulation of *COL4A1*, *COL11A1*, and *HAPLN1* (the latter encodes the hyaluronan and proteoglycan link protein 1) (Burkitt Wright et al. 2011). Moreover, immunofluorescence staining demonstrated the altered deposition of type I collagen, type III collagen, fibronectin, and their receptor α2β1 and α5β1 integrins in *PRDM5* and *ZNF469* in the ECM of fibroblasts from patients (Burkitt Wright et al. 2013; Burkitt Wright et al. 2011). Together, these data suggest a regulatory role for ZNF469 and PRDM5 in the organization of the ECM of several connective tissues, including bone Abu et al. (2008).

3.6 Bone Phenotypes in EDS Caused by Defects in GAG Biosynthesis

Most EDS types are caused by defects in fibrillar types I, III, or V collagen, or their modifying enzymes, but a few rare EDS types have been linked to defects in proteoglycan biosynthesis (see also Chaps. 5 and 6). Proteoglycans are involved in a wide variety of vital functions, including cell-cell and cell-matrix interactions, cell proliferation and migration, cytokine and growth factor signaling, and embryonic and postnatal development. Proteoglycans are composed of a core protein and one or more glycosaminoglycan (GAG) chains that consist of repeating disaccharide units. Depending on the composition of these units, the proteoglycan superfamily is subdivided into heparan sulfate (HS) proteoglycans and chondroitin sulfate (CS)/dermatan sulfate (DS) proteoglycans. The major biological functions of proteoglycans derive from the physicochemical characteristics of the GAG component of the molecules. Defects in enzymes involved in HS and CS/DS biosynthesis, due to genetic alterations in their respective genes, typically give rise to a range of heritable disorders, including diverse osteochondrodysplasias, multiple hereditary exostoses, and specific types of Ehlers-Danlos syndromes (EDS) (50), which reflects the key role of proteoglycans in the development and homeostasis of the cartilage, bone, and skin (Syx et al. 2022).

The biosynthesis of both HS and CS/DS starts with the formation of a tetrasaccharide linker region that is covalently attached to serine residues on the core protein. This linker region is synthesized in a successive and coordinated effort by specific glycosyltransferases. It starts with the transfer of a xylose residue by xylosyltransferases I/II (encoded by *XYLT1/XYLT2*), followed by the addition of two galactose residues by galactosyltransferase I (β4GalT7, encoded by *B4GALT7*) and galactosyltransferase II (β3GalT6, encoded by *B3GALT6*). It is completed by the transfer of glucuronic acid catalyzed by glucuronosyltransferase I (GlcAT-I, encoded by *B3GAT3*). The result is a glucuronic acid-β1,3-galactose-β1,3-galactose-β1,4-xylose-β1-*O*- tetrasaccharide that serves as a primer for HS and CS/DS chain polymerization.

Biallelic (likely) pathogenic variants in *B3GALT7*, coding for galactosyltransferase I, or in *B3GALT6*, coding for galactosyltransferase II, both key enzymes in the biosynthesis of the linker region of GAGs give rise to a spondylodysplastic form of EDS (Malfait et al. 2017; Syx et al. 2022). Both conditions show overlapping phenotypes with craniofacial abnormalities; joint hypermobility, especially of the hands; and a soft hyperextensible skin, placing it within the EDS spectrum (Syx et al. 2022; Van Damme et al. 2018). Most patients with *B4GALT7* mutations have short stature, bowing of the limbs, and radioulnar synostosis, but they lack an overt bone fragility phenotype (Brady et al. 2017), except for one patient who was reported to have low-impact rib and vertebral fractures in infancy. The latter received bisphosphonate treatment, with improvement of bone pain and muscle function (Salter et al. 2016). In patients with *B3GALT6* mutations, kyphoscoliosis (congenital or early onset and progressive), muscle hypotonia, joint contractures, and radiological signs of a spondyloepimetaphyseal dysplasia with metaphyseal widening and bowing of the limb bones, hypoplasia of the iliac bones, acetabular dysplasia, and vertebral body changes are frequently seen. In addition, many of them (25–30%) had severe osteoporosis with multiple fractures after minor trauma, sometimes evoking an initial diagnosis of osteogenesis imperfecta (Van Damme et al. 2018).

Studies on cultured fibroblasts from patients with either *B4GALT7* or *B3GALT6* mutations showed variably reduced galactosyltransferase I and II enzyme activities, respectively, with markedly reduced and/or shorter HS and CS GAG chains, and partial or complete lack of DS on the small leucine-rich proteoglycan decorin (Gotte et al. 2008; Malfait et al. 2013a; Nakajima et al. 2013; Seidler et al. 2006; Van Damme et al. 2018). Ultrastructural analyses of the dermal collagen fibril architecture showed fibrils with variable diameters and some fibrils with irregular contours.

To further study the pathogenic mechanisms underlying spEDS-*B4GALT7* and spEDS-*B3GALT6*, Delbaere et al. generated homozygous *b4galt7* and *b3galt6* knockout (*b4galt7*$^{-/-}$ and *b3galt6*$^{-/-}$) zebrafish (Delbaere et al. 2020a, b). At 4 days postfertilization (dpf), *b4galt7*$^{-/-}$ zebrafish totally lacked HS and CS GAGs, resulting in a severe skeletal phenotype with delayed or absent ossification of several bone structures and aberrant cartilage formation. They died before the age of 10 dpf, suggesting that complete loss of β4GalT7 activity is incompatible with life in zebrafish. This could also be true for humans, as complete loss of β4GalT7 function has not been observed in human patients with spEDS-*B4GALT7*. Therefore, hypomorphic zebrafish were generated, which display only partial loss of β4GalT7 function. Two distinct morpholino-based *b4galt7* knockdown (morphant) and two mosaic *b4galt7*-deficient (crispant) models were created. In these models, defective β4GalT7 activity led to impaired GAG biosynthesis with severely diminished HS and CS proteoglycan levels. Morphant and crispant *b4galt7* embryos showed underdeveloped, misshapen, and partly absent cartilage structures during early development and were smaller than wild-type embryos. Interestingly, spontaneous *B4GALT7* defects have also been identified as a cause of disproportionate dwarfism with joint laxity in Friesian horses (Leegwater et al. 2016).

b3galt6$^{-/-}$ zebrafish, in contrast, survived into adulthood (Delbaere et al. 2020a). They showed progressive craniofacial and skeletal (kyphoscoliosis and extra intramembranous bone) malformations reminiscent of those seen in humans. Quantitative microCT showed that the mutant zebrafish had a lower bone mass and reduced thickness of several parts of the vertebral body. Tissue mineral density was increased in several bony structures (central, hemal, and neural associated elements). Ultrastructural analysis in adult *b3galt6*$^{-/-}$ zebrafish showed an abnormal collagen fibril organization in the intervertebral ligament, in the inner layer of the vertebral bone (that is first deposited during development), and in the dermis. As such, the lower collagen fibril organization in bone probably leads to impaired mineralization, with regions of hypermineralization, thereby compromising the mechanical integrity of the bone. In addition, structural and functional muscle abnormalities were detected in *b3galt6*$^{-/-}$ zebrafish. These phenotypic findings were in accordance with the strongly reduced amounts of CS, DS, and HS disaccharides observed in the bone, skin, and muscle of *b3galt6*$^{-/-}$ zebrafish. Surprisingly, low amounts of GAGs were still produced by *b3galt6*$^{-/-}$ zebrafish, and residual proteoglycans were shown to contain a noncanonical trimeric linker region consisting of only three instead of four sugars (lacking a galactose residue), thereby presenting a possible rescue mechanism for deficiency of one of the linker enzymes. The presence of this noncanonical trimeric linker region was also found to be drastically increased in urine samples of human patients with spEDS-*B3GALT6* (Nikpour et al. 2022).

These findings highlight a key role for β4GalT7 and of β3GalT6 in the development and homeostasis of cartilage and bone (among other tissues) and illustrate that defective enzyme activity alters the biosynthesis of both CS/DS and HS.

Musculocontractural EDS results from (likely) pathogenic variants in genes coding for enzymes specifically involved in dermatan biosynthesis: *CHST14*, encoding dermatan 4-*O*-sulfotransferase 1 (D4ST1), and *DSE*, encoding dermatan sulfate epimerase_1 (DS-epi1). This condition is characterized by a unique set of clinical features related to generalized connective tissue fragility affecting the skin, joints, and internal organs as well as developmental problems leading to typical dysmorphic features and congenital malformations of multiple organ systems. Common musculoskeletal features include joint hypermobility with multiple joint dislocations in combination with multiple congenital contractures; characteristic finger morphologies such as tapering, slender, long, or cylindrical fingers; hypotonia and motor development delay; spinal deformities; and osteoporosis or osteopenia without significantly increased fracture rate (Lautrup et al. 2020; Minatogawa et al. 2021, 2022; Syx et al. 2022). Both D4ST1 and DS-epi1 defects affect consecutive steps of DS biosynthesis and alter the disaccharide composition of CS/DS GAG chains, eventually resulting in the lack of dermatan sulfate and an excess of chondroitin sulfate in dermatan sulfate-rich proteoglycans, e.g., decorin. This leads to altered interfibrillar spacing of collagen fibrils and thus a disorganized collagen bundle organization. No bone studies have been performed in individuals or animal models with *CHST14* of *DSE* defects.

3.7 Bone Phenotypes in EDS Caused by Defects in *TNXB* and *AEBP1*

Some patients with clinical characteristics that strongly resemble those of classical EDS harbor biallelic (likely) pathogenic variants in either *TNXB*, encoding tenascin-X, an ECM glycoprotein that is involved in regulation of collagen fiber assembly (Zweers et al. 2003), or in *AEBP1*, which encodes the ECM-associated adipocyte enhancer-binding protein (AEBP1) or ACLP (aortic carboxypeptidase-like protein), which assists in collagen polymerization (Blackburn et al. 2018). These conditions are classified as classical-like EDS and have an autosomal recessive transmission pattern. Whereas no overt bone phenotype has been reported for tenascin-X-related clEDS, about half of patients with *AEBP1* mutations were noted to have severe osteopenia (Di Giosaffatte et al. 2022).

Tenascin-X has a crucial architectural function in the ECM through its involvement in the regulation of collagen fiber assembly, but it is also an extracellular regulator of several signaling pathways. Indeed, tenascin-X has been shown to regulate the bioavailability of the transforming growth factor (TGF)-β and to modulate epithelial cell plasticity (Valcourt et al. 2015). Kajitani et al. studied the bone phenotype and osteoblastic and osteoclastic differentiation potentials of bone marrow cells in *Tnxb* knockout mice (Kajitani et al. 2019). These mice were shown to exhibit characteristic skin, muscle, and obstetric features that are reminiscent of those seen in human patients with tenascin-X deficiency. They found significant bone loss, which appeared to be caused by enhanced osteoclast maturation, resulting in increased bone resorption activity (Kajitani et al. 2019). The exact mechanisms underlying this bone resorption phenotype remain however poorly understood.

AEBP1 encodes the ECM-associated adipocyte enhancer-binding protein 1 (AEBP1; also known as aortic carboxypeptidase-like protein (ACLP)), which is abundantly present in tissues with high collagen content (Ith et al. 2005; Layne et al. 2001). This protein binds to fibrillar types I, III, and V collagens and assists in type I collagen polymerization (Blackburn et al. 2018). ACLP is also highly expressed in developing skeletal muscle (Layne et al. 1998, 2001), and it was shown to stimulate the differentiation and survival of osteoblasts (Kim et al. 2023). Intriguingly, ACLP was demonstrated to activate the canonical Wnt signaling pathway in hepatic stellate cells through binding frizzled-8 and low-density lipoprotein receptor-related protein 6 (LRP6) and trigger β-catenin-dependent gene expression (Teratani et al. 2018). Activation of the canonical Wnt pathway has been shown to lead to increased bone formation and bone mass and decreased bone resorption (Zhong et al. 2014). Therefore, it is tempting to speculate that a defect in ACLP can contribute, at least in part, to a reduced β-catenin signaling in bone cells, which can result in reduced bone mass and in turn lead to osteoporosis as seen in affected individuals. Nevertheless, deregulation of other pathways that are potentially regulated by ACLP may contribute to this process as well. ACLP was, for instance, also shown to be involved in fibroblast-to-myofibroblast transition through activation of TGFβ receptors (Schissel et al. 2009; Tumelty et al. 2014).

3.8 Conclusions

Despite the mostly anecdotal nature of the information, bone mineral density seems to vary among different types of EDS. In some rare, autosomal recessive types, especially spondylodysplastic type due to *B3GALT6* mutations, but also kyphoscoliotic (*PLOD1, FKBP14*), dermatosparaxis (*ADAMTS2*), classical-like EDS type 2 (*AEBP1*), and BCS (*ZNF469, PRDM5*), low mineral density has been observed. Increased fracture rate has also been reported in some patients with these disorders, including congenital fractures (e.g., congenital skull fractures in dEDS) and low-impact fractures in any stage of life. For classical EDS, mildly reduced bone density and abnormal vertebral morphology seem to be more common at adult age than previously thought, but it remains unclear whether this results from fractures or whether these are dysplastic changes. For vascular EDS, very little information on bone phenotype is available, although studies in a murine model for vEDS suggest a role for type III collagen in trabecular bone development through its effect on osteoblast differentiation and in fracture repair. In order to gain insight into whether EDS should be on the list of differential diagnoses on non-accidental injury during infancy or childhood, Rolfes and colleagues performed a retrospective case-control study on 21 patients with a diagnosis of EDS (14 hEDS, 2 cEDS, 4 vEDS, and 1 aEDS), and they did not find evidence for an increased fracture susceptibility during infancy. According to their observations, older, ambulatory children with these conditions may have a higher fracture rate in childhood (Rolfes et al. 2019) but further studies are needed to confirm this. As such in suspected child abuse cases due increased fractures, EDS is not high on the list of differential diagnoses (Castori 2015, 2016; Rolfes et al. 2019).

No evidence-based clinical guidelines for assessment, follow-up, and/or management of bone health in EDS patients are available, but there seems to be a rationale for recommending assessment of bone mineral density, and evaluation of vertebral morphology, at least at the time of diagnosis. Further studies on the prevalence and incidence of fractures and low bone mineral density among the different EDS types are needed, in order to develop evidence-based guidelines for the assessment and treatment of skeletal complications in EDS.

Funding Information
This work is supported by the Research Foundation—Flanders (FWO), Belgium (1842318N, G041519N, and G0A3322N), and the Special Research Fund Ghent University (GOA019-21).

Compliance with Ethical Standards/Ethical Requirements No conflict of interest.

References

Abu AFM, Marek D, Pras E, Nir U, Reznik-Wolf H, Pras E (2008) Deleterious mutations in the zinc-finger 469 gene cause brittle cornea syndrome. Am J Hum Genet 82:1217–1222

Agrawal P, Kaur H, Kondekar A, Rathi S (2023) A case of Ehlers-Danlos syndrome presenting as short stature: a novel mutation in SLC39A13 causing spondylodysplastic Ehlers-Danlos syndrome. Oxf Med Case Reports 2023:omac107. https://doi.org/10.1093/omcr/omac107

Amar S, Smith L, Fields GB (2017) Matrix metalloproteinase collagenolysis in health and disease. Biochim Biophys Acta Mol Cell Res 1864:1940–1951. https://doi.org/10.1016/j.bbamcr.2017.04.015

Bachinger HP, Bruckner P, Timpl R, Prockop DJ, Engel J (1980) Folding mechanism of the triple helix in type-III collagen and type-III pN-collagen. Role of disulfide bridges and peptide bond isomerization. Eur J Biochem 106:619–632. https://doi.org/10.1111/j.1432-1033.1980.tb04610.x

Basalom S, Rauch F (2020) Bone disease in patients with ehlers-danlos syndromes. Curr Osteoporos Rep 18:95–102. https://doi.org/10.1007/s11914-020-00568-5

Baumann M et al (2012) Mutations in FKBP14 cause a variant of Ehlers-Danlos syndrome with progressive kyphoscoliosis, myopathy, and hearing loss. Am J Hum Genet 90:201–216. https://doi.org/10.1016/j.ajhg.2011.12.004

Bekhouche M, Colige A (2015) The procollagen N-proteinases ADAMTS2, 3 and 14 in pathophysiology. Matrix Biol 44-46:46–53. https://doi.org/10.1016/j.matbio.2015.04.001

Birk DE (2001) Type V collagen: heterotypic type I/V collagen interactions in the regulation of fibril assembly. Micron 32:223–237. https://doi.org/10.1016/s0968-4328(00)00043-3

Birk DE, Fitch JM, Babiarz JP, Doane KJ, Linsenmayer TF (1990) Collagen fibrillogenesis in vitro: interaction of types I and V collagen regulates fibril diameter. J Cell Sci 95(Pt 4):649–657

Blackburn PR et al (2018) Bi-allelic alterations in AEBP1 lead to defective collagen assembly and connective tissue structure resulting in a variant of Ehlers-Danlos syndrome. Am J Hum Genet 102:696–705. https://doi.org/10.1016/j.ajhg.2018.02.018

Boudko SP, Engel J, Bachinger HP (2012) The crucial role of trimerization domains in collagen folding. Int J Biochem Cell Biol 44:21–32. https://doi.org/10.1016/j.biocel.2011.09.009

Bowen CJ et al (2020) Targetable cellular signaling events mediate vascular pathology in vascular Ehlers-Danlos syndrome. J Clin Invest 130:686–698. https://doi.org/10.1172/JCI130730

Bowen JM, Sobey GJ, Burrows NP, Colombi M, Lavallee ME, Malfait F, Francomano CA (2017) Ehlers-Danlos syndrome, classical type. Am J Med Genet C Semin Med Geneta 175:27–39. https://doi.org/10.1002/ajmg.c.31548

Brady AF et al (2017) The Ehlers-Danlos syndromes, rare types. Am J Med Genet C Semin Med Geneta 175:70–115. https://doi.org/10.1002/ajmg.c.31550

Bristow J, Carey W, Egging D, Schalkwijk J (2005) Tenascin-X, collagen, elastin, and the Ehlers-Danlos syndrome. Am J Med Genet C Semin Med Genet 139C:24–30. https://doi.org/10.1002/ajmg.c.30071

Bunel L, Pincet L, Malhotra V, Raote I, Pincet F (2024) A model for collagen secretion by intercompartmental continuities. Proc Natl Acad Sci USA 121:e2310404120. https://doi.org/10.1073/pnas.2310404120

Burkitt Wright EM, Spencer HL, Clayton-Smith J, Au L, Munier FL, Smithson S, Suri M, Rohrbach M, Manson FD, Black GC (2013) Brittle cornea syndrome: recognition, molecular diagnosis and management. Orphanet J Rare Dis 8:6

Burkitt Wright EM et al (2011) Mutations in PRDM5 in brittle cornea syndrome identify a pathway regulating extracellular matrix development and maintenance. Am J Hum Genet 88:767–777. https://doi.org/10.1016/j.ajhg.2011.05.007

Byers PH et al (2017) Diagnosis, natural history, and management in vascular Ehlers-Danlos syndrome. Am J Med Genet C Semin Med Genet 175:40–47. https://doi.org/10.1002/ajmg.c.31553

Byers PH et al (1997) Ehlers-Danlos syndrome type VIIA and VIIB result from splice-junction mutations or genomic deletions that involve exon 6 in the COL1A1 and COL1A2 genes of type I collagen. Am J Med Genet 72:94–105. https://doi.org/10.1002/(sici)1096-8628(19971003)72:1<94::aid-ajmg20>3.0.co;2-o

Cabral WA et al (2007a) Prolyl 3-hydroxylase 1 deficiency causes a recessive metabolic bone disorder resembling lethal/severe osteogenesis imperfecta. Nat Genet 39:359–365. https://doi.org/10.1038/ng1968

Cabral WA et al (2005) Mutations near amino end of alpha1(I) collagen cause combined osteogenesis imperfecta/Ehlers-Danlos syndrome by interference with N-propeptide processing. J Biol Chem 280:19259–19269. https://doi.org/10.1074/jbc.M414698200

Cabral WA et al (2007b) Y-position cysteine substitution in type I collagen (alpha1(I) R888C/p.R1066C) is associated with osteogenesis imperfecta/Ehlers-Danlos syndrome phenotype. Hum Mutat 28:396–405. https://doi.org/10.1002/humu.20456

Cameron JA (1993) Corneal abnormalities in Ehlers-Danlos syndrome type VI. Cornea 12:54–59. https://doi.org/10.1097/00003226-199301000-00009

Cassidy K, Eikenberry EF, Olsen B, Brodsky B (1980) X-ray diffraction investigations of collagen fibril structure in dermatosparactic lamb tissues. Lab Invest 43:542–546

Castori M (2015) Ehlers-Danlos syndrome(s) mimicking child abuse: is there an impact on clinical practice? Am J Med Genet C Semin Med Genet 169:289–292. https://doi.org/10.1002/ajmg.c.31460

Castori M (2016) Ehlers-Danlos syndrome(s) mimicking child abuse: is there an impact on clinical practice? Am J Med Genet A 170:1947. https://doi.org/10.1002/ajmg.a.37733

Castori M, Tinkle B, Levy H, Grahame R, Malfait F, Hakim A (2017) A framework for the classification of joint hypermobility and related conditions. Am J Med Genet C Semin Med Genet 175:148–157. https://doi.org/10.1002/ajmg.c.31539

Chen XD, Dusevich V, Feng JQ, Manolagas SC, Jilka RL (2007) Extracellular matrix made by bone marrow cells facilitates expansion of marrow-derived mesenchymal progenitor cells and prevents their differentiation into osteoblasts. J Bone Miner Res 22:1943–1956. https://doi.org/10.1359/jbmr.070725

Chiodo AA, Hockey A, Cole WG (1992) A base substitution at the splice acceptor site of intron 5 of the COL1A2 gene activates a cryptic splice site within exon 6 and generates abnormal type I procollagen in a patient with Ehlers-Danlos syndrome type VII. J Biol Chem 267:6361–6369

Christensen AEKP, Midtbø M, Gjesdal CG, Mengel-From J, Morling N, Rødahl E, Boman HI (2010) Brittle cornea syndrome associated with a missense mutation in the zinc-finger 469 gene. Invest Ophthalmol Vis Sci 51:47–52

Colige A, Li SW, Sieron AL, Nusgens BV, Prockop DJ, Lapiere CM (1997) cDNA cloning and expression of bovine procollagen I N-proteinase: a new member of the superfamily of zinc-metalloproteinases with binding sites for cells and other matrix components. Proc Natl Acad Sci USA 94:2374–2379

Colige A et al (2005) Domains and maturation processes that regulate the activity of ADAMTS-2, a metalloproteinase cleaving the aminopropeptide of fibrillar procollagens types I-III and V. J Biol Chem 280:34397–34408. https://doi.org/10.1074/jbc.M506458200

Colige AVI, Thiry M, Lambert CA, Van Beeumen J, Li SW, Prockop DJ, Lapiere CM, Nusgens BV (2002) Cloning and characterization of ADAMTS-14, a novel ADAMTS displaying high homology with ADAMTS-2 and ADAMTS-3. J Biol Chem 277:5756–5766

Colman M et al (2022a) Atypical variants in COL1A1 and COL3A1 associated with classical and vascular Ehlers-Danlos syndrome overlap phenotypes: expanding the clinical phenotype based on additional case reports. Clin Exp Rheumatol 40(Suppl 134):46–62. https://doi.org/10.55563/clinexprheumatol/kzkq6y

Colman M, Syx D, De Wandele I, Dhooge T, Symoens S, Malfait F (2021) Clinical and molecular characteristics of 168 probands and 65 relatives with a clinical presentation of classical Ehlers-Danlos syndrome. Hum Mutat 42:1294–1306. https://doi.org/10.1002/humu.24258

Colman M et al (2022b) Kyphoscoliotic Ehlers-Danlos syndrome caused by pathogenic variants in FKBP14: further insights into the phenotypic spectrum and pathogenic mechanisms. Hum Mutat 43:1994–2009. https://doi.org/10.1002/humu.24456

Cundy T, Vincent A, Robertson S (2021) Does brittle cornea syndrome have a bone fragility phenotype? Bone Rep 15:101124. https://doi.org/10.1016/j.bonr.2021.101124

Deak SB, Nicholls A, Pope FM, Prockop DJ (1983) The molecular defect in a nonlethal variant of osteogenesis imperfecta. Synthesis of pro-alpha 2(I) chains which are not incorporated into trimers of type I procollagen. J Biol Chem 258:15192–15197

Delbaere S et al (2020a) b3galt6 Knock-out zebrafish recapitulate beta3GalT6-deficiency disorders in human and reveal a Trisaccharide proteoglycan linkage region. Front Cell Dev Biol 8:597857. https://doi.org/10.3389/fcell.2020.597857

Delbaere S, Van Damme T, Syx D, Symoens S, Coucke P, Willaert A, Malfait F (2020b) Hypomorphic zebrafish models mimic the musculoskeletal phenotype of beta4GalT7-deficient Ehlers-Danlos syndrome. Matrix Biol 89:59–75. https://doi.org/10.1016/j.matbio.2019.12.002

Dhooge T et al (2021) More than meets the eye: expanding and reviewing the clinical and mutational spectrum of brittle cornea syndrome. Hum Mutat 42:711–730. https://doi.org/10.1002/humu.24199

Di Giosaffatte N et al (2022) Congenital defects in a patient carrying a novel homozygous AEBP1 variant: further expansion of the phenotypic Spectrum of Ehlers-Danlos syndrome classical-like type 2? Genes 13. https://doi.org/10.3390/genes13122358

Duran I et al (2017) A chaperone complex formed by HSP47, FKBP65, and BiP modulates Telopeptide Lysyl hydroxylation of type I Procollagen. J Bone Miner Res 32:1309–1319. https://doi.org/10.1002/jbmr.3095

Dusanic M et al (2018) Novel nonsense mutation in SLC39A13 initially presenting as myopathy: case report and review of the literature. Mol Syndromol 9:100–109. https://doi.org/10.1159/000485881

Eller-Vainicher C et al (2016) Bone involvement in adult patients affected with Ehlers-Danlos syndrome. Osteoporos Int 27:2525–2531. https://doi.org/10.1007/s00198-016-3562-2

Fernandes RJ, Hirohata S, Engle JM, Colige A, Cohn DH, Eyre DR, Apte SS (2001) Procollagen II amino propeptide processing by ADAMTS-3. Insights on dermatosparaxis. J Biol Chem 276:31502–31509. https://doi.org/10.1074/jbc.M103466200

Frank S, Boudko S, Mizuno K, Schulthess T, Engel J, Bachinger HP (2003) Collagen triple helix formation can be nucleated at either end. J Biol Chem 278:7747–7750. https://doi.org/10.1074/jbc.C200698200

Fukada T et al (2008) The zinc transporter SLC39A13/ZIP13 is required for connective tissue development; its involvement in BMP/TGF-beta signaling pathways. PLoS One 3:e3642. https://doi.org/10.1371/journal.pone.0003642

Galli GGHLK, Carrara M, Hans W, Wuelling M, Mentz B, Multhaupt HA, Fog CK, Jensen KT, Rappsilber J, Vortkamp A, Coulton L, Fuchs H, Gailus-Durner V, Hrabě de Angelis M, Calogero RA, Couchman JR, Lund AH (2012) Prdm5 regulates collagen gene transcription by association with RNA polymerase II in developing bone. PLoS Genet 8:e1002711. https://doi.org/10.1371/journal.pgen.1002711

Gidalevitz T, Stevens F, Argon Y (2013) Orchestration of secretory protein folding by ER chaperones. Biochim Biophys Acta 1833:2410–2424. https://doi.org/10.1016/j.bbamcr.2013.03.007

Giunta C et al (2018) A cohort of 17 patients with kyphoscoliotic Ehlers-Danlos syndrome caused by biallelic mutations in FKBP14: expansion of the clinical and mutational spectrum and description of the natural history. Genet Med 20:42–54. https://doi.org/10.1038/gim.2017.70

Giunta C et al (2008) Spondylocheiro dysplastic form of the Ehlers-Danlos syndrome–an autosomal-recessive entity caused by mutations in the zinc transporter gene SLC39A13. Am J Hum Genet 82:1290–1305. https://doi.org/10.1016/j.ajhg.2008.05.001

Gjaltema RA, van der Stoel MM, Boersema M, Bank RA (2016) Disentangling mechanisms involved in collagen pyridinoline cross-linking: the immunophilin FKBP65 is critical for dimerization of lysyl hydroxylase 2. Proc Natl Acad Sci U S A 113:7142–7147. https://doi.org/10.1073/pnas.1600074113

Gomez-Navarro N, Miller E (2016) Protein sorting at the ER-Golgi interface. J Cell Biol 215:769–778. https://doi.org/10.1083/jcb.201610031

Gorres KL, Raines RT (2010) Prolyl 4-hydroxylase. Crit Rev Biochem Mol Biol 45:106–124. https://doi.org/10.3109/10409231003627991

Gotte M, Spillmann D, Yip GW, Versteeg E, Echtermeyer FG, van Kuppevelt TH, Kiesel L (2008) Changes in heparan sulfate are associated with delayed wound repair, altered cell migration, adhesion and contractility in the galactosyltransferase I (beta4GalT-7) deficient form of Ehlers-Danlos syndrome. Hum Mol Genet 17:996–1009. https://doi.org/10.1093/hmg/ddm372

Heard ME et al (2016) Sc65-null mice provide evidence for a novel endoplasmic reticulum complex regulating collagen Lysyl hydroxylation. PLoS Genet 12:e1006002. https://doi.org/10.1371/journal.pgen.1006002

Hendershot LM, Buck TM, Brodsky JL (2023) The essential functions of molecular chaperones and folding enzymes in maintaining endoplasmic reticulum homeostasis. J Mol Biol 436:168418. https://doi.org/10.1016/j.jmb.2023.168418

Hennet T (2019) Collagen glycosylation. Curr Opin Struct Biol 56:131–138. https://doi.org/10.1016/j.sbi.2019.01.015

Hopkins DR, Keles S, Greenspan DS (2007) The bone morphogenetic protein 1/Tolloid-like metalloproteinases. Matrix Biol 26:508. https://doi.org/10.1016/j.matbio.2007.05.004

Hudson DM et al (2017) P3h3-null and Sc65-null mice Phenocopy the collagen lysine underhydroxylation and cross-linking abnormality of Ehlers-Danlos syndrome type VIA. J Biol Chem 292:3877–3887. https://doi.org/10.1074/jbc.M116.762245

Hutchings J, Zanetti G (2019) Coat flexibility in the secretory pathway: a role in transport of bulky cargoes. Curr Opin Cell Biol 59:104–111. https://doi.org/10.1016/j.ceb.2019.04.002

Ihme A et al (1984) Ehlers-Danlos syndrome type VI: collagen type specificity of defective lysyl hydroxylation in various tissues. J Invest Dermatol 83:161–165. https://doi.org/10.1111/1523-1747.ep12263502

Ishikawa Y, Bachinger HP (2013) A molecular ensemble in the rER for procollagen maturation. Biochim Biophys Acta 1833:2479–2491. https://doi.org/10.1016/j.bbamcr.2013.04.008
Ishikawa Y, Bachinger HP (2014) A substrate preference for the rough endoplasmic reticulum resident protein FKBP22 during collagen biosynthesis. J Biol Chem 289:18189–18201. https://doi.org/10.1074/jbc.M114.561944
Ishikawa Y, Bonna A, Gould DB, Farndale RW (2023) Local net charge state of collagen triple helix is a determinant of FKBP22 binding to collagen III. Int J Mol Sci 24. https://doi.org/10.3390/ijms242015156
Ishikawa Y, Boudko S, Bachinger HP (2015) Ziploc-ing the structure: triple helix formation is coordinated by rough endoplasmic reticulum resident PPIases. Biochim Biophys Acta 1850:1983–1993. https://doi.org/10.1016/j.bbagen.2014.12.024
Ishikawa Y, Holden P, Bachinger HP (2017a) Heat shock protein 47 and 65-kDa FK506-binding protein weakly but synergistically interact during collagen folding in the endoplasmic reticulum. J Biol Chem 292:17216–17224. https://doi.org/10.1074/jbc.M117.802298
Ishikawa Y, Ito S, Nagata K, Sakai LY, Bachinger HP (2016) Intracellular mechanisms of molecular recognition and sorting for transport of large extracellular matrix molecules. Proc Natl Acad Sci USA 113:E6036–E6044. https://doi.org/10.1073/pnas.1609571113
Ishikawa Y, Mizuno K, Bachinger HP (2017b) Ziploc-ing the structure 2.0: endoplasmic reticulum-resident peptidyl prolyl isomerases show different activities toward hydroxyproline. J Biol Chem 292:9273–9282. https://doi.org/10.1074/jbc.M116.772657
Ishikawa Y et al (2020) The novel missense mutation Met48Lys in FKBP22 changes its structure and functions. Sci Rep 10:497. https://doi.org/10.1038/s41598-019-57374-y
Ishikawa Y et al (2022) Lysyl hydroxylase 3-mediated post-translational modifications are required for proper biosynthesis of collagen alpha1alpha1alpha2(IV). J Biol Chem 298:102713. https://doi.org/10.1016/j.jbc.2022.102713
Ishikawa Y et al (2021) Type I and type V procollagen triple helix uses different subsets of the molecular ensemble for lysine posttranslational modifications in the rER. J Biol Chem 296:100453. https://doi.org/10.1016/j.jbc.2021.100453
Ishikawa Y, Vranka J, Wirz J, Nagata K, Bachinger HP (2008) The rough endoplasmic reticulum-resident FK506-binding protein FKBP65 is a molecular chaperone that interacts with collagens. J Biol Chem 283:31584–31590. https://doi.org/10.1074/jbc.M802535200
Ishikawa Y et al (2012) Mutation in cyclophilin B that causes hyperelastosis cutis in American quarter horse does not affect peptidylprolyl cis-trans isomerase activity but shows altered cyclophilin B-protein interactions and affects collagen folding. J Biol Chem 287:22253–22265. https://doi.org/10.1074/jbc.M111.333336
Ishikawa Y, Wirz J, Vranka JA, Nagata K, Bachinger HP (2009) Biochemical characterization of the prolyl 3-hydroxylase 1.Cartilage-associated protein.Cyclophilin B complex. J Biol Chem 284:17641–17647. https://doi.org/10.1074/jbc.M109.007070
Ith B, Wei J, Yet SF, Perrella MA, Layne MD (2005) Aortic carboxypeptidase-like protein is expressed in collagen-rich tissues during mouse embryonic development. Gene Expr Patterns 5:533–537. https://doi.org/10.1016/j.modgep.2004.11.002
Ito S, Nagata K (2017) Biology of Hsp47 (serpin H1), a collagen-specific molecular. Semin Cell Dev Biol 62:142. https://doi.org/10.1016/j.semcdb.2016.11.005
Ito S, Nagata K (2019) Roles of the endoplasmic reticulum-resident, collagen-specific molecular chaperone Hsp47 in vertebrate cells and human disease. J Biol Chem 294:2133–2141. https://doi.org/10.1074/jbc.TM118.002812
Ito S, Nagata K (2021) Quality control of procollagen in cells. Annu Rev Biochem 90:631–658. https://doi.org/10.1146/annurev-biochem-013118-111603
Izu Y, Birk DE (2023) Collagen XII mediated cellular and extracellular mechanisms in development, regeneration, and disease. Front Cell Dev Biol 11:1129000. https://doi.org/10.3389/fcell.2023.1129000

Jeong J et al (2012) Promotion of vesicular zinc efflux by ZIP13 and its implications for spondylocheiro dysplastic Ehlers-Danlos syndrome. Proc Natl Acad Sci U S A 109:E3530–E3538. https://doi.org/10.1073/pnas.1211775110

Kadler KE, Hill A, Canty-Laird EG (2008) Collagen fibrillogenesis: fibronectin, integrins, and minor collagens as organizers and nucleators. Curr Opin Cell Biol 20:495–501. https://doi.org/10.1016/j.ceb.2008.06.008

Kajitani N, Yamada T, Kawakami K, Matsumoto KI (2019) TNX deficiency results in bone loss due to an increase in multinucleated osteoclasts. Biochem Biophys Res Commun 512:659–664. https://doi.org/10.1016/j.bbrc.2019.03.134

Kalamajski S, Oldberg A (2010) The role of small leucine-rich proteoglycans in collagen fibrillogenesis. Matrix Biol 29:248–253. https://doi.org/10.1016/j.matbio.2010.01.001

Keene DR, Sakai LY, Bachinger HP, Burgeson RE (1987) Type III collagen can be present on banded collagen fibrils regardless of fibril diameter. J Cell Biol 105:2393–2402. https://doi.org/10.1083/jcb.105.5.2393

Keene DR, Sakai LY, Burgeson RE (1991) Human bone contains type III collagen, type VI collagen, and fibrillin: type III collagen is present on specific fibers that may mediate attachment of tendons, ligaments, and periosteum to calcified bone cortex. J Histochem Cytochem 39:59–69. https://doi.org/10.1177/39.1.1983874

Kim H et al (2023) Aortic carboxypeptidase-like protein, a putative myokine, stimulates the differentiation and survival of bone-forming osteoblasts. FASEB J 37:e23104. https://doi.org/10.1096/fj.202300140R

Klaassens M et al (2012) Ehlers-Danlos arthrochalasia type (VIIA-B)–expanding the phenotype: from prenatal life through adulthood. Clin Genet 82:121–130. https://doi.org/10.1111/j.1399-0004.2011.01758.x

Koski MK et al (2017) Assembly of the elongated collagen prolyl 4-hydroxylase alpha(2) beta(2) heterotetramer around a central alpha(2) dimer. Biochem J 474:751–769. https://doi.org/10.1042/BCJ20161000

Lautrup CK et al (2020) Delineation of musculocontractural Ehlers-Danlos Syndrome caused by dermatan sulfate epimerase deficiency. Mol Genet Genomic Med 8:e1197. https://doi.org/10.1002/mgg3.1197

Layne MD et al (1998) Aortic carboxypeptidase-like protein, a novel protein with discoidin and carboxypeptidase-like domains, is up-regulated during vascular smooth muscle cell differentiation. J Biol Chem 273:15654–15660. https://doi.org/10.1074/jbc.273.25.15654

Layne MD, Yet SF, Maemura K, Hsieh CM, Bernfield M, Perrella MA, Lee ME (2001) Impaired abdominal wall development and deficient wound healing in mice lacking aortic carboxypeptidase-like protein. Mol Cell Biol 21:5256–5261. https://doi.org/10.1128/MCB.21.15.5256-5261.2001

Le Goff C, Somerville RP, Kesteloot F, Powell K, Birk DE, Colige AC, Apte SS (2006) Regulation of procollagen amino-propeptide processing during mouse embryogenesis by specialization of homologous ADAMTS proteases: insights on collagen biosynthesis and dermatosparaxis. Development 133:1587–1596. https://doi.org/10.1242/dev.02308

Le Goff CSR, Kesteloot F, Powell K, Birk DE, Colige AC, Apte SS (2007) Regulation of procollagen amino-propeptide processing during mouse embryogenesis by specialization of homologous ADAMTS proteases: insights on collagen biosynthesis and dermatosparaxis. Development 133:1587–1596

Leegwater PA et al (2016) Dwarfism with joint laxity in Friesian horses is associated with a splice site mutation in B4GALT7 BMC. Genomics 17:839. https://doi.org/10.1186/s12864-016-3186-0

Li SW et al (2001) Transgenic mice with inactive alleles for procollagen N-proteinase (ADAMTS-2) develop fragile skin and male sterility. Biochem J 355:271. https://doi.org/10.1042/0264-6021:3550271

Liu X, Wu H, Byrne M, Krane S, Jaenisch R (1997) Type III collagen is crucial for collagen I fibrillogenesis and for normal cardiovascular development. Proc Natl Acad Sci USA 94:1852–1856. https://doi.org/10.1073/pnas.94.5.1852

Liu YA et al (2020) Pathologic skull fracture in a near-term neonate with Arthrochalasia type Ehlers-Danlos syndrome: a case report. Fetal Pediatric pathol 1-6. https://doi.org/10.1080/15513815.2020.1753269

Lopez SG, Bonassar LJ (2022) The role of SLRPs and large aggregating proteoglycans in collagen fibrillogenesis, extracellular matrix assembly, and mechanical function of fibrocartilage. Connect Tissue Res 63:269–286. https://doi.org/10.1080/03008207.2021.1903887

Lund A, Joensen F, Christensen E, Duno M, Skovby F, Schwartz M (2008) A novel arginine-to-cysteine substitution in the triple helical region of the alpha1(I) collagen chain in a family with an osteogenesis imperfecta/Ehlers-Danlos phenotype. Clin Genet 73:97–101. https://doi.org/10.1111/j.1399-0004.2007.00926.x

Ma W, Goldberg J (2016) TANGO1/cTAGE5 receptor as a polyvalent template for assembly of large COPII coats. Proc Natl Acad Sci USA 113:10061–10066. https://doi.org/10.1073/pnas.1605916113

Macsai MS, Lemley HL, Schwartz T (2000) Management of oculus fragilis in Ehlers-Danlos type VI. Cornea 19:104–107. https://doi.org/10.1097/00003226-200001000-00020

Maehata Y et al (2007) Type III collagen is essential for growth acceleration of human osteoblastic cells by ascorbic acid 2-phosphate, a long-acting vitamin C derivative. Matrix Biol 26:371–381. https://doi.org/10.1016/j.matbio.2007.01.005

Malfait F, Vroman R, Colman M, Syx D (2021) The collagen superfamily and Collagenopathies. In: Biology of extracellular matrix, vol 8. Springer, Cham. https://doi.org/10.1007/978-3-030-67592-9_3

Malfait F, Castori M, Francomano CA, Giunta C, Kosho T, Byers PH (2020) The Ehlers-Danlos syndromes. Nat Rev Dis Primers 6:64. https://doi.org/10.1038/s41572-020-0194-9

Malfait F et al (2017) The 2017 international classification of the Ehlers-Danlos syndromes. Am J Med Genet C Semin Med Genet 175:8–26. https://doi.org/10.1002/ajmg.c.31552

Malfait F et al (2013a) Defective initiation of glycosaminoglycan synthesis due to B3GALT6 mutations causes a pleiotropic Ehlers-Danlos-syndrome-like connective tissue disorder. Am J Hum Genet 92:935–945. https://doi.org/10.1016/j.ajhg.2013.04.016

Malfait F, Symoens S, Coucke P, Nunes L, De Almeida S, De Paepe A (2006) Total absence of the alpha2(I) chain of collagen type I causes a rare form of Ehlers-Danlos syndrome with hypermobility and propensity to cardiac valvular problems. J Med Genet 43:e36. https://doi.org/10.1136/jmg.2005.038224

Malfait F et al (2007) Three arginine to cysteine substitutions in the pro-alpha (I)-collagen chain cause Ehlers-Danlos syndrome with a propensity to arterial rupture in early adulthood. Hum Mutat 28:387–395. https://doi.org/10.1002/humu.20455

Malfait F et al (2013b) Helical mutations in type I collagen that affect the processing of the amino-propeptide result in an osteogenesis imperfecta/Ehlers-Danlos syndrome overlap syndrome. Orphanet J Rare Dis 8:78. https://doi.org/10.1186/1750-1172-8-78

Manka SW, Bihan D, Farndale RW (2019) Structural studies of the MMP-3 interaction with triple-helical collagen introduce new roles for the enzyme in tissue remodelling. Sci Rep 9:18785. https://doi.org/10.1038/s41598-019-55266-9

Mattoteia D et al (2023) Identification of regulatory molecular "hot spots" for LH/PLOD collagen glycosyltransferase activity. Int J Mol Sci 24. https://doi.org/10.3390/ijms241311213

Mazziotti G et al (2016) High prevalence of radiological vertebral fractures in adult patients with Ehlers-Danlos syndrome. Bone 84:88–92. https://doi.org/10.1016/j.bone.2015.12.007

McCaughey J, Stephens DJ (2019) ER-to-Golgi transport: a sizeable problem. Trends Cell Biol 29:940–953. https://doi.org/10.1016/j.tcb.2019.08.007

Miedel EL et al (2015) Type III collagen modulates fracture callus bone formation and early remodeling. J Orthop Res 33:675–684. https://doi.org/10.1002/jor.22838

Minatogawa M et al (2022) Clinical and pathophysiological delineation of musculocontractural Ehlers-Danlos syndrome caused by dermatan sulfate epimerase deficiency (mcEDS-DSE): a detailed and comprehensive glycobiological and pathological investigation in a novel patient. Hum Mutat 43:1829–1836. https://doi.org/10.1002/humu.24437

Minatogawa M et al (2021) Clinical and molecular features of 66 patients with musculocontractural Ehlers-Danlos syndrome caused by pathogenic variants in CHST14 (mcEDS-CHST14). J Med Genet 59:865. https://doi.org/10.1136/jmedgenet-2020-107623
Montgomery NT, Zientek KD, Pokidysheva EN, Bachinger HP (2018) Post-translational modification of type IV collagen with 3-hydroxyproline affects its interactions with glycoprotein VI and nidogens 1 and 2. J Biol Chem 293:5987–5999. https://doi.org/10.1074/jbc.RA117.000406
Morello R et al (2006) CRTAP is required for prolyl 3- hydroxylation and mutations cause recessive osteogenesis imperfecta. Cell 127:291–304. https://doi.org/10.1016/j.cell.2006.08.039
Morlino S et al (2020) COL1-related overlap disorder: a novel connective tissue disorder incorporating the osteogenesis imperfecta/Ehlers-Danlos syndrome overlap. Clin Genet 97:396–406. https://doi.org/10.1111/cge.13683
Murthy AV et al (2022) Crystal structure of the collagen prolyl 4-hydroxylase (C-P4H) catalytic domain complexed with PDI: toward a model of the C-P4H alpha(2)beta(2) tetramer. J Biol Chem 298:102614. https://doi.org/10.1016/j.jbc.2022.102614
Musiime M, Chang J, Hansen U, Kadler KE, Zeltz C, Gullberg D (2021) Collagen assembly at the cell surface. Dogmas Revisited Cells 10. https://doi.org/10.3390/cells10030662
Nakajima M et al (2013) Mutations in B3GALT6, which encodes a glycosaminoglycan linker region enzyme, cause a spectrum of skeletal and connective tissue disorders. Am J Hum Genet 92:927–934. https://doi.org/10.1016/j.ajhg.2013.04.003
Nicholls AC et al (1984) The clinical features of homozygous alpha 2(I) collagen deficient osteogenesis imperfecta. J Med Genet 21:257–262. https://doi.org/10.1136/jmg.21.4.257
Nikpour M et al (2022) Glycosaminoglycan linkage region of urinary bikunin as a potentially useful biomarker for beta3GalT6-deficient spondylodysplastic Ehlers-Danlos syndrome. JIMD reports 63:462–467. https://doi.org/10.1002/jmd2.12311
Ogur G, Baykan N, De Paepe A, Steinmann B, Quatacker J, Kuseyri F, Yuksel-Apak M (1994) Clinical, ultrastructural and biochemical studies in two sibs with Ehlers-Danlos syndrome type VI-B-like features. Clin Genet 46:417–422. https://doi.org/10.1111/j.1399-0004.1994.tb04409.x
Ono T, Miyazaki T, Ishida Y, Uehata M, Nagata K (2012) Direct in vitro and in vivo evidence for interaction between Hsp47 protein and collagen triple helix. J Biol Chem 287:6810–6818. https://doi.org/10.1074/jbc.M111.280248
Pihlajaniemi T, Dickson LA, Pope FM, Korhonen VR, Nicholls A, Prockop DJ, Myers JC (1984) Osteogenesis imperfecta: cloning of a pro-alpha 2(I) collagen gene with a frameshift mutation. J Biol Chem 259:12941–12944
Prox J, Arnold P, Becker-Pauly C (2015) Meprin alpha and meprin beta: procollagen proteinases in health and disease. Matrix Biol 44-46:7–13. https://doi.org/10.1016/j.matbio.2015.01.010
Qiu R et al (2023) Cis-trans isomerization of peptoid residues in the collagen triple-helix. Nat Commun 14:7571. https://doi.org/10.1038/s41467-023-43469-8
Raote I et al (2018) TANGO1 builds a machine for collagen export by recruiting and spatially organizing COPII, tethers and membranes. Elife 7. https://doi.org/10.7554/eLife.32723
Raote I et al (2017) TANGO1 assembles into rings around COPII coats at ER exit sites. J Cell Biol 216:901–909. https://doi.org/10.1083/jcb.201608080
Raote I, Saxena S, Malhotra V (2023) Sorting and export of proteins at the endoplasmic reticulum. Cold Spring Harb Perspect Biol 15:15. https://doi.org/10.1101/cshperspect.a041258
Reddi AH, Gay R, Gay S, Miller EJ (1977) Transitions in collagen types during matrix-induced cartilage, bone, and bone marrow formation. Proc Natl Acad Sci USA 74:5589–5592. https://doi.org/10.1073/pnas.74.12.5589
Ricard-Blum S (2011) The collagen family. Cold Spring Harb Perspect Biol 3:a004978. https://doi.org/10.1101/cshperspect.a004978
Ritelli M, Venturini M, Cinquina V, Chiarelli N, Colombi M (2020) Multisystemic manifestations in a cohort of 75 classical Ehlers-Danlos syndrome patients: natural history and nosological perspectives. Orphanet J Rare Dis 15:197. https://doi.org/10.1186/s13023-020-01470-0

Rohrbach M et al (2011) Phenotypic variability of the kyphoscoliotic type of Ehlers-Danlos syndrome (EDS VIA): clinical, molecular and biochemical delineation. Orphanet J Rare Dis 6:46. https://doi.org/10.1186/1750-1172-6-46

Rolfes MC, Deyle DR, King KS, Hand JL, Graff AH, Derauf C (2019) Fracture incidence in Ehlers-Danlos syndrome–A population-based case-control study. Child Abuse Negl 91:95–101. https://doi.org/10.1016/j.chiabu.2019.03.008

Rolvien T, Kornak U, Linke SJ, Amling M, Oheim R (2020) Whole-exome sequencing identifies novel compound heterozygous ZNF469 mutations in two siblings with mild brittle cornea syndrome. Calcif Tissue Int 107:294–299. https://doi.org/10.1007/s00223-020-00721-3

Saito K et al (2009) TANGO1 facilitates cargo loading at endoplasmic reticulum exit sites. Cell 136:891. https://doi.org/10.1016/j.cell.2008.12.025

Salo AM, Myllyharju J (2021) Prolyl and lysyl hydroxylases in collagen synthesis. Exp Dermatol 30:38–49. https://doi.org/10.1111/exd.14197

Salo AM et al (2023) Collagen prolyl 4-hydroxylase isoenzymes I and II have sequence specificity towards different X-pro-Gly triplets. Matrix Biol 125:73. https://doi.org/10.1016/j.matbio.2023.12.001

Salter CG, Davies JH, Moon RJ, Fairhurst J, Bunyan D, DDD S, Foulds N (2016) Further defining the phenotypic spectrum of B4GALT7 mutations. Am J Med Genet A 170:1556–1563. https://doi.org/10.1002/ajmg.a.37604

Schissel SL, Dunsmore SE, Liu X, Shine RW, Perrella MA, Layne MD (2009) Aortic carboxypeptidase-like protein is expressed in fibrotic human lung and its absence protects against bleomycin-induced lung fibrosis. Am J Pathol 174:818–828. https://doi.org/10.2353/ajpath.2009.080856

Schwarze U, Hata R, McKusick VA, Shinkai H, Hoyme HE, Pyeritz RE, Byers PH (2004) Rare autosomal recessive cardiac valvular form of Ehlers-Danlos syndrome results from mutations in the COL1A2 gene that activate the nonsense-mediated RNA decay pathway. Am J Hum Genet 74:917–930. https://doi.org/10.1086/420794

Seidler DG et al (2006) Defective glycosylation of decorin and biglycan, altered collagen structure, and abnormal phenotype of the skin fibroblasts of an Ehlers-Danlos syndrome patient carrying the novel Arg270Cys substitution in galactosyltransferase I (beta4GalT-7). J Mol Med (Berl) 84:583–594. https://doi.org/10.1007/s00109-006-0046-4

Shaw LM, Olsen BR (1991) FACIT collagens: diverse molecular bridges in extracellular matrices. Trends Biochem Sci 16:191–194. https://doi.org/10.1016/0968-0004(91)90074-6

Silver MH, Foidart JM, Pratt RM (1981) Distribution of fibronectin and collagen during mouse limb and palate development. Differentiation 18:141–149. https://doi.org/10.1111/j.1432-0436.1981.tb01115.x

Steinmann BRP, Superti-Furga A (2002) The Ehlers–Danlos syndrome. In: Royce PMSB (ed) Connective tissue and its heritable disorders, 2nd edn. Wiley-Liss, New York

Sun M et al (2020) Collagen XI regulates the acquisition of collagen fibril structure, organization and functional properties in tendon. Matrix Biol 94:77. https://doi.org/10.1016/j.matbio.2020.09.001

Syx D et al (2022) Alterations in glycosaminoglycan biosynthesis associated with the Ehlers-Danlos syndromes. Am J Physiol Cell Physiol 323:C1843–C1859. https://doi.org/10.1152/ajpcell.00127.2022

Taga Y, Kusubata M, Ogawa-Goto K, Hattori S (2016) Developmental stage-dependent regulation of prolyl 3-hydroxylation in tendon type I collagen. J Biol Chem 291:837–847. https://doi.org/10.1074/jbc.M115.686105

Takaluoma K et al (2007) Tissue-specific changes in the hydroxylysine content and cross-links of collagens and alterations in fibril morphology in lysyl hydroxylase 1 knock-out mice. J Biol Chem 282:6588–6596. https://doi.org/10.1074/jbc.M608830200

Takeda R et al (2022) Clinical and molecular features of patients with COL1-related disorders: Implications for the wider spectrum and the risk of vascular complications. Am J Med Genet A 188:2560–2575. https://doi.org/10.1002/ajmg.a.62887

Tarnutzer K, Siva Sankar D, Dengjel J, Ewald CY (2023) Collagen constitutes about 12% in females and 17% in males of the total protein in mice. Sci Rep 13:4490. https://doi.org/10.1038/s41598-023-31566-z
Terajima M et al (2019) Cyclophilin B control of lysine post-translational modifications of skin type I collagen. PLoS Genet 15:e1008196. https://doi.org/10.1371/journal.pgen.1008196
Terajima M et al (2016) Cyclophilin-B modulates collagen cross-linking by differentially affecting lysine hydroxylation in the helical and Telopeptidyl domains of tendon type I collagen. J Biol Chem 291:9501–9512. https://doi.org/10.1074/jbc.M115.699470
Terajima M et al (2022) Lysyl hydroxylase 2 mediated collagen post-translational modifications and functional outcomes. Sci Rep 12:14256. https://doi.org/10.1038/s41598-022-18165-0
Teratani T et al (2018) Aortic carboxypeptidase-like protein, a WNT ligand, exacerbates nonalcoholic steatohepatitis. J Clin Invest 128:1581–1596. https://doi.org/10.1172/JCI92863
Ticho U, Ivry M, Merin S (1980) Brittle cornea, blue sclera, and red hair syndrome (the brittle cornea syndrome). Br J Ophthalmol 64:175–177. https://doi.org/10.1136/bjo.64.3.175
Tinkle B, Castori M, Berglund B, Cohen H, Grahame R, Kazkaz H, Levy H (2017) Hypermobile Ehlers-Danlos syndrome (a.k.a. Ehlers-Danlos syndrome type III and Ehlers-Danlos syndrome hypermobility type): clinical description and natural history. Am J Med Genet C Semin Med Genet 175:48–69. https://doi.org/10.1002/ajmg.c.31538
Tumelty KE, Smith BD, Nugent MA, Layne MD (2014) Aortic carboxypeptidase-like protein (ACLP) enhances lung myofibroblast differentiation through transforming growth factor beta receptor-dependent and -independent pathways. J Biol Chem 289:2526–2536. https://doi.org/10.1074/jbc.M113.502617
Valcourt U, Alcaraz LB, Exposito JY, Lethias C, Bartholin L (2015) Tenascin-X: beyond the architectural function. Cell Adh Migr 9:154–165. https://doi.org/10.4161/19336918.2014.994893
Valli M et al (2012) Deficiency of CRTAP in non-lethal recessive osteogenesis imperfecta reduces collagen deposition into matrix. Clin Genet 82:453–459. https://doi.org/10.1111/j.1399-0004.2011.01794.x
Van Damme T et al (2016) Expanding the clinical and mutational spectrum of the Ehlers-Danlos syndrome, dermatosparaxis type. Genet Med 18:882–891. https://doi.org/10.1038/gim.2015.188
Van Damme T, Colman M, Syx D, Malfait F (2022) The Ehlers-Danlos Syndromes against the Backdrop of Inborn Errors of Metabolism. Genes 13. https://doi.org/10.3390/genes13020265
Van Damme T et al (2018) Biallelic B3GALT6 mutations cause spondylodysplastic Ehlers-Danlos syndrome. Hum Mol Genet 27:3475–3487. https://doi.org/10.1093/hmg/ddy234
Volk SW et al (2014) Type III collagen regulates osteoblastogenesis and the quantity of trabecular bone. Calcif Tissue Int 94:621–631. https://doi.org/10.1007/s00223-014-9843-x
Vranka JA, Sakai LY, Bachinger HP (2004) Prolyl 3-hydroxylase 1, enzyme characterization and identification of a novel family of enzymes. J Biol Chem 279:23615–23621. https://doi.org/10.1074/jbc.M312807200
Wenstrup RJ, Florer JB, Brunskill EW, Bell SM, Chervoneva I, Birk DE (2004) Type V collagen controls the initiation of collagen fibril assembly. J Biol Chem 279:53331–53337. https://doi.org/10.1074/jbc.M409622200
Wenstrup RJ et al (2006) Murine model of the Ehlers-Danlos syndrome. col5a1 haploinsufficiency disrupts collagen fibril assembly at multiple stages. J Biol Chem 281:12888–12895. https://doi.org/10.1074/jbc.M511528200
Wenstrup RJ et al (2011) Regulation of collagen fibril nucleation and initial fibril assembly involves coordinate interactions with collagens V and XI in developing tendon. J Biol Chem 286:20455–20465. https://doi.org/10.1074/jbc.M111.223693
Widmer C et al (2012) Molecular basis for the action of the collagen-specific chaperone Hsp47/SERPINH1 and its structure-specific client recognition. Proc Natl Acad Sci USA 109:13243–13247. https://doi.org/10.1073/pnas.1208072109

Wilhelm D, Wurtz A, Abouelfarah H, Sanchez G, Bui C, Vincourt JB (2023) Tissue-specific collagen hydroxylation at GEP/GDP triplets mediated by P4HA2. Matrix Biol 119:141–153. https://doi.org/10.1016/j.matbio.2023.03.009

Xiao G, Wan Z, Fan Q, Tang X, Zhou B (2014) The metal transporter ZIP13 supplies iron into the secretory pathway in Drosophila melanogaster. eLife 3:e03191. https://doi.org/10.7554/eLife.03191

Yamauchi M, Shiiba M (2008) Lysine hydroxylation and cross-linking of collagen. Methods Mol Biol 446:95–108. https://doi.org/10.1007/978-1-60327-084-7_7

Yamauchi M, Sricholpech M, Terajima M, Tomer KB, Perdivara I (2019a) Glycosylation of type I collagen. Methods Mol Biol 1934:127–144. https://doi.org/10.1007/978-1-4939-9055-9_9

Yamauchi M, Terajima M, Shiiba M (2019b) Lysine hydroxylation and cross-linking of collagen. Methods Mol Biol 1934:309–324. https://doi.org/10.1007/978-1-4939-9055-9_19

Yasukawa S, Uehara M, Suzuki T, Nakano M, Kosho T, Nakamura Y, Takahashi J (2021) The first experience of denosumab therapy on patients with Ehlers-Danlos syndrome and osteoporosis: detailed observation of two patients. Mod Rheumatol Case Rep 5:377–383. https://doi.org/10.1080/24725625.2020.1799493

Yen JL, Lin SP, Chen MR, Niu DM (2006) Clinical features of Ehlers-Danlos syndrome. J Formos Med Assoc 105:475–480. https://doi.org/10.1016/S0929-6646(09)60187-X

Zhong Z, Ethen NJ, Williams BO (2014) WNT signaling in bone development and homeostasis. Wiley Interdiscip Rev Dev Biol 3:489–500. https://doi.org/10.1002/wdev.159

Zweers MCBJ, Steijlen PM, Dean WB, Hamel BC, Otero M, Kucharekova M, Boezeman JB, Schalkwijk J (2003) Haploinsufficiency of TNXB is associated with hypermobility type of Ehlers-Danlos syndrome. Am J Hum Genet 73:214–217

Chapter 4
Collagen II Mutations in Inherited Cartilage Disease: Our Current Understanding of Genotype-Phenotype Correlations

John F. Bateman and Shireen R. Lamandé

Abstract Cartilage contains chondrocytes embedded in a complex extracellular matrix rich in collagen II fibrils that provide the structural framework for other cartilage components such as aggrecan, matrilins 1 and 3, COMP, decorin, fibromodulin, and minor collagens type IX and XI. The central role of the collagen II fibrils is reflected in the large number of mutations in the gene encoding collagen II (*COL2A1*), that cause chondrodysplasias, ranging from premature arthritis through to severe early lethal disorders. Mutations that reduce synthesis of structurally normal protein cause Stickler syndrome, type I, at the milder end of the *COL2A1* chondrodysplasia clinical spectrum. Structural mutations most often cause more severe phenotypes, achondrogenesis/hypochondrogenesis, due to their impacts on procollagen II folding, stability, secretion, and in some cases altered interactions. These are commonly missense mutations, predominantly substitutions for glycine residues in the triple helix. Arginine to cysteine mutations in the helix are generally less severe than glycine substitutions, although p.R989C is a notable exception, causing a severe spondyloepiphyseal dysplasia. How mutant procollagen II engages with the intracellular ER quality control machinery, and if the unfolded/misfolded procollagen elicits an unfolded protein response, is not resolved and awaits further experimentation. Restoring proteostasis by stimulating mutant protein degradation, improving mutant protein folding with chemical chaperones, or modulating

J. F. Bateman
Murdoch Children's Research Institute and Department of Paediatrics, University of Melbourne, Melbourne, VIC, Australia

S. R. Lamandé (✉)
Murdoch Children's Research Institute and Department of Paediatrics, University of Melbourne, Melbourne, VIC, Australia

The Novo Nordisk Foundation Center for Stem Cell Medicine (reNEW), Murdoch Children's Research Institute, Parkville, VIC, Australia
e-mail: shireen.lamande@mcri.edu.au

A. Rossi, F. Zaucke (eds.), *The Extracellular Matrix in Genetic Skeletal Disorders*, Biology of Extracellular Matrix 16,
https://doi.org/10.1007/978-3-031-70835-0_4

unfolded protein response signaling with small molecules offers hope for treating genetic collagen II disorders.

4.1 Introduction

Cartilage formation is the key to endochondral bone formation. Mutations in the many genes affecting the complex pathways of cartilage formation, integrity, and transition to bone via endochondral ossification result in a large group of genetic skeletal disorders. In the 2023 Nosology of Genetic Skeletal Disorders (Unger et al. 2023), 771 clinically defined disorders caused by mutations in 552 genes were described. Cartilage contains chondrocytes embedded within a complex extracellular matrix (ECM) rich in highly organized collagen II fibrils that provide the structural framework for other cartilage components, such as aggrecan, and cartilage-specific interacting components such as matrilins 1 and 3, COMP, decorin, fibromodulin, and minor collagens including type IX and XI. All of these, and other developmentally regulated ECM components, contribute to cartilage structural integrity at all stages of cartilage development and endochondral bone formation and to the resistance to compressive load essential for articular cartilage function. For a detailed overview of the cartilage ECM, see Gentili and Cancedda (2009), Heinegård (2009), Müller et al. (2014), and Wilson et al. (2010). The central role of the cross-linked collagen II fibrillar suprastructure is reflected in the large number of mutations in the collagen II gene (*COL2A1*) that cause inherited cartilage disorders (chondrodysplasias). This article will review the spectrum of *COL2A1* mutations that cause human chondrodysplasias and explore our current knowledge about how these mutations cause cartilage pathology.

4.2 Classification of *COL2A1* Disorders

COL2A1 disorders (Group 2 (Unger et al. 2023)) can be grouped according to their most common clinical severity into those that are 1) lethal, 2) nonlethal with short stature, and 3) nonlethal with normal stature (Table 4.1). Lethal forms include achondrogenesis type 2 (OMIM 200610, AGC2, includes hypochondrogenesis) and platyspondylic dysplasia, Torrance type (OMIM 151210, PLSDT). Nonlethal with short stature forms include spondyloepiphyseal dysplasia congenita (OMIM 183900, SEDC), Kniest dysplasia (OMIM 156550, Kniest), spondyloepimetaphyseal dysplasia, Strudwick type (OMIM 184250, SEMDSTWK), and spondyloperipheral dysplasia (OMIM 121700). Nonlethal with normal stature forms include Stickler syndrome, type I (OMIM 108300, STL1), SED with metatarsal shortening (Czech dysplasia) (OMIM 609162), avascular necrosis of the femoral head (dysplasia of the proximal femoral epiphyses) (OMIM 608805, ANFH1), and Legg-Calve-Perthes disease (OMIM 150600, LCPD).

Table 4.1 Collagen II disorders

Lethal	Inheritance	OMIM abbreviation; no.
Achondrogenesis type 2 (hypochondrogenesis included)	AD	ACG2; 200610
Platyspondylic dysplasia, Torrance type	AD	PLSDT; 151210
Nonlethal with short stature		
Spondyloepiphyseal dysplasia	AD	SEDC; 183900
Kniest dysplasia	AD	Kniest; 156550
Spondyloepimetaphyseal dysplasia, Strudwick type (spondyloepiphyseal dysplasia with marked metaphyseal changes)	AD	SEMDSTWK; 184250
Spondyloperipheral dysplasia	AD	271700
Nonlethal with normal stature		
Stickler syndrome, type I	AD	STL1; 108300
Spondyloepiphyseal dysplasia with metatarsal shortening (formerly Czech dysplasia)	AD	Czech dysplasia; 609162
Spondyloepiphyseal dysplasia, Stanescu type	AD	SEDSTN; 616583
Avascular necrosis of the femoral head	AD	ANFH1; 608805
Legg-Calve-Perthes disease	AD	LCPD; 150600

Table 4.2 Approximate relative frequency

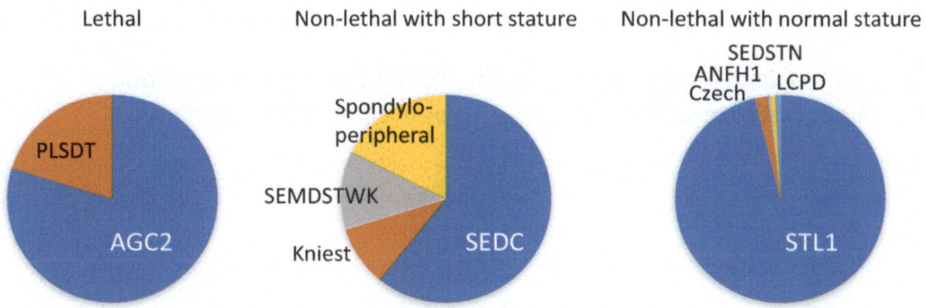

We interrogated the Leiden Open Variation Database v3.0 (LOVD) (https://www.lovd.nl/) and checked designated "pathogenic" or "likely pathogenic" variants against their published clinical phenotypes to generate the approximate relative frequency (number of individual patients) for each disorder within the groups of *COL2A1* disorders (Table 4.2). This revealed that achondrogenesis type 2 (including hypochondrogenesis) is the most common lethal *COL2A1* chondrodysplasia. In the nonlethal, short stature forms, spondyloepiphyseal dysplasia predominates, and almost all the nonlethal with normal stature individuals have Stickler syndrome, Type 1.

4.3 The Spectrum of Reported Pathogenic COL2A1 Mutations

Interrogating LOVD for unique "publicly reported" pathogenic *COL2A1* variants (as defined by LOVD) reveals 578 DNA variants (Fig. 4.1a). These are most commonly substitution mutations (67%) and deletions (24%). These DNA variants result in 520 unique pathogenic collagen II protein variants (Fig. 4.1b). At the protein level, these are most commonly missense amino acid changes (53%), frameshift changes (30%), and premature stop codons (10%). It is important to note that most of the frameshift changes will also generate premature stop codons. Thus, premature termination mutations and missense changes are, by far, the predominant pathogenic changes at the protein level. In this review, we will focus on the molecular pathology of these two protein mutation groups. Further useful information about the spectrum of *COL2A1* mutations and their correlation with clinical phenotype can be found elsewhere (Barat-Houari et al. 2016; Zhang et al. 2020). In Fig. 4.2, we have graphically summarized all the LOVD pathogenic mutations, showing mutation type, location in the collagen II protein structure, and clinical severity. A detailed critical evaluation of the evidence on how these mutations impact collagen biosynthesis, assembly, and function follows, but as background we first present some brief information on collagen biosynthesis.

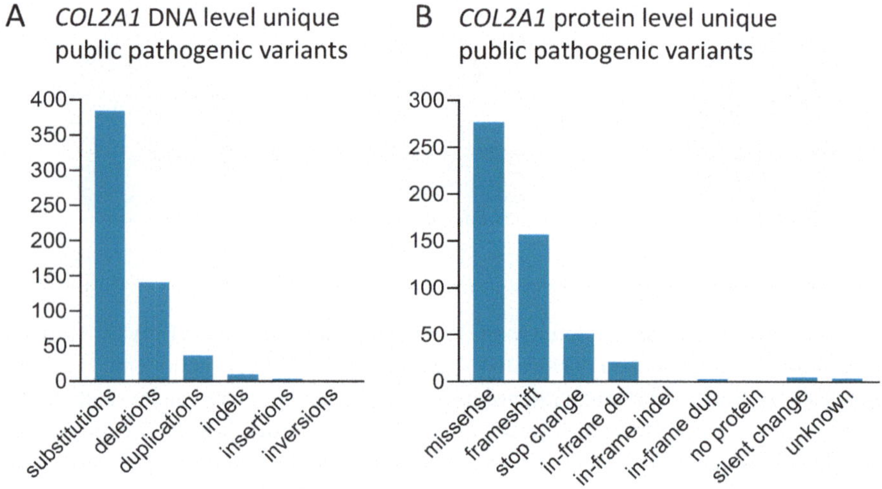

Fig. 4.1 *COL2A1 DNA and protein variants*. Variants designated as "pathogenic" in the LOVD database (May 26, 2023)

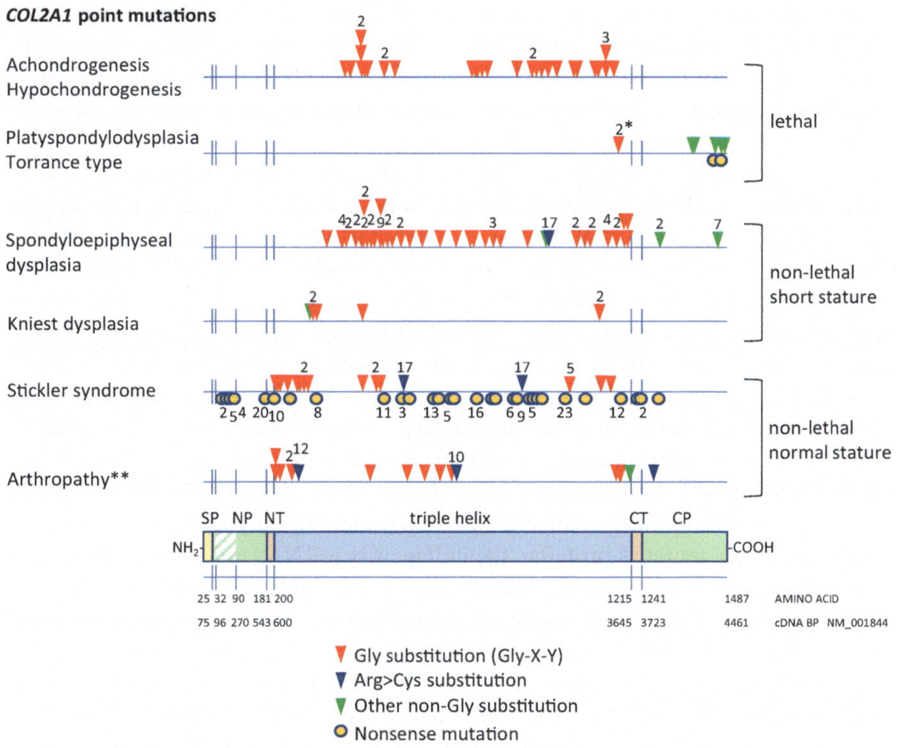

Fig. 4.2 *Position, mutation type, and clinical severity of COL2A1 point mutations*. Point mutations reported in LOVD where the clinical phenotype is described in PubMed-cited publications. Numbers above the designated mutations indicate the number of individuals with the identical mutation in LOVD. Stacked symbols designate different mutations at the same position. *In addition to the missense mutations shown here, frameshift mutations resulting in a premature stop codon also cause platyspondylodysplasia, Torrance type. **The arthropathy category also includes cases of avascular necrosis of the femoral head and Legg-Calves-Perthes disease. Abbreviations for protein domains; *SP* signal peptide; *NP* N-propeptide; *NT* N-telopeptide; *CT* C-telopeptide; *CP* C-propeptide

4.4 Collagen Biosynthesis

Detailed reviews of collagen structure and biosynthesis can be found elsewhere (Canty and Kadler 2005; Ito and Nagata 2021; Myllyharju and Kivirikko 2004; Ruggiero 2021) and will not be presented here; our focus will be on those steps that are vulnerable to *COL2A1* mutations and cause disease. *COL2A1* is a large (>38.5 kb) gene containing 54 exons coding for the collagen proα1(II)-chain protein monomer. Like the genes for other fibril-forming collagens, most exons coding for the central triple-helical region (exons 9–50) are small, commonly 54 or 108 bp. The in-frame arrangement of triple helix coding exons results in exon-skipping mutations maintaining an open reading frame. Exons coding for the N-terminal and

C-terminal protein domains are also mostly in-frame, and only skipping exons 3, 51, or 52 will produce an out-of-frame protein. After pre-mRNA splicing, the mature collagen II mRNA is 4.4 kb and encodes a proα1(II)-chain with a 25-amino acid signal sequence, an N-terminal domain (156 amino acids), an adjoining N-telopeptide (19 amino acids), the triple helix-forming domain (1014 amino acids), a C-terminal telopeptide (27 amino acids), and C-terminal propeptide (246 amino acids). The defining feature of collagen II (and other collagens) is the central region where three pro-α chains form a collagen triple helix. This unique collagen triple helix structure has a strict repetitive sequence where every third amino acid is a glycine such that the helix motif is $(Gly\text{-}X\text{-}Y)_{338}$ where X and Y represent amino acids other than Gly. Glycine is the smallest amino acid, and its occurrence in every third position is critical to fit in a sterically restricted space in the center of the triple helix. The importance of Gly at every third position of the helix for structural integrity is highlighted by the large number of Gly substitutions by bulkier amino acids in *COL2A1* chondrodysplasias, commonly resulting in more severe clinical outcomes than X and Y substitutions. X- and Y-position amino acids are commonly proline and hydroxyproline, which provide additional rigidity to the triple helix.

As with other secreted proteins, the collagen II mRNAs are translated by ribosomes on the rough endoplasmic reticulum (rER), and the free N-terminal ends are extruded into the rER cisternae. Signal peptides at the N-termini are cleaved, and posttranslational modifications such as hydroxylation of prolines in the Y-position of the repeating Gly-X-Y sequence by prolyl hydroxylase are initiated. Likewise, lysine residues in the Y-position are hydroxylated by a lysyl hydroxylase. Hydroxylysine can be further modified by enzymatic glycosylation with galactose or with galactose and glucose. A large mannose-rich oligosaccharide is assembled on the C-terminal propeptide of the proα1(II) chain. The next crucial step is the assembly of proα1(II) chain monomers into the trimers. This is primarily driven by interactions between the C-terminal propeptides which direct the correct chain registration required for subsequent triple helix formation. Emphasizing the importance of this step, mutations that interfere with propeptide sequences cause the lethal chondrodysplasia, platyspondylic dysplasia, Torrance type.

The characteristic component of the collagen molecule, the triple helix, then propagates from nucleated sequences at the C-terminus in a zipper-like manner toward the N-terminus. As the triple helix is the key structural component of collagen II, helix mutations that impair formation or compromise structural stability are a major cause of chondrodysplasias. While triple helix folding is driven primarily by the favorable energetics of the unique amino acid sequences within this domain, and by proline *cis-trans* isomerization, our simplistic preceding overview has not considered the many other enzymes, chaperones, and assembly factors that contribute to correct formation of the mature intracellular procollagen. Readers are referred to an excellent review on the complex "molecular ensemble" that participates in forming mature procollagen in the rER (Ishikawa and Bächinger 2013). It is important to appreciate that mutations in many of these components also cause downstream collagen defects, best demonstrated for collagen I in osteogenesis imperfecta (Jovanovic et al. 2022).

The fully folded procollagen II in the ER then transits the Golgi where the asparagine-linked oligosaccharides in the C-propeptides are further processed. Because the procollagens are large, with rigid helices over 400 nm in length, secretion mechanisms have been problematic to resolve compared to other smaller secretory cargoes. Recently, a model has been proposed involving a specific transmembrane protein, TANGO1, that allows COPII-mediated transport of these bulky proteins (Malhotra and Erlmann 2015). After secretion, two specific proteinases, ADAMTS2 and BMP1, cleave the N-propeptides and C-propeptides, respectively, to process procollagen to collagen. See Canty and Kadler (2005) for further details. The processed collagen, which contains the triple helix and small N- and C- terminal telopeptides, self-assembles into collagen fibrils where the individual collagen monomers are packed axially in a precisely staggered arrangement (Antipova and Orgel 2010) and the fibrils are then further strengthened by cross-linking reactions that form covalent bonds between α chains in adjacent molecules (Eyre 1991). The extracellular collagen II fibrils assemble, not as pure collagen II fibrils, but with a central core of another fibrillar collagen, collagen XI, and with the fibril-associated collagen IX on the outer fibril face (Eyre 1991). Collagen IX has an interrupted helix structure such that flexible collagen IX helix "arms" protrude from the fibril surface providing further sites for interaction with neighboring ECM components. Many other cartilage ECM proteins such as COMP, matrilin 1, and matrilin 3 (among others) interact with the collagen II/XI/IX fibrils and with each other and other ECM components such as aggrecan to produce the mechanical properties critical for cartilage structure and function (Heinegård 2009). Mutations in many of these components (e.g., collagen IX and XI, COMP, matrilin 3, and aggrecan) consequently cause various chondrodysplasias (Dennis et al. 2021).

The preceding superficial summary of collagen biosynthesis is to provide a basis for understanding the molecular pathology of *COL2A1* mutations. Readers should refer to the detailed reviews cited above for a more nuanced understanding of these complex pathways. An overview of the main steps in collagen biosynthesis is provided in Fig. 4.3.

4.5 Mutations that Reduce Collagen II Synthesis

While there are rare mutations, such as large deletions, that prevent *COL2A1* expression, mutations that introduce premature stop codons either via point mutations or frameshifting indels are the common cause of *COL2A1* heterozygous loss of function (Fig. 4.1b). These premature termination mutations result in nonsense-mediated mRNA decay (NMD) of the mutant mRNA. NMD is an evolutionarily conserved mRNA surveillance system that degrades premature stop codon-containing transcripts which would otherwise result in the synthesis of truncated proteins with potentially catastrophic pathophysiological effects. NMD is complex, with several alternative mechanistic models (see review (Fang et al. 2013)), although for most genes, including *COL2A1*, the prevailing model is the "exon junction complex

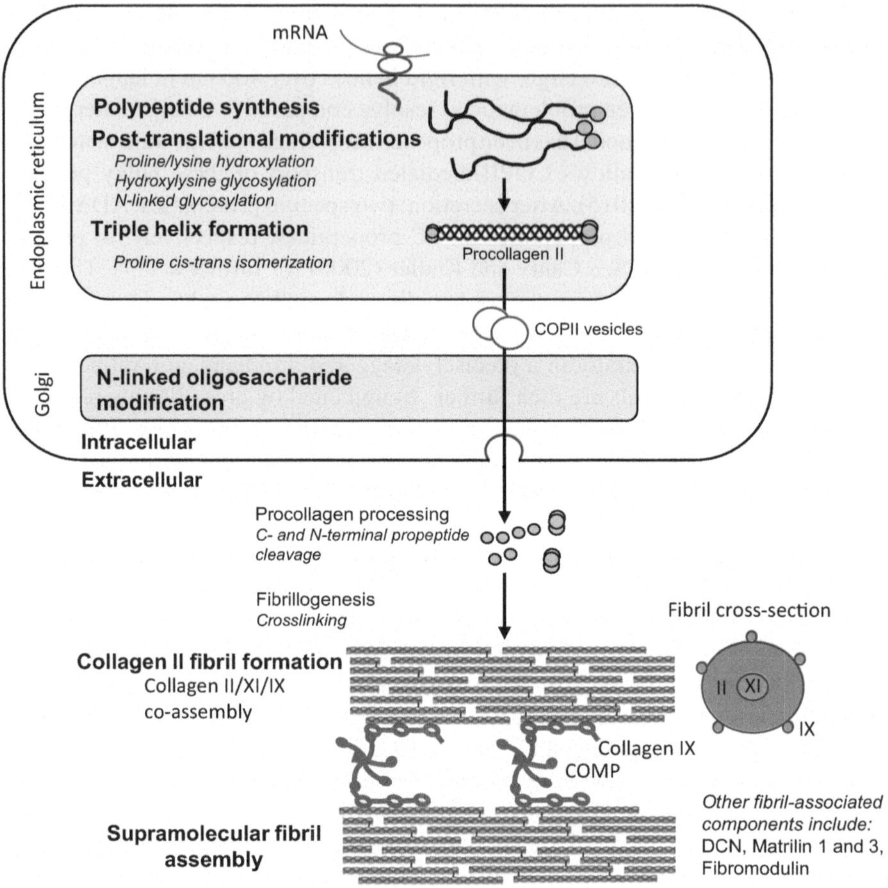

Fig. 4.3 *Overview of the key steps in collagen II biosynthesis and supramolecular assembly.* The collagen II precursor chains are co-translationally translocated into the endoplasmic reticulum (ER) lumen where posttranslational modifications occur. The three proα1(II)-chains associate via their C-terminal domains to form homotrimers. The helical collagens are trafficked via the Golgi network to the plasma membrane and secreted into the extracellular space. The N- and C-terminal propeptide domains are proteolytically removed, and the mature collagen II trimers are assembled into dense covalently cross-linked fibrils with a characteristic D-periodicity. Collagen XI co-assembles with collagen II, forming a collagen XI "core" within the fibril and collagen IX interacts with the fibril surface. COMP homo-pentamers can interact with and facilitate collagen II fibril formation and interactions with other extracellular matrix (ECM) components, including collagen IX, matrilins, and aggrecan

(EJC)-dependent model." In this model, the multi-protein EJC is deposited approximately 20 to 24 nucleotides upstream of each exon-exon junction during pre-mRNA splicing. During an initial "pioneer" round of translation, the translating ribosome displaces the EJCs until the correct termination codon is reached. If a premature stop codon is recognized during this process, the EJCs that are downstream of the PTC recruit NMD factors and the mRNA containing the PTC will be targeted for degradation. Thus, PTC mutations in the final exon (containing the correct termination signal) do not elicit NMD since there are no EJCs downstream that would identify the PTC mutations as premature.

Premature stop codon mutations are the most common cause of Stickler syndrome, type I, which is characterized by ocular, auditory, orofacial, and skeletal abnormalities including mild spondyloepiphyseal dysplasia and early onset osteoarthritis (Fig. 4.2). In vitro transcription and translation of Stickler patient heterozygous mutations confirmed that the Stickler heterozygous premature termination mutations in *COL2A1* result in NMD and collagen II haploinsufficiency (Freddi et al. 2000). These studies demonstrated that the PTC-containing mutant allele mRNA was degraded and could only be detected if NMD was inhibited with cycloheximide. Consistent with PTC-induced NMD as a major cause of Stickler syndrome, there are no Stickler PTC mutations in the final *COL2A1* exon (Fig. 4.2) where NMD would not occur. PTC mutations in the last *COL2A1* exon are predicted to result in the synthesis of a truncated proα1(II) with serious dominant negative impacts on procollagen folding. PTC mutations in the final exon, exon 54, in the lethal platyspondylic dysplasia, Torrance type, are consistent with this model (Fig. 4.2).

4.6 Collagen II Structural Mutations

Mutations that affect collagen II structure cause cartilage disorders with severity varying from mild to severe and lethal conditions. *COL2A1* missense mutations impact cartilage pathology according to their effect on collagen structure, folding, secretion, and ability to form functional fibrils in the ECM. Below we discuss these in three broad groups: (1) mutations in the C-propeptide that are thought to impair initial proα1(II) trimer assembly, (2) missense mutations that substitute Gly residues within the triple helix, and (3) other collagen II helix mutations.

4.6.1 Mutations in the Collagen II C-Propeptide

Missense mutations in the collagen II C-propeptide cause lethal and severe nonlethal chondrodysplasias with short stature, including platyspondylic skeletal dysplasia, Torrance type, and spondyloepiphyseal dysplasia congenita (Fig. 4.2). Nonsense and frameshift mutations in the most 3′ exon coding for this domain causing some cases of platyspondylic dysplasia, Torrance type, and spondyloperipheral dysplasia are predicted to produce truncated and misfolded C-propeptides. However, other nonsense and frameshift mutations in the other exons coding for the C-propeptide, exons 51–53, cause the milder chondrodysplasia, Stickler syndrome type I (Fig. 4.2).

These stop codon-introducing mutations lead to collagen II haploinsufficiency and are not discussed further in this section (see previous section).

There is no direct experimental evidence that shows how missense mutations in this critical region of procollagen II necessary for proα-chain assembly and trimerization impact the protein, but the assumption is that these structural mutations impair or prevent trimeric assembly. This mechanistic assumption is largely based on mutations in the homologous C-propeptide domains of collagen I and collagen X which have been studied in vitro. Several type I collagen C-propeptide structural mutations either prevent or impede proα1(I) chain association (Bateman et al. 1989; Lamandé et al. 1995) and cause lethal perinatal osteogenesis imperfecta. Using the crystal structure of the proα1(III) C-propeptide as a tool to interrogate collagen I osteogenesis imperfecta, C-propeptide mutations (Symoens et al. 2014) predicted that, in general, proα1(I) C-propeptide defects that permit some chain association but severely interfere with procollagen assembly and secretion result in more severe phenotypes than those that completely inhibit chain association. The collagen X NC1 C-terminal domain is critical for the initial association of the chains into the trimeric collagen X structure. Mutations in this domain prevent collagen X trimerization in vitro (Chan et al. 1995b, 1996, 1999) and cause metaphyseal chondrodysplasia, type Schmid (MCDS), in patients (Bateman et al. 2005) and in several mouse models (Rajpar et al. 2009; Tsang et al. 2010). These studies provided the basis for the conclusion that collagen X NC1 domain misfolding resulted in ER stress, and the unfolded protein response (UPR) and the downstream consequences of this caused the MCDS pathology (Cameron et al. 2011; Rajpar et al. 2009; Tsang et al. 2010; Wilson et al. 2005). The possible role of the UPR in the pathogenesis of mutations in other collagen types has been recently reviewed (Bateman et al. 2022). We discuss what is known about *COL2A1* structural mutations and the involvement of the UPR machinery in more detail in the following sections.

Several mouse models have provided the opportunity to explore aspects of the pathophysiology of procollagen II C-propeptide mutations. Cartilage from mouse models of platyspondylic dysplasia, Torrance type, with *Col2a1* p.Y1391S (Kimura et al. 2015) and p.D1469A (Furuichi et al. 2011) mutations in *Col2a1* showed intracellular retention of procollagen II in the ER and upregulation of ER stress markers (*discussed later*), strongly suggesting procollagen misfolding. Two other mouse *Col2a1* C-propeptide mutation chondrodysplasia models, p.S1386P (*Lpk/+*) (Esapa et al. 2012) and a 3 bp deletion in the C-propeptide sequence (*Dmm/+*) (Pace et al. 1997; Ricks et al. 2013), show mutant procollagen II retention with dilated ER, strongly suggesting the mutations impair initial procollagen II assembly.

4.6.2 Glycine Substitutions in the Collagen II Triple Helix

The most common structural mutations in *COL2A1* are missense mutations in the collagen II triple helical domain causing, in general, the more serious clinical phenotypes (Fig. 4.2). Of these, heterozygous mutations that substitute other amino

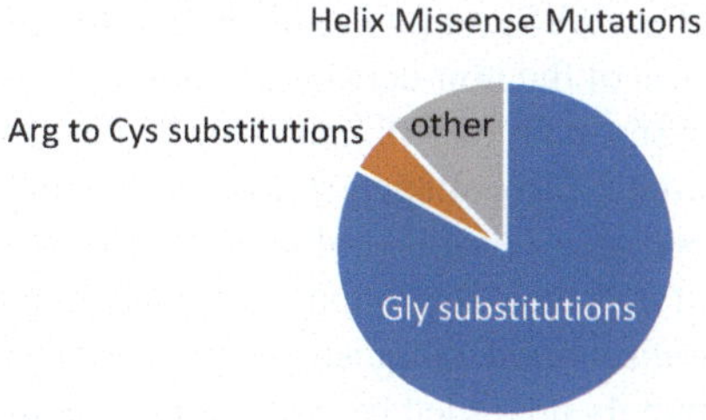

Fig. 4.4 *Collagen II triple helix point mutations.* Missense variants designated as "pathogenic" in the LOVD database (May 26, 2023)

acids for glycine in the helical Gly-X-Y sequence are the most common and most well studied (Fig. 4.4). As discussed earlier, Gly in every third position of the repetitive Gly-X-Y triple sequence is critical for correct triple helix folding and stability; thus, it is unsurprising that Gly substitutions have a major impact on collagen II folding and stability. Amino acid substitutions in the X or Y position can also cause chondrodysplasias (Fig. 4.4), and the only biochemically well-studied of these, Arg to Cys mutations, are discussed in the next section.

Glycine substitution mutations generally cause the more severe chondrodysplasia clinical phenotypes such as achondrogenesis type II (including hypochondrogenesis), spondyloepiphyseal dysplasia congenita, and Kniest dysplasia (Fig. 4.2). There is extensive experimental data on *COL1A1* glycine mutations causing osteogenesis imperfecta, demonstrating that helical glycine substitutions can impair helix formation and destabilize the helix, and as result, the mutant collagen trimers have increased posttranslational modifications and reduced secretion (Bateman et al. 2009; Forlino and Marini 2016; Jovanovic et al. 2022). *COL2A1* helix glycine mutations follow this same general pattern, although there is not as much detailed biochemical information as there is for collagen I glycine mutations. Collagen II is a homo-trimer of three collagen II α1(II)-chains, and thus heterozygous mutant allele expression will result in 7/8 of the collagen II protein trimers containing at least one mutant α1(II)-chain, potentially amplifying the impact compared to mutations in the collagen I heterotrimer.

In patient cartilage or isolated chondrocyte biochemical studies, several achondrogenesis/hypochondrogenesis mutations show impaired collagen II secretion and increased posttranslational modification indicative of delayed helix assembly. These include p.G510D (Bonaventure et al. 1995), p.G774S (Horton et al. 1992), p.G804A (Freisinger et al. 1994), p.G891R (Mortier et al. 1995), p.G1005S (Bonaventure et al. 1995), and p.G1143S (Godfrey and Hollister 1988; Vissing et al. 1989). In some of these, helix melting temperature (Tm) was assessed as an indication of helix instability. In p.G510D Tm was reduced but in p.G1143S Tm was the same as control (Bonaventure et al. 1995). Likewise, the p.G1053E collagen II expressed by transfected SW1353 cells had a normal Tm but confirmed the secretion defect seen with the other mutations (Ito et al. 2005). Cartilage from the p.G1053E patient showed sparse collagen II fibrils in the ECM (Bogaert et al. 1992), and reduced collagen II ECM/sparse fibrils was a feature of p.G804A, p.G510D, p.G1005S, and p.G891R patient cartilage (Bonaventure et al. 1995; Freisinger et al. 1994; Mortier et al. 1995). In p.G804A, in addition to the reduced extracellular collagen II, expression of the non-cartilage collagen I was also apparent (Freisinger et al. 1994). In two cases, p.G969S (Chan et al. 1995a) and p.G1113C (Mundlos et al. 1996), there was little or no collagen II protein in patient cartilage, and it was replaced completely by collagen I despite collagen II mRNA expression by chondrocytes. The trace amount of collagen II protein expressed in p.G1113C chondrocytes was overmodified and not secreted, suggesting that in these two cases, the misfolded collagen II was completely degraded by cellular protein quality control processes. The reduced extracellular collagen determined in many other cases with Gly substitutions also presumably reflects that a proportion of the intracellularly retained mutant collagen II is degraded. The level of degradation may be mutation-dependent, but the essentially complete degradation of collagen II in these two lethal chondrodysplasia patients (Chan et al. 1995a; Mundlos et al. 1996) is striking and clearly a major contribution to the pathology.

Several mouse models support the patient cartilage and chondrocyte analyses. Transgenic mice overexpressing p.G285C (Garofalo et al. 1991) and p.G774S (Maddox et al. 1997) showed similar reduced secretion and sparse collagen II ECM. In a knockin p.G1170S mouse model (Liang et al. 2014), newborn homozygous mutants showed dilated ER, collagen retention, reduced collagen fibrils in the cartilage, and limb and vertebral skeletal defects. Newborn (the latest developmental timepoint analyzed) heterozygous mice had no skeletal phenotype, although dilated ER and reduced extracellular collagen fibrils were obvious, but much less severe than the homozygote.

A comprehensive and systematic analysis of a broad spectrum of collagen II helix Gly substitutions, exploring the effect of position in the triple helix, the type of the amino acid substitution, and the surrounding sequence context, has not been conducted. However, based on the information that is available, and by comparison with the more extensive studies on collagen I helix Gly substitution mutations, it is clear that Gly substitutions can, and probably mostly do, delay helix formation (revealed by overmodification) and reduce secretion, thus engaging and challenging the intracellular proteostasis network. One direct consequence of this is a poor-quality collagen II extracellular matrix. What we know about the other side of the story, how collagen II misfolding influences downstream intracellular consequences such as the ER stress response is discussed in later sections.

4.6.3 Arginine to Cysteine Structural Mutations

There are many non-glycine helix polymorphisms in *COL2A1* in apparently normal individuals (gnomAD, gnomad.broadintstitute.org) indicating that the helix can accommodate some X- and Y-position changes that tend to conserve amino acid charge and size. However, there are a significant number of non-Gly substitutions that cause cartilage disorders (Fig. 4.4). Of these Arg to Cys substitutions are a prominent and relatively well-studied subgroup. Arg to Cys mutations can cause severe but nonlethal forms of spondyloepiphyseal dysplasia congenita but are more commonly found in the milder phenotypes, Stickler syndrome, type I, and precocious osteoarthritis (Fig. 4.2). The triple helix is normally devoid of Cys, and Arg to Cys mutations introduce the possibility of internal disulfide bonding that would be predicted to interfere with helix structure.

There are only a few studies where collagen II Arg to Cys mutations have been examined in patient tissues or patient-derived chondrocytes, and most have been more extensively studied in transfected cell systems. *COL2A1* p.R275C, p.R719C, and p.R904C have normal secretion and normal helix Tm in transfected cells (Ito et al. 2005), suggesting that these mutations occur in regions of the helix that can accommodate the mutation without reducing helix thermal stability. This is supported by in vitro-transfected cell studies on these patient mutations (Chakkalakal et al. 2018). The p.R275C and p.R719C mutations do not induce apoptosis in vitro (Hintze et al. 2008), and they produce milder clinical phenotypes, consistent with their small impact on collagen structure. Apoptosis has not been studied for the p.R904C mutation. In contrast, the p.R989C mutation that causes a severe spondyloepiphyseal epiphyseal dysplasia reduces helix Tm in collagen from transfected cells (Chakkalakal et al. 2018), although in a separate study, collagen II extracted from patient cartilage did not have a reduced Tm (Chan et al. 1993). The p.R989C mutant-containing collagen II had reduced secretion and increased posttranslational modification (Chan et al. 1993), indicative of delayed helix formation such as seen with the Gly substitutions in the helix. In addition to the deleterious impacts of the mutation on collagen folding and secretion, p.R989C mutant collagen interacts with fibronectin inside the cell and extracellularly, and this binding likely inhibits collagen II fibril formation (Ito et al. 2005). Since collagen and fibronectin fibrillogenesis in vivo requires both proteins, strong fibronectin binding to p.R989C mutant collagen II could also be responsible for the disorganized fibronectin matrix and could change the organization of other ECM components. Other studies have shown the p.R989C mutant is sensitive to cleavage with MMPs resulting in dysfunctional truncated fragments (Chakkalakal et al. 2018). In contrast, the less severe p.R275C mutant binds only very weakly to fibronectin and rapidly dissociates (Ito et al. 2005). Fibronectin deposited in the ECM of cells expressing this mutant has a normal fibrillar structure. Furthermore, the p.R989C mutant (and p.G1053E) showed a reduced binding affinity for collagen IX (Steplewski et al. 2005) which could disturb the formation of the collagen II/IX/XI heterotypic fibrils. The clinically milder p.R275C and p.R719C mutants did not have a reduced binding affinity for collagen IX. In the case of the p.R275C mutant, collagen IX binding was unaltered (Steplewski et al. 2005), whereas p.R719C displayed increased collagen IX affinity (Fertala et al. 2001).

The downstream impacts of the p.R989C mutant challenge to the proteostasis machinery ultimately results in apoptosis in transfected cells (Chakkalakal et al. 2018). Likewise, a *Col2a1* p.R1192C mutation in a spondyloepiphyseal mouse model (Donahue et al. 2003) also caused apoptosis in vivo (Hintze et al. 2008). In vitro studies on transfected cells also showed that collagen II had reduced thermal stability, atypical disulfide bond formation, and reduced secretion associated with apoptosis (Chung et al. 2009).

4.7 Do Misfolding Mutations Cause Pathology via an ER Stress Response?

The structural mutations discussed above clearly impact collagen II folding and secretion and thus present a severe challenge to the cellular proteostasis machinery. Many studies have suggested that this misfolding initiates an ER stress response with pathogenic downstream consequences. In this section, we will explore the experimental data underpinning this proposal. The role of ER stress in the pathogenesis of collagenopathies has been recently critically evaluated (Bateman et al. 2022).

The proteostasis network in the ER coordinates protein synthesis, folding, trafficking, and degradation and thus imposes stringent quality control to ensure that proteins leaving the ER are correctly folded and functional. One important component of the proteostasis machinery is the unfolded protein response (UPR) (Walter and Ron 2011) and the overlapping pathway, the integrated stress response (ISR) (Costa-Mattioli and Walter 2020). The UPR is a conserved pathway that is activated in a coordinated attempt to reduce the load of misfolded proteins. The details of the UPR have been discussed in many excellent reviews (Hetz et al. 2020; Walter and Ron 2011) and will not be presented in detail here. The key features of the canonical UPR are that unfolded and misfolded proteins in the ER bind to a molecular chaperone BiP, which recognizes exposed hydrophobic protein sequences normally buried within correctly folded protein structures. This interaction of BiP with misfolded proteins removes BiP from the ER luminal domains of three ER stress sensors, IRE1, ATF6, and PERK, activating them and triggering downstream signaling cascades. IRE1 dimerizes and trans-autophosphorylates via its kinase domain, resulting in a conformational rearrangement that activates a site-selective RNase domain. The activated RNase splices an unconventional intron from the mRNA for an otherwise inactive transcription factor, XBP1. Spliced XBP1 mRNA translation produces an active transcription factor, termed $XBP1_s$. The IRE1-$XBP1_s$ arm targets several secretory pathway-related genes via UPR-responsive promoter elements, particularly upregulating protein trafficking, quality control, and folding machineries. The second UPR arm is mediated by ATF6, a member of the bZIP transcription factor family. Upon activation, ATF6 transits to the Golgi where it is cleaved by S1P and S2P to generate an active cytosolic transcription factor, ATF6p50. ATF6p50 translocates to the nucleus, binds to ER stress-responsive promoter elements, and upregulates transcription of ER protein folding chaperones. The third arm is mediated by PERK, which dimerizes upon activation, activating its kinase domain and leading to trans-autophosphorylation and downstream phosphorylation of the translational initiation factor eIF2α. Phosphorylated eIF2α prevents formation of the translation initiation complex and globally downregulates protein translation, thus reducing the protein folding load. Whereas phosphorylated eIF2α inhibits translation of most mRNAs, it promotes translation of the transcription factor ATF4 that is involved in upregulating genes involved in amino acid metabolism and transport, oxidation-reduction reactions, and the apoptosis-inducing transcription factor CHOP. Thus, long-term PERK pathway activation is cytotoxic. PERK is the initiating unfolded protein sensor for the overlapping stress response network, the integrated stress response (ISR) which can also be activated by GCN2, PKR, and HRI kinases that sense nutrient deprivation, viral infection, and oxidative stress, respectively (Costa-Mattioli and Walter 2020).

As a part of the normal cellular proteostasis machinery, initial UPR activation upregulates chaperones to improve protein folding, reduces protein translation, and stimulates ER-associated degradation pathways to reduce the protein folding burden to attempt to return the cell to proteostasis. However, if the UPR is chronically activated by long-term accumulation of misfolded proteins (such as occurs with collagen misfolding mutations), the transient UPR may be ineffectual and the balance of the three UPR arms may shift away from inducing pro-survival folding and quality control activities to directing the cell to apoptosis via PERK-mediated induction of the pro-apoptotic transcription factor CHOP and IRE1-mediated induction of the JNK pathway (Hetz et al. 2020).

The best evidence for collagen misfolding causing a canonical UPR and resulting in a cartilage disorder comes from studies on metaphyseal chondrodysplasia, type Schmid (MCDS) mutations in the collagen X C-terminal trimerization domain where in vitro data demonstrated impaired collagen X folding and upregulation of several sentinel UPR markers (Wilson et al. 2005). However, in vitro studies with transfected cells may not be a good system for evaluating pathological effects of the UPR, since expression of the mutant collagen at the correct levels and at the correct wild type to mutant ratio in chondrocytes is likely to be critical. The most compelling evidence comes from in vivo mouse studies that show a human collagen X misfolding mutation, p.N617K, activates a UPR in the collagen X expressing hypertrophic cartilage cells, and the resultant pathology is equivalent to human MCDS pathology (Rajpar et al. 2009). It is important to note that the UPR and pathology was evident only in the homozygous mice. In heterozygous mice (equivalent to the heterozygous MCDS patients), little or no pathology was evident and only weak upregulation of several UPR markers was noted (Rajpar et al. 2009). Direct evidence that the UPR is the major contributor to MCDS pathology came from studies on a transgenic mouse where a misfolding thyroglobulin was specifically expressed in hypertrophic cartilage, eliciting a UPR and causing the same pathology (Rajpar et al. 2009). Another mouse model expressing a *Col10a1* transgene with an MCDS 13 bp frameshift deletion causing a 9-amino acid truncation of the C-terminal trimerization domain confirmed collagen X misfolding and ER stress signaling altered chondrocyte maturation in the growth plate, supporting ER stress-related mechanisms for the skeletal abnormalities ((Tsang et al. 2007); see also review articles (Bateman et al. 2009; Tsang et al. 2010)). Detailed transcriptomic studies of the p.N617K MCDS mouse model revealed that activating the PERK pathway was the key pathogenic mechanism (Cameron et al. 2011). This is consistent with separate studies on the 13del MCDS mouse model showing that inhibiting the PERK/integrated stress response pathway with the small molecule ISRIB ameliorated the MCDS phenotype (Wang et al. 2018).

Several studies have attempted to show that intracellularly retained and misfolded collagen II results in a UPR. Based on the induction of a canonical UPR by misfolded collagen X with C-terminal trimerization domain mutations MCDS and procollagen I mutations in OI (Bateman et al. 2022), the collagen II mutations most likely to induce a canonical UPR would be those in the C-terminal propeptide. Two mouse models with collagen II C-propeptide mutations generating a platyspondyly chondrodysplasia phenotype, *Col2a1* p.Y1391S (Kimura et al. 2015) and p.D1469A (Furuichi et al. 2011), show procollagen II retention in the chondrocyte ER, reduced collagen ECM, and increased expression of the downstream markers of ER stress, GRP94 and CHOP. The *Col2a1* p.Y1391S mouse had increased chondrocyte apoptosis, a common consequence of PERK pathway activation. In a p.S1386P mutant modeling spondyloepiphyseal dysplasia congenita, there was also reduced collagen II secretion and a reduced collagen II ECM, but unfortunately ER stress markers were not evaluated (Esapa et al. 2012). The Dmm/+ mouse containing a Lys-Thr to Asn mutation (resulting from a 3 bp deletion) in the C-propeptide demonstrated a small increase in BiP and increased eIF2α phosphorylation also suggesting PERK activation (Ricks et al. 2013). Studies on iPSC-derived chondrocytes with a spondyloperipheral dysplasia mutation (*COL2A1* p.K1447CInsfsX18) that was assumed to result in a truncated, and therefore misfolded, C-propeptide did not reveal an increase in mRNA expression of the UPR markers, BiP, GRP94, or CHOP (Okada et al. 2015). Thus, while procollagen II C-propeptide mutations that impair trimerization and reduce secretion are good candidates for the induction of a canonical UPR, further detailed studies are needed to conclusively demonstrate a UPR and ascertain if this contributes to the severe cartilage phenotype.

For collagen II helix Gly substitution mutations, few studies have looked at the impact on the UPR. In vivo data comes from a mouse chondrodysplasia model with a *Col2a1* p.G1170S mutation (Liang et al. 2014). In the homozygous mutant, there was increased XBP1 splicing, BiP, ATF4, and ATF6 mRNA expression, but these increases were not apparent in the heterozygous mutant, the human disease relevant genotype, with CHOP and ATF6 the only evaluated UPR markers upregulated in both heterozygous and homozygous mice. In contrast, in vitro studies on chondrocytes differentiated from patient fibroblasts with *COL2A1* p. G1182A and p.G450R mutations (achondrogenesis/hypochondrogenesis) did show upregulation of BiP, GRP94, and CHOP mRNA and ATF6 protein activation, eIF2α phosphorylation, and XBP1 cleavage. Similar results were seen for a helical exon-skip mutation (Okada et al. 2015). While these studies suggested that these helical mutations may activate a canonical UPR in this cell model system, there are important caveats around the expression levels of collagen II, the relative efficiency of fibroblast to chondrocyte conversion in control and mutant cultures, and how closely these chondrocytes, directly converted from fibroblasts by transduction with retroviruses expressing c-MYC, KLF4, and SOX9, recapitulate the human chondrocyte phenotype in vitro (Okada et al. 2015).

Several in vitro studies on Arg-Cys helix mutations explore possible UPR activation. Chakkalakal et al. (2018) used HT1080 cell models transfected with *COL2A1* p.R275C, p.R334C, p.R904C, p.R940C, and p.R989C mutant constructs. The more N-terminal helix mutations (p.R275C, p.R334C, p.R904C) did not reduce helix stability (as discussed earlier) or stimulate XBP1 splicing, whereas the more C-terminal mutants that reduced helix stability upregulated XBP1splicing, BiP protein expression, and apoptosis. These data suggested that the more C-terminal arginine to cysteine mutations could provoke a UPR. In other studies on transfected SW1353 cells, the p.R989C mutant was shown by co-precipitation studies to form a complex with BiP (and possibly upregulated BiP expression), while the other mutants tested, p.R275C, p.R719C, and p.G1053E, did not (Hintze et al. 2008). Analysis of the p.R1192C mutant transfected into SW1353 cells showed increased BiP expression (Chung et al. 2009). None of these studies performed a comprehensive analysis of the sentinel markers of a canonical UPR, so even in these in vitro studies, the activation of a canonical UPR has not been unequivocally demonstrated. Further analysis of the *Col2a1* p.R1192C mouse model will be critical to determine if there is a UPR in chondrocytes in vivo and if this is an important contributor to the phenotype. So far analysis has identified BiP-positive cells and apoptosis in the mutant mouse growth plate cartilage (Hintze et al. 2008).

4.8 Conclusion and Future Perspectives

There are several clear correlations between the genotype of *COL2A1* mutations and the resulting clinical chondrodysplasia phenotype. The first is that mutations that introduce premature stop codons, either because of a point mutation or a frameshift mutation, generally result in mutant allele mRNA NMD and collagen II protein haploinsufficiency. Consequently, the cartilage ECM contains half normal amounts of collagen II, but the collagen II produced is not compromised structurally and has unimpeded folding, secretion, and fibril formation, albeit at half the normal level (Fig. 4.5). These mutations cause a phenotype at the milder end of the *COL2A1* chondrodysplasia clinical spectrum, Stickler syndrome, type I. While premature termination codons in the first 53 exons of *COL2A1* are expected to cause NMD and haploinsufficiency, those in the final exon, exon 54, are expected, based on the widely accepted EJC-dependent model of NMD (see earlier), to be NMD-immune. These are predicted to thus result in synthesis of a proα1(II) with an altered amino acid sequence in the C-terminal trimerization domain, impacting protein folding and potentially causing a more severe phenotype. This is borne out by the distribution of premature termination mutations in Stickler syndrome, type I, and the lethal platyspondylic dysplasia where these mutations are clustered in exon 54 (Fig. 4.2).

The second generalization, therefore, is that structural mutations cause more severe chondrodysplasia phenotypes due to their impacts on procollagen II folding,

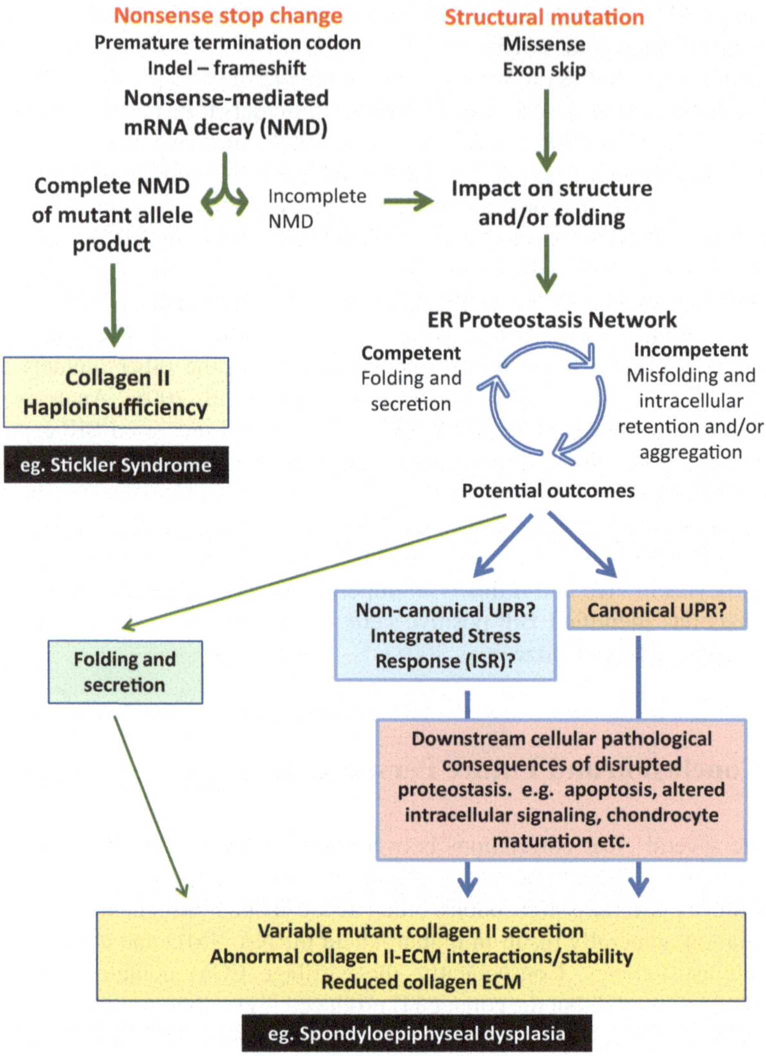

Fig. 4.5 Potential cellular impacts of collagen mutations

stability, secretion, and in some cases altered ECM interactions. These are commonly missense mutations, but also include in-frame exon deletions. The predominant missense mutations in these more severe chondrodysplasias are substitutions for Gly in the triple helix causing achondrogenesis (hypochondrogenesis) (Fig. 4.2). There is abundant evidence that these can disturb collagen folding and structure and engage and challenge the cellular proteostasis machinery (Fig. 4.5).

While a gradient in collagen II thermal stability from thermostable C-terminal to the thermolabile N-terminal regions would suggest that the N-terminus may be more tolerant to structural disruption by the Gly substitutions (Steplewski et al.

2004), a simple relationship between the mutation position in the helix and the clinical severity is not apparent (Fig. 4.2). While studies on how and which mutations affect helix stability are limited for collagen II, for collagen I and osteogenesis imperfecta mutations, it has been concluded that "the effect of a particular substitution on collagen stability is not the primary determinant of clinical outcome but that stability effects do play a modifying role in α1(I) chain substitutions" (Marini et al. 2007). However, it is apparent that the most N-terminal Gly substitutions cause the milder phenotypes, Stickler, type I, and OA-related phenotypes (Fig. 4.2). Gly mutations more C-terminal to helix position ~200 cause the more severe clinical phenotypes (Fig. 4.2), but the mutations in this region do not conform to a continuous gradient of severity and are most likely impacted by the type of substitution, local sequence context, and if disruption of the sequence by the mutation impacts subsequent collagen functionality. Studies on type I collagen mutations in osteogenesis imperfecta support this and provide potential insights (Bodian et al. 2008; Marini et al. 2007). Mutations in the first ~200 amino acids of the triple helix were nonlethal, and beyond that N-terminal region substitutions of Gly with the bulky or changed amino acids, Val, Asp, Glu, and Arg were much more commonly lethal than Cys, Ser, and Ala. The same seems broadly true for collagen II mutations. The most C-terminal lethal mutations (achondrogenesis/hypochondrogenesis) occur at helix position 202 (p.G402E), and more C-terminal to that Val, Asp, Glu, and Arg are the most common changes (64%). The milder phenotypes, Stickler, type I, and OA-related phenotypes also fit with this model with a strong bias against the more disruptive Val, Asp, Glu, and Arg substitutions.

While the Arg-Cys mutations in the helix are generally less severe than Gly substitutions (Fig. 4.2) due to a lesser impact on helix structure, one of these, p.G989C, is a notable exception causing a severe spondyloepiphyseal dysplasia (Fig. 4.2). This may be because of the additional effects of this mutation on fibronectin and collagen IX binding that further impacts fibrillogenesis. Furthermore, this region around the collagenase cleavage site at amino acid 975/976 is a hotspot for protein binding with over 20 different proteins interacting with this region in collagen II peptide studies (Farndale 2019).

Collagen I α1(I) mutations in two regions of the triple helix (positions 691–823 and 910–964) always result in the lethal osteogenesis imperfecta phenotype. These two regions correspond to major ligand binding regions, suggesting crucial interactions of collagen I with integrins, fibronectin, and other cartilage ECM proteins (Marini et al. 2007). In collagen II, with the exception of the p.R989C mutation that has altered interactions with collagen IX and fibronectin (Ito et al. 2005; Steplewski et al. 2005), protein interaction sites on collagen II have not been well mapped and which interactions might be compromised by mutations is not known. While there are mutations in the more severe phenotypes corresponding to the major ligand binding regions of collagen I, there is no evidence of an overrepresentation of severe mutations in these regions (Fig. 4.2). However, there does seem to be a strong overrepresentation of Gly substitution mutations in achondrogenesis/hypochondrogenesis and spondyloepiphyseal dysplasia in an ~200 amino region around helix position 450 (Fig. 4.2). There are no currently known specific ECM interactions within this region. However, collagen II synthetic peptide studies have shown two significant regions binding ~20 proteins within the D2 region of collagen II (Farndale 2019) which contains this possible mutation hotspot. Further studies are required to determine if alterations in mutant collagen II do impact the binding to specific ECM proteins and if this contributes to the clinical phenotype of these mutations.

The extent of mutant collagen II degradation will also contribute to the severity of the clinical phenotype. There are relatively few studies that address this, but in two cases, *COL2A1* p.G969S (Chan et al. 1995a) and p.G1113C (Mundlos et al. 1996), where human cartilage has been studied, there was little or no detectable collagen II, indicating that the mutant containing collagen II trimers are degraded. Presumably this degradation is intracellular, resulting from the reduced folding and secretion and engagement with the ER quality control machinery. In addition, extracellular degradation of mutant collagen II may occur. This is suggested by the studies showing that the p.R989C mutant is sensitive to cleavage with MMPs (Chakkalakal et al. 2018). Collagen I containing an osteogenesis imperfecta Gly substitution mutation has been shown to be selectively degraded within the ECM over time (Bateman and Golub 1994). Based on the many studies, discussed earlier, a common finding in cartilage tissue from patients with collagen II Gly substitutions is a much-reduced collagen II fibrillar matrix, suggesting that degradation (intracellular and/or extracellular) along with reduced mutant procollagen II secretion is a significant contributor to the cartilage pathology. At this point, how this precisely correlates with mutation type and position is unclear, although studies on one mutation offer possible insight. The homozygous p.G1170S mutation results in a dramatic reduction in the collagen II ECM. This interacted poorly with the cellular β1 integrin collagen receptors with a downstream effect on BMP-SMAD signaling that accelerated chondrocyte hypertrophy (Lian et al. 2019). This provides a possible mechanism for how mutations that cause severe collagen II depletion, such as this homozygous p.G1170S mutation, could disrupt ECM-chondrocyte regulatory pathways and have a significant impact of chondrocyte development and maturation.

How mutant procollagen II might engage with the intracellular ER quality control machinery, and if the unfolded/misfolded procollagen elicits unfolded protein response (UPR), is not resolved (Fig. 4.5). The concept that mutations trigger a canonical UPR with pathogenic consequences has been proposed in several studies as an overarching disease mechanism, but the evidence is patchy (as discussed earlier), and it is most likely that the engagement of the UPR is a nuanced story that will show considerable mutation specificity. While several studies show some collagen II mutations activate a few UPR components, the canonical UPR networks involve many more components than those so far identified, and there have been no studies analyzing these broader UPR gene sets. The "UPRome" has 278 components with a "core UPRome" of nine (*HERPUD1, HYOU1, HSPA5, HSP90B1, PDIA4, DNAJB11, SDF2L1, CRELD2, MANF*) (Bergmann et al. 2018). Of these only HSP90B1 (GRP94) and HSPA5 (BiP) have been reported dysregulated with some collagen II mutations. In contrast, all nine of these were dysregulated with a collagen X misfolding mutation that instigated a canonical UPR (Cameron et al. 2011). A key to understanding the possible engagement of collagen II mutations with the canonical UPR will be critical in determining the effect of mutations on, at least, this core UPRome and on the other identified bona fide UPR target genes of the IRE1, ATF6, and PERK activation (Kline et al. 2023).

An interesting consideration that may be of significance to the relevance of the canonical UPR to collagen II mutation pathogenesis comes from studies on collagen I mutations where helix Gly mutation did not bind BiP (Chessler and Byers 1993), considered to be a crucial initiating step of the canonical UPR. In contrast, procollagen I C-propeptide mutations that compromised initial procollagen assembly did bind BiP (Chessler and Byers 1993; Lamandé et al. 1995). This is consistent with later studies on collagen X showing that mutations that impacted C-terminal assembly did initiate a canonical UPR (Cameron et al. 2011; Chan et al. 1996; Wilson et al. 2005) and with the impacts of procollagen I C-propeptide mutants resulting in ER-associated degradation (a key UPR pathway). In contrast collagen I helix Gly mutations resulted in the formation of mutant collagen intracellular aggregates that were subsequently degraded by macroautophagy (Ishida et al. 2009).

While it is too early to conclude that chronic UPR/ISR could be pathogenic in the collagen II collagenopathies, it remains a possibility awaiting further experimentation. However, there are several instances where apoptosis resulting from collagen II misfolding occurs, so either downstream pathological signaling from an ER stress response or from altered ECM-cell signaling due to the compromised collagen ECM must occur. Apoptosis can occur downstream of PERK activation (Hetz et al. 2020), so this arm of the UPR/ISR remains an important pathway to explore. It is also possible that the misfolded collagen II, which has unique features due to its fibrillar rather than prototypical globular structure, may fail to activate the pro-survival aspects of the UPR, and this may be a more significant contributor to pathophysiology than is chronic UPR activation. The accumulation of misfolded collagen II aggregates within the ER (which would be expected for Gly helix mutations) (Ishida et al. 2009) may have a profound impact on the global proteostasis network, thus disrupting protein synthesis, secretion, and cellular metabolic function well beyond the direct impact on collagen proteostasis.

While how the proteostasis challenges of different collagen II mutations cause the pathology is not resolved, it seems likely that approaches that restore proteostasis could be beneficial. Such approaches could involve reducing the mutant collagen load by stimulating degradation of the misfolded collagen or correcting the mutant protein folding. Alternatively, if a pathological UPR/ISR is activated, there may be opportunities to therapeutically manipulate these once the pathways for collagen II misfolding are properly delineated. Increasing breakdown of the mutant misfolded intracellular collagens by stimulating the proteasomal and/or autophagosome-lysosome system is a promising approach. The best example of this comes from collagen X mutations in MCDS. Treating a mouse disease model with carbamazepine, which stimulated autophagy, reduced the clinical severity of the cartilage defect (Mullan et al. 2017), and as a result of these data, a clinical trial treating MCDS patients with carbamazepine is underway (https://MCDS-therapy.eu). However, preclinical trials of carbamazepine and rapamycin, another autophagy stimulator, in an osteogenesis imperfecta mouse model with a collagen I helix Gly mutation did not improve the bone phenotype, and there were deleterious off-target effects (Bateman et al. 2019; Blank et al. 2022). It is worth pointing out that this approach, stimulating mutant degradation, would reduce the amount of collagen II in the ECM, albeit producing a more structurally normal collagen ECM. Given that 7/8 of the collagen II molecules contain one or more mutant chains, degrading mutant-containing collagen is likely to still result in a significant clinical phenotype.

A second approach to reducing the mutant load could involve stimulating mutant collagen II folding with chemical chaperones. In transfected cells expressing the p.R989C and p.R1192C collagen II mutants, the chemical chaperone TMAO (trimethylamine N-oxide) stabilized the thermolabile mutant collagen II, increased secretion, and improved cell survival (Gawron et al. 2010). Other studies on iChon "chondrocyte-like" cells (virally transduced fibroblasts) showed that TMAO reduced apoptosis and BiP and CHOP expression in several collagen II mutants (Okada et al. 2015). Studies with another chemical chaperone, sodium phenylbutyrate (4-PBA) on cells expressing mutant collagen I, have shown rescue of secretion and some markers of the UPR with some helix Gly mutations (Besio et al. 2018). Another approach, which has not been tested with collagen mutations, is using small molecules to stimulate a non-pathological UPR to restore proteostasis (Grandjean and Wiseman 2020). These approaches offer considerable promise, but a possible detrimental effect of rescuing structurally abnormal collagen II secretion on subsequent fibril formation and stability and altered interactions within the ECM formation need to be considered.

A third therapeutic approach being explored in several diseases involves targeting the UPR sensing and downstream signaling pathways (Hetz et al. 2019; Marciniak et al. 2022; Park and Ozcan 2013). While several small molecules are available that can modulate all three arms of the UPR, IRE1α-XBP1, ATF6, and PERK, the appropriate approach for different collagen II mutations will rely on a much more nuanced understanding of any elicited UPR/ISR. One example, however, serves to highlight the promise of this approach. With the collagen X MCDS trimerization mutation (*Col10a1* p.N617K), detailed transcriptomics of mouse cartilage showed that the PERK arm was critical to the pathology (Cameron et al. 2011). Treating another MCDS trimerization mutation mouse model with ISRIB (integrated stress response inhibitor) that reverses the effects of eIF2α phosphorylation downstream of PERK activation improved the cartilage pathology (Wang et al. 2018).

Chondrocytes appear to have a particular susceptibility to ER stress from mutations that cause misfolding of ECM proteins (Briggs et al. 2020), presumably due to their biological role as professional secretory cells. Since their proteostasis networks are tuned to this function, it will be particularly important to dissect the molecular pathology of collagen mutations and their engagement, or lack thereof, with the proteostasis machinery in the right biological context, chondrocytes, or cartilage tissues. Using in vitro chondrocyte models and mouse models that accurately recapitulate cellular pathologies will be vital to understand the collagen II mutation pathophysiology and to provide a platform to evaluate new therapies based on these disease mechanisms.

Acknowledgments Research in the author's laboratory is supported by the National Health and Medical Research Council of Australia through project grants (GNT2003393 and GNT2012936) and Novo Nordisk Foundation grants (NNF21CC0073729, Novo Nordisk Foundation Center for Stem Cell Medicine). Additional infrastructure funding to the Murdoch Children's Research Institute was provided by the Australian Government National Health and Medical Research Council Independent Research Institute Infrastructure Support Scheme and the Victorian Government's Operational Infrastructure Support Program.

Compliance with Ethical Standards/Ethical Requirements No conflicts of interest.

References

Antipova O, Orgel JPRO (2010) In situ D-periodic molecular structure of type II collagen. J Biol Chem 285:7087–7096. https://doi.org/10.1074/jbc.M109.060400

Barat-Houari M, Sarrabay G, Gatinois V, Fabre A, Dumont B, Genevieve D, Touitou I (2016) Mutation update for COL2A1 gene variants associated with type II Collagenopathies. Hum Mutat 37:7–15. https://doi.org/10.1002/humu.22915

Bateman JF, Boot-Handford RP, Lamandé SR (2009) Genetic diseases of connective tissues: cellular and extracellular effects of ECM mutations. Nat Rev Genet 10:173–183. https://doi.org/10.1038/nrg2520

Bateman JF, Golub SB (1994) Deposition and selective degradation of structurally-abnormal type I collagen in a collagen matrix produced by osteogenesis imperfecta fibroblasts in vitro. Matrix Biol 14:251–262. https://doi.org/10.1016/0945-053x(94)90189-9

Bateman JF, Lamande SR, Dahl HH, Chan D, Mascara T, Cole WG (1989) A frameshift mutation results in a truncated nonfunctional carboxyl-terminal pro alpha 1(I) propeptide of type I collagen in osteogenesis imperfecta. J Biol Chem 264:10960–10964

Bateman JF, Sampurno L, Maurizi A, Lamandé SR, Sims NA, Cheng TL, Schindeler A, Little DG (2019) Effect of rapamycin on bone mass and strength in the α2(I)-G610C mouse model of osteogenesis imperfecta. J Cell Mol Med 23:1735–1745. https://doi.org/10.1111/jcmm.14072

Bateman JF, Shoulders MD, Lamandé SR (2022) Collagen misfolding mutations: the contribution of the unfolded protein response to the molecular pathology. Connect Tissue Res 63:210–227. https://doi.org/10.1080/03008207.2022.2036735

Bateman JF, Wilson R, Freddi S, Lamandé SR, Savarirayan R (2005) Mutations of COL10A1 in Schmid metaphyseal chondrodysplasia. Hum Mutat 25:525–534. https://doi.org/10.1002/humu.20183

Bergmann TJ, Fregno I, Fumagalli F, Rinaldi A, Bertoni F, Boersema PJ, Picotti P, Molinari M (2018) Chemical stresses fail to mimic the unfolded protein response resulting from luminal load with unfolded polypeptides. J Biol Chem 293:5600–5612. https://doi.org/10.1074/jbc.RA117.001484

Besio R, Iula G, Garibaldi N, Cipolla L, Sabbioneda S, Biggiogera M, Marini JC, Rossi A, Forlino A (2018) 4-PBA ameliorates cellular homeostasis in fibroblasts from osteogenesis imperfecta patients by enhancing autophagy and stimulating protein secretion. Biochim Biophys Acta Mol basis Dis 1864:1642–1652. https://doi.org/10.1016/j.bbadis.2018.02.002

Blank M, McGregor NE, Rowley L, Kung LH, Crimeen-Irwin B, Poulton IJ, Walker EC, Gooi JH, Lamande SR, Sims NA, Bateman JF (2022) The effect of carbamazepine on bone structure and strength in control and osteogenesis imperfecta ($Col1a2^{+/p.G610C}$) mice (preprint). Pathology. https://doi.org/10.1101/2022.02.23.481692

Bodian DL, Madhan B, Brodsky B, Klein TE (2008) Predicting the clinical lethality of osteogenesis imperfecta from collagen glycine mutations. Biochemistry 47:5424–5432. https://doi.org/10.1021/bi800026k

Bogaert R, Tiller GE, Weis MA, Gruber HE, Rimoin DL, Cohn DH, Eyre DR (1992) An amino acid substitution (Gly 853-->Glu) in the collagen alpha 1(II) chain produces hypochondrogenesis. J Biol Chem 267:22522–22526

Bonaventure J, Cohen-Solal L, Ritvaniemi P, Van Maldergem L, Kadhom N, Delezoide AL, Maroteaux P, Prockop DJ, Ala-Kokko L (1995) Substitution of aspartic acid for glycine at position 310 in type II collagen produces achondrogenesis II, and substitution of serine at position 805 produces hypochondrogenesis: analysis of genotype-phenotype relationships. Biochem J 307(Pt 3):823–830. https://doi.org/10.1042/bj3070823

Briggs MD, Dennis EP, Dietmar HF, Pirog KA (2020) New developments in chondrocyte ER stress and related diseases. F1000Res 9:F1000 Faculty Rev-290. https://doi.org/10.12688/f1000research.22275.1

Cameron TL, Bell KM, Tatarczuch L, Mackie EJ, Rajpar MH, McDermott BT, Boot-Handford RP, Bateman JF (2011) Transcriptional profiling of chondrodysplasia growth plate cartilage reveals adaptive ER-stress networks that allow survival but disrupt hypertrophy. PLoS One 6:e24600. https://doi.org/10.1371/journal.pone.0024600

Canty EG, Kadler KE (2005) Procollagen trafficking, processing and fibrillogenesis. J Cell Sci 118:1341–1353. https://doi.org/10.1242/jcs.01731

Chakkalakal SA, Heilig J, Baumann U, Paulsson M, Zaucke F (2018) Impact of arginine to cysteine mutations in collagen II on protein secretion and cell survival. Int J Mol Sci 19. https://doi.org/10.3390/ijms19020541

Chan D, Cole WG, Chow CW, Mundlos S, Bateman JF (1995a) A COL2A1 mutation in achondrogenesis type II results in the replacement of type II collagen by type I and III collagens in cartilage. J Biol Chem 270:1747–1753

Chan D, Cole WG, Rogers JG, Bateman JF (1995b) Type X collagen multimer assembly in vitro is prevented by a Gly618 to Val mutation in the alpha 1(X) NC1 domain resulting in Schmid metaphyseal chondrodysplasia. J Biol Chem 270:4558–4562. https://doi.org/10.1074/jbc.270.9.4558

Chan D, Freddi S, Weng YM, Bateman JF (1999) Interaction of collagen alpha1(X) containing engineered NC1 mutations with normal alpha1(X) in vitro. Implications for the molecular basis of schmid metaphyseal chondrodysplasia. J Biol Chem 274:13091–13097. https://doi.org/10.1074/jbc.274.19.13091

Chan D, Taylor TK, Cole WG (1993) Characterization of an arginine 789 to cysteine substitution in alpha 1 (II) collagen chains of a patient with spondyloepiphyseal dysplasia. J Biol Chem 268:15238–15245

Chan D, Weng YM, Hocking AM, Golub S, McQuillan DJ, Bateman JF (1996) Site-directed mutagenesis of human type X collagen. Expression of alpha1(X) NC1, NC2, and helical mutations in vitro and in transfected cells. J Biol Chem 271:13566–13572. https://doi.org/10.1074/jbc.271.23.13566

Chessler SD, Byers PH (1993) BiP binds type I procollagen pro alpha chains with mutations in the carboxyl-terminal propeptide synthesized by cells from patients with osteogenesis imperfecta. J Biol Chem 268:18226–18233

Chung HJ, Jensen DA, Gawron K, Steplewski A, Fertala A (2009) R992C (p.R1192C) substitution in collagen II alters the structure of mutant molecules and induces the unfolded protein response. J Mol Biol 390:306–318. https://doi.org/10.1016/j.jmb.2009.05.004

Costa-Mattioli M, Walter P (2020) The integrated stress response: from mechanism to disease. Science 368:eaat 5314. https://doi.org/10.1126/science.aat5314

Dennis EP, Greenhalgh-Maychell PL, Briggs MD (2021) Multiple epiphyseal dysplasia and related disorders: molecular genetics, disease mechanisms, and therapeutic avenues. Dev Dyn 250:345–359. https://doi.org/10.1002/dvdy.221

Donahue LR, Chang B, Mohan S, Miyakoshi N, Wergedal JE, Baylink DJ, Hawes NL, Rosen CJ, Ward-Bailey P, Zheng QY, Bronson RT, Johnson KR, Davisson MT (2003) A missense mutation in the mouse Col2a1 gene causes spondyloepiphyseal dysplasia congenita, hearing loss, and retinoschisis. J Bone Miner Res 18:1612–1621. https://doi.org/10.1359/jbmr.2003.18.9.1612

Esapa CT, Hough TA, Testori S, Head RA, Crane EA, Chan CPS, Evans H, Bassett JHD, Tylzanowski P, McNally EG, Carr AJ, Boyde A, Howell PGT, Clark A, Williams GR, Brown MA, Croucher PI, Nesbit MA, Brown SDM, Cox RD, Cheeseman MT, Thakker RV (2012) A mouse model for spondyloepiphyseal dysplasia congenita with secondary osteoarthritis due to a Col2a1 mutation. J Bone Miner Res 27:413–428. https://doi.org/10.1002/jbmr.547

Eyre DR (1991) The collagens of articular cartilage. Semin Arthritis Rheum 21:2–11. https://doi.org/10.1016/0049-0172(91)90035-X

Fang Y, Bateman JF, Mercer JF, Lamandé SR (2013) Nonsense-mediated mRNA decay of collagen -emerging complexity in RNA surveillance mechanisms. J Cell Sci 126:2551–2560. https://doi.org/10.1242/jcs.120220

Farndale RW (2019) Collagen-binding proteins: insights from the collagen toolkits. Essays Biochem 63:337–348. https://doi.org/10.1042/EBC20180070

Fertala A, Sieron AL, Adachi E, Jimenez SA (2001) Collagen II containing a Cys substitution for Arg-alpha1-519: abnormal interactions of the mutated molecules with collagen IX. Biochemistry 40:14422–14428. https://doi.org/10.1021/bi0109109

Forlino A, Marini JC (2016) Osteogenesis imperfecta. Lancet 387:1657–1671. https://doi.org/10.1016/S0140-6736(15)00728-X

Freddi S, Savarirayan R, Bateman JF (2000) Molecular diagnosis of stickler syndrome: a COL2A1 stop codon mutation screening strategy that is not compromised by mutant mRNA instability. Am J Med Genet 90:398–406

Freisinger P, Ala-Kokko L, LeGuellec D, Franc S, Bouvier R, Ritvaniemi P, Prockop DJ, Bonaventure J (1994) Mutation in the COL2A1 gene in a patient with hypochondrogenesis. Expression of mutated COL2A1 gene is accompanied by expression of genes for type I procollagen in chondrocytes. J Biol Chem 269:13663–13669

Furuichi T, Masuya H, Murakami T, Nishida K, Nishimura G, Suzuki T, Imaizumi K, Kudo T, Ohkawa K, Wakana S, Ikegawa S (2011) ENU-induced missense mutation in the C-propeptide coding region of Col2a1 creates a mouse model of platyspondylic lethal skeletal dysplasia, Torrance type. Mamm Genome 22:318–328. https://doi.org/10.1007/s00335-011-9329-3

Garofalo S, Vuorio E, Metsaranta M, Rosati R, Toman D, Vaughan J, Lozano G, Mayne R, Ellard J, Horton W (1991) Reduced amounts of cartilage collagen fibrils and growth plate anomalies in transgenic mice harboring a glycine-to-cysteine mutation in the mouse type II procollagen alpha 1-chain gene. Proc Natl Acad Sci USA 88:9648–9652. https://doi.org/10.1073/pnas.88.21.9648

Gawron K, Jensen DA, Steplewski A, Fertala A (2010) Reducing the effects of intracellular accumulation of thermolabile collagen II mutants by increasing their thermostability in cell culture conditions. Biochem Biophys Res Commun 396:213–218. https://doi.org/10.1016/j.bbrc.2010.04.056

Gentili C, Cancedda R (2009) Cartilage and bone extracellular matrix. Curr Pharm Des 15:1334–1348. https://doi.org/10.2174/138161209787846739

Godfrey M, Hollister DW (1988) Type II achondrogenesis-hypochondrogenesis: identification of abnormal type II collagen. Am J Hum Genet 43:904–913

Grandjean JMD, Wiseman RL (2020) Small molecule strategies to harness the unfolded protein response: where do we go from here? J Biol Chem 295:15692–15711. https://doi.org/10.1074/jbc.REV120.010218

Heinegård D (2009) Fell-Muir Lecture: Proteoglycans and more–from molecules to biology. Int J Exp Pathol 90:575–586. https://doi.org/10.1111/j.1365-2613.2009.00695.x

Hetz C, Axten JM, Patterson JB (2019) Pharmacological targeting of the unfolded protein response for disease intervention. Nat Chem Biol 15:764–775. https://doi.org/10.1038/s41589-019-0326-2

Hetz C, Zhang K, Kaufman RJ (2020) Mechanisms, regulation and functions of the unfolded protein response. Nat Rev Mol Cell Biol 21:421–438. https://doi.org/10.1038/s41580-020-0250-z

Hintze V, Steplewski A, Ito H, Jensen DA, Rodeck U, Fertala A (2008) Cells expressing partially unfolded R789C/p.R989C type II procollagen mutant associated with spondyloepiphyseal dysplasia undergo apoptosis. Hum Mutat 29:841–851. https://doi.org/10.1002/humu.20736

Horton WA, Machado MA, Ellard J, Campbell D, Bartley J, Ramirez F, Vitale E, Lee B (1992) Characterization of a type II collagen gene (COL2A1) mutation identified in cultured chondrocytes from human hypochondrogenesis. Proc Natl Acad Sci USA 89:4583–4587. https://doi.org/10.1073/pnas.89.10.4583

Ishida Y, Yamamoto A, Kitamura A, Lamandé SR, Yoshimori T, Bateman JF, Kubota H, Nagata K (2009) Autophagic elimination of misfolded procollagen aggregates in the endoplasmic reticulum as a means of cell protection. Mol Biol Cell 20:2744–2754. https://doi.org/10.1091/mbc.e08-11-1092

Ishikawa Y, Bächinger HP (2013) A molecular ensemble in the rER for procollagen maturation. Biochim Biophys Acta 1833:2479–2491. https://doi.org/10.1016/j.bbamcr.2013.04.008

Ito H, Rucker E, Steplewski A, McAdams E, Brittingham RJ, Alabyeva T, Fertala A (2005) Guilty by association: some collagen II mutants alter the formation of ECM as a result of atypical interaction with fibronectin. J Mol Biol 352:382–395. https://doi.org/10.1016/j.jmb.2005.07.019

Ito S, Nagata K (2021) Quality control of procollagen in cells. Annu Rev Biochem 90:631–658. https://doi.org/10.1146/annurev-biochem-013118-111603

Jovanovic M, Guterman-Ram G, Marini JC (2022) Osteogenesis imperfecta: mechanisms and signaling pathways connecting classical and rare OI types. Endocr Rev 43:61–90. https://doi.org/10.1210/endrev/bnab017

Kimura M, Ichimura S, Sasaki K, Masuya H, Suzuki T, Wakana S, Ikegawa S, Furuichi T (2015) Endoplasmic reticulum stress-mediated apoptosis contributes to a skeletal dysplasia resembling platyspondylic lethal skeletal dysplasia, Torrance type, in a novel Col2a1 mutant mouse line. Biochem Biophys Res Commun 468:86–91. https://doi.org/10.1016/j.bbrc.2015.10.160

Kline GM, Paxman RJ, Lin C-Y, Madrazo N, Yoon L, Grandjean JMD, Lee K, Nugroho K, Powers ET, Wiseman RL, Kelly JW (2023) Divergent proteome reactivity influences arm-selective activation of the unfolded protein response by pharmacological endoplasmic reticulum Proteostasis regulators. ACS Chem Biol 18:1719. https://doi.org/10.1021/acschembio.3c00042

Lamandé SR, Chessler SD, Golub SB, Byers PH, Chan D, Cole WG, Sillence DO, Bateman JF (1995) Endoplasmic reticulum-mediated quality control of type I collagen production by cells from osteogenesis imperfecta patients with mutations in the pro alpha 1 (I) chain carboxyl-terminal propeptide which impair subunit assembly. J Biol Chem 270:8642–8649. https://doi.org/10.1074/jbc.270.15.8642

Lian C, Wang X, Qiu X, Wu Z, Gao B, Liu L, Liang G, Zhou H, Yang X, Peng Y, Liang A, Xu C, Huang D, Su P (2019) Collagen type II suppresses articular chondrocyte hypertrophy and osteoarthritis progression by promoting integrin β1-SMAD1 interaction. Bone Res 7:8. https://doi.org/10.1038/s41413-019-0046-y

Liang G, Lian C, Huang D, Gao W, Liang A, Peng Y, Ye W, Wu Z, Su P, Huang D (2014) Endoplasmic reticulum stress-unfolding protein response-apoptosis cascade causes chondrodysplasia in a col 2a1 p.Gly1170Ser mutated mouse model. PLoS One 9:e86894. https://doi.org/10.1371/journal.pone.0086894

Maddox BK, Garofalo S, Keene DR, Smith C, Horton WA (1997) Type II collagen pro-alpha-chains containing a Gly574Ser mutation are not incorporated into the cartilage matrix of transgenic mice. Matrix Biol 16:93–103. https://doi.org/10.1016/s0945-053x(97)90038-x

Malhotra V, Erlmann P (2015) The pathway of collagen secretion. Annu Rev Cell Dev Biol 31:109–124. https://doi.org/10.1146/annurev-cellbio-100913-013002

Marciniak SJ, Chambers JE, Ron D (2022) Pharmacological targeting of endoplasmic reticulum stress in disease. Nat Rev Drug Discov 21:115–140. https://doi.org/10.1038/s41573-021-00320-3

Marini JC, Forlino A, Cabral WA, Barnes AM, San Antonio JD, Milgrom S, Hyland JC, Körkkö J, Prockop DJ, De Paepe A, Coucke P, Symoens S, Glorieux FH, Roughley PJ, Lund AM, Kuurila-Svahn K, Hartikka H, Cohn DH, Krakow D, Mottes M, Schwarze U, Chen D, Yang K, Kuslich C, Troendle J, Dalgleish R, Byers PH (2007) Consortium for osteogenesis imperfecta mutations in the helical domain of type I collagen: regions rich in lethal mutations align with collagen binding sites for integrins and proteoglycans. Hum Mutat 28:209–221. https://doi.org/10.1002/humu.20429

Mortier GR, Wilkin DJ, Wilcox WR, Rimoin DL, Lachman RS, Eyre DR, Cohn DH (1995) A radiographic, morphologic, biochemical and molecular analysis of a case of achondrogenesis type II resulting from substitution for a glycine residue (Gly691-->Arg) in the type II collagen trimer. Hum Mol Genet 4:285–288. https://doi.org/10.1093/hmg/4.2.285

Mullan LA, Mularczyk EJ, Kung LH, Forouhan M, Wragg JM, Goodacre R, Bateman JF, Swanton E, Briggs MD, Boot-Handford RP (2017) Increased intracellular proteolysis reduces disease severity in an ER stress-associated dwarfism. J Clin Invest 127:3861–3865. https://doi.org/10.1172/JCI93094

Müller C, Khabut A, Dudhia J, Reinholt FP, Aspberg A, Heinegård D, Önnerfjord P (2014) Quantitative proteomics at different depths in human articular cartilage reveals unique patterns of protein distribution. Matrix Biol 40:34–45. https://doi.org/10.1016/j.matbio.2014.08.013

Mundlos S, Chan D, McGill J, Bateman JF (1996) An alpha 1(II) Gly913 to Cys substitution prevents the matrix incorporation of type II collagen which is replaced with type I and III collagens in cartilage from a patient with hypochondrogenesis. Am J Med Genet 63:129–136. https://doi.org/10.1002/(SICI)1096-8628(19960503)63:1<129::AID-AJMG23>3.0.CO;2-P

Myllyharju J, Kivirikko KI (2004) Collagens, modifying enzymes and their mutations in humans, flies and worms. Trends Genet 20:33–43. https://doi.org/10.1016/j.tig.2003.11.004

Okada M, Ikegawa S, Morioka M, Yamashita A, Saito A, Sawai H, Murotsuki J, Ohashi H, Okamoto T, Nishimura G, Imaizumi K, Tsumaki N (2015) Modeling type II collagenopathy skeletal dysplasia by directed conversion and induced pluripotent stem cells. Hum Mol Genet 24:299–313. https://doi.org/10.1093/hmg/ddu444

Pace JM, Li Y, Seegmiller RE, Teuscher C, Taylor BA, Olsen BR (1997) Disproportionate micromelia (Dmm) in mice caused by a mutation in the C-propeptide coding region of Col2a1. Dev Dyn 208:25–33. https://doi.org/10.1002/(SICI)1097-0177(199701)208:1<25::AID-AJA3>3.0.CO;2-3

Park SW, Ozcan U (2013) Potential for therapeutic manipulation of the UPR in disease. Semin Immunopathol 35:351–373. https://doi.org/10.1007/s00281-013-0370-z

Rajpar MH, McDermott B, Kung L, Eardley R, Knowles L, Heeran M, Thornton DJ, Wilson R, Bateman JF, Poulsom R, Arvan P, Kadler KE, Briggs MD, Boot-Handford RP (2009) Targeted induction of endoplasmic reticulum stress induces cartilage pathology. PLoS Genet 5:e1000691. https://doi.org/10.1371/journal.pgen.1000691

Ricks ML, Farrell JT, Falk DJ, Holt DW, Rees M, Carr J, Williams T, Nichols BA, Bridgewater LC, Reynolds PR, Kooyman DL, Seegmiller RE (2013) Osteoarthritis in temporomandibular joint of Col2a1 mutant mice. Arch Oral Biol 58:1092–1099. https://doi.org/10.1016/j.archoralbio.2013.02.008

Ruggiero F (2021) The collagen superfamily and collagenopathies, biology of extracellular matrix. Springer International Publishing. https://doi.org/10.1007/978-3-030-67592-9

Steplewski A, Brittingham R, Jimenez SA, Fertala A (2005) Single amino acid substitutions in the C-terminus of collagen II alter its affinity for collagen IX. Biochem Biophys Res Commun 335:749–755. https://doi.org/10.1016/j.bbrc.2005.07.139

Steplewski A, Majsterek I, McAdams E, Rucker E, Brittingham RJ, Ito H, Hirai K, Adachi E, Jimenez SA, Fertala A (2004) Thermostability gradient in the collagen triple helix reveals its multi-domain structure. J Mol Biol 338:989–998. https://doi.org/10.1016/j.jmb.2004.03.037

Symoens S, Hulmes DJS, Bourhis J-M, Coucke PJ, De Paepe A, Malfait F (2014) Type I procollagen C-propeptide defects: study of genotype-phenotype correlation and predictive role of crystal structure. Hum Mutat 35:1330–1341. https://doi.org/10.1002/humu.22677

Tsang KY, Chan D, Bateman JF, Cheah KSE (2010) In vivo cellular adaptation to ER stress: survival strategies with double-edged consequences. J Cell Sci 123:2145–2154. https://doi.org/10.1242/jcs.068833

Tsang KY, Chan D, Cheslett D, Chan WCW, So CL, Melhado IG, Chan TWY, Kwan KM, Hunziker EB, Yamada Y, Bateman JF, Cheung KMC, Cheah KSE (2007) Surviving endoplasmic reticulum stress is coupled to altered chondrocyte differentiation and function. PLoS Biol 5:e44. https://doi.org/10.1371/journal.pbio.0050044

Unger S, Ferreira CR, Mortier GR, Ali H, Bertola DR, Calder A, Cohn DH, Cormier-Daire V, Girisha KM, Hall C, Krakow D, Makitie O, Mundlos S, Nishimura G, Robertson SP, Savarirayan R, Sillence D, Simon M, Sutton VR, Warman ML, Superti-Furga A (2023) Nosology of genetic skeletal disorders: 2023 revision. Am J Med Genet A 191:1164–1209. https://doi.org/10.1002/ajmg.a.63132

Vissing H, D'Alessio M, Lee B, Ramirez F, Godfrey M, Hollister DW (1989) Glycine to serine substitution in the triple helical domain of pro-alpha 1 (II) collagen results in a lethal perinatal form of short-limbed dwarfism. J Biol Chem 264:18265–18267

Walter P, Ron D (2011) The unfolded protein response: from stress pathway to homeostatic regulation. Science 334:1081–1086. https://doi.org/10.1126/science.1209038

Wang C, Tan Z, Niu B, Tsang KY, Tai A, Chan WCW, Lo RLK, Leung KKH, Dung NWF, Itoh N, Zhang MQ, Chan D, Cheah KSE (2018) Inhibiting the integrated stress response pathway prevents aberrant chondrocyte differentiation thereby alleviating chondrodysplasia. elife 7. https://doi.org/10.7554/eLife.37673

Wilson R, Diseberg AF, Gordon L, Zivkovic S, Tatarczuch L, Mackie EJ, Gorman JJ, Bateman JF (2010) Comprehensive profiling of cartilage extracellular matrix formation and maturation using sequential extraction and label-free quantitative proteomics*. Mol Cell Proteomics 9:1296–1313. https://doi.org/10.1074/mcp.M000014-MCP201

Wilson R, Freddi S, Chan D, Cheah KSE, Bateman JF (2005) Misfolding of collagen X chains harboring Schmid metaphyseal chondrodysplasia mutations results in aberrant disulfide bond formation, intracellular retention, and activation of the unfolded protein response. J Biol Chem 280:15544–15552. https://doi.org/10.1074/jbc.M410758200

Zhang B, Zhang Y, Wu N, Li J, Liu H, Wang J (2020) Integrated analysis of COL2A1 variant data and classification of type II collagenopathies. Clin Genet 97:383–395. https://doi.org/10.1111/cge.13680

Chapter 5
Skeletal Disorders Linked to GAG Synthesis

Huber Céline, Guasto Alessandra, Dubail Johanne, and Cormier-Daire Valérie

Abstract Skeletal dysplasias (SD) are a group of rare disorders mainly characterized by cartilage and bone growth anomalies. A significant number of them, including chondrodysplasia with multiple dislocations (CMD), have been related to pathogenic variants in genes encoding proteins involved directly or indirectly in the proteoglycan (PG) biosynthesis. PGs are organic macromolecules composed of a core protein and one or several glycosaminoglycan side chains (GAGs), consisting in repetitive sulfated disaccharides forming the Heparan Sulfate, Chondroitin Sulfate, and Dermatan sulfate. PGs are abundantly expressed in the chondrocyte ExtraCellular Matrix (ECM). They play an important role in chondrocyte maturation and endochondral ossification as co-receptors for growth factors and ensuring the maintenance of mechanical proprieties of cartilage. However, the implication of PG biosynthesis defect in the CMD physiopathology is still poorly understood and a number of patients still remains with unidentified molecular bases.

This review focuses on genetic and glycobiological studies of CMD associated with glycosaminoglycan biosynthesis defects and related animal models. A deep comprehension of the molecular mechanisms leading to those disorders, mostly through extensive phenotypic analyses of in vitro and/or in vivo models, is essential for the development of novel biomarkers for clinical screenings and innovative therapeutics for these diseases.

H. Céline · G. Alessandra · D. Johanne
Paris Cité University, INSERM UMR1163, Imagine Institute, Paris, France

C.-D. Valérie (✉)
Paris Cité University, INSERM UMR1163, Imagine Institute, Paris, France

Service de Génétique clinique, Centre de référence pour les maladies osseuses constitutionnelles Hôpital Necker Enfants Malades, Paris, France
e-mail: valerie.cormier-daire@inserm.fr

A. Rossi, F. Zaucke (eds.), *The Extracellular Matrix in Genetic Skeletal Disorders*, Biology of Extracellular Matrix 16,
https://doi.org/10.1007/978-3-031-70835-0_5

5.1 Introduction

Skeletal dysplasia (SD), also named osteochondrodysplasia, are a group of genetic disorders mainly characterized by cartilage and bone growth anomalies. According to the 11th revision of the "Nosology of Genetic Skeletal Disorders" (Unger et al. 2023), 771 different SD have been recognized and classified in 41 groups depending on their dyadic naming, systematically associating a phenotypic entity with the gene it arises from, maintaining the MIM catalog and its historical clinical data. Thanks to the Next-generation DNA sequencing, pathogenic variants affecting 552 different genes have been identified. These disease-causing genes code for proteins involved not only in bone and cartilage formation but also in several cell biological processes like regulation of gene transcription, cell division, or intracellular transport. In addition, genes that do not code for proteins or alterations in regulatory sequences residing outside the genes have been identified as a cause of skeletal disorders (Guasto and Cormier-Daire 2021). SD can also be caused by alteration in other mechanisms like metabolic processes or the extracellular matrix (ECM) regulation. On this chapter, we will be focusing on the skeletal dysplasias caused by defects in a specific group of ECM proteins, the glycosaminoglycans (GAGs), polysaccharides linked covalently to a core protein to form macromolecules called proteoglycans (PGs).

5.2 Extracellular Matrix and Glycosaminoglycans

5.2.1 Extracellular Matrix

The extracellular matrix (ECM) is a complex micro-environment essential for cell-cell and cell-growth factor interactions (Karamanos et al. 2021). Composition and structure of the ECM vary with the specific functions and needs of a given tissue. The ECM is composed by four major types of macromolecules: collagens, elastin, adhesive glycoproteins, and a heterogeneous family of proteoglycans (PGs). Cartilage is particularly rich in ECM, with matrix making up 90% of the dry weight of the tissue. The network of collagen fibers provides both tensile strength to the tissue and the capacity to contain the swelling pressure of the embedded PGs.

Proteoglycans (PGs)

PGs are organic macromolecules expressed at the cell surface or in the ECM. They are composed of a core protein and one or several covalently bound glycosaminoglycan (GAG) side chains. These chains consist of a common linker region generated by the sequential addition of specific saccharides and repeated disaccharide units formed by amino sugar, either N-acetylglucosamine (GlcNAc) or N-acetylgalactosamine (GalNAc), and an uronic acid, either glucuronic acid (GlcUA) or iduronic acid (IdoUA), except keratan sulfate (KS) consisting of

GlcNAc and galactose. GAGs can be grouped in two families: non-sulfated GAGs, including hyaluronan which is not attached to any core protein, and sulfated GAGs, mainly represented by chondroitin sulfate (CS), dermatan sulfate (DS), KS, heparan sulfate (HS), and heparin (Paganini et al. 2019a). The acidic and hydrophobic properties of GAGs have a major role on tissue hydration and elasticity. In addition, through the negative charge of sulfate groups, PGs interact with ECM molecules and bind to growth factors regulating tissue development, especially in cartilage where PGs are abundant components of the ECM.

PGs can be classified into three main groups according to their location, as they can be secreted, remain associated with the cell surface or found in the cytoplasm. Each group can be further divided into subfamilies according to their gene homology, core protein properties, size and GAG chains composition.

GAGs Biosynthesis

PG biosynthesis is a tightly regulated process involving several enzymes and transporters and starts with the synthesis of the core protein in the rough endoplasmic reticulum (ER). At the ER-Golgi interface, the GAG chain synthesis starts with the transfer of a xylose (Xyl) residue from uridine diphosphate (UDP)-Xyl onto a serine (Ser) residue in the protein core (Fig. 5.1). This action is mediated by β-xylosyltransferases encoded by *XYLT1* or *XYLT2* (Götting et al. 2007). After Xyl addition, a linkage tetrasaccharide is synthetized. Two galactose (Gal) residues are transferred from UDP-Gal and one GlcUA from UDP-GlcUA thanks to the sequential action of β1,4-galactosyltransferase-I (GalT-I), β1,3-galactosyltransferase-II (GalT-II), and β1,3-glucuronosyltransferase-I (GlcAT-I), encoded by *B4GALT7*, *B3GALT6*, and *B3GAT3* respectively (Okajima et al. 1999; Pedersen et al. 2000; Bai et al. 2001). Several modifications may occur on the linkage region, including 2-O-phosphorylation of Xyl residue, catalyzed by a GAG-Xyl kinase, encoded by *FAM20B* gene (Koike et al. 2009), or sulfation of the first Gal residue at the C-6 position and of the second Gal residue at the C-4 or C-6 position (Gulberti et al. 2005). These events regulate the catalytic activity of GalT-I, GalT-II and GlcAT-I and in consequence the linker region assembly and GAG elongation.

Once the linkage region is synthetized, GAG chain elongation continues with two types of reactions which determine the type of GAG being synthetized: addition to the linkage tetrasaccharide of either a β4-linked GalNAc, which will initiate CS/DS assembly, or an α4-linked GlcNAc, which will initiate HS assembly (Lindahl et al. 2017).

In CS, a GalNAc residue is transferred from UDP-GalNAc to GlcUA by β1,4-N-acteylgalactosaminyltransferases, GalNAcT-I and GalNAcT-II, encoded by *CSGALNACT1* and *CSGALNACT2* respectively (Uyama et al. 2002; Sato et al. 2011). Then the CS chains are polymerized by the action of chondroitin synthases (CHSY1-3) which have both β3 glucuronosyltranferase and β4 N- acetylgalactosaminyltransferase activities, and by the chondroitin-polymerizing factor (CHPF) interacting with the CHYS (Kitagawa et al. 2001; Izumikawa et al. 2008). The disaccharides units are then sulfated by different sulfotransferases: chondroitin-4-O-sulfotransferase 1, 2 and 3 (CHST11-13), specific for C-4 sulfation of GalNAc

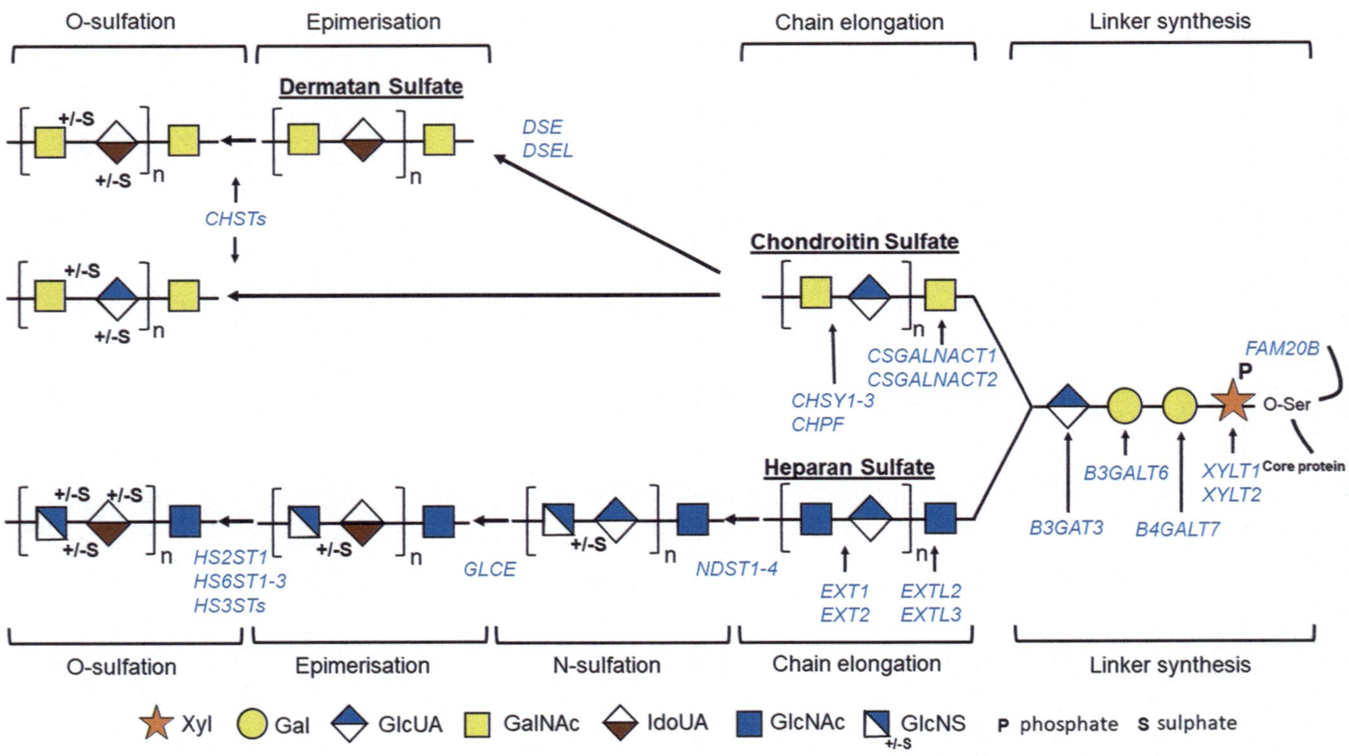

Fig. 5.1 Schematic representation of GAG biosynthesis. From Dubail and Cormier-Daire 2021. GAG biosynthesis starts with the sequential addition to serine residues of PG core protein of one xylose (Xyl), two galactoses (Gal), and one glucuronic acid (GlcUA) constituting the linker region common to three group of GAG, i.e., HS, CS, and DS. GAG chains will then be elongated by binding specific repetitive disaccharides [N-acetylgalactosamine (GalNAc) and glucuronic acid (GlcUA) for CS and N-acetylglucosamine (GlcNAc) and GlcUA for HS]. Some residues are modified: epimerization of GlcUA to iduronic acid (IdoUA) that will generate DS from CS, or N-sulfation/N-deacetylation of GlcNAc to N-sulfoglucosamine (GlcNS) followed by epimerization of GlcUA to IdoUA for HS. Finally, GAGs are further modified by O-sulfation. The enzymes implicated in these processes are indicated in blue

residues, two chondroitin-6-O-sulfotransferase (CHST3 and 7) for C-6 sulfation of GalNAc residues, one GalNAc-4-O-sulfate-6-O-sulfotransferase (CHST15) responsible for the sulfation of disulfated GalNAc and finally one uronosyl-2-O-sulfotransferase for C-2 sulfation of GlcUA (Mizumoto et al. 2013).

The disaccharide unit of DS, containing IdoUA, is formed from CS by the epimerization of the C- 5 position of GlcUA residues thanks to the action of two DS epimerases (DSE1-2), and the sulfation in distinct position by dermatan-4-O-sulfotransferase (CHST14) and uronosyl-2-O- sulfotransferase (Malmström and Aberg 1982).

In HS, the synthesis starts with the addition of a GlcNAc residue to the linkage tetrasaccharide catalyzed by α1,4-N-acetylglucosaminyltransferase-I (GlcNAcT-I) encoded by *EXTL3* (Kim et al. 2001). Later, the HS chain elongation is conducted by HS polymerase complex composed by two enzymes with N-acetylglucosaminyltransferase and glucuronyltransferase activities, and encoded by *EXT1* and *EXT2* genes respectively (McCormick et al. 2000). HS goes through a large number of modifications such as the deacetylation and N-sulfation carried out by N-deacetylase/N-sulfotransferases (NDST1-4), the epimerization of GlcUA in IdoUA by a glucuronyl epimerase, and O-sulfation by several sulfotransferases as 2-O-sulfotransferase (HS2ST1), 6-O-sulfotransferases (HS6ST1-3), and 3-O-sulfotransferases (HS3ST) (Mizumoto et al. 2014). In addition, the sulfate groups can be removed from the 6-O-sulfated GlcNS by an extracellular HS-6-O-endosulfatase (Morimoto-Tomita et al. 2002).

The synthesis of KS follows a different scheme. KS is either N-linked to asparagine residues or O-linked to Ser and threonine residues through D-GalNAc (Vynios 2014). Then a poly-N-acetylactosamine chain is added. This chain, composed by Gal and GlcNAc, is then sulfated by two sulfotransferases: GlcNAc-6-O-sulfotransferases and galactose-6-O-sulfotransferase (Funderburgh 2002).

PGs are abundantly modified by sulfation. The addition of sulfate to their covalently attached GAG chains is required for the PG physiological function (see Chap. 6). The sulfation occurs in the Golgi apparatus and is catalyzed by membrane-bound sulfotransferases which transfer sulfate groups from 3′- phospshoadenosine 5’phosphosulfate (PAPS), the universal sulfate donor, to specific residues of GAG chains. The major source of intracellular sulfate pool comes from the extracellular environment thanks to a specific sulfate/chloride antiporter of the plasma membrane named SLC26A2. A small amount of intracellular sulfate can come also from lysosomal degradation of glycoproteins and GAGs and from the catabolism of sulfur-containing amino acids and thiols. In the cytosol, this sulfate is then activated to PAPS by PAPS synthases (PAPSS1 and PAPSS2), bifunctional enzymes that catalyze the formation of adenosine 5′-phosphosulfate (APS) from sulfate and ATP (Xu et al. 2000). Then PAPS is produced from APS and ATP by the action of the same enzyme. Because most of the sulfation of glycoconjugates occurs in the Golgi apparatus, PAPS is translocated by PAPS transporters (PAPST1 and PAPST2, also named SLC35B2 and SLC35B3) from the cytoplasm into the Golgi lumen where it serves as substrate for the sulfotransferases (Kamiyama et al. 2003). Sulfotransferase activity produces PAP which can inhibit those sulfotransferases via negative feedback. To avoid this event, PAP is rapidly degraded into adenosine monophosphate (AMP) and phosphate by a Golgi resident adenosine 3′-5′-biphosphate 3′-phosphsatase (gPAPP), encoded by *IMPAD1* (Frederick et al. 2008). Finally, newly synthetized PGs are secreted in the ECM through secretory vesicles.

Correct organization of Golgi cisternae and their enzymatic content, as well as properly established pH gradient and concentration of ions like Ca2+, are requirements for a normal GAG formation (Prydz 2015).

5.2.2 Role of GAGs in Cartilage and Bone

PGs together with collagens are the major constituents of the organic matrix of bone, constituting about 5–7% of the non-collagenous matrix components. Cartilage and bone matrix contains different PG families of various structure. The predominate GAG present in cartilage has long been known to be CS. Among the different PGs, aggrecan is the most abundant component of human articular cartilage. It consists of a core protein with attached KSs and CSs. Aggrecans do not exist in isolation within the ECM, but they bind to hyaluronic acid and forms large complexes, which are stabilized by link proteins. These PG aggregates bind water and give cartilage the osmotic properties needed to resist compressive loads. In common with all other connective tissues, cartilages also contain a variety of small leucine-rich repeat PGs (SLRPs). Decorin, biglycan, fibromodulin, and lumican are the major SLRPs present in cartilage, where they help maintain the integrity of the tissue and modulate its metabolism. The function of the SLRPs depends on both their core protein and GAG chains. Their core proteins allow the SLRPs to interact with different macromolecules as collagenes, fibronectin and elastin, and growth factors such as EGF, TGFβ, and TNFα. The GAG chains have been also associated with the interaction with several growth factors, and enable the SLRPs to provide a sink for growth factor accumulation within the ECM. In this manner, the SLRPs can help modulate chondrocyte metabolism by regulating growth factor access to the cells. During bone formation, SLRPs participate in all stages including cell proliferation, osteogenesis, mineral deposition, and bone remodeling. Other PGs expressed during chondrogenesis and in cartilage include the cell surface syndecans and glypican, and the basement membrane HSPG, perlecan (Knudson and Knudson 2001). Perlecans are able to bind several molecules including FGF2, FGF7, fibronectin, heparin, laminin, and integrins and they promotes chondrogenesis in vitro. Thanks to their ability to bind and sequester growth factors such as FGF, they modulate signaling pathways in chondrocyte differentiation and in growth plate development.

5.3 Chondrodysplasia with Multiple Dislocations

Chondrodysplasia with multiple dislocations (CMD) are a group of rare genetic skeletal diseases listed mostly in the groups 5 (dysplasias with multiple joint dislocations), 4 (sulfation disorders), and 26 (bone fragility group) in the International Classification of Genetic Skeletal Disorders (Unger et al. 2023). They are characterized by severe short stature, joint laxity, dislocation of large joints (as hip, knee and shoulder), hand anomalies, and/or vertebral anomalies. The most common radiographic features include advanced carpal and tarsal bone ossification, a monkey wrench appearance of the femoral neck and dislocations of large joints. Additional skeletal features are epiphyseal, metaphyseal and vertebral changes, facial dysmorphism, and cleft palate (Fig. 5.2). The patients can also show a variable combination of extraskeletal features, such as congenital heart defects, intellectual disabilities, obesity, and tooth anomalies. So far, more than 32 distinct human genetic disorders have been described in the CDM group and most of them have been linked to pathogenic variants in 25 genes coding for proteins implicated directly or indirectly in PG biosynthesis (Dubail and Cormier-Daire 2021). Genetic defects have been described in each step of PG biosynthesis. For this reason, they can be divided in impairment of core protein synthesis, GAG chain synthesis, and GAG sulfation-related disorders. Here we focus on disorders caused by defective GAG chain synthesis (Table 5.1).

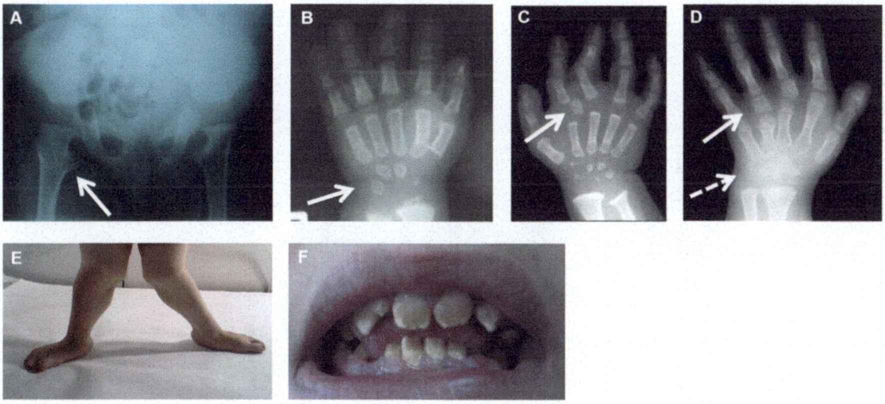

Fig. 5.2 Common and specific clinical features in CMD. From Dubail and Cormier-Daire 2021. (**a**) Hip X-rays showing a monkey wrench appearance of femur (see arrow). (**b**) Hand X-rays of a patient with XYLT1 variants at 8 months of age showing advanced carpal ossification (see arrow). (**c**) Hand X-rays of a patient with CANT1 variants at 1 year of age showing presence of a delta phalanx (see arrow). (**d**) Hand X-rays of a patient with IMPAD1 variants at 5 years of age showing presence of hyperphalangy (see plain arrow) and carpal synostosis (see dashed arrow). (**e**) Genu valgum due to joint laxity. (**f**) Amelogenesis imperfecta in a patient with SLC10A7 variants

Table 5.1 Skeletal disorders caused by variants in genes coding for proteins involved in GAG biosynthesis and related animal models

	Human					Mouse		Zebrafish/xenopus	
	Clinical entities (*MIM / inheritance, group of nosology*)	Main skeletal features	Others clinical features	Specific features	GAG defect	Genotype	Main phenotype	Genotype	Main phenotype
Linker synthesis									
XYLT1	Desbuquois dysplasia type 2 (*MIM: 615777 / AR, group 5*)	Dislocation of large joints, severe prenatal and postnatal growth retardation, flat face, short long bones and advanced carpal and tarsal ossification	Cleft palate, developmental delay, truncal obesity		Reduced total GAG synthesis after incubation with methylumbelliferyl-β-D-xylopyranoside in patient fibroblasts	*pug* mice	Early chondrocyte maturation and early ossification leading to disproportionate dwarfism	*xylt*$^{-}$ mutant zebrafish	Altered craniofacial skeletal morphology, decreased cartilage matrix and increased perichondral bone
XYLT2	Spondylocular syndrome (*MIM: 605822 / AR, group 26*)	Facial dysmorphism, short trunk, platyspondyly and osteoporosis	Ocular defects, cardiac septal defect	Osteoporosis, cataracts, renal detachment, hearing loss	Reduced HS staining in patient fibroblasts and reduced CS and HS chains synthesis in patient fibroblasts	*Xylt2*$^{-/-}$mice	Postnatal liver and kidney cysts, adipose tissue loss, increased heart, spleen and lung weight	N.D	N.D.
FAM20B	Neonatal short limb dysplasia (*MIM: - / AR, N.D.*)	Midface hypoplasia, thoracic hypoplasia, very short stature and multiple dislocations of the large joints	Respiratory failure	Mesomelic shortening, preaxial digital hypoplasia	N.D.	*Fam20b*$^{-/-}$mice *Osr2-Cre;Fam20b*$^{fl/fl}$ mice *Wnt1-Cre;Fam20b* $^{f/fl}$ mice *K14-Cre;Fam20b* $^{f/fl}$ mice *Col1a1-Cre;Fam20b* $^{f/fl}$ mice[4]	Lethal during embryonic period with multiorgan hypoplasia Chondrosarcoma and postnatal ossification defects Multiple craniofacial defects including complete cleft palate leading to postnatal death Supernumerary incisors Growth retardation and spine deformity	*Fam20b*$^{-}$ mutant zebrafish	Altered craniofacial skeletal morphology, decreased cartilage matrix and increased perichondral bone

B4GALT7	Ehlers-Danlos syndrome (EDS) progeroid variant or EDS spondylodysplastic type 1 (EDSSPD1), including Larsen syndrome, la Reunion variant *(MIM: 130070 / AR, group 5)*	Short stature, hypermobile joints, generalized osteopenia, craniofacial dysmorphism	Loose but elastic skin, defective wound healing, hypotonic muscle		Reduced epimerization of decorin GAG chain in patient fibroblasts	N.D.	N.D.	*b4galt7* morphant zebrafish, *b4galt7$^{Cas9/sgRNA}$* crispant zebrafish	Short stature, deformed pectoral fins, craniofacial dysmorphism, reduced mineralization
B3GALT6	Spondyloepimetaphyseal dysplasia with joint laxity *(MIM: 271640 / AR)* or EDS spondylodysplastic type 2 (EDSSPD2) *(MIM: 615349 / AR, group 5)*	Short stature, severe kyphoscoliosis, epimetaphyseal dysplasia, craniofacial dysmorphism, osteopenia, joint laxity	Loose skin, defective wound healing, marked muscular hypotonia		Reduced HS chains and increased CS and DS chains in patient lymphoblastoid cells. Reduced total GAG synthesis in patient fibroblasts Marked reduction of the HS GAG chains ECM organization in patient fibroblasts	N.D.	N.D.	N.D.	N.D.
B3GAT3	Larsen-like syndrome *(MIM: 245600 / AR, group 5)*	Multiple joint dislocations, short stature, craniofacial dysmorphism	Congenital heart defects		Reduced CS, DS and HS chains synthesis in patient fibroblasts	*B3gat3$^{-/-}$* mice	Embryonic lethality before the 8-cell stage due to the failure of cytokinesis	*b3gat3$^{-/-}$* mutant zebrafish	CS synthesis abolished, abnormal pharyngeal cartilage morphogenesis
CS/DS chain elongation									
CSGALNACT1	Joint dislocations and skeletal dysplasia, Desbuquois-like *(MIM: 616615 / AR, group 5)*	Non proportionate short stature, hyperlordosis, advanced bone age, mild joint laxity.			Reduced CS And DS Synthesis in Patient	*Csgalnact1$^{-/-}$* mice	Slight dwafirsm Abnormal perineural net and behavior	N.D.	N.D.

(continued)

Table 5.1 (continued)

	Human					Mouse		Zebrafish/xenopus	
	Clinical entities *(MIM / inheritance, group of nosology)*	Main skeletal features	Others clinical features	Specific features	GAG defect	Genotype	Main phenotype	Genotype	Main phenotype
CHSY1	Temtany preaxial brachydactyl syndrome (TPBS) *(MIM: 605282 / AR, group 19)*	Growth retardation, joint laxity, bilateral and symmetric preaxial brachydactyly and hyperphalangism of digits, facial dysmorphism, dental anomalies	Sensorineural hearing loss, delayed motor and mental development	Hyperphalangism and preaxial brachydactyly	Decreased CS immunostaining in patient skin Low HS levels in patient fibroblasts	*Chsy1*$^{-/-}$ mice	Chondrodysplasia and decreased bone density with severe digit patterning defects	*Chsy* morphant zebrafish	Reduced body length, compromised pectoral fin formation, cranial dysmorphism, inner ear formation defects
DSE	Ehlers-Danlos syndrome musculocontractural type 2 *(MIM: 615539 / AR, group 4)*	Joint dislocation and deformities, distinct craniofacial features	Skin hyperextensibility, bruisability and fragility, multiple congenital contractures		Decreased DS disaccharides in patient fibroblasts	*Dse*$^{-/-}$ mice	Smaller, with a 30% reduced body weight and kinked tail at birth, altered skin morphology and skin tensile strength, abdominal wall defect	*dse* morphant xenopus	Abnormal development of neural crest-derived structures
HS chain elongation									
EXTL3	Immunoskeletal dysplasia with neurodevelopmental abnormalities (ISDNA) *(MIM: 617425 / AR, group 13)*	Severe platysplondyly, brachydactyly, kyphoscoliosis, facial dysmorphisms	Severe motor developmental delay, immunodeficiency linked to T-cell lymphopenia		Lower HS concentration in patient fibroblasts, increased CS and DS concentrations in patient serum and urine Abnormal HS sulfation pattern in patient fibroblasts	*Extl3*$^{-/-}$ mice	Embryonic lethality around 8 days postcoitum	*Extl3*$^{-/-}$ (box) mutant zebrafish	Mildly altered pharyngeal cartilage morphogenesis Abnormal pectoral fin development Defective thymopoiesis

EXT1	Hereditary multiple exostosis type 1 (*MIM: 133700 / AD, group 30)*	Cartilaginous/ bony tumors within the perichondrium in long bones, ribs and other skeletal elements		Multiple exostosis	Reduced HS/CS ratio in patient serum	*Ext1*$^{-/-}$ mice *Ext1*$^{gt/gt}$ mice *Prx-Cre; Ext1*$^{fl/fl}$ mice *Gdf5-Cre; Ext1*$^{fl/fl}$ mice	Embryonic lethality at day 8,5 to 14,5 due to defects in the formation of the mesoderm and failure in egg cylinder elongation Embryonic lethal, delayed hypertrophic chondrocytes differentiation leading to skeletal defects Shortened and malformed limb bones, oligodactyly, and fusion of joints Abnormal joint formation	*ext1* morphant xenopus	Gastrulation defects
EXT2	Hereditary multiple exostosis type 2 (*MIM: 133701 / AD, group 30)*	Cartilaginous/ bony tumors within the perichondrium in long bones, ribs and other skeletal elements		Multiple exostosis	Reduced HS/CS ratio in patient serum	*Ext2*$^{-/-}$ mice *Ext2*$^{+/-}$ mice *Ext1*$^{+/-}$ *Ext2*$^{+/-}$ mice	Embryonic lethality at day 6 Exostoses in ribs Exostoses in ribs and long bones	*Ext2*$^{-/-}$ (dak) mutant zebrafish	Shorter and thicker pharyngeal cartilage elements, severe truncation of pectoral fin, severe tooth formation defects

(continued)

Table 5.1 (continued)

	Human					Mouse		Zebrafish/xenopus	
	Clinical entities *(MIM / inheritance, group of nosology)*	Main skeletal features	Others clinical features	Specific features	GAG defect	Genotype	Main phenotype	Genotype	Main phenotype
Transporter or other									
SLC35D1	Schneckendecken dysplasia (SHNKND) *(MIM: 269250 / AR, group 14)*	Neonatal lethal skeletal dysplasia, oval shaped vertebral bodies, extremely short long bones and small ilia		Snail-shaped ilia	N.D.	*Slc35d1*$^{-/-}$ mice	Neonatal lethality, extremely short limbs, hypoplasia of craniofacial bones, flattening of vertebral bodies and longitudinally short ilia	*slc35d1* morphant xenopus	Lethal form of skeletal dysplasia
SLC35A3	Multiple congenital malformation syndrome including vertebral malsegmentation and joint dislocations *(MIM: - / AR, N.D.)*	Anomalous vertebrae, limb deformities, knee and hip dislocation	Epilepsy		N.D.	N.D.	N.D.	N.D.	N.D.
SLC10A7	Skeletal dysplasia, osteoporosis, multiple dislocations and amelogenesis imperfecta *(MIM: 618363 / AR, group 5)*	Severe pre-and postnatal short stature, microretrognathia, multiple dislocation, advanced carpal ossification, amelogenesis imperfecta		Amelogenesis imperfecta	Low amount of HS and a corresponding increase in CS in patient fibroblasts	*Slc10a7*$^{-/-}$ mice	Skeletal dysplasia, short stature, low bone density, amelogenesis imperfecta	*slc10a7* morphant zebrafish	Abnormal development of several cartilage elements, strong reduction in bone mineralization

CANT1	Desbuquois dysplasia type 1, including Kim variant *(MIM: 251450 / AR)* Recessive multiple epiphyseal dysplasia *(MIM: 617719 / AR, group 5 et 9)*	Short stature, joint laxity, scoliosis, and advanced carpal ossification with a delta phalanx Mild short stature, joint pain, early-onset osteoarthropathy		Bifid distal phalanx of thumb/delta phalanx	Reduced GAG synthesis in presence of β-D-xyloside in patient fibroblasts	*Cant1*$^{-/-}$ mice	Short stature, thoracic kyphosis, delta phalanx	N.D.	N.D.
TGDS	Catel-Manzke syndrome (CATMANS) *(MIM: 616145 / AR, group 19)*	Pierre Robin sequence, bilateral hyperphalangy causing a clinodactyly of the index finger		Accessory bones between second metacarpal and proximal phalanx causing radial deviation of index fingers	N.D.	N.D.	N.D.	N.D.	N.D.
TMEM165	TMEM-CDG (*MIM: 614727 / AR, group 13)*	Postnatal growth deficiency and major spondylo-, epi-, and metaphyseal skeletal dysplasia, joint laxity	Psychomotor retardation, hypotonia		N.D.	N.D.	N.D.	*tmem165* morphant zebrafish	Reduced size and craniofacial defects
KIF22	SEMD with joint laxity (Hall type or leptodactylic type),(*MIM 603546 / AD, group 5*)	Short stature, flat face, generalized joint laxity with multiple dislocations	Progressive scoliosis and limb deformity		N.D.	*Kif22*$^{+/-}$ and *Kif22*$^{-/-}$ mice	No phenotype reported in *Kif22*$^{+/-}$ mice; ~50% of *Kif22*$^{-/-}$ mouse embryos do not survive past the morula stage	*Kif22* morphant zebrafish	Required for the formation of neural progenitors

5.3.1 Disorders Linked to Defective Linker Region Biosynthesis

Disorders linked to variants in genes encoding proteins involved in the synthesis of linker region are now frequently referred as "linkeropathies." This group includes the disorders caused by variants in *XYLT1*, *XYLT2*, *FAM20B*, *B4GALT7*, *B3GALT6*, and *B3GAT3* (Paganini et al. 2019a; Dubail and Cormier-Daire 2021). They all have an autosomal recessive mode of inheritance.

5.3.1.1 XYLT1/XYLT2

Bi-allelic variants in *XYLT1* cause Desbuquois dysplasia type 2 (DD2), also called Baratela-Scott syndrome (Bui et al. 2014; Silveira et al. 2016; Jamsheer et al. 2016; Al-Jezawi et al. 2017; Guo et al. 2017a; LaCroix et al. 2019). DD 2 is characterized by severe pre- and post-natal growth retardation, dislocation of large joints with generalized joint laxity, short, long bones with a monkey wrench appearance to the proximal femur, advanced ossification of carpal and tarsal bones, and facial dysmorphism, including flat face with prominent eyes. Alterations in the fat distribution, variable degree of intellectual disabilities, and cleft palate are also reported. In patient fibroblasts, the defect caused a reduced total GAG synthesis and decorin glycosylation (Bui et al. 2014; Schreml et al. 2014). In the pug mouse characterized by Xylt1 deficiency, due to homozygous missense variant in *Xylt1*, early chondrocyte maturation and early ossification leading to disproportionate dwarfism have been found (Mis et al. 2014). In zebrafish, *xylt1* mutants showed craniofacial anomalies and reduced cartilage matrix (Eames et al. 2011). Variants have been also identified in *XYLT2*, an isoform of *XYLT1* coding for a second xyloxyltransferase, and causing Spondyloocular syndrome characterized by facial dysmorphism, short trunk, platyspondyly, osteoporosis, cardiac septal defect, cataracts, retinal detachment, and hearing loss (Munns et al. 2015; Taylan et al. 2016; Taylan et al. 2017; Umair et al. 2018; Guleray et al. 2019; Kausar et al. 2019). *Xylt2* knock-out mice does not display skeletal defects but post-natal liver and kidney cysts, adipose tissue loss and an increased heart, spleen, and lung weight (Condac et al. 2007; Sivasami et al. 2019; Ferencz et al. 2020).

5.3.1.2 FAM20B

Recently, compound heterozygous variants in *FAM20B* have been identified in patients with a lethal form of neonatal short limb dysplasia characterized by very short stature, multiple dislocations of the large joints, thoracic hypoplasia, respiratory failure, and midface hypoplasia (Kuroda et al. 2019). Xylose phosphorylation catalyzed by FAM20B is crucial for the GalT-II activity, which is essential to complete the synthesis of the linker region. Fam20b- deficient mice exhibited embryonic lethality at embryonic day 13.5 with multiorgan hypoplasia (Vogel et al. 2012). Inactivation of Fam20b in several murine-specific tissues or cell types leading to skeletal defects demonstrated a role of Fam20b in bone development (Tian et al. 2015; Ma et al. 2016; Liu et al. 2018; Saiyin et al. 2019). Similar to xylt1 zebrafish mutant, fam20b zebrafish mutant exhibited altered craniofacial skeletal morphology, decreased cartilage matrix, and increased perichondral bone (Eames et al. 2011).

5.3.1.3 B4GALT7

Biallelic variants in *B4GALT7* cause Ehlers-Danlos syndrome (EDS) progeroid type1, now called EDS spondylodysplastic type 1 characterized by hypermobile joints, an aged appearance with loose yet elastic skin, poor wound healing, hypotonic muscles, craniofacial dysmorphism, short stature, developmental delays, and generalized osteopenia (Okajima et al. 1999; Faiyaz-Ul-Haque et al. 2004; Guo et al. 2013; Salter et al. 2016; Ritelli et al. 2017; Mihalic Mosher et al. 2019). Homozygous variants in *B4GALT7* are also responsible for a variant of Larsen syndrome, frequent on the La Reunion Island, called Larsen of Reunion Island syndrome characterized by facial features, multiple dislocations, short stature, and hyperlaxity (Cartault et al. 2015). In patient fibroblasts, *B4GALT7* variants result in abnormal biosynthesis of mature decorin and biglycan with reduced GAG chain epimerization with a core protein bearing only a Xyl residue (Seidler et al. 2006; Quentin et al. 1990). Recently, a reduced level of bikunin bearing CS chain was detected by Western blot in patient serum compared with healthy controls (Bruneel et al. 2018; Haouari et al. 2020).

Both knockdown (morphant) and mosaic knockdown b4galt7 zebrafish models presented with short stature, deformed pectoral fins, craniofacial dysmorphism, and reduced mineralization (Delbaere et al. 2020).

5.3.1.4 B3GALT6

Variants in *B3GALT6* are responsible for EDS progeroid type 2, also known as EDS spondyloplastic type 2 and spondyloepimetaphyseal dysplasia with joint laxity, Beighton type (Vorster et al. 2015; Alazami et al. 2016; Trejo et al. 2017). The main clinical features for this autosomal recessive syndrome include an aged appearance with loose but elastic skin and defective wound healing, hypermobile joints, developmental delay, short stature, craniofacial disproportion, kyphoscoliosis, epimetaphyseal dysplasia, generalized osteopenia, and hypotonic muscles. Fibroblasts from affected individuals exhibited altered GAG synthesis, with a significant reduction of HS level and an impaired glycation of decorin and reduction of HS synthesis (Malfait et al. 2013; Nakajima et al. 2013; Ritelli et al. 2015). As for patients with B4GALT7 variants, a reduced level of bikunin bearing CS chain was detected by Western blot in patient serum compared to healthy controls (Bruneel et al. 2018; Haouari et al. 2020).

5.3.1.5 B3GAT3

Finally, recessive variants in *B3GAT3* cause Larsen-like syndrome characterized by short stature, multiple joint dislocations, scoliosis, osteopenia, and cranial dysmorphisms such as a flattened midface, hypertelorism, depressed nasal bridge, and prominent forehead (Von Oettingen et al. 2014; Jones et al. 2015; Budde et al. 2015; Job et al. 2016; Bloor et al. 2017; Yauy et al. 2018; Van Damme et al. 2018; Colman et al. 2019; Ritelli et al. 2019). Congenital heart defects, including mitral valve prolapse, ventricular defect, and bicuspid aortic valve, can be observed in those patients. Patient fibroblasts exhibited reduced CS, DS, and HS and an increased level of DS-free decorin as compared with healthy controls (Baasanjav et al. 2011). Moreover, a reduced level of bikunin bearing CS chain was detected in patient serum compared with healthy controls (Bruneel et al. 2018, Haouari et al. 2020).

B3gat3-deficient mice synthesized a smaller CS and HS chain in their blastocysts than that of heterozygous mice and exhibited an embryonic lethality before the eight-cell stage due to the failure of cytokinesis (Izumikawa et al. 2010). On the other hand, *b3gat3* mutant zebrafish presented with abolished CS synthesis and abnormal pharyngeal cartilage morphogenesis (Holmborn et al. 2012).

5.3.2 Disorders Linked to Defective GAG Chain Elongation or Epimerization

5.3.2.1 CS/DS Chain Elongation or Epimerization

CSGALNACT1, *CHSY1*, and *DSE* code for enzymes implicated in CS/DS chain elongation and epimerization of CS to DS. No joint laxity or joint dislocations were described in disorders linked to pathogenic variants encoding for enzymes implicated in HS chains elongation. They will thus not be discussed in this section.

5.3.2.1.1 CSGALNACT1

Pathogenic variants in *CSGALNACT1* have been identified recently in patients with a skeletal dysplasia characterized by a mild micromelic and non-proportioned stature, joint laxity, and advanced bone age. Altered levels of CS, DS, and HS moieties were observed in patient fibroblasts compared with healthy controls (Vodopiutz et al. 2017; Mizumoto et al. 2020). CSGalNAcT1-deficient mice were described several years before and presented with moderate dwarfism and abnormalities in perineural nets and behavior (Sato et al. 2011; Watanabe et al. 2010; Yoshioka et al. 2017).

5.3.2.1.2 CHSY1

Temtamy preaxial brachydactyly syndrome is caused by homozygous variants in *CHSY1* (Li et al. 2010; Sher and Naeem 2014). It is characterized by bilateral and symmetric preaxial brachydactyly and hyperphalangism of digits, growth retardation, facial dysmorphism, deafness, and delayed motor and mental development. Patient skin biopsy exhibited decreased CS- specific immunostaining compared with controls, and reduced molecular weight bikunin bearing CS chain was detected by Western blot in patient serum compared with healthy controls (Tian et al. 2015; Bruneel et al. 2018). Chsy1-deficient mice developed a phenotype mimicking the human pathology, presenting with a chondrodysplasia and decreased bone density with severe digit patterning defects (Wilson et al. 2012). In the same way, chsy morphant zebrafish presented with reduced body length, compromised pectoral fin formation, cranial dysmorphism, and inner ear formation defects (Müller et al. 2013).

5.3.2.1.3 DSE

Homozygous variants in *DSE* cause EDS musculocontractural type 2 characterized by joint hypermobility (finger, elbow, and knee), distinctive facial features, multiple congenital contracture contractures (thumbs and feet), and myopathy. Additional features might include cardiac, valvular, respiratory, gastrointestinal, and ophthalmic complications (Gustafsson et al. 2014; Syx et al. 2015; Lautrup et al. 2020).

Decreased DS level and reduced glycanation of decorin have been evidenced in patient fibroblasts compared with controls (Quentin et al. 1990). Dse-deficient mice were smaller, with a kinky tail at birth, altered skin morphology and skin tensile strength, and abdominal wall defects (Gustafsson et al. 2014; Maccarana et al. 2009; Bartolini et al. 2012). On the other hand, dse morphant xenopus showed abnormal development of neural crest-derived structure (Gouignard et al. 2016).

5.3.2.2 HS Chain Elongation

5.3.2.2.1 EXTL3

Spondyloepimetaphyseal dysplasia (SEMD) with immune deficiency and intellectual disability also known as immunoskeletal dysplasia with developmental abnormalities (ISDNAS) is caused by homozygous variants in *EXTL3* (Guo et al. 2017b; Oud et al. 2017; Volpi et al. 2017). Patients present severe platyspondyly, brachydactyly, kyphoscoliosis, facial dysmorphism, severe motor developmental delay, and immunodeficiency linked to T-cell lymphopenia. *EXTL3* encodes α1,4-N-acetylglucosaminyltransferase-I which catalyze the addition of a GlcNac to the linkage tetrasaccharide in the first step of HS polymerization. Low HS level has been detected in fibroblasts of affected individuals (Oud et al. 2017). Embryonic lethality around 8 days post-coitum is observed in deficient *Extl3* mice (Takahashi et al. 2009), while extl3$^{-/-}$ mutant zebrafish presents mildly altered pharyngeal cartilage morphogenesis, abnormal pectoral fin development, and defective thymopoiesis (Guo et al. 2017b, Oud et al. 2017, Volpi et al. 2017).

5.3.2.2.2 EXT1/EXT2

Hereditary multiple exostosis (HME) type 1 and type 2 are caused by germline monoallelic loss of function variants in *EXT1* and *EXT2* genes. Also known as multiple osteochondromas or exostoses, they are characterized by benign osteocartilaginous tumors, located in the perichondrium flanking the growth plates of long bones, vertebrae, ribs, and cranial base in young and adolescent patients until the end of their puberty. Large number, multiple locations, and size of osteochondromas cause health problems, including skeletal deformities and bowing, joint ankylosis, skeletal growth disparities, and damages to adjacent muscles, nerves and blood vessels (Wuyts and Van Hul 2000; Jennes et al. 2009; Pacifici 2018). EXT1 and EXT2 form a Golgi heterooligomeric complex, in which EXT1 has an enzymatic function and EXT2 a structural and scaffold role, and that catalyzes the polymerization of HS (Busse-Wicher et al. 2014; Stickens et al. 2005). Heterozygous mutations in either gene lead to a HS deficiency of about 50% in serum of patients (Anower-E-Khuda et al. 2013). However a partial HS loss seems not sufficient to cause multiple exostoses and the presence of a somatic "second hit" might lead to local decreases in HS synthesis and/or levels. Cellular studies have shown that the strong decreases in local HS levels disturbed the normal homeostatic signaling pathways and caused excessive BMP signaling. Data from HME mouse models have revealed that treatment with a BMP signaling antagonist reduces osteochondroma formation (Pacifici 2018; Stickens et al. 2005; Lin et al. 2000; Koziel et al. 2004; Matsumoto et al. 2010; Mundy et al. 2011; Zak et al. 2011; Wiweger et al. 2012; Shieh et al. 2014).

5.3.3 Disorders Linked to Defective Activity of Transporters and Other Proteins Localized in the Golgi Apparatus

In this section, we will focus on genes encoding transporter or protein expressed in the Golgi for which pathogenic variants have been identified in patients with CMD and associated with defects in GAG biosynthesis, even if their exact functions on GAG biosynthesis have not been elucidated.

5.3.3.1 SLC35D1

Biallelic variants in *SLC35D1* have been identified in Schneckenbecken dysplasia (SHNKND), a neolethal chondrodysplasia with extremely short long bones, small ilia, and oval-shaped vertebral bodies (Furuichi et al. 2009; Rautengarten et al. 2019). SLC35D1 encodes a UDP-GlcUA/UDP-GalNAc transporter localized in the membrane of Golgi. The *Scl35d1* knock-out mice mimics the phenotype of SHNKND with neonatal lethality, extremely short limbs, flattening of vertebral bodies, hypoplasia of craniofacial bones, and short ilia (Hiraoka et al. 2007). The slc35d1 morphant xenopus presents a lethal form of skeletal dysplasia (De Domenico et al. 2015). Defects in this transporter reduced recruitment of UDP-sugars from the cytoplasm to the Golgi and then their availability for glycosyltransferase reactions for CS synthesis, confirmed in *Scl35d1*$^{-/-}$ mice to a marked reduction of CS chains with reduced length, whereas rate of HS is unaltered (Hiraoka et al. 2007).

5.3.3.2 SLC35A3

Pathogenic variants in *SLC35A3* have been identified in patients with epilepsy, mental retardation, and multiple skeletal defects, including shortened long bones, vertebral anomalies, large joint dislocation, and arthrogryposis (Edmondson et al. 2017; Marini et al. 2017). The skeletal features are similar to those observed in the complex vertebral malformation phenotype observed in cattle and due to homozygous missense variant in bovine *Slc35a3* (Thomsen et al. 2006). *SLC35A3* encodes a carrier that transports UDP-GlcNAc from the cytosol to the Golgi, where it serves as a substrate for (Maszczak-Seneczko et al. 2013). Reduced N-glycan branching has been evidenced in patient cells and plasma compared with control (Edvardson et al. 2013). Moreover, due to its localization and the transporter substrate, it was assumed that SLC35A3 might affect GAG metabolism, and, recently, an abnormal migration profile of bikunin bearing CS chain was observed on Western blot on patient serum (Haouari et al. 2020).

5.3.3.3 SLC10A7

Biallelic variants in *SLC10A7* have been identified in skeletal dysplasia with amelogenesis imperfecta characterized by a pre- and post-natal short stature, large joint dislocations, luxation of knees with genua valga, hypomineralized amelogenesis imperfecta, decreased bone density, "monkey wrench" appearance of the proximal femora, small epiphyses, advanced carpal ossification abnormal vertebrae, hyperlordosis or kyphoscoliosis, and dysmorphic facial features including Pierre–Robin sequence, micrognathia, and flat face (Ashikov et al. 2018; Dubail et al. 2018; Laugel-Haushalter et al. 2019). Additional features included heart defects, hearing loss, and obesity. *SLC10A7* codes for 10-transmembrane-domain transporter of unknown substrate specificity located at the Golgi and plasma membrane. Although its exact function remains unknown, it has been demonstrated that it negatively affects intracellular calcium homeostasis and increased calcium intake that have been measured in patient fibroblasts compared with controls (Dubail et al. 2018; Karakus et al. 2018).

This inadequate intracellular calcium influx most likely disturbs the Golgi ionic environment and in fine glycosyltransferase activities. Significant reduction of HS proportion linked to increased CS proportion was observed in patient fibroblasts compared with control (Dubail et al. 2018=. Interestingly, defective N-glycosylation was also detected in patient serum (Ashikov et al. 2018, Dubail et al. 2018). Slc10a7-deficient mice recapitulated human phenotype with short stature, growth plate disorganization, low bone density, and tooth enamel defects (Dubail et al. 2018). Moreover, alterations of HS/CS content similar to those measured in patient fibroblasts were detected in Slc10a7-deficient mouse cartilage. On the other hand, slc10a7 morphant zebrafish presented with abnormal development of several cartilage elements and a strong reduction in bone mineralization (Ashikov et al. 2018).

5.3.3.4 CANT1

Variants in *CANT1* cause Desbuquois dysplasia type 1 (DBQD1), characterized by severe pre-natal and post-natal growth retardation with short extremities, joint laxity, and progressive scoliosis (Huber et al. 2009; Faden et al. 2010; Inoue et al. 2014; Balasubramanian et al. 2017; Menzies et al. 2019). The main radiologic features include short, long bones with metaphyseal flaring, a "monkey wrench" appearance of the femur neck, and advanced carpal and tarsal ossification with the presence of an extra ossification center, a delta phalanx, between the proximal phalanx of the index and bifid distal thumb phalanx. A milder variant of DBQD1, referred to as "Kim variant," with hands appearing almost normal externally, but characterized on radiographic analyses by elongated phalanges, short metacarpals, and remarkably advanced carpal bone age, has also been linked to pathogenic variants in *CANT1* (Kim et al. 2010). *CANT1* codes for calcium-activated nucleotidase, an ER and Golgi nucleotidase that hydrolyses UDP to UMP and phosphate (Smith et al. 2002). It has been suggested that CANT1 deficiency would lead to inhibition of glycosyltransferase activities and reduced transport in the Golgi of UDP-sugar via negative feedback resulting from increased Golgi UDP level. CANT1 is implicated, through inositol 1,4,5-triphosphate receptor activation, in vesicular trafficking in Golgi cisternae by calcium release, both potentially affecting GAG synthesis. However, its exact function remains unknown (Huber et al. 2009; Nizon et al. 2012). Demonstrating the implication of CANT1 in GAG biosynthesis regulation, reduced GAG synthesis after stimulation by β-D-xyloside has been measured in patient fibroblasts compared with controls (Nizon et al. 2012). Cant1-deficient mice recapitulated the human phenotype of patients with DBQD1, with short stature, thoracic kyphosis, and delta phalanx (Paganini et al. 2019b; Kodama et al. 2020). Cant1 deficiency led to altered GAG synthesis, with reduced chain length and increased sulfation and delayed secretion of PG in the ECM.

5.3.3.5 TGDS

Biallelic variants in *TGDS* gene result in Catel-Manzke syndrome (CATMANS) characterized by Pierre Robin sequence, clinodactyly of index finger to a bilateral hyperphalangy, the radial deviation of index due to the presence of accessory bones between second metacarpal and proximal phalanx (Manzke et al. 2008; Ehmke et al. 2014; Pferdehirt et al. 2015; Schoner et al. 2017; Boschann et al. 2020). *TGDS* encodes for a dTDP-D-glucose 4,6-de-hydrogenase. The closest paralog of *TGDS* is *UXS1*, an UDP-glucuronate decarboxylase 1 which catalyses UDP-xylose synthesis from UDP-glucuronate required for GAG synthesis (Moriarity et al. 2002). Due to overlap between Catel-Manzke syndrome and disorders caused by defects in GAG synthesis and sulfation and its homology with UXS1, TGDS might be involved in nucleotide sugar metabolism (Ehmke et al. 2014).

5.3.3.6 TMEM165

Congenital disorder of glycosylation (CDG), type IIk or TMEM165-CDG, is caused by biallelic variants in *TMEM165* (Foulquier et al. 2012; Schulte Althoff et al. 2016). The most severe phenotypes observed in patients with TMEM165-CDG include growth retardation resistant to human growth hormone, associated with a psychomotor disability, microcephaly, facial hypoplasia, hypotonia, seizures, and hepatosplenomegaly with increased serum transaminases. Skeletal features of these patients include severe dwarfism, osteoporosis, epi-, meta-, and diaphyseal dysplasia, and joint laxity. *TMEM165* encodes the transmembrane protein 165, TMEM165, located in the Golgi membrane and, in a lower proportion, at the plasma membrane, and in late endosomes/lysosomes, and which exact function remains unknown. However, several studies are in favor of a role of TMEM165 in Mn2+ homeostasis in the Golgi, suggesting that TMEM165 could act as a putative Mn2+ transporter. This function on Mn2+ homeostasis is essential for appropriate protein N-glycosylation occurring in the Golgi (Dulary et al. 2017). *TMEM165* deficiency has mostly been associated with N-, O-glycosylation defects, with increased undersialylated and undergalactosylated glycans, as well as in high mannose type N-Glycan detected in patient serum compared with healthy control (Foulquier et al. 2012; Xia et al. 2013; Zeevaert et al. 2013). However, recently, an abnormal migration profile of bikunin bearing CS chain was observed on a Western blot on patient serum (Haouari et al. 2020). Moreover, tmem165 deficiency in zebrafish led to alterations in CS PG expression detected by immunohistochemical staining of cartilage (Bammens et al. 2015). Altogether, those results imply that TMEM165 deficiency can also lead to GAG synthesis defects. Analyses of these *tmem165* morphant zebrafish demonstrated skeletal abnormalities, more particularly in craniofacial structures with altered bone growth and development. Complete *Tmem165* deficiency in mice has not yet been described, but conditional deletion of *Tmem165* in the mammary gland led to abnormal milk production due to defective lactose biosynthesis (Snyder et al. 2019).

5.4 Glycosaminoglycan Synthesis-Related Disorders Form a Subgroup of Congenital Disorders of Glycosylation

CDG are an expanding group of rare, multisystem, underdiagnosed heterogeneous diseases caused by deficient or improper synthesis or attachment of glycans to proteins and lipids. More than 130 inherited disorders have been identified so far, most of them following an autosomal recessive inheritance pattern (Ondruskova et al. 2021). Glycosylation is a very important post-translational modification, as it is estimated that at least 2% of the human genome codes for proteins involved in this vital biochemical pathway (Ng and Freeze 2018). Due to the involvement of glycosylation in various cellular processes, CDG are characterized by multiorgan dysfunction and variable clinical phenotype. Therefore, CDG result in a broad spectrum of pathologies, including skin laxity, skeletal dysplasia, congenital heart defects, neurodevelopmental disorders, and endocrine abnormalities. The most common clinical manifestations include developmental delay, failure to thrive, microcephaly, coagulopathy, and abnormal brain magnetic resonance imaging, including cerebral and/or cerebellar atrophy, cell migration abnormalities, and immune dysfunction (Francisco et al. 2019).

CDG are classified into two groups, CDG I and CDG II. CDG I are defects in the glycan assembly and in the attachment of glycans to proteins in the ER. CDG II are defects in the processing of the already assembled glycans in the Golgi apparatus resulting in truncated and abnormal glycan structures. As it is now considered that glycosylation processes occurring in the ER are required for good folding and stability of the glycoprotein, whereas glycan remodeling in the Golgi will finely regulate the functionality of the protein, CDG I and CDG II will differently affect glycoprotein functions (Freeze 2007).

Most protein glycosylation disorders are due to defects in the N-glycosylation pathways, but they can also be due to defective O-linked glycans (O-Mannose, O-Glucose, O-Fucose, O-GlcNac, and O-GalNac), GAG, glycosylphosphatidylinositol, and glycolipids. As at the Golgi level, the different glycosylation processes coexist, alterations of the Golgi environment/organization, and inadequate supply in common subtrates, will lead to combinations of two or more affected glycosylation pathways. This is well-illustrated by CMD due to variants in genes coding for transporters or Golgi proteins, i.e., *SLC35A3*, *SLC10A7*, and *TMEM165*, for which alterations in both N-glycosylation and GAG biosynthesis were detected (Bruneel et al. 2018; Edvardson et al. 2013; Ashikov et al. 2018; Zeevaert et al. 2013).

5.5 Diagnosis

The group of CMD includes several complex syndromes with overlapping features and other clinical signs that are often non-specific, such as intellectual disability or cardiac defects.

Even if few features are typical for some disorders (Table 5.1), characteristic hand anomalies in *CANT1*, *IMPAD1*, *CHSY1*, and *TGDS*-related conditions or amelogenesis imperfecta in dysplasia due to variants in *SLC10A7*, for example, may help clinicians to orientate the clinical diagnosis. For the vast majority of CMD, phenotype–genotype correlations are still incomplete. Molecular or enzymatic assays to detect specific defects are thus crucial for final diagnostic. However, simple and affordable laboratory tests for use in screening are still missing, and, currently, clinicians must rely on research laboratories to perform the clinical molecular and biochemical tests to confirm the diagnosis. Many research teams have developed and used various techniques to evaluate quantitatively and qualitatively GAG synthesis. With these techniques, they were able to demonstrate in patient biological samples, for some PGs, a reduction of expression, concentration, or molecular weight or an alteration in the concentration of specific disaccharides, sulfated or unsulfated, constituting the GAG (Table 5.1). In most cases, analyses were performed on lysates or culture media from fibroblasts, cultured or not with radiolabeled substrates, after separation by high-performance liquid chromatography or gel electrophoresis (sodium dodecyl sulfate–polyacrylamide gel electrophoresis). Alternatively, altered migration of some low molecular weight PG was detected by sodium dodecyl sulfate–polyacrylamide gel electrophoresis, and immunoblotting or abnormal GAG expression was revealed by immunohistochemistry or immunofluorescence techniques using primary antibodies against specific GAG residues. Being more easily obtained than fibroblasts, blood and urine samples are more convenient for biochemical testing to perform diagnosis in clinics. Only a few studies have studied the GAG in blood or urine from patients using high-performance liquid chromatography. Recently, abnormal electrophoretic migration profiles of bikunin, a serum PG with a single CS chain, were detected after immunoblotting in serum from patients with specific CMD and, more specifically, from patients with "linkeropathies," indicating that bikunin is a potential biomarker, easily detectable, for these pathologies [61–62]. Moreover, for patients with variants in *SLC35A3*, *SLC10A7*, or *TMEM165*, alterations in both N- glycosylation and GAG biosynthesis were detected (Edvardson et al. 2013; Ashikov et al. 2018; Dubail et al. 2018; Foulquier et al. 2012). This demonstrates that it could be useful to look for glycosylation defects in patients with CMD to help refine the diagnosis.

5.6 Therapeutic Approaches

As for most of the skeletal and connective tissue disorders, treatments for CMD patients are still restricted to physiotherapy, orthopedic surgery, symptomatic treatment, or monitoring for potential complications to slow down or modify disease progression (Briggs et al. 2015; Marzin and Cormier-Daire 2020). The development of new innovative therapies requires a strong comprehension of the molecular mechanisms leading to those disorders, mostly through extensive phenotypic analyses of in vitro and/or in vivo models. An example is given by TMEM165. Although the function of TMEM165 is still not completely understood, many observations suggest that TMEM165 has a role in Mn^{2+} homeostasis in the Golgi. As Mn^{2+} is a cofactor of glycosyltransferases, impaired Golgi Mn^{2+} homeostasis in the TMEM165-deficient patient is most likely responsible for the glycosylation defects. Strengthening this hypothesis, experiments performed in vitro have demonstrated that Mn^{2+} supplementation suppressed the glycosylation defects, and Mn^{2+} supplementation was proposed as potential new therapy for TMEM165-CDG patients (Dulary et al. 2017).

5.7 Conclusions and Future Perspectives

The recent evolution of genomic technologies has allowed a big step forward in the comprehension of pathophysiological mechanisms leading to rare genetic disorders, including skeletal disorders and the group formed by CMD. Studies of this latter group of diseases have proven a central role of GAG biosynthesis in the pathogenesis of these rare disorders. GAG biosynthesis has characteristic biochemical properties and affects many processes such as embryonic development and connective tissue formation and functions. It is also a complex and tightly regulated process that is still not yet completely clarified. Studies performed on patient samples, cell cultures, and animal models for human CMD have provided new insights on the GAG synthesis and on the physiologic functions of GAG in cartilage, bone, and connective tissues. However, further comprehensive approaches to the molecular pathogenesis involving GAG chains, in association or not with other glycosylation defects, are required to facilitate the development of new biomarkers for clinical screenings and innovative therapeutics for these diseases.

Compliance with Ethical Standards No conflict of interest.

References

Alazami AM, Al-Qattan SM, Faqeih E, Alhashem A, Alshammari M, Alzahrani F, Al-Dosari MS, Patel N, Alsagheir A, Binabbas B, Alzaidan H, Alsiddiky A, Alharbi N, Alfadhel M, Kentab A, Daza RM, Kircher M, Shendure J, Hashem M, Alshahrani S, Rahbeeni Z, Khalifa O, Shaheen R, Alkuraya FS (2016) Expanding the clinical and genetic heterogeneity of hereditary disorders of connective tissue. Hum Genet 135:525–540

Al-Jezawi NK, Ali BR, Al-Gazali L (2017) Endoplasmic reticulum retention of xylosyltransferase 1 (XYLT1) mutants underlying Desbuquois dysplasia type II. Am J Med Genet A 173:1773–1781

Anower-E-Khuda MF, Matsumoto K, Habuchi H, Morita H, Yokochi T, Shimizu K, Kimata K (2013) Glycosaminoglycans in the blood of hereditary multiple exostoses patients: half reduction of heparan sulfate to chondroitin sulfate ratio and the possible diagnostic application. Glycobiology 23:865–876

Ashikov A, Abu Bakar N, Wen XY, Niemeijer M, Rodrigues Pinto Osorio G, Brand-Arzamendi K, Hasadsri L, Hansikova H, Raymond K, Vicogne D, Ondruskova N, Simon MEH, Pfundt R, Timal S, Beumers R, Biot C, Smeets R, Kersten M, Huijben K, Linders PTA, van den Bogaart G, van Hijum SAFT, Rodenburg R, van den Heuvel LP, van Spronsen F, Honzik T, Foulquier F, van Scherpenzeel M, Lefeber DJ, Mirjam W, Han B, Helen M, Peter VH, Jiddeke VK, Diego M, Lars M, Katja BH, Jozef H, Majid A, Kevin C, Johann TWN, group C (2018) Integrating glycomics and genomics uncovers SLC10A7 as essential factor for bone mineralization by regulating post-Golgi protein transport and glycosylation. Hum Mol Genet 27:3029–3045

Baasanjav S, Al-Gazali L, Hashiguchi T, Mizumoto S, Fischer B, Horn D, Seelow D, Ali BR, Aziz SA, Langer R, Saleh AA, Becker C, Nürnberg G, Cantagrel V, Gleeson JG, Gomez D, Michel JB, Stricker S, Lindner TH, Nürnberg P, Sugahara K, Mundlos S, Hoffmann K (2011) Faulty initiation of proteoglycan synthesis causes cardiac and joint defects. Am J Hum Genet 89:15–27

Bai X, Zhou D, Brown JR, Crawford BE, Hennet T, Esko JD (2001) Biosynthesis of the linkage region of glycosaminoglycans: cloning and activity of galactosyltransferase II, the sixth member of the beta 1,3-galactosyltransferase family (beta 3GalT6). J Biol Chem 276:48189–48195

Balasubramanian K, Li B, Krakow D, Nevarez L, Ho PJ, Ainsworth JA, Nickerson DA, Bamshad MJ, Immken L, Lachman RS, Cohn DH (2017) MED resulting from recessively inherited mutations in the gene encoding calcium-activated nucleotidase CANT1. Am J Med Genet A 173:2415–2421

Bammens R, Mehta N, Race V, Foulquier F, Jaeken J, Tiemeyer M, Steet R, Matthijs G, Flanagan-Steet H (2015) Abnormal cartilage development and altered N-glycosylation in Tmem165-deficient zebrafish mirrors the phenotypes associated with TMEM165-CDG. Glycobiology 25:669–682

Bartolini B, Thelin MA, Rauch U, Feinstein R, Oldberg A, Malmström A, Maccarana M (2012) Mouse development is not obviously affected by the absence of dermatan sulfate epimerase 2 in spite of a modified brain dermatan sulfate composition. Glycobiology 22:1007–1016

Bloor S, Giri D, Didi M, Senniappan S (2017) Novel splicing mutation in. Case Rep Genet 2017:3941483

Boschann F, Stuurman KE, de Bruin C, van Slegtenhorst M, van Duyvenvoorde HA, Kant SG, Ehmke N (2020) TGDS pathogenic variants cause Catel-Manzke syndrome without hyperphalangy. Am J Med Genet A 182:431–436

Briggs MD, Bell PA, Wright MJ, Pirog KA (2015) New therapeutic targets in rare genetic skeletal diseases. Expert Opin Orphan Drugs 3:1137–1154

Bruneel A, Dubail J, Roseau C, Prada P, Haouari W, Huber C, Dupré T, Poüs C, Cormier-Daire V, Seta N (2018) Serum bikunin is a biomarker of linkeropathies. Clin Chim Acta 485:178–180

Budde BS, Mizumoto S, Kogawa R, Becker C, Altmüller J, Thiele H, Rüschendorf F, Toliat MR, Kaleschke G, Hämmerle JM, Höhne W, Sugahara K, Nürnberg P, Kennerknecht I (2015) Skeletal dysplasia in a consanguineous clan from the Island of Nias/Indonesia is caused by a novel mutation in B3GAT3. Hum Genet 134:691–704

Bui C, Huber C, Tuysuz B, Alanay Y, Bole-Feysot C, Leroy JG, Mortier G, Nitschke P, Munnich A, Cormier-Daire V (2014) XYLT1 mutations in Desbuquois dysplasia type 2. Am J Hum Genet 94:405–414

Busse-Wicher M, Wicher KB, Kusche-Gullberg M (2014) The exostosin family: proteins with many functions. Matrix Biol 35:25–33

Cartault F, Munier P, Jacquemont ML, Vellayoudom J, Doray B, Payet C, Randrianaivo H, Laville JM, Munnich A, Cormier-Daire V (2015) Expanding the clinical spectrum of B4GALT7 deficiency: homozygous p.R270C mutation with founder effect causes Larsen of Reunion Island syndrome. Eur J Hum Genet 23:49–53

Colman M, Van Damme T, Steichen-Gersdorf E, Laccone F, Nampoothiri S, Syx D, Guillemyn B, Symoens S, Malfait F (2019) The clinical and mutational spectrum of B3GAT3 linkeropathy: two case reports and literature review. Orphanet J Rare Dis 14:138

Condac E, Silasi-Mansat R, Kosanke S, Schoeb T, Towner R, Lupu F, Cummings RD, Hinsdale ME (2007) Polycystic disease caused by deficiency in xylosyltransferase 2, an initiating enzyme of glycosaminoglycan biosynthesis. Proc Natl Acad Sci USA 104:9416–9421

De Domenico E, Owens ND, Grant IM, Gomes-Faria R, Gilchrist MJ (2015) Molecular asymmetry in the 8-cell stage Xenopus tropicalis embryo described by single blastomere transcript sequencing. Dev Biol 408:252–268

Delbaere S, Van Damme T, Syx D, Symoens S, Coucke P, Willaert A, Malfait F (2020) Hypomorphic zebrafish models mimic the musculoskeletal phenotype of β4GalT7-deficient Ehlers-Danlos syndrome. Matrix Biol 89:59–75

Dubail J, Huber C, Chantepie S, Sonntag S, Tüysüz B, Mihci E, Gordon CT, Steichen-Gersdorf E, Amiel J, Nur B, Stolte-Dijkstra I, van Eerde AM, van Gassen KL, Breugem CC, Stegmann A, Lekszas C, Maroofian R, Karimiani EG, Bruneel A, Seta N, Munnich A, Papy-Garcia D, De La Dure-Molla M, Cormier-Daire V (2018) SLC10A7 mutations cause a skeletal dysplasia with amelogenesis imperfecta mediated by GAG biosynthesis defects. Nat Commun 9:3087

Dubail J, Cormier-Daire V (2021) Chondrodysplasias with multiple dislocations caused by defects in glycosaminoglycan synthesis. Front Genet 12:642097

Dulary E, Potelle S, Legrand D, Foulquier F (2017) TMEM165 deficiencies in congenital disorders of glycosylation type II (CDG-II): clues and evidences for roles of the protein in Golgi functions and ion homeostasis. Tissue Cell 49:150–156

Eames BF, Yan YL, Swartz ME, Levic DS, Knapik EW, Postlethwait JH, Kimmel CB (2011) Mutations in fam20b and xylt1 reveal that cartilage matrix controls timing of endochondral ossification by inhibiting chondrocyte maturation. PLoS Genet 7:e1002246

Edmondson AC, Bedoukian EC, Deardorff MA, McDonald-McGinn DM, Li X, He M, Zackai EH (2017) A human case of SLC35A3-related skeletal dysplasia. Am J Med Genet A 173:2758–2762

Edvardson S, Ashikov A, Jalas C, Sturiale L, Shaag A, Fedick A, Treff NR, Garozzo D, Gerardy-Schahn R, Elpeleg O (2013) Mutations in SLC35A3 cause autism spectrum disorder, epilepsy and arthrogryposis. J Med Genet 50:733–739

Ehmke N, Caliebe A, Koenig R, Kant SG, Stark Z, Cormier-Daire V, Wieczorek D, Gillessen-Kaesbach G, Hoff K, Kawalia A, Thiele H, Altmüller J, Fischer-Zirnsak B, Knaus A, Zhu N, Heinrich V, Huber C, Harabula I, Spielmann M, Horn D, Kornak U, Hecht J, Krawitz PM, Nürnberg P, Siebert R, Manzke H, Mundlos S (2014) Homozygous and compound-heterozygous mutations in TGDS cause Catel-Manzke syndrome. Am J Hum Genet 95:763–770

Faden M, Al-Zahrani F, Arafah D, Alkuraya FS (2010) Mutation of CANT1 causes Desbuquois dysplasia. Am J Med Genet A 152A:1157–1160

Faiyaz-Ul-Haque M, Zaidi SH, Al-Ali M, Al-Mureikhi MS, Kennedy S, Al-Thani G, Tsui LC, Teebi AS (2004) A novel missense mutation in the galactosyltransferase-I (B4GALT7) gene in a family exhibiting facioskeletal anomalies and Ehlers-Danlos syndrome resembling the progeroid type. Am J Med Genet A 128A:39–45

Ferencz B, Condac E, Poudel N, Munteanu MC, Sivasami P, Choudhury B, Naidu NN, Zhang F, Breshears M, Linhardt RJ, Hinsdale ME (2020) Xylosyltransferase 2 deficiency and organ homeostasis. Glycoconj J 37:755–765

Foulquier F, Amyere M, Jaeken J, Zeevaert R, Schollen E, Race V, Bammens R, Morelle W, Rosnoblet C, Legrand D, Demaegd D, Buist N, Cheillan D, Guffon N, Morsomme P, Annaert W, Freeze HH, Van Schaftingen E, Vikkula M, Matthijs G (2012) TMEM165 deficiency causes a congenital disorder of glycosylation. Am J Hum Genet 91:15–26

Francisco R, Marques-da-Silva D, Brasil S, Pascoal C, Dos Reis FV, Morava E, Jaeken J (2019) The challenge of CDG diagnosis. Mol Genet Metab 126:1–5

Frederick JP, Tafari AT, Wu SM, Megosh LC, Chiou ST, Irving RP, York JD (2008) A role for a lithium-inhibited Golgi nucleotidase in skeletal development and sulfation. Proc Natl Acad Sci USA 105:11605–11612

Freeze HH (2007) Congenital disorders of glycosylation: CDG-I, CDG-II, and beyond. Curr Mol Med 7:389–396

Funderburgh JL (2002) Keratan sulfate biosynthesis. IUBMB Life 54:187–194

Furuichi T, Kayserili H, Hiraoka S, Nishimura G, Ohashi H, Alanay Y, Lerena JC, Aslanger AD, Koseki H, Cohn DH, Superti-Furga A, Unger S, Ikegawa S (2009) Identification of loss-of-function mutations of SLC35D1 in patients with Schneckenbecken dysplasia, but not with other severe spondylodysplastic dysplasias group diseases. J Med Genet 46:562–568

Gouignard N, Maccarana M, Strate I, von Stedingk K, Malmström A, Pera EM (2016) Musculocontractural Ehlers-Danlos syndrome and neurocristopathies: dermatan sulfate is required for Xenopus neural crest cells to migrate and adhere to fibronectin. Dis Model Mech 9:607–620

Götting C, Kuhn J, Kleesiek K (2007) Human xylosyltransferases in health and disease. Cell Mol Life Sci 64:1498–1517

Guasto A, Cormier-Daire V (2021) Signaling pathways in bone development and their related skeletal dysplasia. Int J Mol Sci 22

Gulberti S, Lattard V, Fondeur M, Jacquinet JC, Mulliert G, Netter P, Magdalou J, Ouzzine M, Fournel-Gigleux S (2005) Phosphorylation and sulfation of oligosaccharide substrates critically influence the activity of human beta1,4-galactosyltransferase 7 (GalT-I) and beta1,3-glucuronosyltransferase I (GlcAT-I) involved in the biosynthesis of the glycosaminoglycan-protein linkage region of proteoglycans. J Biol Chem 280:1417–1425

Guleray N, Simsek Kiper PO, Utine GE, Boduroglu K, Alikasifoglu M (2019) Intrafamilial variability of XYLT2-related spondyloocular syndrome. Eur J Med Genet 62:103585

Guo MH, Stoler J, Lui J, Nilsson O, Bianchi DW, Hirschhorn JN, Dauber A (2013) Redefining the progeroid form of Ehlers-Danlos syndrome: report of the fourth patient with B4GALT7 deficiency and review of the literature. Am J Med Genet A 161A:2519–2527

Guo L, Elcioglu NH, Iida A, Demirkol YK, Aras S, Matsumoto N, Nishimura G, Miyake N, Ikegawa S (2017a) Novel and recurrent XYLT1 mutations in two Turkish families with Desbuquois dysplasia, type 2. J Hum Genet 62:447–451

Guo L, Elcioglu NH, Mizumoto S, Wang Z, Noyan B, Albayrak HM, Yamada S, Matsumoto N, Miyake N, Nishimura G, Ikegawa S (2017b) Identification of biallelic EXTL3 mutations in a novel type of spondylo-epi-metaphyseal dysplasia. J Hum Genet 62:797–801

Gustafsson R, Stachtea X, Maccarana M, Grottling E, Eklund E, Malmström A, Oldberg A (2014) Dermatan sulfate epimerase 1 deficient mice as a model for human abdominal wall defects. Birth Defects Res A Clin Mol Teratol 100:712–720

Haouari W, Dubail J, Lounis-Ouaras S, Prada P, Bennani R, Roseau C, Huber C, Afenjar A, Colin E, Vuillaumier-Barrot S, Seta N, Foulquier F, Poüs C, Cormier-Daire V, Bruneel A (2020) Serum bikunin isoforms in congenital disorders of glycosylation and linkeropathies. J Inherit Metab Dis 43:1349–1359

Hiraoka S, Furuichi T, Nishimura G, Shibata S, Yanagishita M, Rimoin DL, Superti-Furga A, Nikkels PG, Ogawa M, Katsuyama K, Toyoda H, Kinoshita-Toyoda A, Ishida N, Isono K, Sanai Y, Cohn DH, Koseki H, Ikegawa S (2007) Nucleotide-sugar transporter SLC35D1 is critical to chondroitin sulfate synthesis in cartilage and skeletal development in mouse and human. Nat Med 13:1363–1367

Holmborn K, Habicher J, Kasza Z, Eriksson AS, Filipek-Gorniok B, Gopal S, Couchman JR, Ahlberg PE, Wiweger M, Spillmann D, Kreuger J, Ledin J (2012) On the roles and regulation of chondroitin sulfate and heparan sulfate in zebrafish pharyngeal cartilage morphogenesis. J Biol Chem 287:33905–33916

Huber C, Oulès B, Bertoli M, Chami M, Fradin M, Alanay Y, Al-Gazali LI, Ausems MG, Bitoun P, Cavalcanti DP, Krebs A, Le Merrer M, Mortier G, Shafeghati Y, Superti-Furga A, Robertson SP, Le Goff C, Muda AO, Paterlini-Bréchot P, Munnich A, Cormier-Daire V (2009) Identification of CANT1 mutations in Desbuquois dysplasia. Am J Hum Genet 85:706–710

Inoue S, Ishii A, Shirotani G, Tsutsumi M, Ohta E, Nakamura M, Mori T, Inoue T, Nishimura G, Ogawa A, Hirose S (2014) Case of Desbuquois dysplasia type 1: potentially lethal skeletal dysplasia. Pediatr Int 56:e26–e29

Izumikawa T, Koike T, Shiozawa S, Sugahara K, Tamura J, Kitagawa H (2008) Identification of chondroitin sulfate glucuronyltransferase as chondroitin synthase-3 involved in chondroitin polymerization: chondroitin polymerization is achieved by multiple enzyme complexes consisting of chondroitin synthase family members. J Biol Chem 283:11396–11406

Izumikawa T, Kanagawa N, Watamoto Y, Okada M, Saeki M, Sakano M, Sugahara K, Sugihara K, Asano M, Kitagawa H (2010) Impairment of embryonic cell division and glycosaminoglycan biosynthesis in glucuronyltransferase-I-deficient mice. J Biol Chem 285:12190–12196

Jamsheer A, Olech EM, Kozłowski K, Niedziela M, Sowińska-Seidler A, Obara-Moszyńska M, Latos-Bieleńska A, Karczewski M, Zemojtel T (2016) Exome sequencing reveals two novel compound heterozygous XYLT1 mutations in a Polish patient with Desbuquois dysplasia type 2 and growth hormone deficiency. J Hum Genet 61:577–583

Jennes I, Pedrini E, Zuntini M, Mordenti M, Balkassmi S, Asteggiano CG, Casey B, Bakker B, Sangiorgi L, Wuyts W (2009) Multiple osteochondromas: mutation update and description of the multiple osteochondromas mutation database (MOdb). Hum Mutat 30:1620–1627

Job F, Mizumoto S, Smith L, Couser N, Brazil A, Saal H, Patterson M, Gibson MI, Soden S, Miller N, Thiffault I, Saunders C, Yamada S, Hoffmann K, Sugahara K, Farrow E (2016) Functional validation of novel compound heterozygous variants in B3GAT3 resulting in severe osteopenia and fractures: expanding the disease phenotype. BMC Med Genet 17:86

Jones KL, Schwarze U, Adam MP, Byers PH, Mefford HC (2015) A homozygous B3GAT3 mutation causes a severe syndrome with multiple fractures, expanding the phenotype of linkeropathy syndromes. Am J Med Genet A 167A:2691–2696

Kamiyama S, Suda T, Ueda R, Suzuki M, Okubo R, Kikuchi N, Chiba Y, Goto S, Toyoda H, Saigo K, Watanabe M, Narimatsu H, Jigami Y, Nishihara S (2003) Molecular cloning and identification of 3′-phosphoadenosine 5′-phosphosulfate transporter. J Biol Chem 278:25958–25963

Karakus E, Zahner D, Grosser G, Leidolf R, Gundogdu C, Sánchez-Guijo A, Wudy SA, Geyer J (2018) Estrone-3-sulfate stimulates the proliferation of T47D breast cancer cells stably transfected with the sodium-dependent organic anion transporter SOAT (SLC10A6). Front Pharmacol 9:941

Karamanos NK, Theocharis AD, Piperigkou Z, Manou D, Passi A, Skandalis SS, Vynios DH, Orian-Rousseau V, Ricard-Blum S, Schmelzer CEH, Duca L, Durbeej M, Afratis NA, Troeberg L, Franchi M, Masola V, Onisto M (2021) A guide to the composition and functions of the extracellular matrix. FEBS J 288(24):6850–6912. https://doi.org/10.1111/febs.15776. Epub 2021 Mar 23

Kausar M, Chew EGY, Ullah H, Anees M, Khor CC, Foo JN, Makitie O, Siddiqi S (2019) A novel homozygous frameshift variant in. Front Genet 10:144

Kim BT, Kitagawa H, Tamura J, Saito T, Kusche-Gullberg M, Lindahl U, Sugahara K (2001) Human tumor suppressor EXT gene family members EXTL1 and EXTL3 encode alpha 1,4-N-acetylglucosaminyltransferases that likely are involved in heparan sulfate/ heparin biosynthesis. Proc Natl Acad Sci USA 98:7176–7181

Kim OH, Nishimura G, Song HR, Matsui Y, Sakazume S, Yamada M, Narumi Y, Alanay Y, Unger S, Cho TJ, Park SS, Ikegawa S, Meinecke P, Superti-Furga A (2010) A variant of Desbuquois dysplasia characterized by advanced carpal bone age, short metacarpals, and elongated phalanges: report of seven cases. Am J Med Genet A 152A:875–885

Kitagawa H, Uyama T, Sugahara K (2001) Molecular cloning and expression of a human chondroitin synthase. J Biol Chem 276:38721–38726

Knudson CB, Knudson W (2001) Cartilage proteoglycans. Semin Cell Dev Biol 12:69–78

Kodama K, Takahashi H, Oiji N, Nakano K, Okamura T, Niimi K, Takahashi E, Guo L, Ikegawa S, Furuichi T (2020) CANT1 deficiency in a mouse model of Desbuquois dysplasia impairs glycosaminoglycan synthesis and chondrocyte differentiation in growth plate cartilage. FEBS Open Bio 10:1096–1103

Koike T, Izumikawa T, Tamura J, Kitagawa H (2009) FAM20B is a kinase that phosphorylates xylose in the glycosaminoglycan-protein linkage region. Biochem J 421:157–162

Koziel L, Kunath M, Kelly OG, Vortkamp A (2004) Ext1-dependent heparan sulfate regulates the range of Ihh signaling during endochondral ossification. Dev Cell 6:801–813

Kuroda Y, Murakami H, Enomoto Y, Tsurusaki Y, Takahashi K, Mitsuzuka K, Ishimoto H, Nishimura G, Kurosawa K (2019) A novel gene (FAM20B encoding glycosaminoglycan xylosylkinase) for neonatal short limb dysplasia resembling Desbuquois dysplasia. Clin Genet 95:713–717

LaCroix AJ, Stabley D, Sahraoui R, Adam MP, Mehaffey M, Kernan K, Myers CT, Fagerstrom C, Anadiotis G, Akkari YM, Robbins KM, Gripp KW, Baratela WAR, Bober MB, Duker AL, Doherty D, Dempsey JC, Miller DG, Kircher M, Bamshad MJ, Nickerson DA, Mefford HC, Sol-Church K, UoWCfM G (2019) GGC repeat expansion and exon 1 methylation of XYLT1 is a common pathogenic variant in Baratela-Scott syndrome. Am J Hum Genet 104:35–44

Laugel-Haushalter V, Bär S, Schaefer E, Stoetzel C, Geoffroy V, Alembik Y, Kharouf N, Huckert M, Hamm P, Hemmerlé J, Manière MC, Friant S, Dollfus H, Bloch-Zupan A (2019) A New. Front Genet 10:504

Lautrup CK, Teik KW, Unzaki A, Mizumoto S, Syx D, Sin HH, Nielsen IK, Markholt S, Yamada S, Malfait F, Matsumoto N, Miyake N, Kosho T (2020) Delineation of musculocontractural Ehlers-Danlos syndrome caused by dermatan sulfate epimerase deficiency. Mol Genet Genomic Med 8:e1197

Li Y, Laue K, Temtamy S, Aglan M, Kotan LD, Yigit G, Canan H, Pawlik B, Nürnberg G, Wakeling EL, Quarrell OW, Baessmann I, Lanktree MB, Yilmaz M, Hegele RA, Amr K, May KW, Nürnberg P, Topaloglu AK, Hammerschmidt M, Wollnik B (2010) Temtamy preaxial brachydactyly syndrome is caused by loss-of-function mutations in chondroitin synthase 1, a potential target of BMP signaling. Am J Hum Genet 87:757–767

Lin X, Wei G, Shi Z, Dryer L, Esko JD, Wells DE, Matzuk MM (2000) Disruption of gastrulation and heparan sulfate biosynthesis in EXT1-deficient mice. Dev Biol 224:299–311

Lindahl U, Couchman J, Kimata K, Esko JD (2017) Proteoglycans and sulfated Glycosaminoglycans. In: Varki A, Cummings RD, Esko JD, Stanley P, Hart GW, Aebi M, Darvill AG, Kinoshita T, Packer NH, Prestegard JH, Schnaar RL, Seeberger PH (eds) Essentials of Glycobiology [Internet], 3rd edn. Cold Spring Harbor Laboratory Press, Cold Spring Harbor (NY), pp 2015–2017

Liu X, Li N, Zhang H, Liu J, Zhou N, Ran C, Chen X, Lu Y, Wang X, Qin C, Xiao J, Liu C (2018) Inactivation of Fam20b in the neural crest-derived mesenchyme of mouse causes multiple craniofacial defects. Eur J Oral Sci 126:433–436

Ma P, Yan W, Tian Y, Wang J, Feng JQ, Qin C, Cheng YS, Wang X (2016) Inactivation of Fam20B in joint cartilage leads to chondrosarcoma and postnatal ossification defects. Sci Rep 6:29814

Maccarana M, Kalamajski S, Kongsgaard M, Magnusson SP, Oldberg A, Malmström A (2009) Dermatan sulfate epimerase 1-deficient mice have reduced content and changed distribution of iduronic acids in dermatan sulfate and an altered collagen structure in skin. Mol Cell Biol 29:5517–5528

McCormick C, Duncan G, Goutsos KT, Tufaro F (2000) The putative tumor suppressors EXT1 and EXT2 form a stable complex that accumulates in the Golgi apparatus and catalyzes the synthesis of heparan sulfate. Proc Natl Acad Sci USA 97:668–673

Malfait F, Kariminejad A, Van Damme T, Gauche C, Syx D, Merhi-Soussi F, Gulberti S, Symoens S, Vanhauwaert S, Willaert A, Bozorgmehr B, Kariminejad MH, Ebrahimiadib N, Hausser I, Huysseune A, Fournel-Gigleux S, De Paepe A (2013) Defective initiation of glycosaminoglycan synthesis due to B3GALT6 mutations causes a pleiotropic Ehlers-Danlos-syndrome-like connective tissue disorder. Am J Hum Genet 92:935–945

Malmström A, Aberg L (1982) Biosynthesis of dermatan sulphate. Assay and properties of the uronosyl C-5 epimerase. Biochem J 201:489–493

Manzke H, Lehmann K, Klopocki E, Caliebe A (2008) Catel-Manzke syndrome: two new patients and a critical review of the literature. Eur J Med Genet 51:452–465

Marini C, Hardies K, Pisano T, May P, Weckhuysen S, Cellini E, Suls A, Mei D, Balling R, Jonghe PD, Helbig I, Garozzo D, Guerrini R, group EcAw (2017) Recessive mutations in

SLC35A3 cause early onset epileptic encephalopathy with skeletal defects. Am J Med Genet A 173:1119–1123
Marzin P, Cormier-Daire V (2020) New perspectives on the treatment of skeletal dysplasia. Ther Adv Endocrinol Metab 11:2042018820904016
Maszczak-Seneczko D, Sosicka P, Olczak T, Jakimowicz P, Majkowski M, Olczak M (2013) UDP-N-acetylglucosamine transporter (SLC35A3) regulates biosynthesis of highly branched N-glycans and keratan sulfate. J Biol Chem 288:21850–21860
Matsumoto Y, Matsumoto K, Irie F, Fukushi J, Stallcup WB, Yamaguchi Y (2010) Conditional ablation of the heparan sulfate-synthesizing enzyme Ext1 leads to dysregulation of bone morphogenic protein signaling and severe skeletal defects. J Biol Chem 285:19227–19234
Menzies L, Cullup T, Calder A, Wilson L, Faravelli F (2019) A novel homozygous variant in CANT1 in a patient with Kim-type Desbuquois dysplasia. Clin Dysmorphol 28:219–223
Mihalic Mosher T, Zygmunt DA, Koboldt DC, Kelly BJ, Johnson LR, McKenna DS, Hood BC, Hickey SE, White P, Wilson RK, Martin PT, McBride KL (2019) Expansion of B4GALT7 linkeropathy phenotype to include perinatal lethal skeletal dysplasia. Eur J Hum Genet 27:1569–1577
Mis EK, Liem KF, Kong Y, Schwartz NB, Domowicz M, Weatherbee SD (2014) Forward genetics defines Xylt1 as a key, conserved regulator of early chondrocyte maturation and skeletal length. Dev Biol 385:67–82
Mizumoto S, Ikegawa S, Sugahara K (2013) Human genetic disorders caused by mutations in genes encoding biosynthetic enzymes for sulfated glycosaminoglycans. J Biol Chem 288:10953–10961
Mizumoto S, Yamada S, Sugahara K (2014) Human genetic disorders and knockout mice deficient in glycosaminoglycan. Biomed Res Int 2014:495764
Mizumoto S, Janecke AR, Sadeghpour A, Povysil G, McDonald MT, Unger S, Greber-Platzer S, Deak KL, Katsanis N, Superti-Furga A, Sugahara K, Davis EE, Yamada S, Vodopiutz J (2020) CSGALNACT1-congenital disorder of glycosylation: a mild skeletal dysplasia with advanced bone age. Hum Mutat 41:655–667
Moriarity JL, Hurt KJ, Resnick AC, Storm PB, Laroy W, Schnaar RL, Snyder SH (2002) UDP-glucuronate decarboxylase, a key enzyme in proteoglycan synthesis: cloning, characterization, and localization. J Biol Chem 277:16968–16975
Morimoto-Tomita M, Uchimura K, Werb Z, Hemmerich S, Rosen SD (2002) Cloning and characterization of two extracellular heparin-degrading endosulfatases in mice and humans. J Biol Chem 277:49175–49185
Müller T, Mizumoto S, Suresh I, Komatsu Y, Vodopiutz J, Dundar M, Straub V, Lingenhel A, Melmer A, Lechner S, Zschocke J, Sugahara K, Janecke AR (2013) Loss of dermatan sulfate epimerase (DSE) function results in musculocontractural Ehlers-Danlos syndrome. Hum Mol Genet 22:3761–3772
Mundy C, Yasuda T, Kinumatsu T, Yamaguchi Y, Iwamoto M, Enomoto-Iwamoto M, Koyama E, Pacifici M (2011) Synovial joint formation requires local Ext1 expression and heparan sulfate production in developing mouse embryo limbs and spine. Dev Biol 351:70–81
Munns CF, Fahiminiya S, Poudel N, Munteanu MC, Majewski J, Sillence DO, Metcalf JP, Biggin A, Glorieux F, Fassier F, Rauch F, Hinsdale ME (2015) Homozygosity for frameshift mutations in XYLT2 result in a spondylo-ocular syndrome with bone fragility, cataracts, and hearing defects. Am J Hum Genet 96:971–978
Nakajima M, Mizumoto S, Miyake N, Kogawa R, Iida A, Ito H, Kitoh H, Hirayama A, Mitsubuchi H, Miyazaki O, Kosaki R, Horikawa R, Lai A, Mendoza-Londono R, Dupuis L, Chitayat D, Howard A, Leal GF, Cavalcanti D, Tsurusaki Y, Saitsu H, Watanabe S, Lausch E, Unger S, Bonafé L, Ohashi H, Superti-Furga A, Matsumoto N, Sugahara K, Nishimura G, Ikegawa S (2013) Mutations in B3GALT6, which encodes a glycosaminoglycan linker region enzyme, cause a spectrum of skeletal and connective tissue disorders. Am J Hum Genet 92:927–934
Nizon M, Huber C, De Leonardis F, Merrina R, Forlino A, Fradin M, Tuysuz B, Abu-Libdeh BY, Alanay Y, Albrecht B, Al-Gazali L, Basaran SY, Clayton-Smith J, Désir J, Gill H, Greally MT,

Koparir E, van Maarle MC, MacKay S, Mortier G, Morton J, Sillence D, Vilain C, Young I, Zerres K, Le Merrer M, Munnich A, Le Goff C, Rossi A, Cormier-Daire V (2012) Further delineation of CANT1 phenotypic spectrum and demonstration of its role in proteoglycan synthesis. Hum Mutat 33:1261–1266

Ng BG, Freeze HH (2018) Perspectives on glycosylation and its congenital disorders. Trends Genet 34:466–476

Okajima T, Fukumoto S, Furukawa K, Urano T (1999) Molecular basis for the progeroid variant of Ehlers-Danlos syndrome. Identification and characterization of two mutations in galactosyltransferase I gene. J Biol Chem 274:28841–28844

Ondruskova N, Cechova A, Hansikova H, Honzik T, Jaeken J (2021) Congenital disorders of glycosylation: still "hot" in 2020. Biochim Biophys Acta Gen Subj 1865:129751

Oud MM, Tuijnenburg P, Hempel M, van Vlies N, Ren Z, Ferdinandusse S, Jansen MH, Santer R, Johannsen J, Bacchelli C, Alders M, Li R, Davies R, Dupuis L, Cale CM, Wanders RJA, Pals ST, Ocaka L, James C, Müller I, Lehmberg K, Strom T, Engels H, Williams HJ, Beales P, Roepman R, Dias P, Brunner HG, Cobben JM, Hall C, Hartley T, Le Quesne SP, Mendoza-Londono R, Davies EG, de Sousa SB, Lessel D, Arts HH, Kuijpers TW (2017) Mutations in EXTL3 cause neuro-immuno-skeletal dysplasia syndrome. Am J Hum Genet 100:281–296

Pacifici M (2018) The pathogenic roles of heparan sulfate deficiency in hereditary multiple exostoses. Matrix Biol 71-72:28–39

Paganini C, Costantini R, Superti-Furga A, Rossi A (2019a) Bone and connective tissue disorders caused by defects in glycosaminoglycan biosynthesis: a panoramic view. FEBS J 286:3008–3032

Paganini C, Monti L, Costantini R, Besio R, Lecci S, Biggiogera M, Tian K, Schwartz JM, Huber C, Cormier-Daire V, Gibson BG, Pirog KA, Forlino A, Rossi A (2019b) Calcium activated nucleotidase 1 (CANT1) is critical for glycosaminoglycan biosynthesis in cartilage and endochondral ossification. Matrix Biol 81:70–90

Pedersen LC, Tsuchida K, Kitagawa H, Sugahara K, Darden TA, Negishi M (2000) Heparan/chondroitin sulfate biosynthesis. Structure and mechanism of human glucuronyltransferase I. J Biol Chem 275:34580–34585

Pferdehirt R, Jain M, Blazo MA, Lee B, Burrage LC (2015) Catel-Manzke syndrome: further delineation of the phenotype associated with pathogenic variants in. Mol Genet Metab Rep 4:89–91

Prydz K (2015) Determinants of glycosaminoglycan (GAG) structure. Biomol Ther 5:2003–2022

Quentin E, Gladen A, Rodén L, Kresse H (1990) A genetic defect in the biosynthesis of dermatan sulfate proteoglycan: galactosyltransferase I deficiency in fibroblasts from a patient with a progeroid syndrome. Proc Natl Acad Sci USA 87:1342–1346

Rautengarten C, Quarrell OW, Stals K, Caswell RC, De Franco E, Baple E, Burgess N, Jokhi R, Heazlewood JL, Offiah AC, Ebert B, Ellard S (2019) A hypomorphic allele of SLC35D1 results in Schneckenbecken-like dysplasia. Hum Mol Genet 28:3543–3551

Ritelli M, Chiarelli N, Zoppi N, Dordoni C, Quinzani S, Traversa M, Venturini M, Calzavara-Pinton P, Colombi M (2015) Insights in the etiopathology of galactosyltransferase II (GalT-II) deficiency from transcriptome-wide expression profiling of skin fibroblasts of two sisters with compound heterozygosity for two novel. Mol Genet Metab Rep 2:1–15

Ritelli M, Dordoni C, Cinquina V, Venturini M, Calzavara-Pinton P, Colombi M (2017) Expanding the clinical and mutational spectrum of B4GALT7-spondylodysplastic Ehlers-Danlos syndrome. Orphanet J Rare Dis 12:153

Ritelli M, Cinquina V, Giacopuzzi E, Venturini M, Chiarelli N, Colombi M (2019, 10) Further defining the phenotypic spectrum of B3GAT3 mutations and literature review on linkeropathy syndromes. Genes (Basel)

Saiyin W, Li L, Zhang H, Lu Y, Qin C (2019) Inactivation of FAM20B causes cell fate changes in annulus fibrosus of mouse intervertebral disc and disc defects via the alterations of TGF-β and MAPK signaling pathways. Biochim Biophys Acta Mol basis Dis 1865:165555

Salter CG, Davies JH, Moon RJ, Fairhurst J, Bunyan D, Foulds N, Study D (2016) Further defining the phenotypic spectrum of B4GALT7 mutations. Am J Med Genet A 170:1556–1563

Sato T, Kudo T, Ikehara Y, Ogawa H, Hirano T, Kiyohara K, Hagiwara K, Togayachi A, Ema M, Takahashi S, Kimata K, Watanabe H, Narimatsu H (2011) Chondroitin sulfate N-acetylgalactosaminyltransferase 1 is necessary for normal endochondral ossification and aggrecan metabolism. J Biol Chem 286:5803–5812

Schulte Althoff S, Grüneberg M, Reunert J, Park JH, Rust S, Mühlhausen C, Wada Y, Santer R, Marquardt T (2016) TMEM165 deficiency: postnatal changes in glycosylation. JIMD Rep 26:21–29

Schoner K, Bald R, Horn D, Rehder H, Kornak U, Ehmke N (2017) Mutations in TGDS associated with additional malformations of the middle fingers and halluces: atypical Catel-Manzke syndrome in a fetus. Am J Med Genet A 173:1694–1697

Schreml J, Durmaz B, Cogulu O, Keupp K, Beleggia F, Pohl E, Milz E, Coker M, Ucar SK, Nürnberg G, Nürnberg P, Kuhn J, Ozkinay F (2014) The missing "link": an autosomal recessive short stature syndrome caused by a hypofunctional XYLT1 mutation. Hum Genet 133:29–39

Seidler DG, Faiyaz-Ul-Haque M, Hansen U, Yip GW, Zaidi SH, Teebi AS, Kiesel L, Götte M (2006) Defective glycosylation of decorin and biglycan, altered collagen structure, and abnormal phenotype of the skin fibroblasts of an Ehlers-Danlos syndrome patient carrying the novel Arg270Cys substitution in galactosyltransferase I (beta4GalT-7). J Mol Med (Berl) 84:583–594

Sher G, Naeem M (2014) A novel CHSY1 gene mutation underlies Temtamy preaxial brachydactyly syndrome in a Pakistani family. Eur J Med Genet 57:21–24

Shieh YE, Wells DE, Sater AK (2014) Zygotic expression of Exostosin1 (Ext1) is required for BMP signaling and establishment of dorsal-ventral pattern in Xenopus. Int J Dev Biol 58:27–34

Silveira C, Leal GF, Cavalcanti DP (2016) Desbuquois dysplasia type II in a patient with a homozygous mutation in XYLT1 and new unusual findings. Am J Med Genet A 170:3043–3047

Sivasami P, Poudel N, Munteanu MC, Hudson J, Lovern P, Liu L, Griffin T, Hinsdale ME (2019) Adipose tissue loss and lipodystrophy in xylosyltransferase II deficient mice. Int J Obes 43:1783–1794

Snyder NA, Palmer MV, Reinhardt TA, Cunningham KW (2019) Milk biosynthesis requires the Golgi cation exchanger TMEM165. J Biol Chem 294:3181–3191

Smith TM, Hicks-Berger CA, Kim S, Kirley TL (2002) Cloning, expression, and characterization of a soluble calcium-activated nucleotidase, a human enzyme belonging to a new family of extracellular nucleotidases. Arch Biochem Biophys 406:105–115

Stickens D, Zak BM, Rougier N, Esko JD, Werb Z (2005) Mice deficient in Ext2 lack heparan sulfate and develop exostoses. Development 132:5055–5068

Syx D, Van Damme T, Symoens S, Maiburg MC, van de Laar I, Morton J, Suri M, Del Campo M, Hausser I, Hermanns-Lê T, De Paepe A, Malfait F (2015) Genetic heterogeneity and clinical variability in musculocontractural Ehlers-Danlos syndrome caused by impaired dermatan sulfate biosynthesis. Hum Mutat 36:535–547

Takahashi I, Noguchi N, Nata K, Yamada S, Kaneiwa T, Mizumoto S, Ikeda T, Sugihara K, Asano M, Yoshikawa T, Yamauchi A, Shervani NJ, Uruno A, Kato I, Unno M, Sugahara K, Takasawa S, Okamoto H, Sugawara A (2009) Important role of heparan sulfate in postnatal islet growth and insulin secretion. Biochem Biophys Res Commun 383:113–118

Taylan F, Costantini A, Coles N, Pekkinen M, Héon E, Şıklar Z, Berberoğlu M, Kämpe A, Kıykım E, Grigelioniene G, Tüysüz B, Mäkitie O (2016) Spondyloocular syndrome: novel mutations in XYLT2 gene and expansion of the phenotypic spectrum. J Bone Miner Res 31:1577–1585

Taylan F, Yavaş Abalı Z, Jäntti N, Güneş N, Darendeliler F, Baş F, Poyrazoğlu Ş, Tamçelik N, Tüysüz B, Mäkitie O (2017) Two novel mutations in XYLT2 cause spondyloocular syndrome. Am J Med Genet A 173:3195–3200

Thomsen B, Horn P, Panitz F, Bendixen E, Petersen AH, Holm LE, Nielsen VH, Agerholm JS, Arnbjerg J, Bendixen C (2006) A missense mutation in the bovine SLC35A3 gene, encoding a UDP-N-acetylglucosamine transporter, causes complex vertebral malformation. Genome Res 16:97–105

Tian Y, Ma P, Liu C, Yang X, Crawford DM, Yan W, Bai D, Qin C, Wang X (2015) Inactivation of Fam20B in the dental epithelium of mice leads to supernumerary incisors. Eur J Oral Sci 123:396–402

Trejo P, Rauch F, Glorieux FH, Ouellet J, Benaroch T, Campeau PM (2017) Spondyloepimetaphysial dysplasia with joint laxity in three siblings with. Mol Syndromol 8:303–307

Umair M, Eckstein G, Rudolph G, Strom T, Graf E, Hendig D, Hoover J, Alanay J, Meitinger T, Schmidt H, Ahmad W (2018) Homozygous XYLT2 variants as a cause of spondyloocular syndrome. Clin Genet 93:913–918

Unger S, Ferreira CR, Mortier GR, Ali H, Bertola DR, Calder A, Cohn DH, Cormier-Daire V, Girisha KM, Hall C, Krakow D, Makitie O, Mundlos S, Nishimura G, Robertson SP, Savarirayan R, Sillence D, Simon M, Sutton VR, Warman ML, Superti-Furga A (2023) Nosology of genetic skeletal disorders: 2023 revision. Am J Med Genet A 191:1164–1209

Uyama T, Kitagawa H, Tamura Ji J, Sugahara K (2002) Molecular cloning and expression of human chondroitin N-acetylgalactosaminyltransferase: the key enzyme for chain initiation and elongation of chondroitin/dermatan sulfate on the protein linkage region tetrasaccharide shared by heparin/heparan sulfate. J Biol Chem 277:8841–8846

Van Damme T, Pang X, Guillemyn B, Gulberti S, Syx D, De Rycke R, Kaye O, de Die-Smulders CEM, Pfundt R, Kariminejad A, Nampoothiri S, Pierquin G, Bulk S, Larson AA, Chatfield KC, Simon M, Legrand A, Gerard M, Symoens S, Fournel-Gigleux S, Malfait F (2018) Biallelic B3GALT6 mutations cause spondylodysplastic Ehlers-Danlos syndrome. Hum Mol Genet 27:3475–3487

Vodopiutz J, Mizumoto S, Lausch E, Rossi A, Unger S, Janocha N, Costantini R, Seidl R, Greber-Platzer S, Yamada S, Müller T, Jilma B, Ganger R, Superti-Furga A, Ikegawa S, Sugahara K, Janecke AR (2017) Chondroitin sulfate N-acetylgalactosaminyltransferase-1 (CSGalNAcT-1) deficiency results in a mild skeletal dysplasia and joint laxity. Hum Mutat 38:34–38. https://doi.org/10.1002/humu.23070

Vogel P, Hansen GM, Read RW, Vance RB, Thiel M, Liu J, Wronski TJ, Smith DD, Jeter-Jones S, Brommage R (2012) Amelogenesis imperfecta and other biomineralization defects in Fam20a and Fam20c null mice. Vet Pathol 49:998–1017

Volpi S, Yamazaki Y, Brauer PM, van Rooijen E, Hayashida A, Slavotinek A, Sun Kuehn H, Di Rocco M, Rivolta C, Bortolomai I, Du L, Felgentreff K, Ott de Bruin L, Hayashida K, Freedman G, Marcovecchio GE, Capuder K, Rath P, Luche N, Hagedorn EJ, Buoncompagni A, Royer-Bertrand B, Giliani S, Poliani PL, Imberti L, Dobbs K, Poulain FE, Martini A, Manis J, Linhardt RJ, Bosticardo M, Rosenzweig SD, Lee H, Puck JM, Zúñiga-Pflücker JC, Zon L, Park PW, Superti-Furga A, Notarangelo LD (2017) Mutations cause skeletal dysplasia, immune deficiency, and developmental delay. J Exp Med 214:623–637

Von Oettingen JE, Tan WH, Dauber A (2014) Skeletal dysplasia, global developmental delay, and multiple congenital anomalies in a 5-year-old boy-report of the second family with B3GAT3 mutation and expansion of the phenotype. Am J Med Genet A 164A:1580–1586

Vorster AA, Beighton P, Ramesar RS (2015) Spondyloepimetaphyseal dysplasia with joint laxity (Beighton type); mutation analysis in eight affected south African families. Clin Genet 87:492–495

Vynios DH (2014) Metabolism of cartilage proteoglycans in health and disease. Biomed Res Int 2014:452315

Watanabe Y, Takeuchi K, Higa Onaga S, Sato M, Tsujita M, Abe M, Natsume R, Li M, Furuichi T, Saeki M, Izumikawa T, Hasegawa A, Yokoyama M, Ikegawa S, Sakimura K, Amizuka N, Kitagawa H, Igarashi M (2010) Chondroitin sulfate N-acetylgalactosaminyltransferase-1 is required for normal cartilage development. Biochem J 432:47–55

Wilson DG, Phamluong K, Lin WY, Barck K, Carano RA, Diehl L, Peterson AS, Martin F, Solloway MJ (2012) Chondroitin sulfate synthase 1 (Chsy1) is required for bone development and digit patterning. Dev Biol 363:413–425

Wiweger MI, Zhao Z, van Merkesteyn RJ, Roehl HH, Hogendoorn PC (2012) HSPG-deficient zebrafish uncovers dental aspect of multiple osteochondromas. PLoS One 7:e29734

Wuyts W, Van Hul W (2000) Molecular basis of multiple exostoses: mutations in the EXT1 and EXT2 genes. Hum Mutat 15:220–227

Xia B, Zhang W, Li X, Jiang R, Harper T, Liu R, Cummings RD, He M (2013) Serum N-glycan and O-glycan analysis by mass spectrometry for diagnosis of congenital disorders of glycosylation. Anal Biochem 442:178–185

Xu ZH, Otterness DM, Freimuth RR, Carlini EJ, Wood TC, Mitchell S, Moon E, Kim UJ, Xu JP, Siciliano MJ, Weinshilboum RM (2000) Human 3′-phosphoadenosine 5′-phosphosulfate synthetase 1 (PAPSS1) and PAPSS2: gene cloning, characterization and chromosomal localization. Biochem Biophys Res Commun 268:437–444

Yauy K, Tran Mau-Them F, Willems M, Coubes C, Blanchet P, Herlin C, Taleb Arrada I, Sanchez E, Faure JM, Le Gac MP, Prodhomme O, Boland A, Meyer V, Rivière JB, Duffourd Y, Deleuze JF, Guignard T, Captier G, Barat-Houari M, Genevieve D (2018) B3GAT3-related disorder with craniosynostosis and bone fragility due to a unique mutation. Genet Med 20:269–274

Yoshioka N, Miyata S, Tamada A, Watanabe Y, Kawasaki A, Kitagawa H, Takao K, Miyakawa T, Takeuchi K, Igarashi M (2017) Abnormalities in perineuronal nets and behavior in mice lacking CSGalNAcT1, a key enzyme in chondroitin sulfate synthesis. Mol Brain 10:47

Zak BM, Schuksz M, Koyama E, Mundy C, Wells DE, Yamaguchi Y, Pacifici M, Esko JD (2011) Compound heterozygous loss of Ext1 and Ext2 is sufficient for formation of multiple exostoses in mouse ribs and long bones. Bone 48:979–987

Zeevaert R, de Zegher F, Sturiale L, Garozzo D, Smet M, Moens M, Matthijs G, Jaeken J (2013) Bone dysplasia as a key feature in three patients with a novel congenital disorder of glycosylation (CDG) type II due to a deep intronic splice mutation in TMEM165. JIMD Rep 8:145–152

Chapter 6
Skeletal Dysplasias Caused by Defects in Glycosaminoglycan Sulfation

Chiara Gramegna Tota, Alessandra Leone, Chiara Paganini, Asifa Khan, Antonio Rossi, and Andrea Superti-Furga

Abstract Among the large number of skeletal conditions, there is a group associated with defects in the sulfation of macromolecules. Sulfate metabolism plays important roles in the development and function of different tissues and organs. In cartilage tissue, proteoglycans (PGs) are the most abundant macromolecules of the extracellular matrix. PGs are composed by a core protein to which, linear aminopolysaccharide chains (glycosaminoglycans, GAGs) are covalently bound. GAGs can be heavily sulfated, and the sulfate groups are fundamental, as they confer specific physico-chemical properties (charge and hydrophilicity) to PGs and in addition assure the interactions with other extracellular matrix molecules as well as with growth factors regulating tissue structure, mechanical properties, and cell behavior. To assure PG sulfation, chondrocytes mostly exploit sulfate from the extracellular environment, which is taken up and activated in the cytosol to 3′-phosphoadenosine 5′-phosphosulfate, the universal sulfate donor, used in sulfotransferase reactions. This chapter presents an overview of skeletal dysplasias caused by mutations in genes encoding for transporters or enzymes involved in glycosaminoglycan sulfation. Future research in these dysplasias will hopefully contribute to the deep understanding of the disease pathogenesis in order to allow for the development of targeted pharmacological treatment to alleviate, modify, or prevent the disease progression.

C. Gramegna Tota · A. Khan · A. Rossi (✉)
Department of Molecular Medicine, Unit of Biochemistry, University of Pavia, Pavia, Italy
e-mail: antrossi@unipv.it

A. Leone
Department of Molecular Medicine, Unit of Biochemistry, University of Pavia, Pavia, Italy

University School for Advanced Studies Pavia, Pavia, Italy

C. Paganini
Centre for Inherited Diseases, Department of Research, Fondazione IRCCS Policlinico San Matteo, Pavia, Italy

A. Superti-Furga
University of Lausanne and Genetica AG, Lausanne, Switzerland

A. Rossi, F. Zaucke (eds.), *The Extracellular Matrix in Genetic Skeletal Disorders*, Biology of Extracellular Matrix 16,
https://doi.org/10.1007/978-3-031-70835-0_6

Abbreviations

ACG1B	achondrogenesis type 1B;
AO2	atelosteogenesis type 2;
APS	adenosine 5′-phosphosulfate;
BCYM1A	brachyolmia type 1 Hobaek form;
BCYM1B	brachyolmia type 1 Toledo form;
BCYM4	brachyolmia type 4;
bkn	bikunin;
BPNT2	3′(2′), 5′-bisphosphate nucleotidase 2;
C4ST	chondroitin 4-O-sulfotransferase;
C6ST	chondroitin 6-O-sulfotransferase;
CS	chondroitin sulfate;
D4ST	dermatan 4-O-sulfotransferase;
DHEA	dehydroepiandrosterone;
DHEAS	dehydroepiandrosterone sulfate ester;
DS	dermatan sulfate;
DTD	diastrophic dysplasia;
DTDST	diastrophic dysplasia sulfate transporter;
ECM	extracellular matrix;
EDM4	recessive multiple epiphyseal dysplasia;
EDSMC1	Ehlers-Danlos syndrome musculocontractural type 1;
FGFR	fibroblast growth factor receptor 3;
GAG	glycosaminoglycan;
GalNAc	N-acetylgalactosamine;
GlcNAc	N-acetylglucosamine;
GlcUA	glucuronic acid;
gPAPP	Golgi resident phosphoadenosine phosphate phosphatase;
HS	heparan sulfate;
IdoUA	iduronic acid;
Ihh	Indian hedgehog;
IMPAD1	inositol monophosphatase domain-containing protein 1;
KS	keratan sulfate;
NAC	N-acetylcysteine;
NFSRA	Neurofacioskeletal syndrome with or without renal agenesis;
OCBMD	osteochondrodysplasia, brachydactyly and overlapping malformed digits;
PAP	phosphoadenosine phosphate;
PAPS	3′-phosphoadenosine 5′-phosphosulfate;
PAPSS	PAPS synthetase;
PAPST	PAPS transporter;
PG	proteoglycan;
SEDCJD	spondyloepiphyseal dysplasia with congenital joint dislocations;
SEMD	spondyloepimetaphyseal dysplasia;

ST	carbohydrate sulfotransferase;
SULT	cytosolic sulfotransferase;
Tcf7	Transcription factor 7;
TPST	tyrosylprotein sulfotransferase;
UPR	unfolded protein response.

6.1 Introduction

Sulfation and phosphorylation are two of the most prominent conjugating systems; they are ubiquitous phenomena carried out in all organisms. The phosphorylation process has always received extensive coverage, since it plays a major role in regulatory mechanisms. Conversely, the importance of sulfation is less appreciated, despite its essential role in growth, development, and maintenance of cellular and tissue functions. Sulfated endogenous compounds include a wide array of molecules. Among small substances, conjugation of xenobiotics with sulfate in the liver allows their detoxification and excretion with the bile. Sulfation also occurs in the biotransformation of catecholamines, iodothyronines, cholesterol, bile acids, and steroids (Aksoy et al. 1993; Falany et al. 1994; Richard et al. 2001; Weinshilboum et al. 1997). Sulfoglycolipids such as sphingolipids and galactoglycerolipids are abundant in myelin as well as spermatozoa, kidney, and small intestine (Farooqui and Horrocks 1985). Tyrosine sulfation is a post-translation modification of many secretory and membrane proteins and peptides that impacts their functions (Stone et al. 2009).

Among macromolecules, proteoglycans (PGs) present the highest density of sulfate groups. PGs are a heterogeneous family of macromolecules characterized by a core protein to which one to more linear polysaccharide chains, named glycosaminoglycans (GAGs), are attached (Iozzo 1998). To date 43 distinct PG encoding genes and many alternatively spliced variants have been characterized (Iozzo and Schaefer 2015). PGs present different cellular localization including the nucleus, the cytoplasm and the cell membrane, but they are particularly abundant in the extracellular matrix (ECM) of connective tissues. Based on their location and function, they have been grouped in four major classes: the intracellular, cell surface, pericellular and extracellular PGs (Iozzo and Schaefer 2015). PG molecular diversity is largely due to GAG different structures and characteristics. The biological activity of GAGs depends on several properties including the molecular weight, the monosaccharide species and most important, their negative charge associated to sulfate groups. Indeed, tissue hydration, elasticity, cation composition and the binding of GAGs with high-affinity ECM proteins, growth factors, enzymes, and cell surface receptors are mainly influenced by the highly acidic and hydrophilic properties of GAGs (Bowman and Bertozzi 1999; Gallagher 1994; Honke and Taniguchi 2002). The highly negative charge of hyaluronic acid, the only unsulfated GAG, is due to glucuronic acid, while sulfate groups importantly contribute to the negative charge of all other sulfated GAGs. Sulfation can occur at various positions within the glycan backbone generating specific sulfation motifs that function as molecular recognition elements (Gama et al. 2006). This functional information provided in a sequence-specific manner, referred to as the "sulfation code," is analogous to DNA, RNA, and protein sequences. Thus, it is conceivable that the genetic disturbance of the fine structures of GAGs both at the saccharide and/or sulfate levels may cause a variety of diseases. Among them several heritable disorders caused by defects in the biosynthesis (see Chap. 5) and catabolism of sulfated compounds are associated with abnormal development of the skeleton (Paganini et al. 2019). In the skeleton GAG sulfation plays structural and signalling roles. In articular cartilage, it generates the osmotic swelling pressure, which enables cartilage to withstand compressive loads and support cell-proteoglycan interactions with specific growth factors and signalling molecules (Plaas et al. 2023).

This book chapter reviews the consequences of defects in sulfation of PGs during their synthesis, which are associated mainly with abnormal skeletal development.

6.2 The Different Types of Sulfated GAGs

GAGs are linear polysaccharide chains that are composed of repeated disaccharide units. GAGs are attached to the PG core protein through a linkage tetrasaccharide region composed of a xylose (Xyl), 2 galactose (Gal) residues and a glucuronic acid (GlcUA). Xylose binds a serine residue of the core protein by O-glycosylation. Keratan sulfate and hyaluronic acid are exceptions: the former does not bear the tetrasaccharide region and is O-linked to serine/threonine residues or N-linked to asparagine, the latter is a polysaccharide without a core protein (Mizumoto and Yamada 2021; Soares da Costa et al. 2017). According to their monosaccharide constituents and sulfation patterns, four families of sulfated GAGs have been described (Soares da Costa et al. 2017) (Fig. 6.1).

6.2.1 Chondroitin Sulfate

Chondroitin sulfate (CS) consists of repeating disaccharide units composed of glucuronic acid (GlcUA) and N-acetylgalactosamine (GalNAc) linked by a β1–3 glycosidic bond. Disaccharide units are joined together by β1–4 glycosidic bonds. CSs may be sulfated on C4 or C6 of the GalNAc residues forming CS-A and CS-C, respectively but it may contain sulfate groups on both C4 and C6 forming CS-E. The GlcUA residue can also be sulfated on C2 in CS-A and CS-C, forming CS-B and CS-D, respectively.

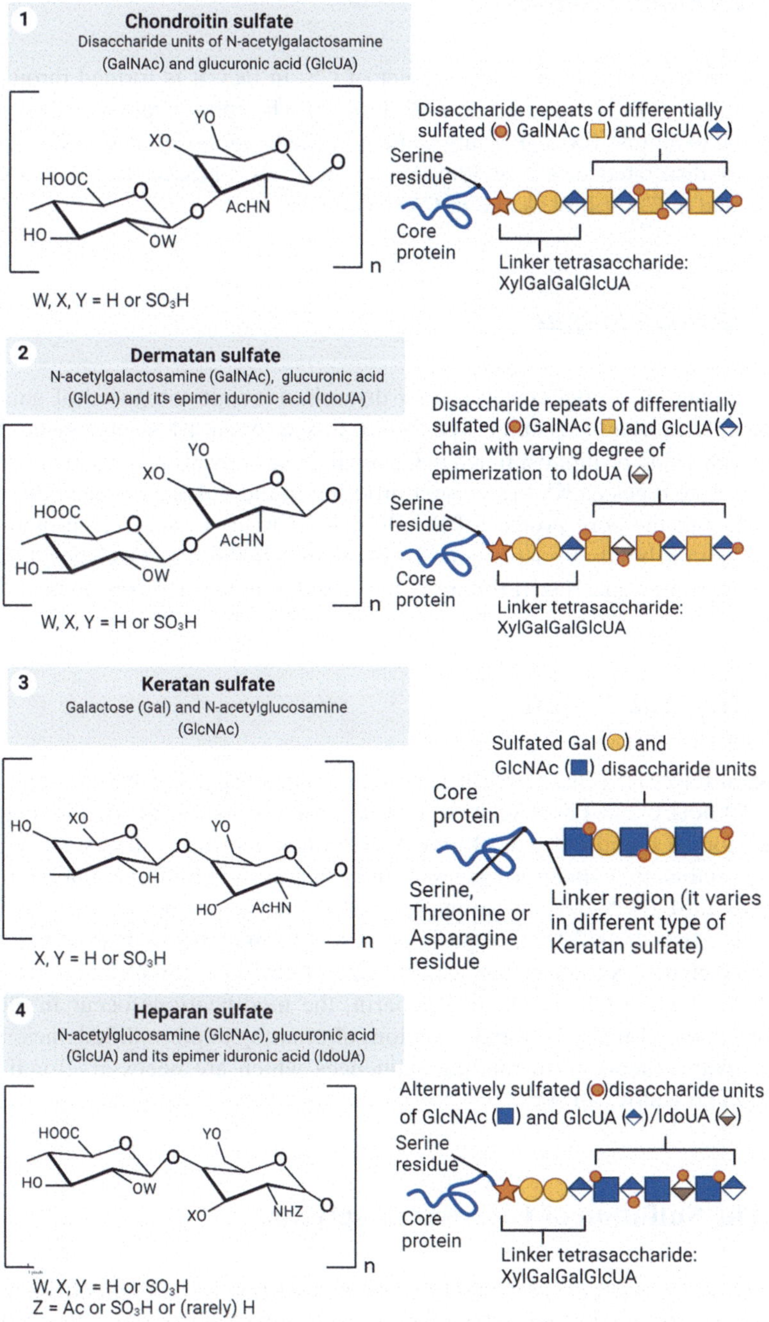

Fig. 6.1 Schematic representation of key GAG structures. (1) Chondroitin sulfate (CS) disaccharide unit, linker tetrasaccharide structure, and CS chains. (2) Dermatan sulfate (DS) disaccharide unit, linker tetrasaccharide structure and DS chains. (3) Keratan sulfate (KS) disaccharide unit, linker tetrasaccharide structure and KS chains. (4) Heparan sulfate (HS) disaccharide unit, linker tetrasaccharide structure and HS chains. GalNAc N-acetylgalactosamine, GlcUA glucuronic acid, GlcNAc N-acetylglucosamine, IdoUA iduronic acid, Gal galactose. Figure created with BioRender.

6.2.2 Dermatan Sulfate

Dermatan sulfate (DS) is the stereoisomer of CS; in fact, it is formed through the epimerization of GlcUA to iduronic acid (IdoUA). The epimerization can vary from few percent to almost 100% in a single chain. It can be monosulfated in C4 or C6 of GalNAc or disulfated in C2 of IdoUA and in C6 of GalNAc, in C4 and C6 of GalNAc, and in C2 of IdoUA and in C4 of GalNAc.

6.2.3 Keratan Sulfate

Keratan sulfate (KS) contains a peculiar disaccharide unit composed of galactose (Gal) and N-acetylglucosamine (GlcNAc) linked by a β1-4 glycosidic bond. Disaccharides may be monosulfated in C6 of GlcNAc or disulfated in C6 of GlcNAc and Gal. Three types of KS are present differing in the linkage between the oligosaccharide and the core protein of the PGs. KS-I binds to the core protein with N-glycosylation bonds to asparagine residues, KS-II through O-glycosylation to serine or threonine and KS-III through O-glycosylation to serine residues.

6.2.4 Heparan Sulfate

Heparan sulfate (HS) and heparin consists of repeating disaccharide units composed of GlcUA and GlcNAc linked by a β1-4 glycosidic bond. Disaccharide units are joined together by α1-4 glycosidic bonds. During their assembly in the Golgi apparatus, they undergo extensive modifications including epimerization of GlcUA to IdoUA, N-deacetylation, N- and O- sulfation at different positions (Soares da Costa et al. 2017). Heparin point to a form of HS with short and heavily sulfated chains. Sulfation can occur in C2 of GlcUA or IdoUA and in C6 or C3 of GlcNAc. Unlike CS, in HS and heparin, the modifications occur in clusters along the chain with highly sulfated regions alternating to unmodified ones. These specific arrangements form binding sequences which are necessary for ligands (Townley and Bülow 2018).

6.3 The Sulfation of Glycosaminoglycans

The GAG sulfation occurs in the Golgi apparatus starting from 3′-phophoadenosine 5′-phophosulfate (PAPS), the sulfate donor for all sulfation reactions (Fig. 6.2).

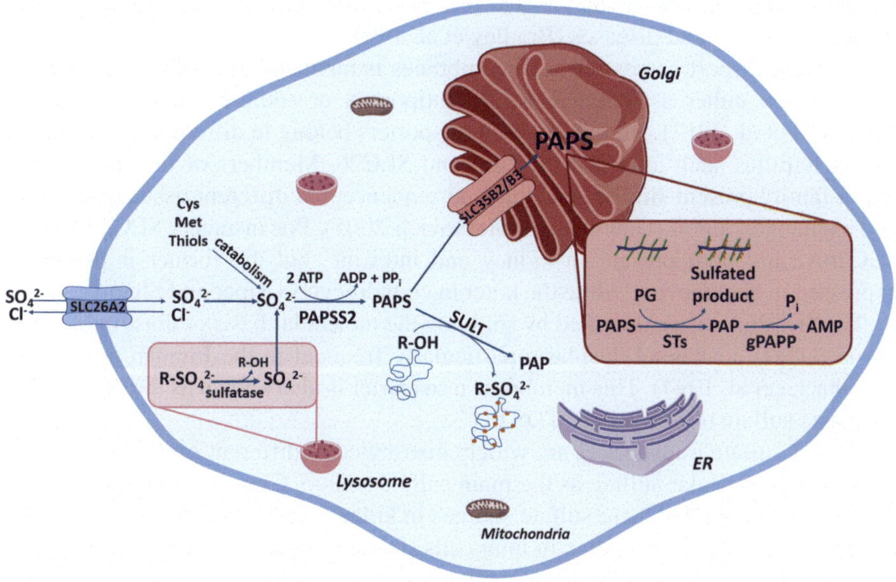

Fig. 6.2 Schematic view of sulfate metabolism in cells. Intracellular level of sulfate depends mainly by extracellular uptake through membrane transporters, such as the SLC26A2. A small amount of sulfate comes also from the catabolism of sulfur-containing amino acids and thiols or from sulfatase reactions in lysosomes. Once in cell, sulfate is activated to 3′-phosphoadenosine 5′-phosphosulfate (PAPS) by PAPS synthetase (PAPSS). PAPS represent the universal sulfate donor and it is used by cytosolic sulfotransferases (SULTs) for hormone and xenobiotics sulfation or by Golgi sulfotransferases (STs) for sulfation of macromolecules such as PGs. The SLC35B2 and SLC35B3 transporters allow Golgi uptake of PAPS. During sulfotransferase reactions, phosphoadenosine phosphate (PAP) is produced as by-product and it is hydrolyzed to AMP and phosphate by a Golgi resident phosphoadenosine phosphate phosphatase (gPAPP, also known as IMPAD1 or BPNT2)

6.3.1 Intracellular Sulfate Origin

The cytoplasmic sulfate pool may be derived from (i) the extracellular fluids by specific transporters that mediate its transfer through the cell membrane, (ii) the catabolism of sulfur containing amino acids and other thiols via intermediary metabolism with production of sulfite, which is then oxidized to sulfate, and (iii) the degradation of sulfated molecules through sulfatases in lysosomes (Langford et al. 2017; Paganini et al. 2020) (Fig. 6.2).

In healthy adults, one third of daily sulfate need comes from the diet (Dawson 2011) through intestinal absorption. The plasma level spans a broad range in humans (250–400 μM) and it is maintained by renal reabsorption in proximal tubule through SLC13A1 and SLC26A1 transporters (Dawson et al. 2017; Langford et al. 2017). The sulfate physiological homeostasis in the body fluids (Dawson 2011) can be altered by several factors such as age (Cole and Scriver 1980), gestational period (Cole et al. 1985) and diseases (Bradley et al. 1994).

Sulfate transport across the cell membranes is mediated by sulfate transporters that function either as sulfate/chloride antiporters or sodium/sulfate symporters (Langford et al. 2017). Human sulfate transporters belong to different solute linked carrier families such as SLC4, SLC13, and SLC26. Members of the same transporter family present similar structure and sequence, but different tissue expression and exchanged anion (Dawson and Markovich 2005). For instance, SLC26A1 and SLC26A2 are both present in kidney and intestine, but the former is primarily expressed in hepatocytes, while the latter in chondrocytes (Alper and Sharma 2013).

The SLC26A2 was identified by studying the molecular basis of diastrophic dysplasia (DTD), a skeletal dysplasia particularly frequent in the Finnish population (Hästbacka et al. 1994). This membrane transporter is also known as the diastrophic dysplasia sulfate transporter (DTDST).

Even if sulfate transporters are widely distributed in different organs, not all tissues use extracellular sulfate as the main sulfate source. Cysteine, glutathione, and N-acetylcysteine (NAC) are sulfate sources in kidney cells, whereas cysteine is the most efficient source of sulfate in lung cells (Dawson et al. 1983). In human fibroblast cultures, at less than physiological extracellular sulfate concentrations, cysteine is a major contributor to the intracellular sulfate pool (Elgavish and Meezan 1991).

A small amount of intracellular sulfate can come from lysosomal degradation of the carbohydrate moiety of glycoproteins and GAGs, generating monosaccharides and sulfate that must efflux from the lysosomes before re-entering the biosynthetic pathways (Rome and Hill 1986).

In conclusion, in each cell type, the three distinct sources of sulfate have a different incidence and work together to maintain the intracellular sulfate pool.

6.3.2 The Sulfate Activation Pathway

In sulfation reactions, PAPS is the activated form of sulfate used by sulfotransferases. The concerted action of two enzymes ATP sulfurylase and adenosine 5′-phosphosulfate (APS) kinase that, in animals, are present in a bifunctional protein named PAPS synthetase (PAPSS), is required for the activation of sulfate to PAPS (Lyle et al. 1994; Paganini et al. 2020). The amino-terminal region of PAPSS contains the catalytic domain of APS kinase, whereas the carboxy-terminal bears the ATP sulfurylase domain. ATP sulfurylase catalyzes the activation of sulfate with ATP to form APS and pyrophosphate, then APS kinase catalyzes the conversion of APS and another ATP molecule into PAPS and ADP. The first reaction is the rate limiting step in PAPS formation, because it is not energetically favored; however, subsequent hydrolysis of pyrophosphate and the rapid utilization of APS in the second reaction drives the overall reaction in the forward direction. Two PAPSS isozymes encoded by genes located on separate chromosomes have been reported. In humans, the genes for PAPS synthetase 1 (PAPSS1) and PAPS synthetase 2 (PAPSS2) are located on chromosome 4q25–26 and 10q23–24, respectively (Xu et al. 2000). As occurred for the SLC26A2, the mouse and human PAPSS2 isozymes were discovered through investigation of specific developmental dwarfism disorders, i.e., brachymorphism in mice and a form of spondyloepimetaphyseal dysplasia in humans (Kurima et al. 1998; ul Haque et al. 1998). More recently, defects in PAPSS2 have been linked to altered steroid metabolism (Noordam et al. 2009). No human pathologies have been linked so far to mutations in PAPSS1 suggesting embryonic lethality since it is the main isoform in developing nervous system (Langford et al. 2017; Strott 2002).

PAPSS1 and 2 are 76% identical both in humans and in mice (Venkatachalam 2003), but the catalytic activity of PAPSS2 is 10- to 15-fold higher than that of PAPSS1 (Fuda et al. 2002). Moreover, the two isoforms differ for cellular and tissue localization. PAPSS1 localizes mainly to the nucleus and is expressed in skin and brain, while PAPSS2 is mainly present in the cytosol of liver and cartilage cells (Schroder et al. 2012; Venkatachalam 2003).

Sulfation of small molecules such as hormones and neurotransmitters as well as drugs and xenobiotics occurs in the cytosol, whereas sulfation of carbohydrates, peptides, and proteins occurs in the Golgi apparatus (Fig. 6.2). For this purpose, PAPS is transported to the Golgi by specific PAPS transporters (PAPSTs). Two different PAPST isoforms have been cloned, PAPST1 and PAPST2 (Kamiyama et al. 2006; Kamiyama et al. 2003), also known as SLC35B2 and SLC35B3, respectively (Günal et al. 2019; Parker and Newstead 2019). Little is known about these two transporters; they are closely related and show similar transport activity; they are both widely expressed among tissues, in particular PAPST2 is highly expressed in colon (Kamiyama et al. 2006). Kinetic studies have suggested that PAPSTs work through an antiport mechanism that might involve phosphoadenosine phosphate (PAP) or AMP (Parker and Newstead 2019). Interestingly, variants in *SLC35B2* have been reported in patients with skeletal defects and intellectual disability (Guasto et al. 2022) and cartilage defects have been described in *slc35b2* null zebrafish mutants (Clément et al. 2008), while no human genetic disorders have been linked so far to mutations in the *SLC35B3* gene.

In humans the PAPS concentration ranges between 3.6 and 22.6 nmol/g tissue (Klaassen and Boles 1997); the highest PAPS level is present in the liver. The PAPS concentration is low compared with the amount of PAPS needed to sustain high rates of sulfation suggesting that high rates of PAPS synthesis are required (Klaassen and Boles 1997).

6.3.3 The Sulfation Pathway

Sulfotransferases are the enzymes that catalyze the transfer of sulfate from PAPS to a variety of acceptor molecules. In eukaryotes two classes of sulfotransferases are present differing for cellular localization: cytosolic and Golgi resident sulfotransferases. As described above, cytosolic sulfotransferases are involved in sulfation of small molecules such as hormones (catecholamines, steroid and thyroid hormones) and xenobiotics including drugs, while Golgi resident sulfotransferases sulfate carbohydrates and proteins involved in the secretory pathway (Bojarova and Williams 2008; Strott 2002).

Human cytosolic sulfotransferases (SULTs), classified in four families (SULT1, SULT2, SULT4 and SULT6) based on distribution and substrate preference, even if SULTs may show overlapping substrate specificity, are encoded by 13 genes (Coughtrie 2016). In the Golgi, carbohydrate sulfotransferases (STs) are membrane bound; up to now, 37 genes have been identified in humans (Langford et al. 2017). STs are stereoselective, exhibit strict substrate specificity, and are differently distributed (Günal et al. 2019; Honke and Taniguchi 2002).

During protein maturation in the Golgi, specific tyrosine residues of secreted proteins and peptides are also sulfated by tyrosylprotein sulfotransferases (TPST1 and TPST2) (Hartmann-Fatu et al. 2015). Sulfated tyrosines are important to boost protein-protein or protein-biopolymer interactions (Stewart and Ronald 2022).

Sulfotransferase reactions using PAPS as universal sulfate donor produce PAP that inhibits sulfotransferase activity as a reaction product (Günal et al. 2019). Thus, cytoplasmic and Golgi PAP phosphatases named BPNT1 and 2 (previously known as IMPAD1), respectively, rapidly degrade PAP in AMP and phosphate at sites of its production preventing sulfotransferases inhibition (Günal et al. 2019) (Fig. 6.2).

In contrast to sulfotransferases, sulfatases catalyze the hydrolysis of sulfate ester bonds from a wide variety of substrates including GAGs, sulfolipids, and steroid sulfates (Diez-Roux and Ballabio 2005). Sulfatases are grouped in two main categories based on their subcellular localization. Lysosomal sulfatases act at acidic pH and are involved in catabolism, while those present in the endoplasmic reticulum, Golgi apparatus and at the cell surface, acting at neutral pH, are mainly involved in biosynthetic pathways (Diez-Roux and Ballabio 2005).

6.4 Sulfate Metabolism and Skeletal Dysplasias

Genetic defects have been identified in several steps of PG biosynthesis including the core protein and GAGs. The number of disorders caused by defects in GAG synthesis definitely exceed the number of disorders caused by defects in genes encoding for the core proteins. This is due to the enormous structural diversity of the polysaccharide chains in term of sugar composition and extent of sulfation, leading to a huge number of diseases that affect primarily cartilage and endochondral bone (Paganini et al. 2019; Paganini et al. 2020). The cartilage phenotype results from the high amount of sulfated PGs that are synthesized and deposited in the cartilage ECM. The skeletal diseases caused by alterations in sulfate uptake, metabolic activation, and sulfation of GAGs are discussed in this session and are summarized in Fig. 6.3 and Table 6.1.

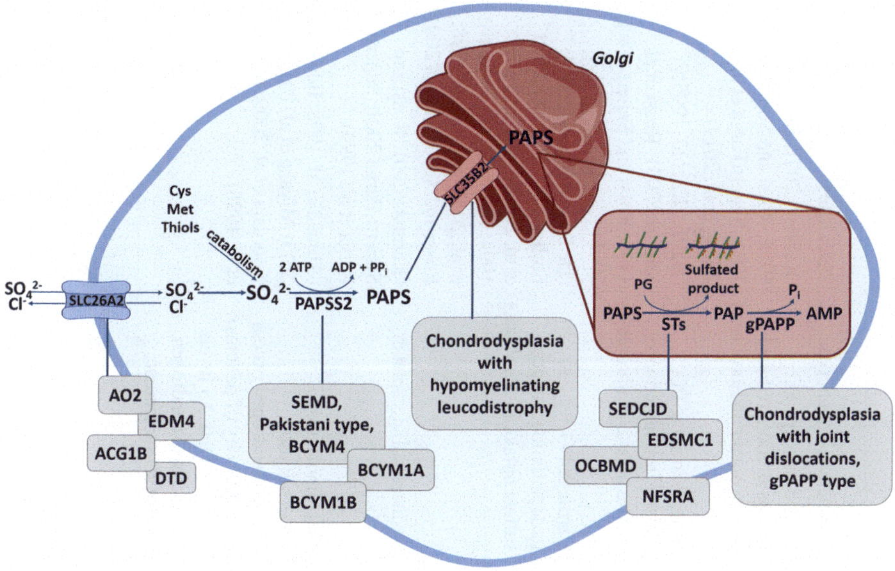

Fig. 6.3 Schematic diagram of the defects in the sulfation pathway causing skeletal dysplasias. Mutations in genes encoding for transporters and enzymes involved in sulfate metabolism cause several skeletal dysplasias (grey box). ACG1B, achondrogenesis type 1B; AO2, atelosteogenesis type 2; DTD, diastrophic dysplasia; EDM4, recessive multiple epiphyseal dysplasia; SEMD, spondyloepimetaphyseal dysplasia; BCYM4, brachyolmia type 4; BCYM1A, brachyolmia type 1, Hobaek form; BCYM1B, brachyolmia type 1, Toledo form; SEDCJD, spondyloepiphyseal dysplasia with congenital joint dislocations; EDSMC1, Ehlers-Danlos syndrome musculocontractural type 1 and OCBMD, osteochondrodysplasia, brachydactyly and overlapping malformed digits; NFSRA, neurofacioskeletal syndrome with or without renal agenesis

6.4.1 Skeletal Dysplasias Linked to Proteins Involved in Sulfate Metabolism

SLC26A2 Linked Chondrodysplasias

Achondrogenesis type 1B (ACG1B, MIM 600972), atelosteogenesis type 2 (AO2, MIM 256050), diastrophic dysplasia (DTD, MIM 222600), and recessive multiple epiphyseal dysplasia (EDM4, MIM 226900) are four chondrodysplasias, reported in decreasing order of severity, that belong to the heterogeneous group of skeletal diseases known as “SLC26A2 linked chondrodysplasias” (Rossi and Superti-Furga 2001). These disorders are caused by mutations in the *SLC26A2* gene encoding for a sulfate transporter of the cell membrane, also known as diastrophic dysplasia sulfate transporter. The gene was characterized at the end of a long-term project aimed at the identification of the molecular basis of DTD (Hästbacka et al. 1994).

The heterogeneity of SLC26A2 linked chondrosysplasias is demonstrated by a wide range of clinical phenotypes with key clinical features including lethality in the fetal period or immediately after birth (ACG1B and AO2), shortened limbs (all

Table 6.1 Skeletal dysplasias caused by defects in sulfate metabolism

Pathology	MIM/ inheritance	Causative gene	Protein product and function	Biochemical phenotype	References
Achondrogenesis type 1B (ACG1B)	600972/AR	*SLC26A2*	Sulfate/chloride antiporter of the cell membrane	Severe cartilage PG undersulfation; reduced sulfate uptake in fibroblasts	(Karniski 2001; Rossi et al. 1998; Rossi and Superti-Furga 2001; SupertiFurga et al. 1996; Zheng et al. 2019)
Atelosteogenesis type 2 (AO2)	256050/AR	*SLC26A2*	Sulfate/chloride antiporter of the cell membrane	Severe cartilage PG undersulfation; reduced sulfate uptake in fibroblasts	(Hästbacka et al. 1996; Karniski 2001; Rossi et al. 1998; Rossi and Superti-Furga 2001)
Diastrophic dysplasia (DTD)	222600/AR	*SLC26A2*	Sulfate/chloride antiporter of the cell membrane	Cartilage PG undersulfation; reduced sulfate uptake in fibroblasts	(De Leonardis et al. 2014; Forlino et al. 2005; Gualeni et al. 2010; Hästbacka et al. 1994; Karniski 2001; Monti et al. 2015; Rossi et al. 1998; Rossi and Superti-Furga 2001)
Recessive multiple epiphyseal dysplasia (EDM4)	226900/AR	*SLC26A2*	Sulfate/chloride antiporter of the cell membrane	Reduced sulfate uptake	(Karniski 2001; Rossi and Superti-Furga 2001; Superti-Furga et al. 1999)
Spondyloepimetaphyseal dysplasia, SEMD, Pakistani type or Brachyolmia type 4 (BCYM4)	612847/AR	*PAPSS2*	PAPS synthetase 2 catalyzes the synthesis of PAPS	Macromolecular undersulfation; Signs of androgen excess (in a minority of patients); Very low DHEAS levels and increased androgen levels (in one patient)	(Ahmad et al. 1998; Iida et al. 2013; Miyake et al. 2012; Noordam et al. 2009; ul Haque et al. 1998)

Brachyolmia type 1 (includes Hobaek form and Toledo form, BCYM1A and 1B respectively)	271530/AR 271630/AR	*PAPSS2*	PAPS synthetase 2 catalyzes the synthesis of PAPS	Undersulfation of CS; low activity of PAPS-CS sulfotransferase; signs of androgen excess (in a minority of patients)	(Iida et al. 2013; Miyake et al. 2012; Mourao et al. 1981; Toledo et al. 1978)
Chondrodysplasia with hypomyelinating leucodistrophy	620269 / AR	*SLC35B2*	PAPS transporter 1 transports PAPS from the cytosol to the Golgi	CS hyposulfation	(Guasto et al. 2022)
Chondrodysplasia with joint dislocations, gPAPP type	614078/AR	*BPNT2*	Golgi resident PAP phosphatase hydrolyses PAP to AMP and phosphate	Not yet described (see mouse phenotype in Table 6.2)	(Frederick et al. 2008; Nizon et al. 2012; Sohaskey et al. 2008; Vissers et al. 2011)
Spondyloepiphyseal dysplasia with congenital joint dislocations (SEDCJD or SED Omani type)	143095/AR	*CHST3*	Carbohydrate sulfotransferase-3 catalyzes the transfer of sulfate from PAPS to GalNAc residues of CS in C6	Depletion of 6-O-sulfated GalNAc residues in CS chains in fibroblasts and urine	(Hall 1997; Hermanns et al. 2008; Kozlowski et al. 1974; Thiele et al. 2004; Tuysuz et al. 2009; Unger et al. 2010a; van Roij et al. 2008)
Ehlers-Danlos syndrome musculocontractural type 1 (EDSMC1)	601776/AR	*CHST14*	Carbohydrate sulfotransferase-14 catalyzes the transfer of sulfate from PAPS to GalNAc residues of DS in C4	Reduction of 4-O-sulfation in GalNAc residues in DS chains; decrease of DS and increase of CS chain synthesis	(Dundar et al. 2009; Hirose et al. 2018; Janecke et al. 2016; Malfait et al. 2010; Miyake et al. 2010)
Osteochondrodysplasia, brachydactyly and overlapping malformed digits (OCBMD)	618167/AR	*CHST11*	Carbohydrate sulfotransferase-11 catalyzes the transfer of sulfate from PAPS to GalNAc residues of CS in C4	Not yet reported	(Chopra et al. 2015; Kluppel et al. 2005; Shabbir et al. 2018)
Neurofacioskeletal syndrome with or without renal agenesis (NFSRA)	619194/AR	*HS2ST1*	Heparan sulfate 2-O-sulfotransferase 1 catalyzes the transfer of sulfate from PAPS to IdoUA (rarely GlcNAc) residues of HS in C2	Not yet reported	(Schneeberger et al. 2020)

but not EDM4), joint pain, joint contractures, "hitchhiker" thumb (AO2 and DTD), cystic swelling of the external ear (DTD), cleft palate (DTD and AO2), scoliosis, clubfoot, and double-layered patellae (EDM4).

Interestingly, the linkage between the different clinical phenotypes and the mutation severity, the resulting residual activity of the sulfate transporter and the extent of cartilage PG undersulfation have been demonstrated (Karniski 2001; Rossi et al. 1998; Rossi and Superti-Furga 2001). Phenotypic differences observed in unrelated and even first-degree relatives with the same variant suggest that phenotypic severity is not only determined by the nature of the *SLC26A2* mutations.

Spondyloepimetaphyseal Dysplasia, PAPSS2 Related
Spondyloepimetaphyseal dysplasia (SEMD), PAPSS2 related, formerly known as SEMD Pakistani type (Ahmad et al. 1998) is characterized by short stature, short and bowed lower limbs with enlarged knee joints, kyphoscoliosis, mild brachydactyly, delayed ossification of the epiphyses and osteoarthritis (Ahmad et al. 1998). This disorder is caused by mutations in the *PAPSS2* gene, encoding for the 3′-phosphoadenosine 5′-phosphosulfate synthetase 2 (Miyake et al. 2012; ul Haque et al. 1998). Thus, the impairment of PAPSS2 reduces intracellular synthesis of PAPS leading to macromolecular undersulfation. SEMD Pakistani type is also known as Brachyolmia type 4 (BCYM4, MIM 612847), even if this name is obsolete and rarely used, but reported in OMIM. In 2012, *PAPSS2* variants have been identified also in patients affected by Brachyolmia type 1 including Hobaek and Toledo forms (BCYM1A and 1B, MIM 271530 and 271630, respectively) (Miyake et al. 2012), making these disorders allelic to the SEMD Pakistani type. Further studies confirmed that brachyolmia type 1A, 1B and 4 are indeed the same disorder (Iida et al. 2013).

Chondrodysplasia with Hypomyelinating Leucodistrophy
Chondrodysplasia with hypomyelinating leucodistrophy has been recently described in two patients with pre- and postnatal growth retardation, scoliosis, severe motor and intellectual disabilities, and hypomyelinating leukodystrophy. In both patients homozygous variants have been described in the *SLC35B2* gene encoding for the PAPS transporter 1 (PAPST1) that transports PAPS from the cytosol to the Golgi (Guasto et al. 2022). No variant has been described in the homologous *SLC35B3* gene.

Chondrodysplasia with Joint Dislocations, gPAPP Type
Chondrodysplasia with joint dislocations, gPAPP type (MIM 614078) presents short stature, chondrodysplasia with brachydactyly, congenital joint dislocations, micrognathia, cleft palate, and facial dysmorphism as the main clinical features. In patients, variants in the *BPNT2* (formerly *IMPAD1*) gene encoding for a Golgi PAP phosphatase (gPAPP) have been identified (Vissers et al. 2011). When gPAPP activity is impaired, the hydrolysis of PAP, the by-product of the sulfotransferase reactions, is reduced. This reduction causes PAP accumulation in the Golgi, leading to the inhibition of sulfotransferase reactions and of PAPS transport inside the Golgi through the PAPS/AMP antiporter. Thus, undersulfation of macromolecules is the final consequence of gPAPP impairment (Frederick et al. 2008; Sohaskey et al. 2008).

6.4.2 Skeletal Dysplasias Linked to Proteins Involved in GAG Sulfation

Spondyloepiphyseal Dysplasia with Congenital Joint Dislocations, CHST3 Type

Spondyloepiphyseal dysplasia with congenital joint dislocations, CHST3 type (SEDCJD, MIM 143095), also named recessive Larsen syndrome, is characterized by dislocation of the knees and/or hips at birth, clubfoot, elbow joint dysplasia with subluxation and limited extension, short stature, and progressive kyphosis developing in late childhood with marked flattening of intervertebral spaces (Unger et al. 2010a). During childhood, the dislocations become more severe causing arthritis of the hips and spine with intervertebral disc degeneration. At this stage, the clinical features are those previously identified as Spondyloepiphyseal dysplasia, Omani type (Thiele et al. 2004). Thus, spondyloepiphyseal dysplasia, Omani type, humero-spinal dysostosis and recessive Larsen syndrome represent different age-related descriptions of the same condition. Variants in *CHST3* gene encoding for carbohydrate sulfotransferase-3, also known as chondroitin 6-O-sulfotransferase 1 (C6ST-1), cause these disorders and its wide mutational spectrum could be the explanation for the clinical spectrum (Hermanns et al. 2008; Thiele et al. 2004; Tuysuz et al. 2009; Unger et al. 2010b; van Roij et al. 2008). The C6ST-1 is the sulfotransferase that catalyzes 6-O-sulfation of GalNAc residues in CS. Its functional impairment leads to the almost complete depletion of 6-O-sulfated GalNAc residues in CS chains from patients' fibroblasts and urine (Hermanns et al. 2008; Thiele et al. 2004; van Roij et al. 2008).

Ehlers-Danlos Syndrome (EDS) Musculocontractural Type 1

Ehlers-Danlos syndrome (EDS) musculocontractural type 1 (EDSMC1, MIM 601776), formerly EDS type VIB or adducted thumb-clubfoot syndrome, is characterized by craniofacial dysmorphism, congenital contractures of thumbs and fingers, clubfeet, severe kyphoscoliosis, muscular hypotonia, hyperextensible thin skin with easy bruisability and atrophic scarring, wrinkled palms, joint hypermobility and, atrial septal defects (Janecke et al. 2016; Malfait et al. 2010) (see Chap. 3). The disorder is caused by defects in the *CHST14* gene encoding for carbohydrate sulfotransferase-14 (also named dermatan 4-O-sulfotransferase 1, D4ST-1), responsible for sulfate transfer to C4 of GalNAc residues in DS (Dundar et al. 2009; Janecke et al. 2016; Malfait et al. 2010). Patients' fibroblasts show, beyond the reduced 4-O-sulfation in GalNAc residues of DS chains, decreased DS and increased CS chain level. The lack of 4-O-sulfation in GalNAc residues allows the reverse epimerization of iduronic acid present in DS into glucuronic acid present in CS explaining the CS chains increase (Dundar et al. 2009). Thus, decorin DS chains are completely replaced by CS chains leading to reduction of GAG chain flexibility, that in turn affects collagen fibril assembly (Hirose et al. 2018; Miyake et al. 2010).

Osteochondrodysplasia, Brachydactyly, and Overlapping Malformed Digits
Osteochondrodysplasia, brachydactyly, and overlapping malformed digits (OCBMD, MIM 618167) presents with bilateral and symmetrical skeletal defects affecting mainly the limbs and hand and foot malformations such as short and adducted thumbs, overriding fingers and broad halluces. Moreover, scoliosis, dislocated patellae and fibulae, pectus excavatum, and mild osteoarthritis have been reported. Variants in the *CHST11* gene encoding for carbohydrate sulfotransferase 11 have been identified (Shabbir et al. 2018). The enzyme, also named chondroitin 4-O-sulfotrasferase 1 (C4ST-1), is a Golgi sulfotransferase that catalyzes sulfate transfer from PAPS to C4 of the GalNAc residues in CS. Previously, a deletion spanning exon 2 of the *CHST11* gene and an embedded microRNA, MIR3922, was reported in a woman with short stature, brachydactyly with disproportionately short index finger and congenital malformations of hand and foot digits and malignant lymphoproliferative disease (Chopra et al. 2015).

Neurofacioskeletal Syndrome With or Without Renal Agenesis
Neurofacioskeletal syndrome with or without renal agenesis (NFSRA, MIM 619194) is a syndrome characterized by developmental delay with corpus callosum, skeletal and renal abnormalities, recently described in four individuals from three unrelated families (Schneeberger et al. 2020). Affected individuals showed facial dysmorphism, hypoplasia or agenesis of corpus callosum, unilateral or bilateral renal agenesis and skeletal abnormalities, including short stature, scoliosis, flexion contractures, and brachydactyly. Homozygous or compound heterozygous variants have been detected in the *HS2ST1* gene encoding for the heparan sulfate 2-O-sulfotransferase 1 that catalyzes the transfer of the sulfate group from PAPS to C2 of IdoUA and, rarely, of GlcUA of HS.

6.5 Animal Models with Defects in GAG Sulfation

The skeletal phenotypes of different in vivo models with defects in the sulfation pathway have provided insights on the role of sulfation in skeletal development. Therefore, some of these mutant animal strains, described in Table 6.2, have been validated as models of human skeletal dysplasias.

Diastrophic Dysplasia Mouse
The *diastrophic dysplasia* (*dtd)* mouse is an *Slc26a2* knock-in mouse with an Ala386Val variant previously characterized in a patient with DTD. The mutant mouse has been validated as model of human DTD based on morphological and biochemical results. The skeletal phenotype in mutant mice includes reduced skeletal growth, shortened limbs, long bones deformities, reduced toluidine blue staining of cartilage, impaired uptake of sulfate, and cartilage PG undersulfation (Forlino et al. 2005); PG undersulfation has also been demonstrated in the different zones of the growth plate (Gualeni et al. 2010). Even if chondrocytes are mainly dependent on extracellular sulfate for PG sulfation, using the *dtd* mouse it has been

Table 6.2 Animal models with defects in GAG sulfation

Animal model	Generation	Gene	Protein product and function	Morphological phenotype	Biochemical phenotype	References
dtd mouse	Knock-in for a Ala386Val variant by gene targeting	*Slc26a2*	Sulfate/chloride antiporter of the cell membrane	Reduced skeletal growth; shortened limbs; deformities of long bones; reduced toluidine blue staining of cartilage; altered collagen orientation in articular cartilage; smaller and less organized collagen fibers in trabecular bone	Impaired sulfate uptake; cartilage PG undersulfation; altered Ihh distribution; reduced chondrocyte proliferation	(Forlino et al. 2005, Gualeni et al. 2010; Mertz et al. 2009, 2013)
Slc26a2 knock-out mouse	Knock-out by gene targeting	*Slc26a2*	Sulfate/chloride antiporter of the cell membrane	Death at birth; extremely short limbs; bowed tibias; small thoracic cage and protuberant abdomen; reduced collagen in cartilage ECM	Retention of collagen in ER; activation of UPR (ATF6 pathway); overactivation of FGFR3 signalling	(Zheng et al. 2019)
Brachymorphic (*bm*) mouse	Gly79Arg spontaneous variant	*Papss2*	PAPS synthetase 2, enzyme that catalyzes the synthesis of PAPS	Short-limb dwarfism; short spine and tail; domed shaped skull; structurally abnormal growth plate with short chondrocyte columns	Undersulfated CS; normal sulfation of HS; disturbed Ihh signalling; reduced chondrocyte proliferation	(Kurima et al. 1998)
Pinscher (pic) zebrafish	Knock-out generated by N-ethyl-N-nitrosourea mutagenesis	*Slc35b2*	PAPS transporter 1 transports PAPS from the cytosol to the Golgi	Shorter and thicker cartilage elements; HS, CS and KS undersulfated	Normal cartilage differentiation; reduced ossification; lack of hypertrophic markers	(Clément et al. 2008)
Bpnt2 knock-out mice (two animal models generated independently)	Knock-out by gene trapping	*Bpnt2*	Golgi resident PAP phosphatase hydrolyses PAP to AMP and phosphate	Neonatal lethality; severe skeletal abnormalities; reduced trunk; markedly short appendicular bones of the limbs; severe respiratory distress; reduced growth plate area; chondrocyte columns irregularly distributed	CS undersulfation; altered Ihh distribution	(Frederick et al. 2008; Sohaskey et al. 2008)

(Continued)

Table 6.2 (continued)

Animal model	Generation	Gene	Protein product and function	Morphological phenotype	Biochemical phenotype	References
C4st1 knock-out mouse	Knock-out by gene trapping	*C4st1*	Carbohydrate sulfotransferase-11, enzyme that catalyzes the transfer of sulfate from PAPS to GalNAc residues of CS in C4	Perinatal death; severe chondrodysplasia; abnormal localization of CS; altered growth plate morphogenesis; abnormal chondrocyte differentiation and orientation within the growth plate	Altered CS sulfation; upregulation of TGF-β signalling; downregulation of BMP signalling	(Kluppel et al. 2005)
C6st knock-out mouse	Knock-out by gene targeting	*C6st*	Carbohydrate sulfotransferase-3, enzyme that catalyzes the transfer of sulfate from PAPS to GalNAc residues of CS in C6	Lower number of naive T lymphocytes in the spleen; no skeletal phenotype		(Uchimura et al. 2002)
Chst14 knock-out mouse	Knock-out by gene targeting	*Chst14*	Carbohydrate sulfotransferase-14, enzyme that catalyzes the transfer of sulfate from PAPS to GalNAc residues of DS in C4	Decreased birth rate; abnormally increased tooth growth; tail deformation (kinked tail); skin fragility; thoracic kyphosis		(Bian et al. 2011; Tang et al. 2010; Mikami et al. 2003; Akyüz et al. 2013; Nitahara-Kasahara et al. 2021; Yoshizawa and Kosho 2023)
Chst15 knock-out mouse	Knock-out by gene targeting	*Chst15*	Carbohydrate sulfotransferase-15, enzyme that transfers sulfate from PAPS to of the 4-O-sulfated GalNAc residues of CS in C-6	Cortical and trabecular bone loss		(Koike et al. 2015)

demonstrated that intracellular catabolism of thiol compounds can provide sulfate for PG sulfation suggesting potential pharmacological approaches to DTD (Monti et al. 2015; Paganini et al. 2021).

Slc26a2 Knock-out Mouse
Recently the *Slc26a2* gene has been knocked out generating a murine strain which has been validated as a model of AO2 and ACG1B. Homozygous mutant mice die at birth and manifest extremely short limbs, bowed tibias, small thoracic cage, and protuberant abdomen. As a consequence of Slc26a2 null function an overactivation of fibroblast growth factor receptor 3 (FGFR3) signalling has been described (Zheng et al. 2019). This observation suggests the FGFR3 signalling could be targeted in SLC26A2 chondrodysplasias by tyrosine kinase inhibitors to ameliorate the phenotype (Li et al. 2024).

Brachymorphic Mouse
The *brachymorphic (bm)* mouse has been generated by N-ethyl-N-nitrosourea mutagenesis and bears the Gly79Arg variant in the *Papss2* gene (Kurima et al. 1998). The *bm* phenotype is characterized by normal life span, short-limb dwarfism, short spine and tail and domed shaped skull. Abnormal growth plate with a structurally abnormal cartilage ECM and short chondrocyte columns have been observed by histological analysis (Orkin et al. 1977). Mutant mice produced undersulfated CS, but not HS and show disturbed Indian hedgehog (Ihh) signalling due to abnormal distribution in the ECM resulting in reduced chondrocyte proliferation (Cortes et al. 2009; Orkin et al. 1976).

Slc35b2 Knock-out Zebrafish
In an attempt to identify novel genes involved in skeletal development that might cause hereditary multiple hexostosis, Clément et al. have generated the *pinscher* (*pic*) zebrafish knock-out (Clément et al. 2008). The *pic* gene encodes for PAPST1 and is homologous to the *SLC35B2*. In *pic*$^{-/-}$ larvae, cartilage elements are shorter and thicker than wild types and do not stain with Alcian blue at pH 1.0 due to loss of sulfation. This observation is confirmed by immunostaining with antibodies for HS, CS and KS. The onset of cartilage differentiation is normal based on the expression of *gdf5, col2a1* and *sox9,* but ossification is reduced demonstrated by reduced Alizarin red staining and lack of expression of the hypertrophic marker *col10a1*.

Bpnt2 Knock-out Mouse
The *Bpnt2* mutant mouse has been generated by a gene trap approach leading to loss of functional gPAPP, the PAP phosphatase of the Golgi (Frederick et al. 2008). Homozygous mutant mice are neonatally lethal, with severe skeletal abnormalities of the bones formed by endochondral ossification. The length of the trunk is reduced, and the appendicular bones of the limbs are markedly shorter compared to heterozygous or wild type mice. Mutant mice, alive at birth, die within minutes because of severe respiratory distress caused by the deformity and reduced size of the thoracic cage. The areas of the growth plate zones are reduced and the chondrocyte columns are irregularly distributed. CS sulfation analysis in whole embryo has shown marked undersulfation, while no difference has been detected in HS sulfation except for

moderate HS undersulfation in the lungs (Frederick et al. 2008). In the same year, another *Bpnt2* gene trap mutant has been generated (Sohaskey et al. 2008). Also in this mouse strain, homozygous mutant animals die perinatally and further studies have demonstrated that Bpnt2 is required for the orientation and positioning of interphalangeal joints within the endochondral skeleton.

C4st1 Knock-out Mouse

A mouse deficient in the activity of the chondroitin 4-O-sulfotranferase encoded by the *C4st1* gene, also named *Chst11* gene, has been generated by a gene trap approach. Homozygous mutant mice die within hours after birth and display a severe chondrodysplasia which is restricted to bones formed through endochondral ossification (Kluppel et al. 2005). Detailed analysis of the developing skeleton and the cartilage growth plate have demonstrated that loss of C4st1 disturbs the balance of chondroitin sulfation and causes abnormal chondroitin sulfate localization. These defects result in a strong upregulation of TGF-β signalling with concomitant down-regulation of BMP signalling causing severe disturbances in growth plate morphogenesis due to abnormal chondrocyte differentiation and orientation within the growth plate.

C6st Knock-out Mouse

The *C6st* knock-out mouse has been generated by deletion of the *Chst3* gene encoding for chondroitin 6-O-sulfotransferase. Knock-out mice are born at the expected frequency and are viable through adulthood. No apparent differences between normal and knock-out mice have been reported, except a lower number of naive T lymphocytes in the spleen of mutant mice (Uchimura et al. 2002). Considering the skeletal phenotype in patients with defects in the homologous human gene, it remains unclear why humans develop exclusively a skeletal phenotype despite the widespread expression of C6ST-1 in tissues and why there are obvious differences between mouse and human due to enzyme deficiency.

Chst14 Knock-out Mouse

In 2010 and 2011, two strains of *Chst14* knock-out (*Chst14*$^{-/-}$) mice were generated through homologous recombination targeting the single coding exon of *Chst14* by two independent groups (Bian et al. 2011; Tang et al. 2010). *Chst14* encodes for dermatan 4-O-sulfotransferase-1 a critical enzyme in the biosynthesis of DS and catalyzes the 4-O-sulfation of GalNAc in the sequence IdoUA-GalNAc which occurs immediately after epimerization of D-glucuronic acid (GlcUA) to IdoUA by dermatan sulfotransferase 1 and/or 2 (Mikami et al. 2003). In addition, Nitahara-Kasahara et al. generated *Chst14* gene-disrupted mice through CRISPR/Cas9-mediated gene editing (Nitahara-Kasahara et al. 2021). *Chst14*$^{-/-}$ mice show decreased birth rate because of lethality at E16.5-E18.5, but after the perinatal period, the lifespan of mutant mice is similar to heterozygous or wild type animals. *Chst14*$^{-/-}$ mice have abnormally increased tooth growth, tail deformation (kinked tail), and skin fragility (Akyüz et al. 2013; Nitahara-Kasahara et al. 2021). Thoracic kyphosis has been detected in 1-year-old *Chst14* mutant mice (Nitahara-Kasahara et al. 2021), but it is unclear whether the cause of the thoracic kyphosis is an influence of skeletal muscle or a direct effect of DS deficiency in the bone. The phenotypic similarities and differences of patients affected by musculocontractural Ehlers-Danlos syndrome with mutations in *CHST14* gene and the mouse models have been recently reviewed (Yoshizawa and Kosho 2023).

Chst15 Knock-out Mouse

The *Chst15* gene encodes for the sulfotransferase that transfers sulfate from PAPS to the C-6 hydroxyl group of the 4-O-sulfated GalNAc residues forming GalNAc(4,6-SO_4) unit, a CS subtype named CS-E. The *Chst15* knock-out mice show no readily visible developmental abnormalities, but both cortical and trabecular bone loss has been observed in 4 months old male mice. These findings demonstrate that CS-E production is critical to proper bone anabolism and that the CS-E synthesis is regulated by estrogens. This work suggests the novel idea that post-menopausal osteopenia and osteoporosis are caused by both excessive osteoclast formation and reduced anabolic bone formation, that is mediated by estrogen-induced CS-E (Koike et al. 2015).

6.6 Alterations of Extracellular Matrix and Cell Homeostasis Due to Defects in PG Sulfation

PGs are prominent components of ECMs that modulate the biochemical, physical, and mechanical properties of tissues. Several genetic disorders affecting primarily the skeleton are caused by altered PG synthesis that impacts on cartilage structure, development, and homeostasis (Paganini et al. 2019).

Mutations in genes encoding for enzymes and transporters involved in the sulfate activation pathway or in sulfotransferase reactions cause altered PG sulfation which is the first evidence of impaired sulfate metabolism. This condition may affect cartilage homeostasis at different levels impairing the synthesis of whole GAG chains and altering the ECM organization, the ECM mechanical properties and a variety of biological processes including interactions with growth factors, cytokines and morphogens that modulate cell behaviour.

In the extracellular environment, PGs are fundamental for the organization of the complex ECM network. For instance, the correct formation of collagen fibers is regulated by the flexible DS chain of decorin and biglycan. The GAG chains are curved and in close contact with the outer of collagen fibers in healthy skin samples as observed by TEM analysis. Conversely, skin biopsies of EDSMC1 patients show dispersed collagen fibers and GAG chains that appear to be straight since DS is converted to CS (Hirose et al. 2018; Janecke et al. 2016; Miyake et al. 2010). In fact, in DS biosynthesis, the 4-O-sulfation of GalNAc prevents the reconversion of DS to CS. The CHST14 sulfotransferase is the enzyme that catalyzes the 4-O-sulfation of GalNAc in DS chain. Thus, its impairment causes GAG biosynthetic defects leading to the complete replacement of DS with CS on the decorin core protein as demonstrated in fibroblasts from EDSMC1 patients (Dundar et al. 2009; Janecke et al. 2016; Miyake et al. 2010).

Moreover, the non-collagenous and collagenous fractions have been studied in articular cartilage of the animal model of diastrophic dysplasia, the *dtd* mouse, by infrared hyperspectral imaging. The non-collagenous protein and collagen fractions are reduced in the articular cartilage layer of *dtd* animals compared with wild types as well as the collagen orientation. These alterations may contribute to the early cartilage degradation progression with age (Mertz et al. 2009). Altered collagen distribution in the ECM of the *dtd* mouse has been confirmed by the presence of collagen fibers with smaller diameter and less organized in the trabecular bone matrix compared with wild type animals. All these defects in collagen organization are related with chondroitin undersulfation due to impairment of the SLC26A2 transporter (Gualeni et al. 2013).

In the cartilage ECM of ACG1B patients, the collagen fibers are particularly dense around chondrocytes forming the so-called collagen rings (Superti-Furga 1996). This peculiar structure has been reported also in the cartilage growth plate of the *Slc26a2* knock-out mouse (*Slc26a2*$^{-/-}$), a murine model of human ACG1B (Zheng et al. 2019). In particular, in proliferating chondrocytes from this mouse strain, a large amount of collagen has been detected inside the cell within the ER that is enlarged, whereas relatively small amount of collagen is present around chondrocytes and no collagen is detected in acellular areas of cartilage. Overexpression to various degree of the unfolded protein response (UPR) markers has been observed, in particular of the ATF6 arm that causes overexpression of the FGFR3 and activation of this pathway leading to negative regulation of chondrocyte proliferation, thereby protecting chondrocytes from further increased matrix protein load (Zheng et al. 2019). The link between SLC26A2 impairment, PG undersulfation and defective collagen secretion will require further investigations in the future.

Fibronectin is another abundant component of ECMs; it has been demonstrated that PG sulfation is important for fibronectin matrix assembly. In fact, sulfation inhibition either by chlorate, a competitive inhibitor of ATP sulfurylase, or by SLC26A2 knockdown by RNA interference, decreases fibronectin assembly in the ECM of different cell types (Galante and Schwarzbauer 2007). Since fibronectin fibril formation is dependent on macromolecular sulfation, it is possible that the perturbation of this process contributes to skeletal defects in patients with sulfation disorders. Indeed, fibronectin is a major component of the chondrocyte microenvironment during cartilage development and in adult tissues colocalizes with type VI collagen (Chang et al. 1997; Poole et al. 1992). During cartilage differentiation, fibronectin is replaced by a more collagen-rich matrix suggesting that fibronectin matrix may present a framework for collagen deposition during cartilage development.

PGs directly influence cell activity either through cell-ECM interactions or through the binding of specific growth factors in the ECM. Extensive modifications of the glycans such as sulfation, carboxylation, deacetylation, and epimerization generate specific motifs that can function as protein binding sites for signalling molecules (Vallet et al. 2022).

HS and CS are sulfated GAGs involved in interactions with growth factors. They play different roles due to their different localization. CS in the ECM binds growth factors protecting them from proteases and regulating their diffusion in the matrix, while HS on the cell membrane works as co-receptor. The importance of the correct sulfation of CS for growth factor signalling is demonstrated by the altered Ihh distribution in the cartilage ECM of the *bm* mouse and the *dtd* mouse causing reduced chondrocyte proliferation (Cortes et al. 2009; Gualeni et al. 2010). Indeed, Ihh signalling, a fundamental pathway for cartilage development, relies on sulfated PGs for its activity. The binding of Ihh to different GAG chains has been measured in vitro (Cortes et al. 2009); the affinity of Ihh for different glycans in decreasing order is HS, CS-4S, CS-6S, and unsulfated CS (chondroitin). Since in the *bm* mouse HS sulfation is normal as demonstrated by disaccharide analysis (Cortes et al. 2009), altered Ihh distribution in the ECM is mainly linked to CS undersulfation. Moreover, another evidence of the link between sulfated PGs and Ihh is the finding of altered Ihh distribution and expression of its downstream targets in the *Bpnt2* knock-out mouse (Sohaskey et al. 2008).

A complex network of signalling pathways has been described along with the Ihh pathway to control cartilage proliferation and maturation including FGF, BMP, TGF-β, and Wnt pathways (Guasto and Cormier-Daire 2021). In an attempt to further elucidate the molecular basis of reduced chondrocyte proliferation in the growth plate of the *dtd* mouse, activation of the Wnt pathway has been detected causing an increase of the non-phosphorylated form of β-catenin that translocated to the nucleus and increase the expression of Transcription factor 7 (Tcf7) resulting in an increased expression of cyclin D1 in the G1 phase of the cell cycle (De Leonardis et al. 2014). Notwithstanding cyclin D1 expression increase, cell cycle progression is compromised due to reduced phosphorylation of the pocket protein p130 that inhibit the transcription factors of the E2F family crucial for DNA replication and cell proliferation. Thus, the pocket proteins play a regulatory role in cell cycle control as mediators between the cyclin activities regulated by extracellular signals and the transcription factors controlling expression of genes involved in DNA replication and cell cycle progression.

Interestingly, in growth plate chondrocytes of *C4st1* knock-out mice reduced downstream effector of BMP signalling and increased TGF-β effector is observed. This finding demonstrates an altered balance between BMP and TGF-β pathway when PG sulfation is impaired (Kluppel et al. 2005), and further support the relevance of proper chondroitin sulfation in different signalling pathways crucial for chondrocyte proliferation and differentiation.

6.7 Conclusions and Perspectives

In this chapter, the complexity of skeletal defects caused by mutations in genes encoding for enzymes and transporters involved in sulfate metabolism has been deeply reviewed. Despite the great step forward in the identification of causative genes allowed by next-generation genomic technologies, a detailed molecular picture of the cross talk between ECM components, transcription factors, and signalling molecules is still far from complete. Cartilage and bone are the tissues mainly affected by defects in PG sulfation, but defective sulfation could involve other organs such as cardiac tissue in SEDCJD (Hall 1997; Tuysuz et al. 2009) or lymphoid tissue leading to tumor progression in OCBMD (Chopra et al. 2015). The progress in the early diagnosis as well as the targeted treatments either by physiotherapy or orthopedic interventions has led to the amelioration of the patients' quality of life and increased life expectancy which are expected to increase by the potential development of personalized therapies. Thus, patients now reach an age at which the phenotype progresses to a multi-organ impairment resembling accelerated chronological ageing. Indeed, many skeletal dysplasias develop osteoarthritis and/or osteoporosis early in life compared to the normal population. The deep phenotyping of adult in vitro and in vivo models is a powerful mean to define disease mechanisms during ageing and provide important insights into common complex disease forms.

Variants in different genes cause skeletal dysplasias with overlapping features that may be misdiagnosed as occurs in condrodysplasia with joint dislocation, gPAPP type, Catel-Manzke syndrome, and Desbuquois dysplasia type 1. Finally, variants in the same gene present a huge variability in the clinical phenotypes suggesting the importance also of environmental and epigenetic factors.

On the other hand, translational activities including the identification and validation of biomarkers and validation of new therapeutic targets to ultimately pave the way for innovative therapies are important to approach the patient's bedside. The identification of non-invasive biomarkers is an urgent need in view of monitoring the efficacy of any therapy. Recently, the analysis of urinary GAG sulfation as a marker for any treatment targeted at increasing GAG sulfation has been suggested in *SLC26A2* disorders. Indeed, urinary chondroitin sulfate is undersulfated both in the *dtd* mouse and in DTD patients compared to age matched controls (Karousou et al. 2014; Monti et al. 2015; Paganini et al. 2023). Moreover, hyposulfation of CS chains has been detected also in urine from a CHST3 patient with reduction in 6-O-sulfated disaccharide and an increase in the non-sulfated unit (Thiele et al. 2004).

A further potential biomarker for GAG sulfation might involve the sulfation analysis of bikunin (bkn) a serum chondroitin sulfate PG synthetized by the liver. Analysis of bkn isoforms has been suggested as marker of linkeropathies, chondrodysplasias caused by mutations in genes encoding for enzymes involved in the synthesis of the tetrasaccharide linkage region of GAGs (see Chap. 5). By western blot, Haouari et al. have demonstrated an altered ratio between the bkn core protein and the bkn linked to the CS chain in linkeropathies patients compared to age matched controls (Haouari et al. 2020). Moreover, in the two patients with variants in the *SCL35B2* gene encoding for the PAPS transporter, bkn sulfation has been tested by 2-dimensional gel electrophoresis (2-DE) followed by western blot. Serum samples have been digested with chondroitinase ABC in order to generate an hexasaccharide linked bkn form carrying a reduced number of sulfation sites. Under this condition,

2-DE patterns of controls differ from the patients' ones demonstrating a loss of negative charges consistent with CS hyposulfation (Guasto et al. 2022).

Recently, significant advances in the development of therapeutic approaches for some skeletal disorders have been achieved with the aim of alleviating, preventing, or modifying the disease progression (Besio and Forlino 2015; Marzin and Cormier-Daire 2020; Mullan et al. 2017; Ornitz and Legeai-Mallet 2017; Paganini et al. 2021; Sabir and Irving 2021). These progresses have been made thank to deep phenotyping of in vitro and/or in vivo models that has allowed the identification of new molecules or already existing ones through a drug repositioning approach. Thus, in the next future, it is expected that the number of disorders with at least a therapeutic option will increase thanks to the deep understanding of the disease molecular mechanisms and/or the identification, among different phenotypes, of common pathogenic mechanisms that might be targeted with common therapeutic strategies.

Funding This work was supported by MUR "Progetti di ricerca di rilevante interesse nazionale" (PRIN 2022ETJJES_001) and MUR "Dipartimenti di Eccellenza 2023-2027" and the European Community's Horizon-MSCA-DN-01-01 under grant agreement n. 101,072,766 (CHANGE).

Conflicts of Interest The authors declare no conflict of interest.

References

Ahmad M, Haque MF, Ahmad W, Abbas H, Haque S, Krakow D, Rimoin DL, Lachman RS, Cohn DH (1998) Distinct, autosomal recessive form of spondyloepimetaphyseal dysplasia segregating in an inbred Pakistani kindred. Am J Med Genet 78(5):468–473

Aksoy IA, Otterness DM, Weinshilboum RM (1993) Cholesterol sulfation in human liver. Catalysis by dehydroepiandrosterone sulfotransferase. Drug Metab Dispos 21(2):268–276

Akyüz N, Rost S, Mehanna A, Bian S, Loers G, Oezen I, Mishra B, Hoffmann K, Guseva D, Laczynska E, Irintchev A, Jakovcevski I, Schachner M (2013) Dermatan 4-O-sulfotransferase1 ablation accelerates peripheral nerve regeneration. Exp Neurol 247:517

Alper SL, Sharma AK (2013) The SLC26 gene family of anion transporters and channels. Mol Asp Med 34(2–3):494–515

Besio R, Forlino A (2015) Treatment options for osteogenesis imperfecta. Expert Opin Orphan Drugs 3(2):165–181

Bian S, Akyüz N, Bernreuther C, Loers G, Laczynska E, Jakovcevski I, Schachner M (2011) Dermatan sulfotransferase Chst14/D4st1, but not chondroitin sulfotransferase Chst11/C4st1, regulates proliferation and neurogenesis of neural progenitor cells. J Cell Sci 124(Pt 23):4051

Bojarova P, Williams SJ (2008) Sulfotransferases, sulfatases and formylglycine-generating enzymes: a sulfation fascination. Curr Opin Chem Biol 12(5):573–581

Bowman KG, Bertozzi CR (1999) Carbohydrate sulfotransferases: mediators of extracellular communication. Chem Biol 6(1):R9–R22

Bradley H, Gough A, Sokhi RS, Hassell A, Waring R, Emery P (1994) Sulfate metabolism is abnormal in patients with rheumatoid arthritis. Confirmation by in vivo biochemical findings. J Rheumatol 21(7):1192–1196

Chang J, Nakajima H, Poole CA (1997) Structural colocalisation of type VI collagen and fibronectin in agarose cultured chondrocytes and isolated chondrons extracted from adult canine tibial cartilage. J Anat, Pt 4 190
Chopra SS, Leshchiner I, Duzkale H, McLaughlin H, Giovanni M, Zhang C, Stitziel N, Fingeroth J, Joyce RM, Lebo M, Rehm H, Vuzman D, Maas R, Sunyaev SR, Murray M, Cassa CA (2015) Inherited CHST11/MIR3922 deletion is associated with a novel recessive syndrome presenting with skeletal malformation and malignant lymphoproliferative disease. Mol Genet Genomic Med 3(5):413–423
Clément A, Wiweger M, von der Hardt S, Rusch MA, Selleck SB, Chien CB, Roehl HH (2008) Regulation of zebrafish skeletogenesis by ext2/dackel and papst1/pinscher. PLoS Genet 4(7):e1000136
Cole DE, Baldwin LS, Stirk LJ (1985) Increased serum sulfate in pregnancy: relationship to gestational age. Clin Chem 31(6):866–867
Cole DE, Scriver CR (1980) Age-dependent serum sulfate levels in children and adolescents. Clin Chim Acta 107(1–2):135–139
Cortes M, Baria AT, Schwartz NB (2009) Sulfation of chondroitin sulfate proteoglycans is necessary for proper Indian hedgehog signaling in the developing growth plate. Development 136(10):1697–1706
Coughtrie MWH (2016) Function and organization of the human cytosolic sulfotransferase (SULT) family. Chem Biol Interact 259(Pt A):2–7
Dawson JR, Norbeck K, Moldeus P (1983) The effectiveness of different sulfate precursors in supporting extrahepatic sulfate conjugation. Biochem Pharmacol 32(11):1789–1791
Dawson PA (2011) Sulfate in fetal development. Semin Cell Dev Biol 22(6):653–659
Dawson PA, Markovich D (2005) Pathogenetics of the human SLC26 transporters. Curr Med Chem 12(4):385–396
Dawson PA, Richard K, Perkins A, Zhang Z, Simmons DG (2017) Review: nutrient sulfate supply from mother to fetus: placental adaptive responses during human and animal gestation. Placenta 54:45–51
De Leonardis F, Monti L, Gualeni B, Tenni R, Forlino A, Rossi A (2014) Altered signaling in the G1 phase deregulates chondrocyte growth in a mouse model with proteoglycan Undersulfation. J Cell Biochem 115(10):1779–1786
Diez-Roux G, Ballabio A (2005) Sulfatases and human disease. Annu Rev Genomics Hum Genet 6:355–379
Dundar M, Muller T, Zhang Q, Pan J, Steinmann B, Vodopiutz J, Gruber R, Sonoda T, Krabichler B, Utermann G, Baenziger JU, Zhang L, Janecke AR (2009) Loss of dermatan-4-sulfotransferase 1 function results in adducted thumb-clubfoot syndrome. Am J Hum Genet 85(6):873–882
Elgavish A, Meezan E (1991) Sulfation by human lung fibroblasts: SO4(2-) and sulfur- containing amino acids as sources for macromolecular sulfation. Am J Phys 260(6 Pt 1):L450–L456
Falany CN, Wheeler J, Oh TS, Falany JL (1994) Steroid sulfation by expressed human cytosolic sulfotransferases. J Steroid Biochem Mol Biol 48(4):369–375
Farooqui AA, Horrocks LA (1985) On the role of sulfolipids in mammalian metabolism. Mol Cell Biochem 66(1):87–95
Forlino A, Piazza R, Torre SD, Tatangelo L, Bonafe L, Gualeni B, Romano A, Pecora F, Superti-Furga A, Cetta G, Rossi A (2005) A diastrophic dysplasia sulfate transporter (SLC26A2) mutant mouse: morphological and biochemical characterization of the resulting chondrodysplasia phenotype. Hum Mol Genet 14(6):859–871
Frederick JP, Tafari AT, Wu SM, Megosh LC, Chiou ST, Irving RP, York JD (2008) A role for a lithium-inhibited Golgi nucleotidase in skeletal development and sulfation. Proc Natl Acad Sci USA 105(33):11605–11612
Fuda H, Shimizu C, Lee YC, Akita H, Strott CA (2002) Characterization and expression of human bifunctional 3′-phosphoadenosine 5′-phosphosulphate synthase isoforms. Biochem J 365(Pt 2):497–504

Galante LL, Schwarzbauer JE (2007) Requirements for sulfate transport and the diastrophic dysplasia sulfate transporter in fibronectin matrix assembly. J Cell Biol 179(5):999

Gallagher JT (1994) Heparan sulphates as membrane receptors for the fibroblast growth factors. Eur J Clin Chem Clin Biochem 32(4):239–247

Gama CI, Tully SE, Sotogaku N, Clark PM, Rawat M, Vaidehi N, Goddard WA, Nishi A, LC, H.-W. (2006) Sulfation patterns of glycosaminoglycans encode molecular recognition and activity. Nat Chem Biol 2(9):467–473

Gualeni B, de Vernejoul MC, Marty-Morieux C, De Leonardis F, Franchi M, Monti L, Forlino A, Houillier P, Rossi A, Geoffroy V (2013) Alteration of proteoglycan sulfation affects bone growth and remodeling. Bone 54(1):83–91

Gualeni B, Facchini M, De Leonardis F, Tenni R, Cetta G, Viola M, Passi A, Superti-Furga A, Forlino A, Rossi A (2010) Defective proteoglycan sulfation of the growth plate zones causes reduced chondrocyte proliferation via an altered Indian hedgehog signalling. Matrix Biol 29(6):453–460

Guasto A, Cormier-Daire V (2021) Signaling pathways in bone development and their related skeletal dysplasia. Int J Mol Sci 22(9)

Guasto A, Dubail J, Aguilera-Albesa S, Paganini C, Vanhulle C, Haouari W, Gorría-Redondo N, Aznal-Sainz E, Boddaert N, Planas-Serra L, Schlüter A, Vélez-Santamaría V, Verdura E, Bruneel A, Rossi A, Huber C, Pujol A, Cormier-Daire V (2022) Biallelic variants in SLC35B2 cause a novel chondrodysplasia with hypomyelinating leukodystrophy. Brain 145(10):3711–3722

Günal S, Hardman R, Kopriva S, Mueller JW (2019) Sulfation pathways from red to green. J Biol Chem 294(33):12293–12312

Hall BD (1997) Humero-spinal dysostosis: report of the fourth case with emphasis on generalized skeletal involvement, abnormal craniofacial features, and mitral valve thickening. J Pediatr Orthop B 6(1):11–14

Haouari W, Dubail J, Lounis-Ouaras S, Prada P, Bennani R, Roseau C, Huber C, Afenjar A, Colin E, Vuillaumier-Barrot S, Seta N, Foulquier F, Poüs C, Cormier-Daire V, Bruneel A (2020) Serum bikunin isoforms in congenital disorders of glycosylation and linkeropathies. J Inherit Metab Dis 43(6):1349

Hartmann-Fatu C, Trusch F, Moll CN, Michin I, Hassinen A, Kellokumpu S, Bayer P (2015) Heterodimers of tyrosylprotein sulfotransferases suggest existence of a higher organization level of transferases in the membrane of the trans-Golgi apparatus. J Mol Biol 427(6 Pt B):1404–1412

Hermanns P, Unger S, Rossi A, Perez-Aytes A, Cortina H, Bonafe L, Boccone L, Setzu V, Dutoit M, Sangiorgi L, Pecora F, Reicherter K, Nishimura G, Spranger J, Zabel B, Superti-Furga A (2008) Congenital joint dislocations caused by carbohydrate sulfotransferase 3 deficiency in recessive Larsen syndrome and humero-spinal dysostosis. Am J Hum Genet 82(6):1368–1374

Hirose T, Takahashi N, Tangkawattana P, Minaguchi J, Mizumoto S, Yamada S, Miyake N, Hayashi S, Hatamochi A, Nakayama J, Yamaguchi T, Hashimoto A, Nomura Y, Takehana K, Kosho T, Watanabe T (2018) Structural alteration of glycosaminoglycan side chains and spatial disorganization of collagen networks in the skin of patients with mcEDS-CHST14. Biochim Biophys Acta Gen Subj, vol 1863, p 623

Honke K, Taniguchi N (2002) Sulfotransferases and sulfated oligosaccharides. Med Res Rev 22(6):637–654

Hästbacka J, de la Chapelle A, Mahtani MM, Clines G, Reeve Daly MP, Daly M, Hamilton BA, Kusumi K, Trivedi B, Weaver A, Coloma A, Lovett M, Buckler A, Kaitila I, Lander ES (1994) The diastrophic dysplasia gene encodes a novel sulfate transporter: positional cloning by fine-structure linkage disequilibrium mapping. Cell 78(6):1073–1087

Hästbacka J, Superti-Furga A, Wilcox WR, Rimoin DL, Cohn DH, Lander ES (1996) Atelosteogenesis type II is caused by mutations in the diastrophic dysplasia sulfate-transporter gene (DTDST): evidence for a phenotypic series involving three chondrodysplasias. Am J Hum Genet 58(2):255–262

Iida A, Simsek-Kiper PO, Mizumoto S, Hoshino T, Elcioglu N, Horemuzova E, Geiberger S, Yesil G, Kayserili H, Utine GE, Boduroglu K, Watanabe S, Ohashi H, Alanay Y, Sugahara K, Nishimura G, Ikegawa S (2013) Clinical and radiographic features of the autosomal recessive form of brachyolmia caused by PAPSS2 mutations. Hum Mutat 34(10):1381–1386

Iozzo RV (1998) Matrix proteoglycans: from molecular design to cellular function. Annu Rev Biochem 67:609–652

Iozzo RV, Schaefer L (2015) Proteoglycan form and function: A comprehensive nomenclature of proteoglycans. Matrix Biol 42:11–55

Janecke AR, Li B, Boehm M, Krabichler B, Rohrbach M, Muller T, Fuchs I, Golas G, Katagiri Y, Ziegler SG, Gahl WA, Wilnai Y, Zoppi N, Geller HM, Giunta C, Slavotinek A, Steinmann B (2016) The phenotype of the musculocontractural type of Ehlers-Danlos syndrome due to CHST14 mutations. Am J Med Genet A 170a(1):103–115

Kamiyama S, Sasaki N, Goda E, Ui-Tei K, Saigo K, Narimatsu H, Jigami Y, Kannagi R, Irimura T, Nishihara S (2006) Molecular cloning and characterization of a novel 3′-phosphoadenosine 5′-phosphosulfate transporter, PAPST2. J Biol Chem 281(16):10945–10953

Kamiyama S, Suda T, Ueda R, Suzuki M, Okubo R, Kikuchi N, Chiba Y, Goto S, Toyoda H, Saigo K, Watanabe M, Narimatsu H, Jigami Y, Nishihara S (2003) Molecular cloning and identification of 3′-phosphoadenosine 5′-phosphosulfate transporter. J Biol Chem 278(28):25958–25963

Karniski LP (2001) Mutations in the diastrophic dysplasia sulfate transporter (DTDST) gene: correlation between sulfate transport activity and chondrodysplasia phenotype. Hum Mol Genet 10(14):1485–1490

Karousou E, Asimakopoulou A, Monti L, Zafeiropoulou V, Afratis N, Gartaganis P, Rossi A, Passi A, Karamanos NK (2014) FACE analysis as a fast and reliable methodology to monitor the Sulfation and Total amount of chondroitin sulfate in biological samples of clinical importance. Molecules 19(6):7959–7980

Klaassen CD, Boles JW (1997) Sulfation and sulfotransferases 5: the importance of 3′- phosphoadenosine 5′-phosphosulfate (PAPS) in the regulation of sulfation. FASEB J 11(6):404–418

Kluppel M, Wight TN, Chan C, Hinek A, Wrana JL (2005) Maintenance of chondroitin sulfation balance by chondroitin-4-sulfotransferase 1 is required for chondrocyte development and growth factor signaling during cartilage morphogenesis. Development 132(17):3989–4003

Koike T, Mikami T, Shida M, Habuchi O, Kitagawa H (2015) Chondroitin sulfate-E mediates estrogen-induced osteoanabolism. Scientific reports 5:5

Kozlowski KS, Celermajer JM, Tink AR (1974) Humero-spinal dysostosis with congenital heart disease. Am J Dis Child 127(3):407–410

Kurima K, Warman ML, Krishnan S, Domowicz M, Krueger RC Jr, Deyrup A, Schwartz NB (1998) A member of a family of sulfate-activating enzymes causes murine brachymorphism [published erratum appears in proc Natl Acad Sci U S A 1998 Sep 29;95(20):12071]. Proc Natl Acad Sci USA 95(15):8681–8685

Langford R, Hurrion E, Dawson PA (2017) Genetics and pathophysiology of mammalian sulfate biology. J Genet Genomics 44(1):7–20

Li P, Wang D, Lu W, He X, Hu J, Yun H, Zhao C, Yang L, Jie Q, Luo Z (2024) Targeting FGFR3 signaling and drug repurposing for the treatment of SLC26A2-related chondrodysplasia in mouse model. J Orthop Translat 44

Lyle S, Stanczak J, Ng K, Schwartz NB (1994) Rat chondrosarcoma ATP sulfurylase and adenosine 5′-phosphosulfate kinase reside on a single bifunctional protein. Biochemistry 33(19):5920–5925

Malfait F, Syx D, Vlummens P, Symoens S, Nampoothiri S, Hermanns-Le T, Van Laer L, De Paepe A (2010) Musculocontractural Ehlers-Danlos syndrome (former EDS type VIB) and adducted thumb clubfoot syndrome (ATCS) represent a single clinical entity caused by mutations in the dermatan-4-sulfotransferase 1 encoding CHST14 gene. Hum Mutat 31(11):1233–1239

Marzin P, Cormier-Daire V (2020) New perspectives on the treatment of skeletal dysplasia. Ther Adv Endocrinol Metab 11:2042018820904016

Mertz E, Rossi A, Facchini M, Gualeni B, Forlino A (2009) Matrix disruptions, growth and degradation of diastrophic dysplasia cartilage with impaired Sulfation. Glycobiology 19(11):1291–1292

Mikami T, Mizumoto S, Kago N, Kitagawa H, Sugahara K (2003) Specificities of three distinct human chondroitin/dermatan N-acetylgalactosamine 4-O-sulfotransferases demonstrated using partially desulfated dermatan sulfate as an acceptor: implication of differential roles in dermatan sulfate biosynthesis. J Biol Chem 278(38):36115

Miyake N, Elcioglu NH, Iida A, Isguven P, Dai J, Murakami N, Takamura K, Cho TJ, Kim OH, Hasegawa T, Nagai T, Ohashi H, Nishimura G, Matsumoto N, Ikegawa S (2012) PAPSS2 mutations cause autosomal recessive brachyolmia. J Med Genet 49(8):533–538

Miyake N, Kosho T, Mizumoto S, Furuichi T, Hatamochi A, Nagashima Y, Arai E, Takahashi K, Kawamura R, Wakui K, Takahashi J, Kato H, Yasui H, Ishida T, Ohashi H, Nishimura G, Shiina M, Saitsu H, Tsurusaki Y, Doi H, Fukushima Y, Ikegawa S, Yamada S, Sugahara K, Matsumoto N (2010) Loss-of-function mutations of CHST14 in a new type of Ehlers-Danlos syndrome. Hum Mutat 31(8):966–974

Mizumoto S, Yamada S (2021) Congenital disorders of deficiency in glycosaminoglycan biosynthesis. Front Genet 12

Monti L, Paganini C, Lecci S, De Leonardis F, Hay E, Cohen-Solal M, Villani S, Superti-Furga A, Tenni R, Forlino A, Rossi A (2015) N-acetylcysteine treatment ameliorates the skeletal phenotype of a mouse model of diastrophic dysplasia. Hum Mol Genet 24(19):5570–5580

Mourao PA, Kato S, Donnelly PV (1981) Spondyloepiphyseal dysplasia, chondroitin sulfate type: a possible defect of PAPS—chondroitin sulfate sulfotransferase in humans. Biochem Biophys Res Commun 98(2):388–396

Mullan LA, Mularczyk EJ, Kung LH, Forouhan M, Wragg JM, Goodacre R, Bateman JF, Swanton E, Briggs MD, Boot-Handford RP (2017) Increased intracellular proteolysis reduces disease severity in an ER stress-associated dwarfism. J Clin Invest 127(10):3861–3865

Nitahara-Kasahara Y, Mizumoto S, Inoue YU, Saka S, Posadas-Herrera G, Nakamura-Takahashi A, Takahashi Y, Hashimoto A, Konishi K, Miyata S, Masuda C, Matsumoto E, Maruoka Y, Yoshizawa T, Tanase T, Inoue T, Yamada S, Nomura Y, Takeda S, Watanabe A, Kosho T, Okada T (2021) A new mouse model of Ehlers-Danlos syndrome generated using CRISPR/Cas9-mediated genomic editing. Dis Model Mech 14(12)

Nizon M, Alanay Y, Tuysuz B, Kiper PO, Genevieve D, Sillence D, Huber C, Munnich A, Cormier-Daire V (2012) IMPAD1 mutations in two Catel-Manzke like patients. Am J Med Genet A 158A(9):2183–2187

Noordam C, Dhir V, McNelis JC, Schlereth F, Hanley NA, Krone N, Smeitink JA, Smeets R, Sweep FC, Claahsen-van der Grinten HL, Arlt W (2009) Inactivating PAPSS2 mutations in a patient with premature pubarche. N Engl J Med 360(22):2310–2318

Orkin RW, Pratt RM, Martin GR (1976) Undersulfated chondroitin sulfate in the cartilage matrix of brachymorphic mice. Dev Biol 50(1):82–94

Orkin RW, Williams BR, Cranley RE, Poppke DC, Brown KS (1977) Defects in the cartilaginous growth plates of brachymorphic mice. J Cell Biol 73:287–299

Ornitz DM, Legeai-Mallet L (2017) Achondroplasia: development, pathogenesis, and therapy. Dev Dyn 246(4):291–309

Paganini C, Carroll RS, Gramegna Tota C, Schelhaas AJ, Leone A, Duker AL, O'Connell DA, Coghlan RF, Johnstone B, Ferreira CR, Peressini S, Albertini R, Forlino A, Bonafé L, Campos-Xavier AB, Superti-Furga A, Zankl A, Rossi A, Bober MB (2023) Identification of potential non-invasive biomarkers in diastrophic dysplasia. Bone 175:116838

Paganini C, Costantini R, Superti-Furga A, Rossi A (2019) Bone and connective tissue disorders caused by defects in glycosaminoglycan biosynthesis: a panoramic view. FEBS J 286(15):3008–3032

Paganini C, Gramegna Tota C, Monti L, Monti I, Maurizi A, Capulli M, Bourmaud M, Teti A, Cohen-Solal M, Villani S, Forlino A, Superti-Furga A, Rossi A (2021) Improvement of the

skeletal phenotype in a mouse model of diastrophic dysplasia after postnatal treatment with N-acetylcysteine. Biochem Pharmacol 185:114452
Paganini C, Gramegna Tota C, Superti-Furga A, Rossi A (2020) Skeletal Dysplasias caused by Sulfation defects. Int J Mol Sci 21(8)
Parker JL, Newstead S (2019) Gateway to the Golgi: molecular mechanisms of nucleotide sugar transporters. Curr Opin Struct Biol 57:127–134
Plaas AHK, Moran MM, Sandy JD, Hascall VC (2023) Aggrecan and Hyaluronan: the infamous cartilage polyelectrolytes–then and now. Adv Exp Med Biol 1402
Poole CA, Ayad S, Gilbert RT (1992) Chondrons from articular cartilage. V. Immunohistochemical evaluation of type VI collagen organisation in isolated chondrons by light, confocal and electron microscopy. J Cell Sci 103(Pt 4)
Richard K, Hume R, Kaptein E, Stanley EL, Visser TJ, Coughtrie MW (2001) Sulfation of thyroid hormone and dopamine during human development: ontogeny of phenol sulfotransferases and arylsulfatase in liver, lung, and brain. J Clin Endocrinol Metab 86(6):2734–2742
Rome LH, Hill DF (1986) Lysosomal degradation of glycoproteins and glycosaminoglycans. Efflux and recycling of sulphate and N-acetylhexosamines. Biochem J 235(3):707–713
Rossi A, Kaitila I, Wilcox WR, Rimoin DL, Steinmann B, Cetta G, Superti-Furga A (1998) Proteoglycan sulfation in cartilage and cell cultures from patients with sulfate transporter chondrodysplasias: relationship to clinical severity and indications on the role of intracellular sulfate production. Matrix Biol 17(5):361–369
Rossi A, Superti-Furga A (2001) Mutations in the diastrophic dysplasia sulfate transporter (DTDST) gene (SLC26A2): 22 novel mutations, mutation review, associated skeletal phenotypes, and diagnostic relevance. [erratum appears in hum Mutat 2001;18(1):82]. Hum Mutat 17(3):159–171
Sabir A, Irving M (2021) Clinical trials in skeletal dysplasia: a paradigm for treating rare diseases. Br Med Bull 139(1):16
Schneeberger PE, von Elsner L, Barker EL, Meinecke P, Marquardt I, Alawi M, Steindl K, Joset P, Rauch A, Zwijnenburg PJG, Weiss MM, Merry CLR, Kutsche K (2020) Bi-allelic pathogenic variants in HS2ST1 cause a syndrome characterized by developmental delay and corpus callosum, skeletal, and renal abnormalities. Am J Hum Genet 107(6):1044
Schroder E, Gebel L, Eremeev AA, Morgner J, Grum D, Knauer SK, Bayer P, Mueller JW (2012) Human PAPS synthase isoforms are dynamically regulated enzymes with access to nucleus and cytoplasm. PLoS One 7(1):e29559
Shabbir RMK, Nalbant G, Ahmad N, Malik S, Tolun A (2018) Homozygous CHST11 mutation in chondrodysplasia, brachydactyly, overriding digits, clino-symphalangism and synpolydactyly. J Med Genet 55(7):489–496
Soares da Costa D, Reis RL, Pashkuleva I (2017) Sulfation of Glycosaminoglycans and its implications in human health and disorders. Annu Rev Biomed Eng 19:1–26
Sohaskey ML, Yu J, Diaz MA, Plaas AH, Harland RM (2008) JAWS coordinates chondrogenesis and synovial joint positioning. Development 135(13):2215–2220
Stewart V, Ronald PC (2022) Sulfotyrosine residues: interaction specificity determinants for extracellular protein-protein interactions. J Biol Chem 298(8):102232
Stone MJ, Chuang S, Hou X, Shoham M, Zhu JZ (2009) Tyrosine sulfation: an increasingly recognised post-translational modification of secreted proteins. New Biotechnol 25(5):299–317
Strott CA (2002) Sulfonation and molecular action. Endocr Rev 23(5):703–732
Superti-Furga A (1996) Achondrogenesis type 1B. J Med Genet 33(11):957–961
Superti-Furga A, Neumann L, Riebel T, Eich G, Steinmann B, Spranger J, Kunze J (1999) Recessively inherited multiple epiphyseal dysplasia with normal stature, club foot, and double layered patella caused by a DTDST mutation. J Med Genet 36(8):621–624
SupertiFurga A, Hastbacka J, Wilcox WR, Cohn DH, van der Harten HJ, Rossi A, Blau N, Rimoin DL, Steinmann B, Lander ES, Gitzelmann R (1996) Achondrogenesis type IB is caused by mutations in the diastrophic dysplasia sulphate transporter gene. Nat Genet 12(1):100–102

Tang T, Li L, Tang J, Li Y, Lin WY, Martin F, Grant D, Solloway M, Parker L, Ye W, Forrest W, Ghilardi N, Oravecz T, Platt KA, Rice DS, Hansen GM, Abuin A, Eberhart DE, Godowski P, Holt KH, Peterson A, Zambrowicz BP, de Sauvage FJ (2010) A mouse knockout library for secreted and transmembrane proteins. Nat Biotechnol 28(7):749

Thiele H, Sakano M, Kitagawa H, Sugahara K, Rajab A, Hohne W, Ritter H, Leschik G, Nurnberg P, Mundlos S (2004) Loss of chondroitin 6-O-sulfotransferase-1 function results in severe human chondrodysplasia with progressive spinal involvement. Proc Natl Acad Sci USA 101(27):10155–10160

Toledo SP, Mourao PA, Lamego C, Alves CA, Dietrich CP, Assis LM, Mattar E (1978) Recessively inherited, late onset spondylar dysplasia and peripheral corneal opacity with anomalies in urinary mucopolysaccharides: a possible error of chondroitin-6-sulfate synthesis. Am J Med Genet 2(4):385–395

Townley RA, Bülow HE (2018) Deciphering functional glycosaminoglycan motifs in development. Curr Opin Struct Biol 50:144

Tuysuz B, Mizumoto S, Sugahara K, Celebi A, Mundlos S, Turkmen S (2009) Omani-type spondyloepiphyseal dysplasia with cardiac involvement caused by a missense mutation in CHST3. Clin Genet 75(4):375–383

Uchimura K, Kadomatsu K, Nishimura H, Muramatsu H, Nakamura E, Kurosawa N, Habuchi O, El-Fasakhany FM, Yoshikai Y, Muramatsu T (2002) Functional analysis of the chondroitin 6-sulfotransferase gene in relation to lymphocyte subpopulations, brain development, and oversulfated chondroitin sulfates. J Biol Chem 277(2):1443–1450

ul Haque MF, King LM, Krakow D, Cantor RM, Rusiniak ME, Swank RT, Superti-Furga A, Haque S, Abbas H, Ahmad W, Ahmad M, Cohn DH (1998) Mutations in orthologous genes in human spondyloepimetaphyseal dysplasia and the brachymorphic mouse. Nat Genet 20(2):157–162

Unger S, Lausch E, Rossi A, Megarbane A, Sillence D, Alcausin M, Aytes A, Mendoza-Londono R, Nampoothiri S, Afroze B, Hall B, Lo IF, Lam ST, Hoefele J, Rost I, Wakeling E, Mangold E, Godbole K, Vatanavicharn N, Franco LM, Chandler K, Hollander S, Velten T, Reicherter K, Spranger J, Robertson S, Bonafe L, Zabel B, Superti-Furga A (2010a) Phenotypic features of carbohydrate sulfotransferase 3 (CHST3) deficiency in 24 patients: congenital dislocations and vertebral changes as principal diagnostic features. Am J Med Genet A 152A(10):2543–2549

Unger S, Lausch E, Rossi A, Megarbane A, Sillence D, Alcausin M, Aytes A, Mendoza-Londono R, Nampoothiri S, Afroze B, Hall B, Lo IFM, Lam STS, Hoefele J, Rost I, Wakeling E, Mangold E, Godbole K, Vatanavicharn N, Franco LM, Chandler K, Hollander S, Velten T, Reicherter K, Spranger J, Robertson S, Bonafe L, Zabel B, Superti-Furga A (2010b) Phenotypic features of carbohydrate sulfotransferase 3 (CHST3) deficiency in 24 patients: congenital dislocations and vertebral changes as principal diagnostic features. Am J Med Genet A 152A(10):2543–2549

Vallet SD, Berthollier C, Ricard-Blum S (2022) The glycosaminoglycan interactome 2.0. Am J Physiol Cell Physiol 322(6)

van Roij MH, Mizumoto S, Yamada S, Morgan T, Tan-Sindhunata MB, Meijers-Heijboer H, Verbeke JI, Markie D, Sugahara K, Robertson SP (2008) Spondyloepiphyseal dysplasia, Omani type: further definition of the phenotype. Am J Med Genet A 146A(18):2376–2384

Venkatachalam KV (2003) Human 3′-phosphoadenosine 5′-phosphosulfate (PAPS) synthase: biochemistry, molecular biology and genetic deficiency. IUBMB Life 55(1):1–11

Vissers L, Lausch E, Unger S, Campos-Xavier AB, Gilissen C, Rossi A, Del Rosario M, Venselaar H, Knoll U, Nampoothiri S, Nair M, Spranger J, Brunner HG, Bonafe L, Veltman JA, Zabel B, Superti-Furga A (2011) Chondrodysplasia and abnormal joint development associated with mutations in IMPAD1, encoding the Golgi-resident nucleotide phosphatase, gPAPP. Am J Hum Genet 88(5):608–615

Weinshilboum RM, Otterness DM, Aksoy IA, Wood TC, Her C, Raftogianis RB (1997) Sulfation and sulfotransferases 1: sulfotransferase molecular biology: cDNAs and genes. FASEB J 11(1):3–14

Xu ZH, Otterness DM, Freimuth RR, Carlini EJ, Wood TC, Mitchell S, Moon E, Kim UJ, Xu JP, Siciliano MJ, Weinshilboum RM (2000) Human 3′-phosphoadenosine 5′-phosphosulfate synthetase 1 (PAPSS1) and PAPSS2: gene cloning, characterization and chromosomal localization. Biochem Biophys Res Commun 268(2):437–444

Yoshizawa T, Kosho T (2023) Mouse models of Musculocontractural Ehlers-Danlos syndrome. Genes 14(2)

Zheng C, Lin X, Xu X, Wang C, Zhou J, Gao B, Fan J, Lu W, Hu Y, Jie Q, Luo Z, Yang L (2019) Suppressing UPR-dependent overactivation of FGFR3 signaling ameliorates SLC26A2-deficient chondrodysplasias. EBioMedicine 40:695–709

Chapter 7
The Fibrillinopathies: Fibrillin-Dependent Skeletal Dysplasias Affecting Long Bone Growth

Yousef A. T. Morcos and Gerhard Sengle

Abstract Fibrillin-1 and -2 form supramolecular networks in the connective tissue space that are crucial for development and homeostasis of bone tissues. These so called fibrillin microfibrils with their associated ligands form an intricate cellular microenvironment that controls cellular behavior of bone resident cells. This is illustrated by the skeletal features of congenital connective disorders caused by *FBN1* or *FBN2* mutations that are characterized by long bone over- or undergrowth and are collectively termed as fibrillinopathies. The skeletal features of the fibrillinopathies suggest an involvement of both fibrillins in the regulation of growth and differentiation processes of bone tissues; however, a clear understanding of how fibrillins control cellular processes of bone resident cells is still lacking. Fibrillins were shown to target and sequester growth factors of the TGF-β superfamily, but how they control growth factor bioavailability and thereby fine tune signal transduction requires further investigation. However, several studies employing biochemical experiments on the molecular protein level and *Fbn1* knock-in mouse models have provided new insightful mechanistic information, over the last years. Here, we will give an overview about the current concepts of how fibrillin deficiency impacts bone

Y. A. T. Morcos
Department of Pediatrics and Adolescent Medicine, Faculty of Medicine and University Hospital Cologne, University of Cologne, Cologne, Germany

Center for Biochemistry, Faculty of Medicine, University Hospital of Cologne, University of Cologne, Cologne, Germany

G. Sengle (✉)
Department of Pediatrics and Adolescent Medicine, Faculty of Medicine and University Hospital Cologne, University of Cologne, Cologne, Germany

Center for Biochemistry, Faculty of Medicine, University Hospital of Cologne, University of Cologne, Cologne, Germany

Center for Molecular Medicine Cologne (CMMC), University of Cologne, Cologne, Germany

Cologne Center for Musculoskeletal Biomechanics (CCMB), Cologne, Germany
e-mail: gsengle@uni-koeln.de

A. Rossi, F. Zaucke (eds.), *The Extracellular Matrix in Genetic Skeletal Disorders*, Biology of Extracellular Matrix 16,
https://doi.org/10.1007/978-3-031-70835-0_7

physiology and how this information may be utilized to explore new therapeutic avenues for patients with skeletal dysplasias due to fibrillin deficiency.

7.1 Introduction

Fibrillins are large, 350 kDa extracellular glycoproteins that form supramolecular structures within the microenvironment of connective tissue resident cells (Sakai et al. 1986). Although being not major components of the bone matrix, the phenotypic outcomes caused by fibrillin-1 or -2 deficiency are striking. Skeletal features characteristic to patients with *FBN1* or *FBN2* mutations range from tall to short stature, from arachnodactyly to brachydactyly, from hypermobile to contractured joints. This clinical variability implicates fibrillins as important regulators of growth and differentiation processes in bone tissues (Smaldone and Ramirez 2016). Until now, more than 3000 *FBN1* mutations have been described to be causative for Marfan Syndrome characterized by long bone overgrowth while only a limited number of rare mutations were reported to be causative for long bone undergrowth characteristic to the fibrillin-dependent acromicric dysplasias (Sakai and Keene 2019; Sakai et al. 2016). This implicates that a detailed understanding of the assembled tissue form of fibrillins is required to understand why a specific mutation leads to aberrant increase or loss of growth factor activity. When extracted, the assembled form of fibrillin can be visualized via electron microscopy as "beads-on-a-string" microfibrils with a diameter of 10–12 nm and an interbead periodicity of 56 nm (Keene et al. 1991a). In addition to their function as targeting scaffold for growth factors, fibrillin microfibrils (FMF) contribute significantly to the biomechanical properties and physiology of bone tissues (Nistala et al. 2010b; Sengle and Sakai 2015; Smaldone and Ramirez 2016; Zimmermann et al. 2021). Therefore, understanding not only the molecular requirements for FMF assembly and stability but also their interaction repertoire is crucial in determining the specific pathomechanisms in bone triggered by a particular *FBN1* mutation.

7.2 Characteristic Skeletal Features of Connective Tissue Disorders Caused by Fibrillin Mutations

7.2.1 Marfan Syndrome and Beals Syndrome (CCA)

FBN1 and *FBN2* mutations are causative for several skeletal features which can be mostly attributed to dysregulated long bone growth (Fig. 7.1). Clinical features that contribute to the diagnostic criteria include tall stature (Nelle et al. 1994), long fingers (arachnodactyly) (Feingold 1968; Grahame and Hakim 2013; Walker and Murdoch 1970), disproportionally long extremities (dolichostenomelia) (Walker

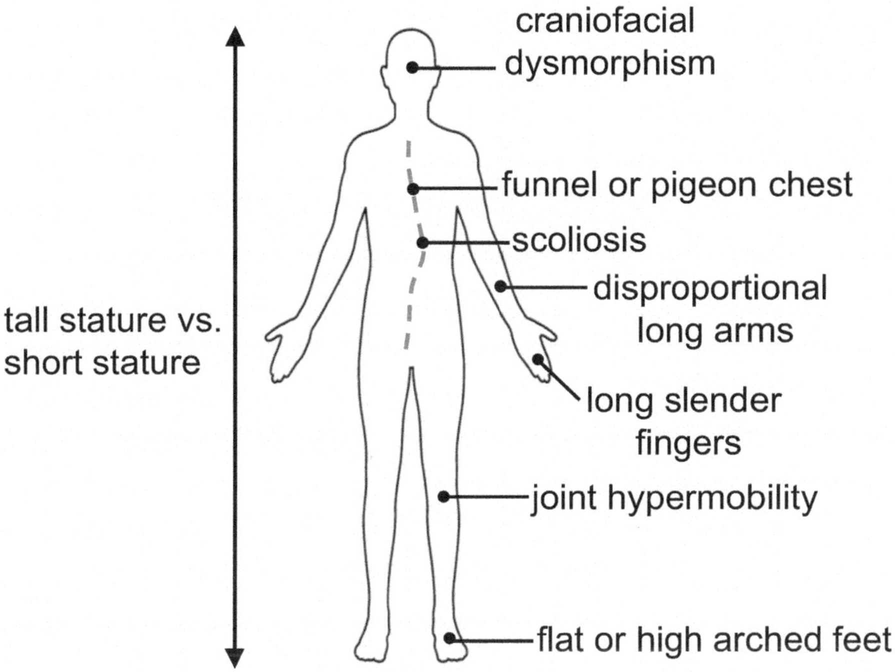

Fig. 7.1 Overview of major skeletal features caused by mutations in the fibrillin genes *FBN1* or *FBN2*. A comprehensive overview of the skeletal features in the fibrillin genes is given in Tables 7.1, 7.2, and 7.3

and Murdoch 1970), an abnormal curvature of the spine (scoliosis/kyphosis) (Birch and Herring 1987; Glard et al. 2009; Sponseller et al. 1997), and chest wall deformities (Desmarais and Keller 2013; Fraser et al. 2017) (Table 7.1). MFS patients also display craniofacial abnormalities and were reported to have a reduced bone mineral density and fracture risk (Carter et al. 2000; Folkestad et al. 2021; Kohlmeier et al. 1995; Moura et al. 2006).

Beals syndrome or congenital contractural arachnodactyly (CCA, MIM#121050) is caused by extremely rare mutations in *FBN2* and shares most skeletal abnormalities with MFS but also displays distinct features (Table 7.2). CCA patients display a Marfan-like habitus with a tall and slender build (Beals and Hecht 1971). They exhibit fixation of certain joints, including the fingers, elbows, knees, and hips, which are in a flexed position, known as contractures. Arachnodactyly, characterized by abnormally elongated and slender fingers and toes, as well as camptodactyly, which is a permanent flexion of the fingers, is also common. Additionally, individuals with CCA may present with abnormally shaped ears that appear crumpled, front-to-back and side-to-side curvature of the spine, talipes equinovarus (clubfoot), ulnar deviation of the fingers, and an unusually short neck (Meerschaut et al. 2020). Despite these distinct skeletal features, CCA patients have a

Table 7.1 Overview of the skeletal features of Marfan syndrome (OMIM#154700)

Skeletal feature	*Description*	*Reference*
Arachnodactyly	Long, slender fingers	(Feingold 1968; Grahame and Hakim 2013; Walker and Murdoch 1970)
Dolichostenomelia	Long arms, legs, fingers in proportion to the rest of the body	(Walker and Murdoch 1970)
Joint hypermobility	Excessive flexibility and looseness of joints	(Grahame and Pyeritz 1995)
Pectus excavatum	Depression of sternum ("funnel chest")	(Arn et al. 1989), (De Maio et al. 2016), (Fraser et al. 2017)
Pectus carinatum	Protrusion of the sternum and ribs ("pigeon chest")	(Ho et al. 2005), (Loeys et al. 2010a), (Desmarais and Keller 2013)
Pes planus	Flat foot	(Lindsey et al. 1998)
Pes cavus	High arch of the foot	(Coelho and Almeida 2020)
Protrusio acetabuli	Deformity medial wall of the acetabulum that allows the femoral head to migrate progressively towards the pelvic cavity	(Steel 1996),(Van de Velde et al. 2006), (Yule et al. 1999)
Scoliosis or kyphosis	Abnormal curvature of the spine	(Robins et al. 1975), (Birch and Herring 1987), (Demetracopoulos and Sponseller 2007), (Glard et al. 2008; Glard et al. 2009), (Sponseller et al. 1997)
Tall stature	> two standard deviations above mean height for age and sex	(Benoist et al. 2018; Nelle et al. 1994)
Craniofacial dysmorphism		(Zarate et al. 2022)
Dolichocephaly	Relatively long skull	(Wozniak-Mielczarek et al. 2022)
Enophthalmos	Abnormally retro-positioned globe with respect to bony orbit	(Ting et al. 2010), (Gehle et al. 2017)
Micrognathia	An abnormally small jaw	(Judge and Dietz 2005)
Highly arched palate	Highly arched roof of the mouth	(Fitch and Secord 1963), (De Coster et al. 2002)
Mandibular retrognathia	Lower jaw is set further back than the upper jaw	(De Coster et al. 2004)
Malar hypoplasia	Underdeveloped cheek bones	(Ting et al. 2010)
Prognathism	Protruding jaw	(Bade and Spirgimn 1965)
Malocclusion	Crowded teeth, overbite, or underbite	(Utreja and Evans 2009)

significantly reduced risk to develop major cardiovascular complications such as aortic aneurysm formation or mitral valve prolapse (Tuncbilek and Alanay 2006). This suggests that while both fibrillins have similar functions in the skeletal system, fibrillin-1 plays a more significant role in cardiovascular tissues.

Table 7.2 Unique skeletal features of Beals syndrome also known as congenital contractural arachnodactyly (CCA, MIM#121050)

Skeletal features	Description	Reference
Joint contractures	Fingers, elbows, knees, and hips are fixed in a flexed position	(Viljoen 1994)
Camptodactyly	Permanent flexion of the fingers	(Deng et al. 2016), (Deng et al. 2016)
Ear malformation	Crumpled ears	(Tuncbilek and Alanay 2006), (Viljoen 1994)
Talipes equinovarus	Clubfoot, feet are abnormally positioned	(Ramos Arroyo et al. 1985)
Ulnar deviation	Outward displacement of the fingers	(Ogino and Kato 1993)
Cervical shortening	Neck is abnormally short	(Inbar-Feigenberg et al. 2014)

7.2.2 *Short Stature Syndromes Caused by* FBN1 *Mutations and Related Dysplasias*

Interestingly, autosomal dominantly inherited *FBN1* mutations are also causative for long bone undergrowth (Faivre et al. 2003; Le Goff et al. 2011b; Sengle et al. 2012) as illustrated by a more recently discovered group of fibrillin-dependent short syndromes that are also characterized by short hands, stiff joints, and thickened skin (Sakai and Keene 2019) (Table 7.3). The recessive forms of these connective tissue disorders are caused by mutations in ECM components that were found to be targeted to the FMF scaffold such as ADAMTSL2 (a disintegrin and metalloproteinase with thrombospondin motifs like-2) (Le Goff et al. 2008), ADAMTS10 (Dagoneau et al. 2004), ADAMTS17 (Karoulias et al. 2020; Morales et al. 2009), LTBP2 (latent TGF-β binding protein-2) (Haji-Seyed-Javadi et al. 2012), and LTBP3 (McInerney-Leo et al. 2016). Members of the ADAMTS and ADAMTSL families have been shown to facilitate proper FMF assembly and therefore likely contribute to the structural integrity of FMF (Hubmacher et al. 2017; Hubmacher et al. 2015; Kutz et al. 2011; Sengle et al. 2012; Tsutsui et al. 2010). LTBP3 serves as a transport molecule for TGF-β growth factors (TGF-β-1,-2, and -3) (Saharinen and Keski-Oja 2000) and is tethered to FMF via TG2 mediated crosslinks (Lockhart-Cairns et al. 2022), while LTBP2 binds to FMF and interacts with FGF-2 (Hirani et al. 2007; Menz et al. 2015), a growth factor known to be important for bone homeostasis. Two related syndromes are stiff skin (SSKS) and Myhre syndrome (MYHRS). SSkS is caused by *FBN1* mutations in the domain containing the integrin binding site suggesting an involvement of the integrin mediated mechanosignal transduction of bone resident cells (Loeys et al. 2010b). MYHRS is caused by mutations of the intracellular signaling intermediate of BMP/TGF-β signaling SMAD4 indicating a link to dysregulated growth factor signal transduction (Le Goff et al. 2011a).

Recently, two other related connective tissue disorder were described that are caused by mutations in LTBP1 (Pottie et al. 2021) and EMILIN1

Table 7.3 Rare mutations in *FBN1* lead to WMS, GPHYSD, ACMICD, and SSkS characterized by long bone undergrowth. Also, mutations in the signaling intermediate SMAD4 lead to Myhre syndrome (MYHRS) which is characterized by short stature, short hands, and stiff joints

Dysplasia	*Skeletal feature*	*Acronym/ subtype*	*MIM#*	*Affected gene*	*Inheritance*	*Reference*
Weill-Marchesani syndrome (WMS)	Brachycephaly: Unusually short, wide head shape	WMS1	27760	*ADAMTS10*	AR	(Dagoneau et al. 2004)
	Brachydactyly: Short fingers	WMS2	608328	*FBN1*	AD	(Faivre et al. 2003), (Sengle et al. 2012)
	Joint stiffness in hands, shoulder, elbows, knees, ankles					
	Short, broad hands /feet, broad metacarpals / phalanges	WMS3	614819	*LTBP2*	AR	(Haji-Seyed-Javadi et al. 2012)
	Scoliosis	WMS4	613195	*ADAMTS17*	AR	(Morales et al. 2009)
	Short stature					
Geleophysic dysplasia (GPHYSD) Acromicric dysplasia (ACMICD)	Cone-shaped epiphyses	GPHYSD1	231050	*ADAMTSL2*	AR	(Le Goff et al. 2008)
	Delayed bone age					
	Joint stiffness					
	Ovoid vertebral bodies	GPHYSD2	614185	*FBN1*	AD	(Le Goff et al. 2011b)
	Short long tubular bones					
	Hypertelorism	GPHYSD3	617809	*LTBP3*	AD	(McInerney-Leo et al. 2016)
	Shortened nose					
	Short hands and feet					
	Short stature	ACMICD	102370	*FBN1*	AD	(Le Goff et al. 2011b)
	Toe walking					
Stiff skin syndrome (SSKS)	Cutaneous nodules at distal interphalangeal joints	SSkS	184900	*FBN1*	AD	(Loeys et al. 2010b)
	Flexion contractures of limb joints					
	Limited joint mobility					
	Short stature					

(continued)

Table 7.3 (continued)

Dysplasia	*Skeletal feature*	*Acronym/ subtype*	*MIM#*	*Affected gene*	*Inheritance*	*Reference*
Myhre syndrome (MYHRS)	Brachydactyly: Hands/ feet	MYHRS	139210	*SMAD4*	AD	(Le Goff et al. 2011a)
	Broad ribs					
	Camptodactyly					
	Clinodactyly: Curvature of a digit					
	Cone-shaped epiphyses					
	Decreased joint mobility					
	Large, flattened vertebrae with large pedicles					
	Microcephaly					
	Overlapping toes					
	Short long bones					
	Thickened calvarium					
	Toe syndactyly, 2–3					
	Vertebral fusions					

(elastin-microfibril-interface-located-protein-1) (Adamo et al. 2022) that are both targeted to the FMF network in the skeletal system (Dallas et al. 2000; Schiavinato et al. 2017). Affected patients of both disorders were identified due to defects in elastic tissues such as cutis laxa; however, they also display skeletal features similar to the fibrillinopathies such as short stature, brachydactyly, joint hypermobility, and highly arched palate (Table 7.4). Also, patients with mutations in other FMF associated ligands of the fibulin and LTBP families show clinical features similar to the fibrillinopathies reflecting their functional interdependence to FMF (Table 7.4).

7.3 Domain Structure of Fibrillin and Its Supramolecular Organization into Microfibrils

7.3.1 Fibrillin Domain Structure

Fibrillins have a conserved domain structure mostly composed of repeating units of calcium binding epidermal growth factor like domains (cbEGF). The fibrillin family comprises three members, namely fibrillins 1-3, which share the highest sequence

Table 7.4 Skeletal features of patients with *LTBP1* and *EMILIN1* mutations

Dysplasia	*Skeletal feature*	*Acronym/ subtype*	*MIM#*	*Affected gene*	*Inheritance*	*Reference*
Autosomal recessive Cutis laxa type IIE	Brachydactyly	ARCL2E	619451	*LTBP1*	*AR*	(Pottie et al. 2021)
	Clinodactyly					
	Craniosynostosis					
	Joint hypmobility					
	Short stature					
	Syndactyly					
Distal hereditary motor neuronopathy, type X	Dolicocephaly		620080	*EMILIN1*	AR	(Adamo et al. 2022)
	Downslanted palpebral fissures					
	Fractures					
	Highly arched palate					
	Joint hypermobility					
	Short palpebral fissures					
Autosomal recessive Cutis laxa type IB	Arachnodactyly	ARCL1B	614437	*EFEMP2 alias FBLN4*	AR	(Hucthagowder et al. 2006)
	Contractures of 3rd to 5th fingers					
	Fractures at birth					
	Highly-arched palate					
	Joint hypermobility					
	Micrognathia					
	Pectus excavatum					
	Prominent forehead					
	Prominent premaxilla					
Autosomal recessive Cutis laxa type IC	Flat midface	ARCL1C	613177	*LTBP4*	AR	(Urban et al. 2009)
	Hypertelorism					
	Joint hypermobility					
	Micrognathia					
	Retrognathia					
	Wide fontanels					
	Wide sutures					
	Widely spaced first and second toes					
	Plantar crease					

(continued)

Table 7.4 (continued)

Dysplasia	*Skeletal feature*	*Acronym/ subtype*	*MIM#*	*Affected gene*	*Inheritance*	*Reference*
Autosomal recessive cutis laxa type IA	Arachnodactyly	ARCL1A	219100	*FBNL5*	AR	(Loeys et al. 2002)
	Congenital fractures					
	Joint hypermobility					
	Microcephaly					
	Pectus excavatum					

homology with four related proteins known as latent transforming growth factor TGF-β binding proteins 1-4 (LTBPs 1-4) (Charbonneau et al. 2004; Robertson et al. 2011). The domain structures of fibrillins and LTBPs mostly contain cbEGF repeats, interspersed by eight-cysteine (8-cys) domains, also referred to as TGF-β-binding like (TB) domains (Fig. 7.2). In addition to these domains, fibrillins contain hybrid domains that share structural similarity with both EGF and TB domains and host LTBP and fibulin-binding epitopes (El-Hallous et al. 2007; Ono et al. 2009). TB domains are specific to fibrillins and LTBPs and are not found in other proteins (Robertson et al. 2011). The third TB domains of LTBPs bind to TGF-β prodomains (PDs), also known as latency-associated peptide (LAP), to which they become covalently attached via two disulphide bridges (Saharinen et al. 1996). However, the TB domains of fibrillins have been demonstrated to be not involved in binding to TGF-β growth factors (Isogai et al. 2003; Saharinen and Keski-Oja 2000). Fibrillin possesses 47 EGF domains, out of which 43 have the ability to bind calcium, known as cbEGF. This calcium-binding capability provides stability against proteolytic degradation and influences the extensibility of fibrillin (Reinhardt et al. 2000). The connecting regions between TB and cbEGF domains, known as interdomain linkers, might function as a molecular spring that stretches when subjected to force (Lee et al. 2004). Under normal physiological conditions, these domain connections are fully occupied by calcium molecules. However, during stretching, these connections unfold, resulting in a reduced binding affinity and the release of calcium (Lee et al. 2004). Fibrillin attaches to the cell surface through $\alpha 5\beta 1$, $\alpha v\beta 3$, and $\alpha v\beta 6$ integrins, facilitated by the presence of the RGD site within the fourth TB domain (Bax et al. 2003; Pfaff et al. 1996). Furin-mediated proteolytic processing of the C-terminus of fibrillin aids in efficient secretion and the assembly of fibrillin into microfibrils (Jensen et al. 2014). The resulting C-terminal propeptide was recently reported to act as a hormone regulating glucose release from the liver and controlling appetite in the brain (Janoschek et al. 2020; Romere et al. 2016). However, asprosin is also produced by chondrocytes and asprosin levels were also shown to correlate with COMP (cartilage oligomeric matrix protein), a known marker for cartilage degradation, upon running exercise and hip replacement surgery (Morcos et al. 2022). In addition, the C-terminal propeptide of fibrillin-2 was reported to

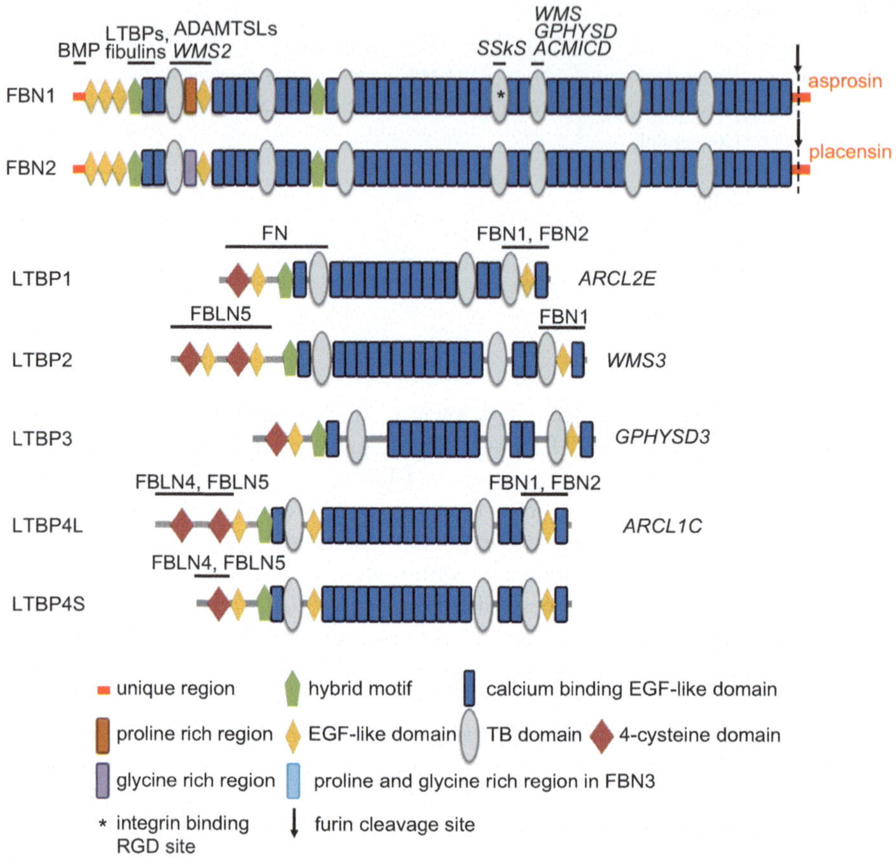

Fig. 7.2 Domain structure of the members of the fibrillin/LTBP protein family. Functional interaction sites are indicated as well as the location of rare *FBN1* mutations leading to short stature syndromes. Mutations in LTBPs also lead to short stature syndromes and forms of cutis laxa. MIM numbers and references for each disorder can be found in Tables 7.3 and 7.4

function as a hormone and stimulates cAMP production, protein kinase A (PKA) activity, glucose secretion, and increased expression of gluconeogenesis enzymes in hepatocytes (Yu et al. 2020).

7.3.2 *Supramolecular Organization of Fibrillin into Microfibrils*

The arrangement of fibrillin molecules into mature microfibrils with a diameter of 10–12 nm is still not fully understood. Electron microscopy (EM) reveals that microfibrils have a "beads-on-a-string" appearance, with the N- and C-termini of the fibrillin molecules located in the bead region (Baldock et al. 2001; Reinhardt et al. 1996). However, it is unclear how the monomers are correctly aligned to stabilize the interbead region. Biochemical analysis suggests that transglutaminase-derived cross-links help to achieve this (Qian and Glanville 1997). The periodicity of microfibrils extracted from tissues is 50–60 nm, but upon stretching, the interbead distance can increase to over 100 nm (Keene et al. 1991a; Wang et al. 2009). Three structural models have been proposed to reconcile the difference between the periodicity and the length of extracted and recombinant fibrillin-1 monomers (Baldock et al. 2001; Kuo et al. 2007; Lee et al. 2004). New imaging approaches, such as serial block-face scanning electron microscopy (SBF-SEM) and atomic force microscopy (AFM), have provided further insights into the ultrastructure of microfibrils. These approaches have shown that microfibrils contain a central cavity in the interbead region where eight symmetrically arranged fibrillin monomers can be accommodated (Godwin et al. 2018). Recently, the first detailed cryo-EM structure of FMF has been reported (Godwin et al. 2023). The major bead region showed pseudo eight-fold symmetry and a buried protease resistant N-terminal core (Fig. 7.3a and b). Based on this structure, it could be shown that a deletion of three domains that is causative for WMS induces a rearrangement with long-range effects blocking interaction with LTBP1 at a remote site (Godwin et al. 2023). Separate deletion of this binding site resulted in the assembly of shorter FMF with structural alterations. These results demonstrated that fibrillin-1 mutations may affect the ultrastructure of FMF with long-range structural consequences on other binding ligands such as growth factors and thereby trigger disease pathways.

7.4 Localization of Fibrillin in Bone and Functional Interactions Within the Bone ECM

7.4.1 Localization of Fibrillin in Bone Tissues

Despite the striking skeletal features of patients with *FBN1* or *FBN2* mutations, fibrillins are not major components of the bone matrix. Previous localization studies in human tissues showed that fibrillin-1 resembles the distribution of type III collagen in fetal bone, but already at 7 years of age fibrillin-1 was found to be absent from the interior of the cortex except for the canaliculi and cement lines, but could be detected in discrete fibers at the bone surface (Keene et al. 1991b). In the skeletal system, elastic fibers are most abundant in ligaments and tendons (Hill et al. 2020); however they are also found in the articular fibrous tissue of various joints, and in the fibrous perichondria of human hyaline and fibrous cartilages (Keene et al. 1997). Within bones, elastic fibers are found in the outer fibrous layer of the periosteum, but not in the inner cellular (cambial) layer. It was proposed a long time ago that the molecular defect being causative for long bone overgrowth in Marfan syndrome might be located in the periosteum, based on the observation that cutting the

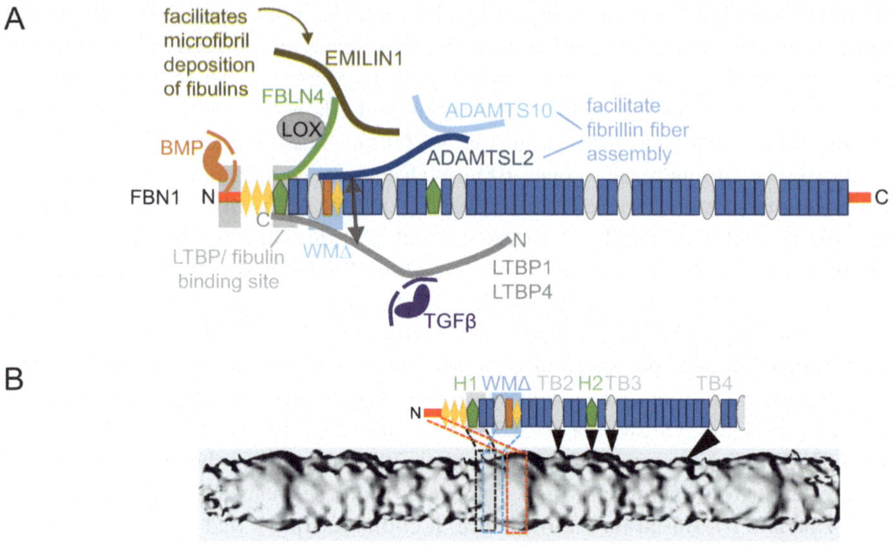

Fig. 7.3 Functional binding sites within fibrillin-1 and their localization within the ultrastructure of the microfibril. (**a**) Fibrillin-1 contains several functional binding region for growth factor of the TGF-β superfamily and the LTBP and fibulin family members. Thereby, bone morphogenetic proteins (BMPs) are directly targeted to the fibrillin N-terminal unique domain via interactions of the prodomains (Sengle et al. 2008). TGF-β is targeted to fibrillin-1 in the form of a large latent complex (LLC) that consists of a prodomain-growth factor small latent complex (SLC) that is tethered to latent TGF-β binding proteins (LTBPs) (Isogai et al. 2003; Saharinen and Keski-Oja 2000). LTBPs compete with fibulins for the same fibrillin-1 binding site containing the first hybrid motif (H1) (Ono et al. 2009). Fibulin-4 is an important transport molecule for lysyl oxidase (LOX) (Horiguchi et al. 2009), and requires EMILIN1 (Elastic Microfibril Interface Located Protein 1) for ECM deposition (Adamo et al. 2022). The WMΔ region in fibrillin-1 is composed of three domains: the first TGF-β binding module (TB1), the proline rich region, and the fourth EGF-like domain. The WMΔ deletion mutation in *FBN1* causes familial autosomal dominant Weill-Marchesani syndrome (WMS2) (Sengle et al. 2012). The WMΔ regions contains a binding site for members of the ADAMTS/L (A Disintegrin And Metalloproteinase with Thrombospondin Motifs) family (Sengle et al. 2012) that facilitate fibrillin microfibril assembly (Hubmacher et al. 2019; Kutz et al. 2011). Mutations in ADAMTSL2 lead to dysregulated TGF-β activity, as it also exerts direct interactions with LTBP1 (Le Goff et al. 2008). (**b**) Functional regions and domains in fibrillin-1 are marked on a fibrillin microfibril negative stain class average (Godwin et al. 2023)

periosteal cuff can stimulate long bone growth. The periosteum is loosely attached to the bone, but strongly connected to the growth plate at the perichondrial ring. During long bone growth, the periosteum is thought to exert a mechanical restraint on elongation of the bone at the perichondrial ring by being stretched. Cutting the periosteum releases this restraint and stimulates long bone growth (Houghton and Dekel 1979). FMF that are detected in the perichondrium and periosteum are highly organized into long elastic fibers that uniformly align parallel to the long axis of the bone (Wlodarski 1989). It is possible that FMF serve a mechanical function in the perichondrium and periosteum, restraining growth of long bones. Continuous pressure against defective, more compliant FMF in the perichondrium may lead to overgrowth of long bones. However, recent binding studies identified FMF as targeting platform of growth factors of the TGF-β superfamily (which include the bone morphogenetic proteins and growth and differentiation factors) which implicated them in the modulation of bone growth (Sengle and Sakai 2015; Zimmermann et al. 2021). Furthermore, interactions with components with the elastic fiber network such as LTBPs and fibulins were identified to be crucial for bone development and stability (Mahajan et al. 2021; Rifkin et al. 2018).

7.4.2 Functional Interactions of Fibrillins in Skeletal Tissues

7.4.2.1 Interactions with the TGF-β/BMP Signal Transduction System

The functional interdependence of TGF-β (TGFβ-1, -2, and -3) growth factors and the FMF network for bone development and homeostasis becomes evident by the clinical overlap of genetic disorders or animal models caused by mutations in the TGF-β genes, their receptors, downstream signaling mediators, as well as extracellular targeting and sequestration proteins. For instance, mutations in TGF-β type I and type II receptors TGF-β2 or SMAD3 lead to similar skeletal features as observed in MFS such as tall stature, arachnodactyly, pectus deformities, or club foot (Boileau et al. 1993, 1995; Lindsay et al. 2012; Loeys et al. 2005) (Fig. 7.4). For a better understanding of the underlying pathological mechanisms, it is important to understand the mechanisms by which TGF-β growth factors are secreted, targeted to the ECM, and utilized by cells.

Until now, three main regulatory steps have been identified for controlling TGF-β bioavailability. TGF-β growth factors are secreted as small latent complexes (SLC) consisting of a dimeric growth factor non-covalently associated with its PD, also known as LAP (latency associated peptide) (Saharinen and Keski-Oja 2000). The PD is proteolytically processed from the growth factor domain during the secretory pathway, and confers latency by either blocking receptor access to the growth factor or by altering its conformation to inhibit receptor recognition (De Crescenzo et al. 2001). Secondly, TGF-β is secreted as a large latent complex (LLC) to the extracellular space in which LAP is covalently tethered via disulfide bridges to LTBPs (Saharinen et al. 1996). Lastly, TGF-β·LLC is targeted to the ECM through specific direct interactions of LTBPs with fibrillin-1 (Isogai et al. 2003). These molecular interactions are required to control the bioavailability of TGF-β, since mutations in LAP, in LTBPs, or in fibrillin-1 correlate with aberrant TGF-β activity in humans and mice. Mutations in LAP lead to Camurati-Engelmann disease (CAEND, MIM# 131300), a skeletal condition that is characterized by abnormally thick bones (hyperostosis) in the arms, legs, and skull. Additional abnormalities include joint deformities (contractures), and scoliosis. Mutations in LTBPs, the carriers of TGF-β to the ECM, have been also implicated in bone disease. Mutations in *LTBP3* lead to acromicric and geleophysic dysplasia (McInerney-Leo et al. 2016), characterized by severe short stature, short hands and feet, and joint limitations (Table 7.3). Similar to affected patients *Ltbp3* knock-out mice display reduced body size, craniofacial malformations, due to early ossification of the skull base synchondroses.

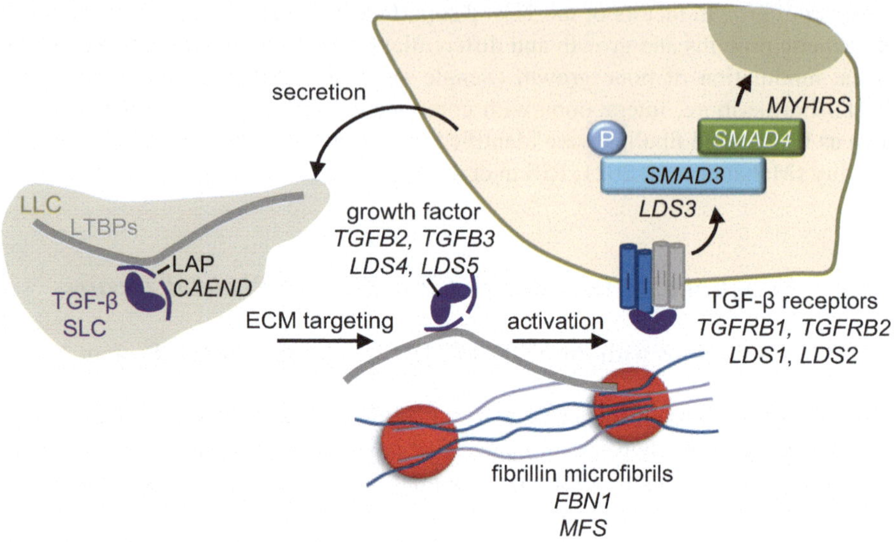

Fig. 7.4 Mutations in TGF-β prodomains or growth factors, their receptors, intracellular signaling mediators, or ECM targeting scaffolds such as fibrillin microfibrils lead to syndromes with common skeletal features. MIM numbers and references for each disorder can be found in Tables 7.3 and 7.4

In addition, histological examination of *Ltbp3*$^{-/-}$ skeletons revealed an increase in bone mass and kyphosis (Dabovic et al. 2002a, b). Craniofacial abnormalities and reduced long bone length were also observed in *Ltbp1*$^{-/-}$ mice (Drews et al. 2008). Patients affected by *LTBP1* mutations present with connective tissue features (cutis laxa and inguinal hernia), craniofacial dysmorphology, variable heart defects, and prominent skeletal features (craniosynostosis, short stature, brachydactyly, and syndactyly) (Pottie et al. 2021) (Table 7.4). Also, zebrafish lines carrying mutations in *ltbp1* showed abnormal collagen fibrillogenesis and intervertebral ligaments and ectopic bone formation on the vertebrae (Pottie et al. 2021). Collectively, it is suggested that all of the reported skeletal features are caused by a reduced bioavailability of TGF-β. Interestingly,

BMPs belong to the TGF-β superfamily of GFs and form TGF-β-like GF complexes (CPLXs) with their processed PDs, but interact with fibrillin-1 and fibrillin-2 through direct interactions with their PDs (Gregory et al. 2005; Sengle et al. 2008; Sengle et al. 2011; Zimmermann et al. 2021). The BMP PD binding site on fibrillin-1 and fibrillin-2 is located within the "fibrillin unique N-terminal domain" (FUN) (Spanou et al. 2023; Wohl et al. 2016). BMP PD binding to fibrillin results in BMP sequestration (Wohl et al. 2016). Although BMP PD-GF CPLXs are active and adopt an open V-shape conformation, their targeting to FMF induces a conformational change into a closed ring-shape conformation (Wohl et al. 2016). This structural re-arrangement renders the GF latent since the PD prevents access of BMP type II receptors to its binding site on the GF moiety. Furthermore, we recently described a new proteolytic activation mechanism of BMPs from latent FMF-bound pools (Furlan et al. 2021). It involves the specific cleavage of the BMP PD within the alpha-helix 2 by MMPs which leads to release of the active growth factor (Furlan et al. 2021). Moreover, BMP PDs maybe also concentrated to FMF via the GAG chains of FMF-associated proteoglycans such as perlecan (Sengle et al. 2011). We recently showed that BMPs targeting to GAGs such as heparin/ heparan sulfate via specific PD binding sites allows its concentration to the ECM in bioactive V-shape conformation (Spanou et al. 2023). This suggests that FMF can fine-tune the bioavailability of active growth factors during skeletal growth by targeting different binding sites within their PDs. It was biochemically shown that *FBN1* mutations occurring in fifth TB-module being causative for short stature and long bone undergrowth impact the heparin/heparan sulfate binding site within this domain (Cain et al. 2012). Our recent findings may suggest that these mutations limit the spatial concentration of bioactive growth factors bound to FMF within the cellular microenvironment of bone (Spanou et al. 2023).

Further mechanistic insight on how FMF control TGF-β/BMP signals in skeletal tissues came from investigations in fibrillin mutant mouse models. *Fbn2*$^{-/-}$ mice show syndactyly and contractures that were shown to be caused by aberrantly increased BMP signaling (Arteaga-Solis et al. 2001; Sengle et al. 2015). *Fbn2*$^{-/-}$ mice have a low bone mass phenotype due to reduced bone formation in vivo and impaired osteoblast maturation in vitro. This phenotype was shown to be caused by improper activation of latent TGF-β, which selectively reduces the expression of osterix and collagen I (Nistala et al. 2010b). Similarly, osteoblasts from *Fbn1*$^{-/-}$ mice also exhibit improper activation of latent TGF-β but mature faster due to increased availability of BMPs (Nistala et al. 2010b). This suggests that FMF do not have a direct role in supporting mineral deposition. These findings indicate that FMF play a critical role in regulating bone formation by modulating endogenous TGF-β and BMP signaling.

7.4.2.2 Fibrillin Interacts with Functional Components of the Elastic Fiber Network Required for Bone Stability

The formation of elastic fibers occurs in a hierarchical manner, beginning with the deposition of fibronectin to create a proper fibrous FMF network (Kinsey et al. 2008; Sabatier et al. 2009). This network serves as a structural scaffold for the targeting and cross-linking of tropoelastin into mature elastin. Specific binding events between fibronectin, fibrillins, fibulin-4, fibulin-5, LTBP4, tropoelastin, and lysyl oxidase (LOX) facilitate this complex process (Adamo et al. 2021; Godwin et al. 2019; Schmelzer et al. 2020). Tropoelastin forms small aggregates and interacts with fibulin-4 and fibulin-5 before being targeted to FMF-bound LTBP4, which facilitates linear deposition along the FMF core structure. Fibulin-4 and fibulin-5 interact with LOX and lysyl oxidase like-1 (LOXL-1), respectively (Horiguchi et al. 2009; Liu et al. 2004), serving as carrier molecules for their extracellular targeting to FMF. Fibulin-4 interacts with the LOX propeptide to mediate extracellular targeting of LOX to FMF, where its activation occurs by BMP-1–mediated propeptide removal (Horiguchi et al. 2009). LOX generates the unique elastin-specific desmosine cross-links, in addition to establishing collagen cross-links in the extracellular matrix (ECM) (Vallet and Ricard-Blum 2019).

The functional consequences of absence of elastic fiber components for bone stability are illustrated by the clinical features of patients and investigations of knock-out mouse models. Apart from forelimb contractures and unusually thick collagen fibrils, no morphological abnormalities were observed in the skeleton of *Fbln4*$^{-/-}$ mice compared to their wild type littermates (Sasaki et al. 2016). However, the deficiency of fibulin-4 led to an increased susceptibility of bone collagen to acid extraction, indicating a significant reduction in lysylpyridinoline and hydroxylysylpyridinoline cross-links in bone (Sasaki et al. 2016). This was further supported by a strong decrease in the amount of lysyl oxidase in long bones and calvaria and a reduced proteolytic activation of lysyl oxidase in fibulin-4 deficient osteoblasts (Sasaki et al. 2016). The addition of recombinant fibulin-4 rescued the activation, suggesting that fibulin-4 plays a crucial role in the proteolytic activation of LOX (Sasaki et al. 2016). Further evidence was provided that fibulin-5 plays a vital role in the development of the craniofacial skeleton in mice (Noda et al. 2015). Fibulin-5 is present as a fibrous matrix in mesenchymal tissues derived from the cranial neural crest, including craniofacial sutures. Fibulin-5 deficient mice exhibited reduced outgrowth of the premaxillary bone during postnatal stages. Although mesenchymal cells in the premaxillo-maxillary suture of fibulin-5 deficient mice were capable of differentiating into osteoblasts, they were less proliferative than those in wild-type mice (Noda et al. 2015). These findings suggest that fibulin-5 is essential for regulating the proliferation of facial suture mesenchymal cells, which is necessary for craniofacial skeletal morphogenesis.

7.5 Insights into Underlying Pathways Controlling Bone Stability and Long Bone Growth in the Fibrillinopathies by Employing Mouse Models

Over the recent years, new molecular insight has been gained by generating functional knock-out or knock-in mouse lines resembling the characteristic clinical features of the fibrillinopathies such as long bone over growth or undergrowth. In Table 7.5 an overview of the investigated mouse lines and their skeletal phenotypes is provided.

7.5.1 Fibrillins as Negative Regulators of Bone Resorption

Similar to reports about Marfan patients, a loss or reduction of fibrillins was shown to be causative for a reduction in bone mass and bone mineral density in *Fbn1* or *Fbn2* mutant mice. In addition to forelimb syndactyly and reduced muscle mass in *Fbn2*$^{-/-}$ mice, features that were shown to be caused by dysregulated BMP

Table 7.5 List of mouse models to study long bone over- or undergrowth in the fibrillinopathies

Line	Type of mutant allele	Skeletal features	Resembling disorder	Reference
Fbn1$^{mgR/mgR}$	Hypomorph	Severe kyphosis, overgrowth of the ribs, reduced bone content and reduction in BMD	MFS	(Pereira et al. 1999), (Nistala et al. 2010a)
Fbn2$^{-/-}$	Knock-out	Reduced femur length, decreased BMD, decreased bone mass, fewer trabeculae, reduced BFR	CCA	(Arteaga-Solis et al. 2001), (Nistala et al. 2010b; Sengle et al. 2015),
Fbn1$^{Prx1-/-}$	Limb-specific knock-out	Longer limb bones, reduced cancellous bone, incremental bone loss, degeneration of trabeculae		(Smaldone et al. 2016),(Sedes et al. 2022)
Fbn1 $^{WM\Delta/WM\Delta}$	Knock-in	Reduction of long bones and digits until P150	WMS2	(Sengle et al. 2012)
Fbn1$^{Scx-/-}$	Tendon-specific knock-out	Elongated femurs		(Sedes et al. 2022)
Fbn1$^{Osx-/-}$	Bone-specific Knock-out	Normal femurs		(Sedes et al. 2022)
Fbn1$^{Col2-/-}$	Cartilage-specific knock-out	Normal femurs		(Sedes et al. 2022)
Fbn1$^{Y1698C/Y1698C}$	Knock-in	Reduced stature, reduced femur and tibia length, decreased size of hypertrophic chondrocyte surface area, impaired hypertrophic chondrocyte columnar organization, reduction in growth plate length, abnormal chondrocyte differentiation, reduced *Col10a1* expression in hypertrophic zone, increased *Sox9* expression in prehypertrophic zone	GPHYSD2	(Delhon et al. 2022)

(continued)

Table 7.5 (continued)

Line	Type of mutant allele	Skeletal features	Resembling disorder	Reference
Adamtsl2$^{-/-}$	Knock-out	Reduced body length, shortened fore- and hindlimbs, impaired digit elongation, ovoid deformation of vertebrae, shortened snout and calvaria, abnormal differentiation/proliferation of chondrocytes, no regular column formation, delayed ossification center in phalanx bones	GPHYSD1	(Delhon et al. 2019)
Adamtsl2 $^{Col2-/-}$	Cartilage-specific Knock-out	Dwarfisms, reduced femur length, disrupted in growth plate of column formation in long bones		(Delhon et al. 2019)
Adamtsl2$^{Prx1-/-}$	Limb-specific knock-out	Reduced long bone length		(Hubmacher et al. 2019)
Adamtsl2 $^{Scx-/-}$	Tendon-specific knock-out	Reduced long bone length		(Hubmacher et al. 2019)
Adamts10$^{S236X/S236X}$	Knock-in	Shortened long bones, shorter resting zone and expanded hypertrophic zone	WMS1	(Mularczyk et al. 2018)

signaling (Arteaga-Solis et al. 2001; Sengle et al. 2015), further analysis of *Fbn2*$^{-/-}$ bones suggests a role in postnatal bone homeostasis. *Fbn2*$^{-/-}$ femurs show a length reduction of 4.5% at 4 month of age (P120) (Nistala et al. 2010b). Additionally, mid-diaphyseal cross sections of mutant femurs showed a 15% smaller bone width and a 25% narrower marrow cavity. However, *Fbn2*$^{-/-}$ femurs exhibited a typical shape at P4 and did not show any notable morphological anomalies in growth plates (Nistala et al. 2010b). This suggests that fibrillin-2 is required for postnatal bone remodeling rather than for endochondral bone development. In line with this, micro-computed tomography (μCT) and histomorphometric analyses detected an appreciable deficit of cancellous bone in mutant vertebrae at P90 and P180 (Nistala et al. 2010b). The analyses revealed a 58% decrease in BMD, a 27% decrease in bone mass, and 52% fewer trabeculae. In vivo analyses also showed a 55% reduction in bone formation rate (BFR) that was associated with a normal number of surface osteoblasts (Nistala et al. 2010b).

Fbn1$^{mgR/mgR}$ mice produce around 15% of normal fibrillin-1 amounts and were established as mouse model resembling MFS-like skeletal manifestations of long bone overgrowth such as severe kyphosis and overgrowth of the ribs (Pereira et al. 1999). Histomorphometry of *Fbn1*$^{mgR/mgR}$ bone tissue along with μCT scans of lumbar vertebrae at P90 revealed a 15% decline in bone content (quantified as bone volume over total volume; BV/TV), a 19.7% decrease in apparent bone mineral density (BMD), and changes in the structure of trabeculae, characterized by diminished trabecular thickness and augmented trabecular space (Nistala et al. 2010a). However, the number of trabecular structures was normal. The observed osteopenia in *Fbn1*$^{mgR/mgR}$ mice may likely be caused by an increased bone resorption, that is supported by elevated levels of urinary deoxypyridinoline (Dpd) collagen cross-links. BFR appeared normal with a normal presence of surface osteoblasts and osteoclasts (Nistala et al. 2010a).

Both adult *Fbn2*$^{-/-}$ and *Fbn1*$^{mgR/mgR}$ mice exhibited an enhanced osteolytic response to locally implanted lipopolysaccharide-coated titanium particles, as observed in studies by Nistala et al. in 2010 (Nistala et al. 2010a, 2010c) which implies the presence of an osteoclastogenic defect in absence of *Fbn1* or *Fbn2*. The increased osteoclastogenic potential of *Fbn2*$^{-/-}$ osteoblasts was mainly attributed to the upregulation of *Rankl* due to increased TGFβ activity (Nistala et al. 2010c). This was drawn from the finding that blocking TGFβ signaling reduced the elevation of Rankl in *Fbn2*$^{-/-}$ osteoblasts and bones, and that antagonizing systemic TGFβ improved locally induced osteolysis in *Fbn2*$^{-/-}$ mice (Nistala et al. 2010c). Similarly, elevated TGFβ activity led to abnormally high *Rankl* expression in cultured osteoblasts from *Fbn1*$^{-/-}$ mice (Nistala et al. 2010c). Together, these findings underscore that extracellular microfibrils play a pivotal role in balancing local catabolic and anabolic signals during bone remodeling, while also suggesting distinct mechanisms of bone loss in MFS and CCA.

7.5.2 *Conditional Genetic Ablation of* **Fbn1** *During Limb Development*

Further studies employing mice with a limb-specific ablation of *Fbn1* employing Prx1-Cre that directs Cre recombinase expression to early limb bud mesenchyme and a subset of craniofacial mesenchyme *Fbn1*$^{Prx1-/-}$ allowed further insight into the reasons of bone loss caused by *Fbn1* ablation (Smaldone et al. 2016). Thereby, it was found that a continuous reduction of cancellous bone in *Fbn1*$^{Prx1-/-}$ is caused by premature depletion of mesenchymal stem cells (MSCs) and osteoprogenitor cells combined with constitutively increased bone resorption. Analyses of *Fbn1*$^{Prx1-/-}$ mice at 1, 3, and 6 months of age showed incremental bone loss and trabecular microarchitecture degeneration accompanied by not only a gradual decrease in MSC number but also a reduction of their colony-forming capability. Marrow stromal cells derived from *Fbn1*$^{Prx1-/-}$ limb bones showed aberrantly increased TGFβ activity, and systemic administration of a TGFβ neutralizing monoclonal antibody (1D11) improved bone mass and trabecular microarchitecture as well as normalized the number of MSCs, and osteoprogenitor cells. Overall, this suggests that fibrillin-1 regulates MSC activity by controlling TGFβ bioavailability within the microenvironment of marrow niches.

7.5.3 Fibrillin-1 Deficiency in Perichondrium Contributes to Long Bone Growth

Further mechanistic insight into long bone overgrowth in MFS was made by employing several tissue-specific genetic ablations of *Fbn1* in the using the Cre-LoxP recombination system. Inactivation of the *Fbn1* gene in the early limb mesenchyme using the *Prx1-Cre* transgene resulted in $Fbn1^{Prx-/-}$ mice with longer limb bones (Sedes et al. 2022). This suggests that proper fibrillin-1 production by limb bud cells is required to maintain continuous limb growth. To better dissect the contributions of bone and cartilage cells for limb growth, the authors ablated fibrillin-1 synthesis in bone and cartilage cells using *Osx-Cre* and *Col2-Cre* transgenes, respectively. In addition, a tendon- and ligament-specific deletion of *Fbn1* was achieved by utilizing the *Scx-Cre* transgene. Interestingly, *Scx* was found to have a broader expression pattern including the perichondrium and the aorta as $Fbn1^{Scx-/-}$ were found to die at early postnatal stages due to dissecting aortic aneurysms. Interestingly, a statistically significant femur overgrowth was only observed in $Fbn1^{Scx-/-}$ mice but not in $Fbn1^{Osx-/-}$ or $Fbn1^{Col2-/-}$ animals at P90.

Analyses of differentially expressed genes in the fibrillin-1-deficient perichondrium predicted that loss of TGFβ signaling may influence chondrogenesis in the neighboring epiphyseal growth plate. Immunohistochemistry revealed that fibrillin-1 deficiency in the outer perichondrium is associated with decreased accumulation of latent TGFβ-binding proteins (LTBPs)-3 and -4, and reduced levels of TGFβ activity. Moreover, mutant metatarsal bones grown in vitro were longer and released less TGFβ. Moreover, addition of recombinant TGFβ1 normalized linear growth of mutant metatarsal bones. Therefore, it was concluded that longitudinal bone overgrowth in MFS is due to a diminished targeting of LTBP3 and LTBP4 into the fibrillin-1 deficient ECM of the outer perichondrium, which results in less TGFβ signaling locally and improper growth plate differentiation distally.

7.5.4 Molecular Insights into Mechanisms Causative for Long Bone Undergrowth in the Fibrillinopathies by Studying Corresponding Mouse Models

7.5.4.1 A three-domain deletion mutation in *FBN1* causes WMS

In 2012, we were able to demonstrate that certain *FBN1* mutations do not impose a dominant negative impact on fibrillin microfibril stability and are therefore not causative for MFS. The identified *FBN1* exon 9 to 11 deletion mutation (WMΔ) leads to autosomal the dominant form of WMS in a family with opposite features of MFS including thick skin, short stature, and brachydactyly (Sengle et al. 2012). The WMΔ mutation causes deletion of the first eight-cysteine domain, the proline rich region, and the adjacent EGF-like domain. When WMΔ was replicated in mice, growth of long bones appeared to be normal in the first two weeks of postnatal life but were found to be reduced by P21-30 in *Fbn1*$^{WM\Delta/WM\Delta}$ mice. Measurements of P30 *Fbn1*$^{WM\Delta/WM\Delta}$ long bones using μCT showed a significant 6–10% length reduction of the radius, ulna, and tibia. Also, metacarpals and proximal and distal phalanges in fore- and hindpaws were reduced by 2–23% in the *Fbn1*$^{WM\Delta/+}$ and *Fbn1*$^{WM\Delta/WM\Delta}$ mice at P21–30. However, these differences in bone length were normalized at P150.

Since it was previously reported that autosomal recessive WMS is caused by mutations in *ADAMTS10* (Dagoneau et al. 2004), and autosomal dominant WMS by mutations in *FBN1* (Faivre et al. 2003), a common genetic pathway and a direct interaction of both proteins was assumed. Further, it was shown that mutations in *ADAMTSL2* and *FBN1* lead to GD with similar features of WMS (Le Goff et al. 2008, 2011b). This suggests that ADAMTSL2 and ADAMTS10 mutually control skeletal growth and differentiation in concert with fibrillin-1. To test whether both proteins exert a direct physical interaction with fibrillin-1, protein-protein interaction studies with recombinantly generated proteins were conducted. The results showed that the WMΔ mutation deletes a binding site that targets ADAMTSLIKE-2, -3, -6, and papilin to fibrillin-1 (Sengle et al. 2012). This finding confirmed the initial assumption that ADAMTSLIKE proteins are involved in a common molecular pathway together with fibrillin-1 and ADAMTS-10. Further studies showed that ADAMTS10 is able to cleave fibrillin-1 and both ADAMTSL2 and ADAMTS10 assist in fibrillin fiber assembly in in vitro cultures (Hubmacher et al. 2019; Kutz et al. 2011).

These findings suggest that distinct tissue-specific microenvironments, that are perturbed in WMS, require a functionally intact fibrillin microfibril network that is structurally maintained and regulated by ADAMTSLIKE proteins in concert with ADAMTS enzymes.

7.5.4.2 *Adamts10* mutant mouse model resembling autosomal recessive WMS

To further gain molecular insight into autosomal recessive WMS, a mouse model was generated that resembles the autosomal recessive R237X mutation in *ADAMTS10* identified in WMS patients (Mularczyk et al. 2018). *Adamts10*$^{S236X/S236X}$ homozygous mice are smaller as assessed by their reduced weight and body length at P30 and P90. Measurements of *Adamts10*$^{S236X/S236X}$ femur and tibia were found to be 6–10% shorter compared to littermate controls. Histological analysis of the growth plate showed an overall normal organization but with differences in the relative proportions of the zones with shorter resting zone and expanded hypertrophic zone indicating altered bone development in the mutant mice. ADAMTS10 was further shown to be expressed in the knee joint, mostly articular cartilage, perichondrium, and meniscus but also in the growth plate and secondary ossification center. Further analysis of *Adamts10*$^{S236X/S236X}$ growth plates showed a heightened chondrocyte proliferation as indicated by the increased presence of BrdU positive cells within the proliferative zone. Embryonic fibroblasts from *Adamts10*$^{S236X/S236X}$ mice showed a reduction of BMP-2, -4, -5-, -7 transcript levels that corresponded with a decrease in BMP activity. These findings associate decreased BMP activity with autosomal recessive WMS.

7.5.4.3 *Adamtsl2* deletion causes impairment of chondrogenesis and tendon development

To further investigate the role of ADAMTSL2 in murine skeletal development, *Adamtsl2* was deleted in all cells or only in chondrocytes (Delhon et al. 2019). *Adamtsl2*$^{-/-}$ mice displayed multiple skeletal abnormalities similar to GD patients such as a 12.8% reduced naso-anal length as well as significant shortened fore- and hindlimbs, impaired digit elongation, a slight ovoid deformation of the vertebrae, and a shortened snout and calvaria (Delhon et al. 2019). Absence of *Adamtsl2* also affects growth plate formation as indicated by abnormal differentiation and proliferation of chondrocytes. *Adamtsl2* null chondrocytes do not form regular columns, are increased in size, and show a scattered localization throughout the growth plate. Also, *Adamtsl2* ablation delayed the formation of the ossification center in phalanx bones and impaired TGF-β signaling impairment in limbs. Further, chondrocyte-specific ablation of *Adamtsl2* (*Adamtsl2*$^{Col2-/-}$) also leads to dwarfism with shorted femur length and a disruption of the column formation in the growth plate of long bones. Interestingly, the formation of the ossification center in phalanx bones was normal in *Adamtsl2*$^{Col2-/-}$ mice. *Adamtsl2*$^{Col2-/-}$ mice also showed no defects in skull formation, suggesting that intramembranous ossification was not affected. Interestingly, primary *Adamtsl2* null chondrocytes failed to establish a microfibrillar network composed by fibrillin-1 and LTBP1 fibrils (Delhon et al. 2019). These findings suggest that ADAMTL2 plays an important role in the growth plate by maintaining a functional ECM microenvironment required for proper TGF-β utilization.

Furthermore, it was shown that selective removal of *Adamtsl2* in the limb mesenchyme by utilizing Prx1-Cre causes abnormalities in tendons and a length reduction of the distal limb (Hubmacher et al. 2019). However interestingly, *Adamtsl2* deletion using Scx-Cre, a tendon-specific Cre-deleter strain, which does not delete in cartilage, also led to impairment of skeletal growth (Hubmacher et al. 2019). Therefore, it can be hypothesized that ADALTSL2 not only regulates limb-growth both by maintaining ECM integrity in the growth plate cartilage, but also via it impact on the ECM architecture in tendons and soft tissues.

7.5.4.4 The TB5 domain of fibrillin-1 plays a role in endochondral ossification

Previously, it was shown that all known *FBN1* mutations causative for GPHYSD2 are localized within the region encoding the fifth transforming growth factor-β binding protein-like (TB5) domain. To investigate the specific role of this domain in endochondral ossification, a new *Fbn1* knock-in mouse model was generated by introducing the p.Tyr1696Cys mutation from a patient with GD into the corresponding murine FBN1 TB5 domain. Both $Fbn1^{+/Y1698C}$ and $Fbn1^{Y1698C/Y1698C}$ mice exhibit a reduced stature (reduction of 8.9% in homozygous and 4.2% in heterozygous P30 mice) reminiscent of the human GD phenotype. Histology of the growth plates showed a decreased size of the hypertrophic chondrocyte surface area in both *Fbn1Y1698C* mutant mice at P1 and P30. Moreover, *Fbn1Y1698C* mutant mice exhibited an impaired hypertrophic chondrocyte columnar organization and global reductions in growth plate length. Further analyses showed that presence of the *Fbn1Y1698C* knock-in mutation impacts cellular behavior in the growth plate characterized by abnormal chondrocyte differentiation, reduced *Col10a1* expression in the hypertrophic zone, but increased *Sox9* expression in the prehypertrophic zone, while no change in TGF-β activity was observed. Moreover, *Fbn1Y1698C* chondrocytes failed to establish a dense microfibrillar network which appears to be required for proper chondrocyte differentiation (Delhon et al. 2022).

7.6 Conclusions and Perspectives

Clinical evidence from connective tissue disorders combined with data from transgenic mouse models and biochemical investigations clearly imply the fibrillin proteins in control of growth and differentiation processes of skeletal tissues. However so far, the underlying pathways explaining how a certain *FBN1* mutation leads to long bone over- or undergrowth remain largely obscure. This is understandable as fibrillins exert their biological function as a microfibrillar multimeric scaffold to which several adaptor molecules growth factors and proteases are targeted. Until now several binding sites of the fibrillin microfibril network have been identified. The challenge for future studies will be to understand how the ensemble of fibrillin-bound ligands work as a machinery in three-dimensional fashion to maintain post-natal growth and balance tissue homeostasis. Thereby, it will be instrumental to understand how fibrillin-1 mutations alter the ultrastructure of the microfibril. Depending on the structural change imposed by the respective mutation, a bound growth factor may become less sequestered and easily accessible by a cell surface receptor, or less utilizable by cells due to its improper presentation to the cell surface. Fibrillin mutations may change the activation status of embedded growth factors due to structural rearrangement of its ultrastructure. Therefore, a comprehensive understanding on how fibrillin-1 mutations impact binding sites of functional adaptor proteins and growth factors will allow us to better understand why over 3000 mutations lead to Marfan syndrome characterized by long bone overgrowth, while only a hand full of rare mutations are causative for the opposite features such as short stature and brachydactyly.

Recent advances have provided a better structural understanding of the three-dimensional assembled form of fibrillin by employing cryogenic electron microscopy on native fibrillin microfibrils in mammalian tissue (Godwin et al. 2023).

Thereby, a pseudo eightfold symmetry of the primary bead region was determined which contains both the amino and carboxy termini. Building upon this structural insight, it was demonstrated that the WMΔ deletion mutation induces a rearrangement that obstructs the interaction with latent LTBP1 at a distant site. These findings confirm that mutations within assembled tissue fibrillin microfibrils can potentially result in long-range structural alterations that disrupt growth factor signaling and contribute to the onset of disease.

Better knowledge about the molecular requirements for the stability of fibrillin microfibrils will lead to a deeper understanding of the molecular pathways initiated by fibrillin-1 mutations and will further open up new therapeutic avenues for patients.

Acknowledgements Funding was provided by Deutsche Forschungsgemeinschaft (DFG; German Research Foundation), project number 384170921 (FOR2722/C2 and FOR2722/M2), and 397484323 (TRR259/B09) to G.S.

Compliance with Ethical StandardsAuthors have no conflict of interest.

References

Adamo CS, Beyens A, Schiavinato A, Keene DR, Tufa SF, Morgelin M, Brinckmann J, Sasaki T, Niehoff A, Dreiner M, Pottie L, Muino-Mosquera L, Gulec EY, Gezdirici A, Braghetta P, Bonaldo P, Wagener R, Paulsson M, Bornaun H, De Rycke R, De Bruyne M, Baeke F, Devine WP, Gangaram B, Tam A, Balasubramanian M, Ellard S, Moore S, Symoens S, Shen J, Cole S, Schwarze U, Holmes KW, Hayflick SJ, Wiszniewski W, Nampoothiri S, Davis EC, Sakai LY, Sengle G, Callewaert B (2022) EMILIN1 deficiency causes arterial tortuosity with osteopenia and connects impaired elastogenesis with defective collagen fibrillogenesis. Am J Hum Genet 109:2230–2252

Adamo CS, Zuk AV, Sengle G (2021) The fibrillin microfibril/elastic fibre network: a critical extracellular supramolecular scaffold to balance skin homoeostasis. Exp Dermatol 30:25–37

Arn PH, Scherer LR, Haller JA Jr, Pyeritz RE (1989) Outcome of pectus excavatum in patients with Marfan syndrome and in the general population. J Pediatr 115:954–958

Arteaga-Solis E, Gayraud B, Lee SY, Shum L, Sakai L, Ramirez F (2001) Regulation of limb patterning by extracellular microfibrils. J Cell Biol 154:275–281

Bade, Spirgimn E (1965) Oral manifestations of Marfan's Syndrome. Oral Surg Oral Med Oral Pathol 19:757–770

Baldock C, Koster AJ, Ziese U, Rock MJ, Sherratt MJ, Kadler KE, Shuttleworth CA, Kielty CM (2001) The supramolecular organization of fibrillin-rich microfibrils. J Cell Biol 152:1045–1056

Bax DV, Bernard SE, Lomas A, Morgan A, Humphries J, Shuttleworth CA, Humphries MJ, Kielty CM (2003) Cell adhesion to fibrillin-1 molecules and microfibrils is mediated by alpha 5 beta 1 and alpha v beta 3 integrins. J Biol Chem 278:34605–34616

Beals RK, Hecht F (1971) Congenital contractural arachnodactyly. A heritable disorder of connective tissue. J Bone Joint Surg Am 53:987–993

Benoist G, Tubach F, Roy C, Rioux S, Michelon-Jouneaux M, Chevallier B, Jondeau G, Stheneur C (2018) Skeletal evolution in Marfan syndrome: growth curves from a French national cohort. Pediatr Res 83:71–77

Birch JG, Herring JA (1987) Spinal deformity in Marfan syndrome. J Pediatr Orthop 7:546–552

Boileau C, Jondeau G, Babron MC, Coulon M, Alexandre JA, Sakai L, Melki J, Delorme G, Dubourg O, Bonaiti-Pellie C et al (1993) Autosomal dominant Marfan-like connective-tissue disorder with aortic dilation and skeletal anomalies not linked to the fibrillin genes. Am J Hum Genet 53:46–54

Boileau C, Junien C, Collod G, Jondeau G, Dubourg O, Bourdarias JP, Bonaiti-Pellie C, Frezal J, Maroteaux P (1995) Reply to "the question of heterogeneity in Marfan syndrome". Nat Genet 9:230–231

Cain SA, McGovern A, Baldwin AK, Baldock C, Kielty CM (2012) Fibrillin-1 mutations causing Weill-Marchesani syndrome and acromicric and geleophysic dysplasias disrupt heparan sulfate interactions. PLoS One 7:e48634

Carter N, Duncan E, Wordsworth P (2000) Bone mineral density in adults with Marfan syndrome. Rheumatology (Oxford) 39:307–309

Charbonneau NL, Ono RN, Corson GM, Keene DR, Sakai LY (2004) Fine tuning of growth factor signals depends on fibrillin microfibril networks. Birth Defects Res C Embryo Today 72:37–50

Coelho SG, Almeida AG (2020) Marfan syndrome revisited: from genetics to the clinic. Rev Port Cardiol (Engl Ed) 39:215–226

Dabovic B, Chen Y, Colarossi C, Obata H, Zambuto L, Perle MA, Rifkin DB (2002a) Bone abnormalities in latent TGF-[beta] binding protein (Ltbp)-3-null mice indicate a role for Ltbp-3 in modulating TGF-[beta] bioavailability. J Cell Biol 156:227–232

Dabovic B, Chen Y, Colarossi C, Zambuto L, Obata H, Rifkin DB (2002b) Bone defects in latent TGF-beta binding protein (Ltbp)-3 null mice; a role for Ltbp in TGF-beta presentation. J Endocrinol 175:129–141

Dagoneau N, Benoist-Lasselin C, Huber C, Faivre L, Megarbane A, Alswaid A, Dollfus H, Alembik Y, Munnich A, Legeai-Mallet L, Cormier-Daire V (2004) ADAMTS10 mutations in autosomal recessive Weill-Marchesani syndrome. Am J Hum Genet 75:801–806

Dallas SL, Keene DR, Bruder SP, Saharinen J, Sakai LY, Mundy GR, Bonewald LF (2000) Role of the latent transforming growth factor beta binding protein 1 in fibrillin-containing microfibrils in bone cells in vitro and in vivo. J Bone Miner Res 15:68–81

De Coster P, De Pauw G, Martens L, De Paepe A (2004) Craniofacial structure in Marfan syndrome: a cephalometric study. Am J Med Genet A 131:240–248

De Coster PJ, Martens LC, De Paepe A (2002) Oral manifestations of patients with Marfan syndrome: a case-control study. Oral Surg Oral Med Oral Pathol Oral Radiol Endod 93:564–572

De Crescenzo G, Grothe S, Zwaagstra J, Tsang M, O'Connor-McCourt MD (2001) Real-time monitoring of the interactions of transforming growth factor-beta (TGF-beta) isoforms with latency-associated protein and the ectodomains of the TGF-beta type II and III receptors reveals different kinetic models and stoichiometries of binding. J Biol Chem 276:29632–29643

De Maio F, Fichera A, De Luna V, Mancini F, Caterini R (2016) Orthopaedic aspects of Marfan syndrome: the experience of a Referral Center for Diagnosis of Rare Diseases. Adv Orthop 2016:8275391

Delhon L, Mahaut C, Goudin N, Gaudas E, Piquand K, Le Goff W, Cormier-Daire V, Le Goff C (2019) Impairment of chondrogenesis and microfibrillar network in Adamtsl2 deficiency. FASEB J 33:2707–2718

Delhon L, Mougin Z, Jonquet J, Bibimbou A, Dubail J, Bou-Chaaya C, Goudin N, Le Goff W, Boileau C, Cormier-Daire V, Le Goff C (2022) The critical role of the TB5 domain of fibrillin-1 in endochondral ossification. Hum Mol Genet 31:3777–3788

Demetracopoulos CA, Sponseller PD (2007) Spinal deformities in Marfan syndrome. Orthop Clin North Am 38:563–572, vii

Deng H, Lu Q, Xu H, Deng X, Yuan L, Yang Z, Guo Y, Lin Q, Xiao J, Guan L, Song Z (2016) Identification of a novel missense FBN2 mutation in a Chinese family with congenital Contractural arachnodactyly using exome sequencing. PLoS One 11:e0155908

Desmarais TJ, Keller MS (2013) Pectus carinatum. Curr Opin Pediatr 25:375–381

Drews F, Knobel S, Moser M, Muhlack KG, Mohren S, Stoll C, Bosio A, Gressner AM, Weiskirchen R (2008) Disruption of the latent transforming growth factor-beta binding protein-1 gene causes alteration in facial structure and influences TGF-beta bioavailability. Biochim Biophys Acta 1783:34–48

El-Hallous E, Sasaki T, Hubmacher D, Getie M, Tiedemann K, Brinckmann J, Batge B, Davis EC, Reinhardt DP (2007) Fibrillin-1 interactions with fibulins depend on the first hybrid domain and provide an adaptor function to tropoelastin. J Biol Chem 282:8935–8946

Faivre L, Gorlin RJ, Wirtz MK, Godfrey M, Dagoneau N, Samples JR, Le Merrer M, Collod-Beroud G, Boileau C, Munnich A, Cormier-Daire V (2003) In frame fibrillin-1 gene deletion in autosomal dominant Weill-Marchesani syndrome. J Med Genet 40:34–36

Feingold M (1968) The "thumb sign" in children. Clin Pediatr (Phila) 7:423–424

Fitch HB, Secord RT (1963) The Marfan syndrome. Oral Surg Oral Med Oral Pathol 16:373–377

Folkestad L, Stochholm K, Groth K, Hove H, Andersen NH, Gravholt CH (2021) Fracture rates and fracture risk in patients with Marfan syndrome: a nationwide register-based cohort study. J Bone Miner Res 36:901–909

Fraser S, Child A, Hunt I (2017) Pectus updates and special considerations in Marfan syndrome. Pediatr Rep 9:7277

Furlan AG, Spanou CES, Godwin ARF, Wohl AP, Zimmermann LA, Imhof T, Koch M, Baldock C, Sengle G (2021) A new MMP-mediated prodomain cleavage mechanism to activate bone morphogenetic proteins from the extracellular matrix. FASEB J 35:e21353

Gehle P, Goergen B, Pilger D, Ruokonen P, Robinson PN, Salchow DJ (2017) Biometric and structural ocular manifestations of Marfan syndrome. PLoS One 12:e0183370

Glard Y, Launay F, Edgard-Rosa G, Collignon P, Jouve JL, Bollini G (2008) Scoliotic curve patterns in patients with Marfan syndrome. J Child Orthop 2:211–216

Glard Y, Pomero V, Collignon P, Skalli W, Jouve JL, Bollini G (2009) Three-dimensional analysis of the vertebral rotation associated with the lateral deviation in Marfan syndrome spinal deformity. J Pediatr Orthop B 18:51–56

Godwin ARF, Dajani R, Zhang X, Thomson J, Holmes DF, Adamo CS, Sengle G, Sherratt MJ, Roseman AM, Baldock C (2023) Fibrillin microfibril structure identifies long-range effects of inherited pathogenic mutations affecting a key regulatory latent TGFbeta-binding site. Nat Struct Mol Biol 30:608–618

Godwin ARF, Singh M, Lockhart-Cairns MP, Alanazi YF, Cain SA, Baldock C (2019) The role of fibrillin and microfibril binding proteins in elastin and elastic fibre assembly. Matrix Biol 84:17–30

Godwin ARF, Starborg T, Smith DJ, Sherratt MJ, Roseman AM, Baldock C (2018) Multiscale imaging reveals the hierarchical Organization of Fibrillin Microfibrils. J Mol Biol 430:4142–4155

Grahame R, Hakim AJ (2013) Arachnodactyly–a key to diagnosing heritable disorders of connective tissue. Nat Rev Rheumatol 9:358–364

Grahame R, Pyeritz RE (1995) The Marfan syndrome: joint and skin manifestations are prevalent and correlated. Br J Rheumatol 34:126–131

Gregory KE, Ono RN, Charbonneau NL, Kuo CL, Keene DR, Bachinger HP, Sakai LY (2005) The prodomain of BMP-7 targets the BMP-7 complex to the extracellular matrix. J Biol Chem 280:27970–27980

Haji-Seyed-Javadi R, Jelodari-Mamaghani S, Paylakhi SH, Yazdani S, Nilforushan N, Fan JB, Klotzle B, Mahmoudi MJ, Ebrahimian MJ, Chelich N, Taghiabadi E, Kamyab K, Boileau C, Paisan-Ruiz C, Ronaghi M, Elahi E (2012) LTBP2 mutations cause Weill-Marchesani and Weill-Marchesani-like syndrome and affect disruptions in the extracellular matrix. Hum Mutat 33:1182–1187

Hill JR, Eekhoff JD, Brophy RH, Lake SP (2020) Elastic fibers in orthopedics: form and function in tendons and ligaments, clinical implications, and future directions. J Orthop Res 38:2305–2317

Hirani R, Hanssen E, Gibson MA (2007) LTBP-2 specifically interacts with the amino-terminal region of fibrillin-1 and competes with LTBP-1 for binding to this microfibrillar protein. Matrix Biol 26:213–223

Ho NC, Tran JR, Bektas A (2005) Marfan's syndrome. Lancet 366:1978–1981

Horiguchi M, Inoue T, Ohbayashi T, Hirai M, Noda K, Marmorstein LY, Yabe D, Takagi K, Akama TO, Kita T, Kimura T, Nakamura T (2009) Fibulin-4 conducts proper elastogenesis via interaction with cross-linking enzyme lysyl oxidase. Proc Natl Acad Sci USA 106:19029–19034

Houghton GR, Dekel S (1979) The periosteal control of long bone growth. An experimental study in the rat. Acta Orthop Scand 50:635–637

Hubmacher D, Schneider M, Berardinelli SJ, Takeuchi H, Willard B, Reinhardt DP, Haltiwanger RS, Apte SS (2017) Unusual life cycle and impact on microfibril assembly of ADAMTS17, a secreted metalloprotease mutated in genetic eye disease. Sci Rep 7:41871

Hubmacher D, Taye N, Balic Z, Thacker S, Adams SM, Birk DE, Schweitzer R, Apte SS (2019) Limb- and tendon-specific Adamtsl2 deletion identifies a role for ADAMTSL2 in tendon growth in a mouse model for geleophysic dysplasia. Matrix Biol 82:38–53

Hubmacher D, Wang LW, Mecham RP, Reinhardt DP, Apte SS (2015) Adamtsl2 deletion results in bronchial fibrillin microfibril accumulation and bronchial epithelial dysplasia–a novel mouse model providing insights into geleophysic dysplasia. Dis Model Mech 8:487–499

Hucthagowder V, Sausgruber N, Kim KH, Angle B, Marmorstein LY, Urban Z (2006) Fibulin-4: a novel gene for an autosomal recessive cutis laxa syndrome. Am J Hum Genet 78:1075–1080

Inbar-Feigenberg M, Meirowitz N, Nanda D, Toi A, Okun N, Chitayat D (2014) Beals syndrome (congenital contractural arachnodactyly): prenatal ultrasound findings and molecular analysis. Ultrasound Obstet Gynecol 44:486–490

Isogai Z, Ono RN, Ushiro S, Keene DR, Chen Y, Mazzieri R, Charbonneau NL, Reinhardt DP, Rifkin DB, Sakai LY (2003) Latent transforming growth factor beta-binding protein 1 interacts with fibrillin and is a microfibril-associated protein. J Biol Chem 278:2750–2757

Janoschek R, Hoffmann T, Morcos YAT, Sengle G, Dotsch J, Hucklenbruch-Rother E (2020) Asprosin in pregnancy and childhood. Mol Cell Pediatr 7:18

Jensen SA, Aspinall G, Handford PA (2014) C-terminal propeptide is required for fibrillin-1 secretion and blocks premature assembly through linkage to domains cbEGF41-43. Proc Natl Acad Sci USA 111:10155–10160

Judge DP, Dietz HC (2005) Marfan's syndrome. Lancet 366:1965–1976

Karoulias SZ, Beyens A, Balic Z, Symoens S, Vandersteen A, Rideout AL, Dickinson J, Callewaert B, Hubmacher D (2020) A novel ADAMTS17 variant that causes Weill-Marchesani syndrome 4 alters fibrillin-1 and collagen type I deposition in the extracellular matrix. Matrix Biol 88:1–18

Keene DR, Jordan CD, Reinhardt DP, Ridgway CC, Ono RN, Corson GM, Fairhurst M, Sussman MD, Memoli VA, Sakai LY (1997) Fibrillin-1 in human cartilage: developmental expression and formation of special banded fibers. J Histochem Cytochem 45:1069–1082

Keene DR, Maddox BK, Kuo HJ, Sakai LY, Glanville RW (1991a) Extraction of extendable beaded structures and their identification as fibrillin-containing extracellular matrix microfibrils. J Histochem Cytochem 39:441–449

Keene DR, Sakai LY, Burgeson RE (1991b) Human bone contains type III collagen, type VI collagen, and fibrillin: type III collagen is present on specific fibers that may mediate attachment of tendons, ligaments, and periosteum to calcified bone cortex. J Histochem Cytochem 39:59–69

Kinsey R, Williamson MR, Chaudhry S, Mellody KT, McGovern A, Takahashi S, Shuttleworth CA, Kielty CM (2008) Fibrillin-1 microfibril deposition is dependent on fibronectin assembly. J Cell Sci 121:2696–2704

Kohlmeier L, Gasner C, Bachrach LK, Marcus R (1995) The bone mineral status of patients with Marfan syndrome. J Bone Miner Res 10:1550–1555

Kuo CL, Isogai Z, Keene DR, Hazeki N, Ono RN, Sengle G, Bachinger HP, Sakai LY (2007) Effects of fibrillin-1 degradation on microfibril ultrastructure. J Biol Chem 282:4007–4020

Kutz WE, Wang LW, Bader HL, Majors AK, Iwata K, Traboulsi EI, Sakai LY, Keene DR, Apte SS (2011) ADAMTS10 protein interacts with fibrillin-1 and promotes its deposition in extracellular matrix of cultured fibroblasts. J Biol Chem 286:17156–17167

Le Goff C, Mahaut C, Abhyankar A, Le Goff W, Serre V, Afenjar A, Destree A, di Rocco M, Heron D, Jacquemont S, Marlin S, Simon M, Tolmie J, Verloes A, Casanova JL, Munnich A, Cormier-Daire V (2011a) Mutations at a single codon in mad homology 2 domain of SMAD4 cause Myhre syndrome. Nat Genet 44:85–88

Le Goff C, Mahaut C, Wang LW, Allali S, Abhyankar A, Jensen S, Zylberberg L, Collod-Beroud G, Bonnet D, Alanay Y, Brady AF, Cordier MP, Devriendt K, Genevieve D, Kiper PO, Kitoh H,

Krakow D, Lynch SA, Le Merrer M, Megarbane A, Mortier G, Odent S, Polak M, Rohrbach M, Sillence D, Stolte-Dijkstra I, Superti-Furga A, Rimoin DL, Topouchian V, Unger S, Zabel B, Bole-Feysot C, Nitschke P, Handford P, Casanova JL, Boileau C, Apte SS, Munnich A, Cormier-Daire V (2011b) Mutations in the TGFbeta binding-protein-like domain 5 of FBN1 are responsible for acromicric and geleophysic dysplasias. Am J Hum Genet 89:7–14

Le Goff C, Morice-Picard F, Dagoneau N, Wang LW, Perrot C, Crow YJ, Bauer F, Flori E, Prost-Squarcioni C, Krakow D, Ge G, Greenspan DS, Bonnet D, Le Merrer M, Munnich A, Apte SS, Cormier-Daire V (2008) ADAMTSL2 mutations in geleophysic dysplasia demonstrate a role for ADAMTS-like proteins in TGF-beta bioavailability regulation. Nat Genet 40:1119–1123

Lee SS, Knott V, Jovanovic J, Harlos K, Grimes JM, Choulier L, Mardon HJ, Stuart DI, Handford PA (2004) Structure of the integrin binding fragment from fibrillin-1 gives new insights into microfibril organization. Structure 12:717–729

Lindsay ME, Schepers D, Bolar NA, Doyle JJ, Gallo E, Fert-Bober J, Kempers MJ, Fishman EK, Chen Y, Myers L, Bjeda D, Oswald G, Elias AF, Levy HP, Anderlid BM, Yang MH, Bongers EM, Timmermans J, Braverman AC, Canham N, Mortier GR, Brunner HG, Byers PH, Van Eyk J, Van Laer L, Dietz HC, Loeys BL (2012) Loss-of-function mutations in TGFB2 cause a syndromic presentation of thoracic aortic aneurysm. Nat Genet 44:922–927

Lindsey JM, Michelson JD, MacWilliams BA, Sponseller PD, Miller NH (1998) The foot in Marfan syndrome: clinical findings and weight-distribution patterns. J Pediatr Orthop 18:755–759

Liu X, Zhao Y, Gao J, Pawlyk B, Starcher B, Spencer JA, Yanagisawa H, Zuo J, Li T (2004) Elastic fiber homeostasis requires lysyl oxidase-like 1 protein. Nat Genet 36:178–182

Lockhart-Cairns MP, Cain SA, Dajani R, Steer R, Thomson J, Alanazi YF, Kielty CM, Baldock C (2022) Latent TGFbeta complexes are transglutaminase cross-linked to fibrillin to facilitate TGFbeta activation. Matrix Biol 107:24–39

Loeys B, Van Maldergem L, Mortier G, Coucke P, Gerniers S, Naeyaert JM, De Paepe A (2002) Homozygosity for a missense mutation in fibulin-5 (FBLN5) results in a severe form of cutis laxa. Hum Mol Genet 11:2113–2118

Loeys BL, Chen J, Neptune ER, Judge DP, Podowski M, Holm T, Meyers J, Leitch CC, Katsanis N, Sharifi N, Xu FL, Myers LA, Spevak PJ, Cameron DE, De Backer J, Hellemans J, Chen Y, Davis EC, Webb CL, Kress W, Coucke P, Rifkin DB, De Paepe AM, Dietz HC (2005) A syndrome of altered cardiovascular, craniofacial, neurocognitive and skeletal development caused by mutations in TGFBR1 or TGFBR2. Nat Genet 37:275–281

Loeys BL, Dietz HC, Braverman AC, Callewaert BL, De Backer J, Devereux RB, Hilhorst-Hofstee Y, Jondeau G, Faivre L, Milewicz DM, Pyeritz RE, Sponseller PD, Wordsworth P, De Paepe AM (2010a) The revised Ghent nosology for the Marfan syndrome. J Med Genet 47:476–485

Loeys BL, Gerber EE, Riegert-Johnson D, Iqbal S, Whiteman P, McConnell V, Chillakuri CR, Macaya D, Coucke PJ, De Paepe A, Judge DP, Wigley F, Davis EC, Mardon HJ, Handford P, Keene DR, Sakai LY, Dietz HC (2010b) Mutations in fibrillin-1 cause congenital scleroderma: stiff skin syndrome. Sci Transl Med 2:23ra20

Mahajan D, Kancharla S, Kolli P, Sharma AK, Singh S, Kumar S, Mohanty AK, Jena MK (2021) Role of Fibulins in embryonic stage development and their involvement in various diseases. Biomol Ther 11

McInerney-Leo AM, Le Goff C, Leo PJ, Kenna TJ, Keith P, Harris JE, Steer R, Bole-Feysot C, Nitschke P, Kielty C, Brown MA, Zankl A, Duncan EL, Cormier-Daire V (2016) Mutations in LTBP3 cause acromicric dysplasia and geleophysic dysplasia. J Med Genet 53:457–464

Meerschaut I, De Coninck S, Steyaert W, Barnicoat A, Bayat A, Benedicenti F, Berland S, Blair EM, Breckpot J, de Burca A, Destree A, Garcia-Minaur S, Green AJ, Hanna BC, Keymolen K, Koopmans M, Lederer D, Lees M, Longman C, Lynch SA, Male AM, McKenzie F, Migeotte I, Mihci E, Nur B, Petit F, Piard J, Plasschaert FS, Rauch A, Ribai P, Pacheco IS, Stanzial F, Stolte-Dijkstra I, Valenzuela I, Varghese V, Vasudevan PC, Wakeling E, Wallgren-Pettersson C, Coucke P, De Paepe A, De Wolf D, Symoens S, Callewaert B (2020) A clinical scoring system for congenital contractural arachnodactyly. Genet Med 22:124–131

Menz C, Parsi MK, Adams JR, Sideek MA, Kopecki Z, Cowin AJ, Gibson MA (2015) LTBP-2 has a single high-affinity binding site for FGF-2 and blocks FGF-2-induced cell proliferation. PLoS One 10:e0135577

Morales J, Al-Sharif L, Khalil DS, Shinwari JM, Bavi P, Al-Mahrouqi RA, Al-Rajhi A, Alkuraya FS, Meyer BF, Al Tassan N (2009) Homozygous mutations in ADAMTS10 and ADAMTS17 cause lenticular myopia, ectopia lentis, glaucoma, spherophakia, and short stature. Am J Hum Genet 85:558–568

Morcos YAT, Lutke S, Tenbieg A, Hanisch FG, Pryymachuk G, Piekarek N, Hoffmann T, Keller T, Janoschek R, Niehoff A, Zaucke F, Dotsch J, Hucklenbruch-Rother E, Sengle G (2022) Sensitive asprosin detection in clinical samples reveals serum/saliva correlation and indicates cartilage as source for serum asprosin. Sci Rep 12:1340

Moura B, Tubach F, Sulpice M, Boileau C, Jondeau G, Muti C, Chevallier B, Ounnoughene Y, Le Parc JM, G. Multidisciplinary Marfan Syndrome Clinic (2006) Bone mineral density in Marfan syndrome. A large case-control study. Joint Bone Spine 73:733–735

Mularczyk EJ, Singh M, Godwin ARF, Galli F, Humphreys N, Adamson AD, Mironov A, Cain SA, Sengle G, Boot-Handford RP, Cossu G, Kielty CM, Baldock C (2018) ADAMTS10-mediated tissue disruption in Weill-Marchesani syndrome. Hum Mol Genet 27:3675–3687

Nelle M, Troger J, Rupprath G, Bettendorf M (1994) Metacarpal index in Marfan's syndrome and in constitutional tall stature. Arch Dis Child 70:149–150

Nistala H, Lee-Arteaga S, Carta L, Cook JR, Smaldone S, Siciliano G, Rifkin AN, Dietz HC, Rifkin DB, Ramirez F (2010a) Differential effects of alendronate and losartan therapy on osteopenia and aortic aneurysm in mice with severe Marfan syndrome. Hum Mol Genet 19:4790–4798

Nistala H, Lee-Arteaga S, Smaldone S, Siciliano G, Carta L, Ono RN, Sengle G, Arteaga-Solis E, Levasseur R, Ducy P, Sakai LY, Karsenty G, Ramirez F (2010b) Fibrillin-1 and -2 differentially modulate endogenous TGF-beta and BMP bioavailability during bone formation. J Cell Biol 190:1107–1121

Nistala H, Lee-Arteaga S, Smaldone S, Siciliano G, Ramirez F (2010c) Extracellular microfibrils control osteoblast-supported osteoclastogenesis by restricting TGFbeta stimulation of RANKL production. J Biol Chem 285:34126–34133

Noda K, Nakamura T, Komatsu Y (2015) Fibulin-5 deficiency causes developmental defect of premaxillary bone in mice. Biochem Biophys Res Commun 466:585–591

Ogino T, Kato H (1993) Clinical features and treatment of congenital fusion of the small and ring finger metacarpals. J Hand Surg Am 18:995–1003

Ono RN, Sengle G, Charbonneau NL, Carlberg V, Bachinger HP, Sasaki T, Lee-Arteaga S, Zilberberg L, Rifkin DB, Ramirez F, Chu ML, Sakai LY (2009) Latent transforming growth factor beta-binding proteins and fibulins compete for fibrillin-1 and exhibit exquisite specificities in binding sites. J Biol Chem 284:16872–16881

Pereira L, Lee SY, Gayraud B, Andrikopoulos K, Shapiro SD, Bunton T, Biery NJ, Dietz HC, Sakai LY, Ramirez F (1999) Pathogenetic sequence for aneurysm revealed in mice underexpressing fibrillin-1. Proc Natl Acad Sci USA 96:3819–3823

Pfaff M, Reinhardt DP, Sakai LY, Timpl R (1996) Cell adhesion and integrin binding to recombinant human fibrillin-1. FEBS Lett 384:247–250

Pottie L, Adamo CS, Beyens A, Lutke S, Tapaneeyaphan P, De Clercq A, Salmon PL, De Rycke R, Gezdirici A, Gulec EY, Khan N, Urquhart JE, Newman WG, Metcalfe K, Efthymiou S, Maroofian R, Anwar N, Maqbool S, Rahman F, Altweijri I, Alsaleh M, Abdullah SM, Al-Owain M, Hashem M, Houlden H, Alkuraya FS, Sips P, Sengle G, Callewaert B (2021) Bi-allelic premature truncating variants in LTBP1 cause cutis laxa syndrome. Am J Hum Genet 108:1095–1114

Qian RQ, Glanville RW (1997) Alignment of fibrillin molecules in elastic microfibrils is defined by transglutaminase-derived cross-links. Biochemistry 36:15841–15847

Ramos Arroyo MA, Weaver DD, Beals RK (1985) Congenital contractural arachnodactyly. Report of four additional families and review of literature. Clin Genet 27:570–581

Reinhardt DP, Keene DR, Corson GM, Poschl E, Bachinger HP, Gambee JE, Sakai LY (1996) Fibrillin-1: organization in microfibrils and structural properties. J Mol Biol 258:104–116
Reinhardt DP, Ono RN, Notbohm H, Muller PK, Bachinger HP, Sakai LY (2000) Mutations in calcium-binding epidermal growth factor modules render fibrillin-1 susceptible to proteolysis. A potential disease-causing mechanism in Marfan syndrome. J Biol Chem 275:12339–12345
Rifkin DB, Rifkin WJ, Zilberberg L (2018) LTBPs in biology and medicine: LTBP diseases. Matrix Biol 71-72:90–99
Robertson I, Jensen S, Handford P (2011) TB domain proteins: evolutionary insights into the multifaceted roles of fibrillins and LTBPs. Biochem J 433:263–276
Robins PR, Moe JH, Winter RB (1975) Scoliosis in Marfan's syndrome. Its characteristics and results of treatment in thirty-five patients. J Bone Joint Surg Am 57:358–368
Romere C, Duerrschmid C, Bournat J, Constable P, Jain M, Xia F, Saha PK, Del Solar M, Zhu B, York B, Sarkar P, Rendon DA, Gaber MW, LeMaire SA, Coselli JS, Milewicz DM, Sutton VR, Butte NF, Moore DD, Chopra AR (2016) Asprosin, a fasting-induced Glucogenic protein hormone. Cell 165:566–579
Sabatier L, Chen D, Fagotto-Kaufmann C, Hubmacher D, McKee MD, Annis DS, Mosher DF, Reinhardt DP (2009) Fibrillin assembly requires fibronectin. Mol Biol Cell 20:846–858
Saharinen J, Keski-Oja J (2000) Specific sequence motif of 8-Cys repeats of TGF-beta binding proteins, LTBPs, creates a hydrophobic interaction surface for binding of small latent TGF-beta. Mol Biol Cell 11:2691–2704
Saharinen J, Taipale J, Keski-Oja J (1996) Association of the small latent transforming growth factor-beta with an eight cysteine repeat of its binding protein LTBP-1. EMBO J 15:245–253
Sakai LY, Keene DR (2019) Fibrillin protein pleiotropy: Acromelic dysplasias. Matrix Biol 80:6–13
Sakai LY, Keene DR, Engvall E (1986) Fibrillin, a new 350-kD glycoprotein, is a component of extracellular microfibrils. J Cell Biol 103:2499–2509
Sakai LY, Keene DR, Renard M, De Backer J (2016) FBN1: the disease-causing gene for Marfan syndrome and other genetic disorders. Gene 591:279–291
Sasaki T, Stoop R, Sakai T, Hess A, Deutzmann R, Schlotzer-Schrehardt U, Chu ML, von der Mark K (2016) Loss of fibulin-4 results in abnormal collagen fibril assembly in bone, caused by impaired lysyl oxidase processing and collagen cross-linking. Matrix Biol 50:53–66
Schiavinato A, Keene DR, Imhof T, Doliana R, Sasaki T, Sengle G (2017) Fibulin-4 deposition requires EMILIN-1 in the extracellular matrix of osteoblasts. Sci Rep 7:5526
Schmelzer CEH, Hedtke T, Heinz A (2020) Unique molecular networks: formation and role of elastin cross-links. IUBMB Life 72:842–854
Sedes L, Wondimu E, Crockett B, Hansen J, Cantalupo A, Asano K, Iyengar R, Rifkin DB, Smaldone S, Ramirez F (2022) Fibrillin-1 deficiency in the outer perichondrium causes longitudinal bone overgrowth in mice with Marfan syndrome. Hum Mol Genet 31:3281–3289
Sengle G, Carlberg V, Tufa SF, Charbonneau NL, Smaldone S, Carlson EJ, Ramirez F, Keene DR, Sakai LY (2015) Abnormal activation of BMP signaling causes myopathy in Fbn2 null mice. PLoS Genet 11:e1005340
Sengle G, Charbonneau NL, Ono RN, Sasaki T, Alvarez J, Keene DR, Bachinger HP, Sakai LY (2008) Targeting of bone morphogenetic protein growth factor complexes to fibrillin. J Biol Chem 283:13874–13888
Sengle G, Ono RN, Sasaki T, Sakai LY (2011) Prodomains of transforming growth factor beta (TGFbeta) superfamily members specify different functions: extracellular matrix interactions and growth factor bioavailability. J Biol Chem 286:5087–5099
Sengle G, Sakai LY (2015) The fibrillin microfibril scaffold: a niche for growth factors and mechanosensation? Matrix Biol 47:3–12
Sengle G, Tsutsui K, Keene DR, Tufa SF, Carlson EJ, Charbonneau NL, Ono RN, Sasaki T, Wirtz MK, Samples JR, Fessler LI, Fessler JH, Sekiguchi K, Hayflick SJ, Sakai LY (2012) Microenvironmental regulation by fibrillin-1. PLoS Genet 8:e1002425

Smaldone S, Clayton NP, del Solar M, Pascual G, Cheng SH, Wentworth BM, Schaffler MB, Ramirez F (2016) Fibrillin-1 regulates skeletal stem cell differentiation by modulating TGFbeta activity within the marrow niche. J Bone Miner Res 31:86–97

Smaldone S, Ramirez F (2016) Fibrillin microfibrils in bone physiology. Matrix Biol 52-54:191–197

Spanou CES, Wohl AP, Doherr S, Correns A, Sonntag N, Lutke S, Morgelin M, Imhof T, Gebauer JM, Baumann U, Grobe K, Koch M, Sengle G (2023) Targeting of bone morphogenetic protein complexes to heparin/heparan sulfate glycosaminoglycans in bioactive conformation. FASEB J 37:e22717

Sponseller PD, Sethi N, Cameron DE, Pyeritz RE (1997) Infantile scoliosis in Marfan syndrome. Spine (Phila Pa 1976) 22:509–516

Steel HH (1996) Protrusio acetabuli: its occurrence in the completely expressed Marfan syndrome and its musculoskeletal component and a procedure to arrest the course of protrusion in the growing pelvis. J Pediatr Orthop 16:704–718

Ting BL, Mathur D, Loeys BL, Dietz HC 3rd, Sponseller PD (2010) The diagnostic value of the facial features of Marfan syndrome. J Child Orthop 4:545–551

Tsutsui K, Manabe R, Yamada T, Nakano I, Oguri Y, Keene DR, Sengle G, Sakai LY, Sekiguchi K (2010) ADAMTSL-6 is a novel extracellular matrix protein that binds to fibrillin-1 and promotes fibrillin-1 fibril formation. J Biol Chem 285:4870–4882

Tuncbilek E, Alanay Y (2006) Congenital contractural arachnodactyly (Beals syndrome). Orphanet J Rare Dis 1:20

Urban Z, Hucthagowder V, Schurmann N, Todorovic V, Zilberberg L, Choi J, Sens C, Brown CW, Clark RD, Holland KE, Marble M, Sakai LY, Dabovic B, Rifkin DB, Davis EC (2009) Mutations in LTBP4 cause a syndrome of impaired pulmonary, gastrointestinal, genitourinary, musculoskeletal, and dermal development. Am J Hum Genet 85:593–605

Utreja A, Evans CA (2009) Marfan syndrome-an orthodontic perspective. Angle Orthod 79:394–400

Vallet SD, Ricard-Blum S (2019) Lysyl oxidases: from enzyme activity to extracellular matrix cross-links. Essays Biochem 63:349–364

Van de Velde S, Fillman R, Yandow S (2006) Protrusio acetabuli in Marfan syndrome. History, diagnosis, and treatment. J Bone Joint Surg Am 88:639–646

Viljoen D (1994) Congenital contractural arachnodactyly (Beals syndrome). J Med Genet 31:640–643

Walker BA, Murdoch JL (1970) The wrist sign. A useful physical finding in the Marfan syndrome. Arch Intern Med 126:276–277

Wang MC, Lu Y, Baldock C (2009) Fibrillin microfibrils: a key role for the interbead region in elasticity. J Mol Biol 388:168–179

Wlodarski KH (1989) Normal and heterotopic periosteum. Clin Orthop Relat Res 241:265–277

Wohl AP, Troilo H, Collins RF, Baldock C, Sengle G (2016) Extracellular regulation of bone morphogenetic protein activity by the microfibril component Fibrillin-1. J Biol Chem 291:12732–12746

Wozniak-Mielczarek L, Osowicka M, Radtke-Lysek A, Drezek-Nojowicz M, Gilis-Malinowska N, Sabiniewicz A, Mielczarek M, Sabiniewicz R (2022) How to distinguish Marfan Syndrome from Marfanoid habitus in a physical examination-comparison of external features in patients with Marfan Syndrome and Marfanoid habitus. Int J Environ Res Public Health 19

Yu Y, He JH, Hu LL, Jiang LL, Fang L, Yao GD, Wang SJ, Yang Q, Guo Y, Liu L, Shang T, Sato Y, Kawamura K, Hsueh AJ, Sun YP (2020) Placensin is a glucogenic hormone secreted by human placenta. EMBO Rep 21:e49530

Yule SR, Hobson EE, Dean JC, Gilbert FJ (1999) Protrusio acetabuli in Marfan's syndrome. Clin Radiol 54:95–97

Zarate YA, Morris SA, Blackshare A, Algaze CA, Connor BS, Kim AJ, Yutzey KE, Miller EM, Weaver KN, Collins RT 2nd (2022) A clinical scoring system for early onset (neonatal) Marfan syndrome. Genet Med 24:1503–1511

Zimmermann LA, Correns A, Furlan AG, Spanou CES, Sengle G (2021) Controlling BMP growth factor bioavailability: the extracellular matrix as multi skilled platform. Cell Signal 85:110071

Part II
ECM Signaling and Skeletal Disorders

Chapter 8
FGF Signaling: A Key Pathway During Skeletal Development

Chantal Fayad and Laurence Legeai-Mallet

Abstract Receptor tyrosine kinases (RTKs) control fundamental cellular processes and their signal transduction is essential in skeletal, nervous, and vascular development. RTKs are activated by dimerization in the plasma membrane. Among the RTK family, the Fibroblast Growth Factor Receptors (FGFRs) play a crucial role in both development and ageing. They are expressed on the cell membrane and this family consists of five FGFRs (FGFR1-5). The native ligands for these receptors are the Fibroblast Growth Factor (FGF). This family is composed of 22 FGFs divided into three subfamilies (paracrine, endocrine, and intracellular subfamily). FGFs and FGFRs are widely expressed in both the appendicular the axial skeleton and skull and their interaction activates several signaling pathways to regulate various cell functions and metabolization of extracellular matrix (ECM) proteins, involved in skeletal development. Mutations in FGFR/FGF genes lead to a deregulation of the FGF/FGFR signaling and consequently to several skeletal dysplasias such as chondrodysplasia and craniosynostoses. Different therapeutic approaches are being established to target the abnormal functioning of FGF/FGFR signaling, the therapeutic strategies are various and associated with different mechanisms of actions. Recent progress in understanding the pathophysiological bases of these osteochondrodysplasias has enabled the first treatments to emerge.

8.1 Introduction

Receptor tyrosine kinases (RTK) control fundamental cellular processes, such as proliferation, differentiation, and their signal transduction, which are essential in skeletal, nervous, and vascular development, in homeostasis and in disease. RTKs'

C. Fayad · L. Legeai-Mallet (✉)
Université de Paris Cité, Imagine Institute, Laboratory of Molecular and Physiopathological Bases of Osteochondrodysplasia. INSERM UMR, Paris, France
e-mail: laurence.legeai-mallet@inserm.fr

A. Rossi, F. Zaucke (eds.), *The Extracellular Matrix in Genetic Skeletal Disorders*, Biology of Extracellular Matrix 16,
https://doi.org/10.1007/978-3-031-70835-0_8

activation is ligand-induced dimerization in the plasma membrane; RTKs form homodimers and heterodimers and each dimer plays a unique role in cell signaling (Schlessinger 2000).

Many of the 58 human RTKs are targeted with activators and inhibitors to counteract their dysregulation in development and cancers. RTK are activated in a tissue specific manner (cell type or tissue) and by families of ligands (Lemmon and Schlessinger 2010).

Among the RTK family, the Fibroblast Growth Factor Receptors (FGFRs) play a crucial role in both development and ageing (Wilkie et al. 1995). Dysregulation of FGFRs has been implicated in skeletal disease and a wide variety of cancers (Webster and Donoghue 1997).

The human FGFR family consists of five members: FGFR1 to FGFR5. Despite being encoded by separate genes, the FGFR1-4 members share high homology, with their sequence identity varying from 56% to 71%. These four FGFRs share a canonical RTK architecture, all four FGFR members contains an intracellular tyrosine kinase domain, a single transmembrane domain, and large extracellular ligand-binding domain that comprises three immunoglobulin (Ig)-like subunits (D1, D2, and D3) (Belov and Mohammadi 2013; Lemmon and Schlessinger 2010). FGFR5 is a unique member of the family lacking the functional tyrosine kinase domain (Zhuang et al. 2010).

Similar to other RTKs, FGFRs are expressed on the cell membrane and can be stimulated and activated by extracellular signals. Fibroblast growth factor (FGF) is the native ligand for this family of RTK. Through its extracellular domain, FGFR recognizes and is stimulated by specific FGFs (Lemmon and Schlessinger 2010; Schlessinger 2000). FGFRs are activated by FGF ligand and this activation will induce the trans auto-phosphorylation of tyrosine residues, phosphorylation of many proteins and activation of many signaling pathways such as Mapk, AKT, PLCγ, and JAK/STAT pathways. The dysregulation of these pathways is responsible for a family of skeletal disorders affecting both endochondral and membranous ossification process (Ornitz and Itoh 2015). FGFR/FGF signaling regulates various cell functions and metabolization of extracellular matrix (ECM) proteins. ECM is composed of a complex protein network including several collagens (e.g., Collagen type II, type IX, X, XI in cartilage and Collagen type I in bone), proteoglycan and matrix metalloproteinase. By preserving the intricate architecture around the cells and the form of the organs, the ECM gives tissues their structural strength (Gentili and Cancedda 2009).

In this chapter, we describe the role of the various family of FGF ligands and their interaction with FGFR during development, the interaction of FGF/FGFR with ECM proteins, and the role of the FGFR family involved in skeletal disease in particular osteochondrodysplasia, and the possible therapeutic approaches for the FGFR-related disorders.

8.2 FGF Ligands

FGFs are part of a broad family of growth factors. FGFs utilize different signaling pathways to regulate several physiological and pathophysiological processes (Hui et al. 2018). Here, a classification of the FGF family members and a list of the known functions are described.

FGF family is constituted of 22 FGFs (Fig. 8.1); Depending on developmental features and sequence homology, 18 FGFs were divided into 5 paracrine subfamilies and 1 endocrine subfamily (Belov and Mohammadi 2013). The five paracrine subfamilies are as follows: FGF1 subfamily (FGF1 and FGF2), FGF4 subfamily (FGF4, FGF5, and FGF6), FGF7 subfamily (FGF3, FGF10, FGF7, and FGF22), FGF8 subfamily (FGF8, FGF17, and FGF18), and FGF9 subfamily (FGF9, FGF16, and FGF20) and the endocrine subfamily FGF 15/19 subfamily. The paracrine and endocrine subfamilies bind and interact with FGFR through co-factors heparan sulfate (HS) and Klotho respectively (Belov and Mohammadi 2013). Other FGF members, FGF11–FGF14, are called the intracellular FGFs and they are not included in the previously six subfamilies due to the fact that they cannot activate FGF receptors (Itoh and Ornitz 2008).

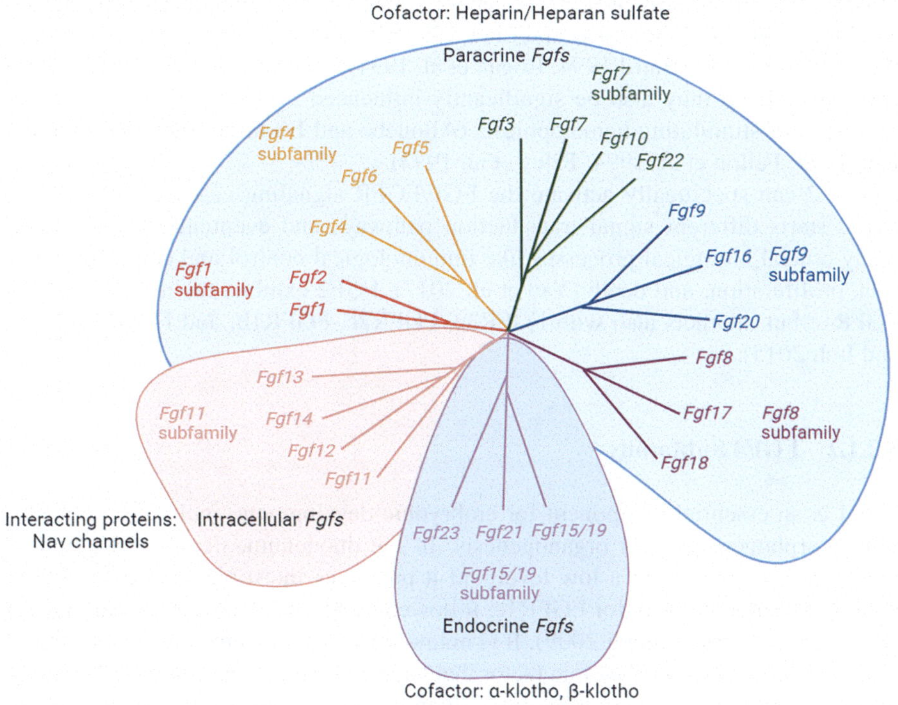

Fig. 8.1 Fibroblast Growth Factor families. *Fgf* family is constituted of 22 *Fgf genes*. They were divided into five paracrine subfamilies, one endocrine subfamily and one intracellular family that cannot activate FGFR. *Fgf15* is the mouse ortholog of human *FGF19*

8.2.1 Paracrine FGFs

8.2.1.1 FGF1 Subfamily

FGF1 is one of the FGF superfamily's members with the best characterization. FGF1 is a potent mitogen with powerful effects on a wide variety of cell types. It participates in many morphogenetic and developmental stages, as well as angiogenesis and wound healing (Zakrzewska et al. 2008). FGF1 exerts its biological activity by binding to four cell surface receptors with high affinity (FGFR1-4), which causes the receptor to dimerize and become trans-autophosphorylated in its tyrosine kinase domain (Ornitz et al. 1996; Spivak-Kroizman et al. 1994). The activated FGFR triggers cellular response by starting a number of signaling cascades, such as the phosphoinositide 3-kinase/Akt, mitogen-activated protein kinase (MAPK), and phospholipase C- (PLC-) pathways (Mohammadi et al. 2005b).

FGF2 is one of the earliest members of the FGF family to be identified. FGF2 is dispersed in different organs, tissues, and cells. Due to its widespread body distribution and potent cell-promoting properties, FGF2 is crucial for tendon to bone mending, cartilage repair, bone repair, and nerve regeneration (Park et al. 2014). In vivo, it is an angiogenic molecule and in vitro, it promotes the proliferation of smooth muscle cells, the healing of wounds, and the restoration of damaged tissue (Basilico and Moscatelli 1992; Schwartz and Liaw 1993).The development and/or function of the nervous system (Baird 1994; Logan et al. 1991), the eye (McAVOY et al. 1991) and the skeleton may also be significantly influenced by FGF2 in addition to its potential for stimulating hematopoiesis (Allouche and Bikfalvi 1995; Bikfalvi and Han 1994; Fallon et al. 1994; Riley et al. 1993).

FGF2 can specifically activate the FGF/FGFR signaling cascade. As a result, FGF2 starts different signal transduction pathways and eventually takes part in many crucial biological processes like immunological control and cell differentiation, proliferation, and death (Yao et al. 2017). FGF2 exhibits the most affinity to FGFR1c but interacts also with FGFR3c, FGFR2c, FGFR1b, and FGFR4 (Ornitz and Itoh 2015).

8.2.1.2 FGF4 Subfamily

FGF4 is an essential component for embryonic development, including implantation, morphogenesis, and organogenesis. In the duodenum, ileum, and colon of adults, it is expressed at a low level and it preserves intestinal stem cells. FGF4 exhibits the most affinity for FGFR1c, followed by FGFR2c, FGFR3c, and FGFR4 (Beenken and Mohammadi 2009). It is noteworthy that a recent study found FGF4 to be a strong antihyperglycemic factor that has no effect on appetite. FGF4 directly affects macrophages, where it inhibits inflammatory responses in the liver, muscle, and adipose tissue (Ying et al. 2021).

FGF5 regulates hair development and more precisely is known to inhibit the hair growth phase (Cho et al. 2003; Pena 1999; Rosenquist and Martin 1996). FGF5 activate either the FGFR1 or FGFR2 receptor (Clements et al. 1993).

FGF6 contributes to myogenesis, primarily in skeletal muscle (Armand et al. 2006). It has potential clinical usefulness for treating muscle injuries since it encourages the development, proliferation, and differentiation of skeletal muscle (Jump et al. 2009). FGFR1 expressed by osteoblasts and osteoclasts and FGFR4 expressed by primary osteoblasts and osteogenic precursor cells are preferentially responsible for FGF6-induced signal transduction. FGF6 is thought to trigger MAP kinase signaling via RANKL to activate osteoclasts and finish bone formation (Bosetti et al. 2010; Ornitz et al. 1996).

8.2.1.3 FGF7 Subfamily

FGF3 is necessary for the development of the inner ear (Tekin et al. 2007). FGF3 usually binds and interacts with FGFR2b (Mason et al. 1994; Zhang et al. 2006).

FGF7 is necessary for the development of the kidney, thymus, and hippocampus (Alpdogan et al. 2006; Lee et al. 2012; Qiao et al. 1999; Terauchi et al. 2010). FGF7, also known as keratinocyte growth factor, is particularly expressed in mesenchyme (Beenken and Mohammadi 2009). FGFR2b, are the primary partners of the FGF7 subfamily, which is the only subfamily exclusively expressed in the mesenchyme (Mason et al. 1994; Zhang et al. 2006).

The receptor tyrosine kinase FGFR2IIIb is selectively activated by the paracrine action of FGF7 (Finch et al. 1989; Knights and Cook 2010). The RAS-mitogen-activated protein kinase (MAPK), phosphoinositide 3-kinase (PI3K), and Akt signaling pathways, which are crucial in the growth of tumors, are among the signal transduction pathways that are quickly activated as a result of FGF7 and FGFR2 binding (Grose and Dickson 2005). In addition, it has been shown that FGF7 plays a critical role in the regulation of liver progenitor cells that are important in supporting the liver regeneration (Takase et al. 2013).

FGF10 is crucial for development and adult tissue homeostasis (Itoh 2016). FGF10 is necessary for the development of the limb, lung, thyroid, pituitary, lacrimal, and salivary gland (Bellusci et al. 1997; Hoffman et al. 2002; Izvolsky et al. 2003; Makarenkova et al. 2000; Min et al. 1998; Moerlooze et al. 2000; Sekine et al. 1999). FGF10 binds and activates FGFR2b, and consequently, induces vital intracellular signaling pathways in a variety of cell types, modifying organ branching, cell proliferation, differentiation, and migration during development; tissue repair and wound healing; preservation of the stem cell compartment; and invasion and proliferation of cancer cells (Itoh 2016).

FGF22 is involved in presynaptic neural differentiation (Umemori et al. 2004). It has been shown to be essential for the development of excitatory synapses in the developing brain, including the cerebellum, hippocampus, and lateral geniculate nucleus. FGF22 mainly binds to FGFR2 (Singh et al. 2012; Terauchi et al. 2010; Umemori et al. 2004).

8.2.1.4 FGF9 Subfamily

FGF9 plays a role in lung development and in the kidney and in particular in the maintenance of stemness in nephrons. This factor is involved in bone repair, as well as in mammalian sex determination (Barak et al. 2012; Colvin et al. 2001; Y. Kim et al. 2006). FGF9 has the highest affinity to bind and activate FGFR3 but exhibits no action toward FGFR1b or -2b; FGF9 can bind and activate FGFR3IIIc, FGFR3IIIb, and FGFR2IIIc; however, it can also activate other receptors with lesser affinity, such as FGFR1IIIc and FGFR4 (Hecht et al. 1995; Zhang et al. 2006).

FGF16 is mostly expressed in cardiomyocytes. It functions as a paracrine signaling molecule and is effectively secreted (Hotta et al. 2008). Through its favorable impact on myocardial development, FGF-16 was shown to be necessary for embryonic heart development in mid-gestation. It competes with FGF2 for FGFR1c binding (Matsumoto et al. 2013).

FGF20, similarly to FGF9, has been shown to play a role in the kidney and in particular in the maintenance of stemness in nephrons (Barak et al. 2012). This factor is also expressed in inner ear development and is most likely the ligand for FGFR1 during sensory specification phase of cochlear development, where it is produced at the appropriate location and timing to regulate sensory cell specification (Hayashi et al. 2008). Studies highlight also the significance of FGF20 in the development and treatment of Parkinson's disease as well as the differentiation and survival of dopaminergic neurons. In fact, FGF20 exerts a paracrine effect on dopaminergic neurons. It interacts with FGFR1c and induces its phosphorylation. The MAPK pathway, will be then activated which is crucial to the survival of dopaminergic neurons (Itoh and Ohta 2013).

8.2.1.5 FGF8 Subfamily

FGF8 plays a crucial part in the control of embryonic development, cell division, cell migration, and cell proliferation. It is necessary for healthy embryonic development of the limbs, ears, eyes, and brain as well as for the appropriate function of gonadotropin-releasing hormone (GnRH) neuronal system (Olsen et al. 2006; Ornitz et al. 1996; Zhang et al. 2006). It is also involved in hippocampus cells' neurite development (Vantaggiato et al. 2011). FGF8 has strong binding affinities for the non-spliced FGFR4 and the c spliced variants of FGFR1-3 (Estienne and Price 2018).

FGF17 is crucial for vascular expansion and typical brain development. It is expressed during embryogenesis and in the adult cerebellum and cortex (Cholfin and Rubenstein 2007). FGF17 has stronger binding affinities to FGFR3c and FGFR4 than FGFR1c (Estienne and Price 2018).

FGF18 appears to control cell division and proliferation positively during osteogenesis and unfavorably during chondrogenesis (Ohbayashi et al. 2002). It contributes to the establishment of the lung's alveoli in the late stages of embryonic lung development (Usui et al. 2004). FGF18 binds to FGFR3c rather than FGFR2c (Estienne and Price 2018; Ornitz and Itoh 2015).

8.2.2 *Endocrine FGFs*

8.2.2.1 FGF 15/19 Subfamily

FGF19 was initially identified in the human brain during fetal and embryonic development. FGF15 is an ortholog of FGF19 in animals and it is found in rodents' central nervous systems where it promotes the differentiation of mature neurons. FGF15/19 have proliferative and metabolic properties. FGF19 can bind and activate FGFR1c, FGFR2c, FGFR3c, and FGFR4 (Wu et al. 2010).

FGF21 is produced by a number of organs and controls the homeostasis of energy (Fisher and Maratos-Flier 2016). Unlike other FGFs, FGF21 has metabolic activity rather than proliferative activity (Itoh et al. 2015; Ornitz and Itoh 2015). FGF21 is mostly expressed in the liver and adipose tissue, with little expression being found in the bones, muscle, heart, kidneys, and testes (Kokkinos et al. 2017). FGF21 was demonstrated to regulate energy balance, glucose, and lipid homeostasis, through its binding to its specific receptor FGFR1 (Geng et al. 2020). FGF21 is able to bind and activate FGFR1c and FGFR3c (Suzuki et al. 2008).

FGF23 is a phosphotropic hormone (Château et al. 2010). It has also proliferative and metabolic properties. The primary role of FGF23 is to control plasma phosphate levels. In fact, it accomplishes this by reducing the kidney's ability to reabsorb phosphate, causing phosphate to be expelled in the urine (Jüppner 2011). Osteoblasts and osteocytes produce FGF23 into the bloodstream, which has effects on the kidney, parathyroid, heart, bone, and possibly other organs as well (Faul et al. 2011; Krajisnik et al. 2007; Murali et al. 2016; Shimada et al. 2004). While FGFR1c seems to be the main endocrine receptor for FGF23, FGFR3 and FGFR4 can also be activated by FGF23 in a Klotho-dependent (Kurosu et al. 2006) or -independent manner (Grabner et al. 2015; Murali et al. 2016).

8.3 Intracellular FGF

8.3.1 *FGF11 Subfamily*

These FGFs are not secreted and do not interact with FGFR (Olsen et al. 2003). They interact with the cytosolic carboxy terminal tail of voltage-gated sodium channels. This interaction may aid to regulate the subcellular localization of voltage-gated sodium channels at the axon beginning segment throughout development and the channel's ion-gating capabilities in adult neurons and other excitable cells like cardiomyocytes (Goldfarb et al. 2007; Hsu et al. 2014; Lou et al. 2005; Wang et al. 2011; Xiao et al. 2013).

FGF11 seems to have signaling functions during tooth development (Kettunen et al. 2011).

FGF13 is colocalized with sodium channels and is expressed in primary sensory neurons. FGF13 is also expressed in cerebral cortical neurons in the late stages of development and at the nodes of Ranvier along with myelinated afferent axons (Zhang et al. 2012).

FGF12 and FGF14 interact with voltage-gated sodium channels to control their activity in neurons (Zhang et al. 2012).

8.4 Cartilage and Bone ECM

Cartilage and bone ECM metabolism is essential to the formation of skeletal tissues and their repair or regeneration in different skeletal diseases, including bone fractures, cartilage damage, arthritis, and hereditary skeletal deformities. Other processes such as tumor invasion, metastasis, fibrosis, wound healing, and tissue expansion depend also on the formation of the ECM (Liotta et al. 1980).

Cartilage is a robust, slightly flexible, and elastic tissue consisting of extracellular fibers and chondrocytes embedded in the ECM (Aigner et al. 2006). This ECM is made of self-assembled macromolecules: collagen, non-collagenous glycoproteins, hyaluronan, and proteoglycans (PG) make up the majority of its composition. The collagenous network gives the cartilage ECM its tensile strength, and the proteoglycans, primarily aggrecan are the reason of the osmotic swelling and the cartilage tissue's elastic qualities (Gentili and Cancedda 2009). In addition to providing support for the cells, ECM also acts as a reservoir for growth factors and cytokines, regulating the turnover and activation state of the cells and controlling cell activity by influencing both proliferation and differentiation (Gentili and Cancedda 2009; Velleman 2000) (Fig. 8.2).

Numerous studies have demonstrated the presence of several collagen and proteoglycan types that are variably expressed in mature bone ECM, the transition zone between mature cartilage and bone (Velleman 2000). The various ECM components are directly involved in a number of processes that turn cartilage into bone.

In summary, complex systems that regulate ECM turnover and remodeling are responsible for maintaining the homeostasis of cartilage and bone. The ECM resident cells, in bone and cartilage, generate matrix-degrading enzymes, inflammatory mediators, and local factors. The reactions of the local cell to auto and paracrine anabolic and catabolic pathways determine the turnover and degradation of ECM (Gentili and Cancedda 2009).

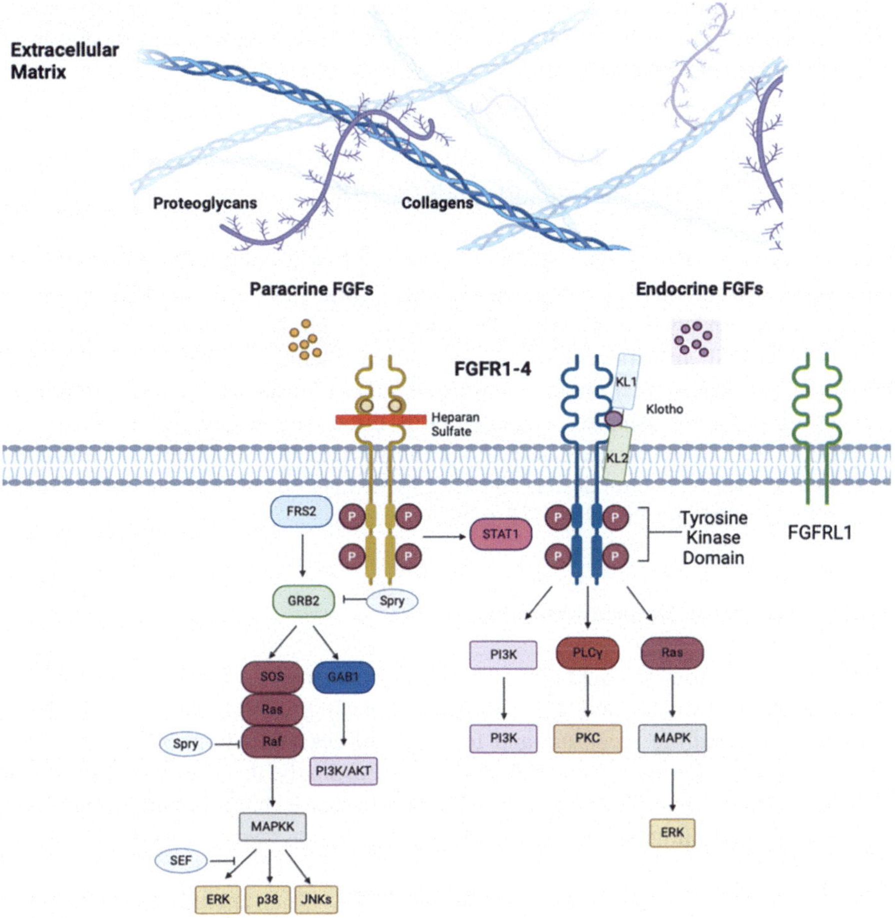

Fig. 8.2 Intracellular pathways triggered by FGF in the cells. FGFs bind to tyrosine kinase FGFRs associated to ECM co-receptor (HS or klotho) and activate many protein kinases and transcription factors

8.5 FGFRs Family

FGFRs are four conventional members, namely FGFR1 to FGFR4, which consist of an extracellular domain, a single transmembrane domain, and an intracellular domain containing a tyrosine kinase domain. FGFR1 to FGFR3 are selectively spliced to produce b and c isoforms and this splicing event modulates the specificity of FGFR ligand binding. FGFR5 also known as FGFR like 1 consists of an extracellular domain with no functional tyrosine kinase domain. FGFR5 is often considered as a decoy receptor (Trueb 2011).

FGFR pathway plays an active role in the development of almost all tissues and organs. FGFRs control both cell proliferation and differentiation and their role is different depending on the tissues.

8.5.1 FGFR Dimerization and Signaling

Many studies have revealed the existence of multiple ways of ligand-FGFR interaction. Through alternative splicing, the FGFR1, FGFR2, FGFR3, and FGFR4 genes produce the seven key FGFR proteins (FGFRs 1b, 1c, 2b, 2c, 3b, 3c, and 4). Different FGFRs bind FGF in different ways. For stables interactions between FGFs and FGFRs, a cofactor, which interacts with FGFs and FGFRs separately, is required. FGF-FGFR-cofactor complex triggers FGFR dimerization and the activation of FGFR intracellular tyrosine kinase regions and subsequently the induction and activation of several key intracellular signaling pathways (Brewer et al. 2016; Carter et al. 2015; Ornitz and Itoh 2015) (Fig. 8.2).

8.5.1.1 Co-factor Heparan Sulfate

Many growth factors (e.g., FGFs, VEGFs) bind actively to HS, a component of cartilage ECM. HS is needed as a cofactor for certain growth factors (e.g., FGF) to bind to their signaling receptors and form a well-described ternary FGF–FGFR–heparin complex (Mohammadi et al. 2005a). Accordingly, the general consensus is that HS may help create stable gradients of growth factors bound to the ECM; these gradients of morphogens are essential for regulating developmental processes (Hynes 2009).

The secreted proteins known as paracrine FGFs, which include FGF1-FGF10, FGF16-FGF18, FGF20, and FGF22, are the ones that have HS-binding sites at their carboxyl termini. By attaching to cell surface FGFRs with HS as a co-factor, paracrine FGFs carry out biological functions. By acting as local signals via diffusion on neighboring target cells, paracrine FGFs primarily function in a variety of physiological and developmental processes. Specific cell surface transmembrane-type proteins are covalently coupled to HS, which has long linear carbohydrate chains of repeated sulfated glucuronic acid linked to N-acetylglucosamine disaccharides. It modifies the diffusion of paracrine FGFs, and this modality directs paracrine FGFs as local signals (Brewer et al. 2016; Ornitz and Itoh 2015).

High affinity ligand binding leads the tyrosine kinase domains of a FGFR to dimerize and tyrosine residues to be trans-phosphorylated, which makes binding sites for adaptor proteins and activates tyrosine kinase activity (Farrell and Breeze 2018). The conventional FGF/FGFR downstream signaling pathways include Ras/Raf-MEK-MAPKs (mitogen-activated protein kinases), phosphatidylinositol-3 kinase/protein kinase B (PI3K/AKT), PLCγ, and signal transducer and activator of transcription (STAT), depending on the cellular composition in different cells and tissues (Huang et al. 2016; Ornitz and Itoh 2015; Turner and Grose 2010). The canonical downstream signaling molecules including the plasma membrane-localized FGFR substrate 2α (FRS2α) dock at the phosphorylated tyrosine residues (Gotoh 2008). The adaptor protein GRB2 is attracted by the phosphorylation of FRS2 and activated MAP kinase signaling pathways (including ERK MAPK, p38 MAPK, and JNK MAPK) (Kouhara et al. 1997; Lamothe et al. 2004). Inhibitors of FGFR signaling include Sprouty (Spry), which acts at the level of GRB2 and/or Raf to prevent MAPK and PI3K signaling (Hanafusa et al. 2002; Timsah et al. 2014), and includes, downstream of FGFR, SEF (Similar Expression to FGF) which plays a role of antagonist for the MAPK pathway by interacting with MEK (Torii et al. 2004) and DUSP6 (Dual-specificity phosphatase 6), which leads to the dephosphorylating ERK1/2 and thus the inhibition of MAPK signaling (Camps et al. 1998).

8.5.1.2 Co-factor Klotho

The endocrine FGFs FGF15/19, FGF21, and FGF23 have Klotho-binding sites at their carboxyl termini and need α Klotho or β Klotho as a co-factor for inducing FGFRs dimerization. In numerous metabolic processes, endocrine FGFs primarily serve as endocrine signals. The single-pass transmembrane proteins α Klotho and β Klotho, each of which has 1000 amino acids and a brief cytoplasmic region, share structural similarities. They are mainly expressed in the target tissues of endocrine FGFs (Ornitz and Itoh 2015). β Klotho is the cofactor of FGF15/19 and FGFR4 while FGF21 primarily binds FGFR1c and β Klotho. FGFR1c is mostly bound by FGF23 via α Klotho. In turn, this binding results in the activation of the FGFR's intracellular signaling pathways. The two parts of Klotho receptor, KL1 and KL2, interact with FGFR to produce a groove that FGF can fit into (G. Chen et al. 2018). The activation of FGFR triggers signaling pathways such as PI3K/Akt, phospholipase C (PLC), and Ras/MAPK/ERK (Prud'homme et al. 2022).

It has been shown that α-Klotho is very sensitive to biophysical signals from the surrounding ECM. For instance, reducing matrix stiffness enhanced α-Klotho expression and enhanced chondrocyte health in vitro, while increasing matrix stiffness decreased α-Klotho expression. The cause of this stiffness-dependent modulation of α-Klotho was the recruitment of epigenetic regulators, such as DNA methyltransferase 1, at the Klotho promoter (Iijima et al. 2023).

8.5.2 FGF/FGFR Expression in Skeleton

8.5.2.1 Expression of FGF and FGF Receptors in the Developing Appendicular Skeleton

FGF ligands and receptors are expressed and work throughout the whole development of the appendicular skeleton, from the formation of the limb bud to the growth, remodeling, homeostasis, and repair of mature bone. Before any morphological or molecular evidence of a mesenchymal condensation (also known as a chondrogenic condensation) in the distal limb bud of mouse and chicken, *Fgfr1* and *Fgfr2* are present in mesenchymal cells (Orr-Urtreger et al. 1991; Sheeba et al. 2010). The expression of *Fgfr3* and *Fgfr4* is not observed at the pre-condensation stage in chicken (Sheeba et al. 2010).

At the distal border of the limb bud, there is a signaling center called the apical ectodermal ridge (AER). Several FGFs are expressed by the AER, mainly FGF4 and FGF8, but also FGF2, FGF9, and FGF17. The FGFs signal to limb mesenchymal FGFRs and slow down cell differentiation, giving cells more time to proliferate and inducing the formation of limb buds (Martin 1998; Sun et al. 2000; Tabin and Wolpert 2007). Mesenchymal cells start to differentiate and are activated by FGF signal from the AER, which will lead to the formation of a mesenchymal (chondrogenic) condensation, the primary event that initiates appendicular skeletogenesis (Bénazet and Zeller 2009; Tabin and Wolpert 2007).

At this step in the formation of a chondrogenic condensation process, Sox9 is expressed, the expression of *FGFR2* or *Fgfr2* is increased (Delezoide et al. 1998; Eswarakumar et al. 2002; Orr-Urtreger et al. 1991; Peters et al. 1992; Sheeba et al. 2010; Szebenyi et al. 1995; Yu and Ornitz 2008). In the mesenchyme of the limb bud, *Fgfr1* expression continues to be more consistent. *FGFR3* or *Fgfr3 and FGFR4* or *Fgfr4* are not expressed in the mesenchyme of the distal limb buds, but they are expressed in developing muscle tissue in more proximal regions of the growing limb (Delezoide et al. 1998; Orr-Urtreger et al. 1991; Peters et al. 1992; Sheeba et al. 2010; Szebenyi et al. 1995). The cells in the condensation's periphery that give rise to the perichondrium and periosteum exhibit *FGFR1* or *Fgfr1* and *FGFR2* or *Fgfr2* expression (Delezoide et al. 1998; Eswarakumar et al. 2002). Cells express *Fgfr3*, Sox9, and type II collagen centrally when they commit to a chondrogenic fate (Peters et al. 1993; Peters et al. 1992; Purcell et al. 2009). Later, *Fgfr1* is expressed by mesenchymal progenitors in the perichondrium and periosteum, which will give rise to the bone collar and cortical bone, whereas FGFR2 is expressed by differentiated osteoblasts (Britto et al. 2001; Coutu et al. 2011; Jacob et al. 2006; Molténi et al. 1999; Ohbayashi et al. 2002).

Chondrocytes in the center of the mesenchymal condensation start to hypertrophy and vascular invasion causes the chondrocytes in the hypertrophic zone to form a primary ossification center, the growth plate is created. These immature chondroprogenitors have FGFR3 expression. *Fgfr3* expression is still strong in chondrocytes in the proliferating and prehypertrophic zones of the established growth plate.

(Delezoide et al. 1998; Eswarakumar et al. 2002; Hamada et al. 1999; Jacob et al. 2006; Karolak et al. 2015; Karuppaiah et al. 2016; Lazarus et al. 2007; Ornitz and Marie 2002; Peters et al. 1993; Yu et al. 2003). *Fgfr3* expression decreases and *Fgfr1* expression increases as chondrocytes start to hypertrophy near the middle of the growing skeletal segment along the proximal-distal axis (Deng et al. 1996; Jacob et al. 2006; Karolak et al. 2015; Naski et al. 1998; Peters et al. 1993; Peters et al. 1992). In the groove of Ranvier and ring of LaCroix chondroprogenitor cells, *Fgfr3* is expressed more strongly (Robinson et al. 1999), and articular chondrocytes from mice and humans express FGFR1 and FGFR3 (Weng et al. 2012; Yan et al. 2011).

8.5.2.2 Expression of FGF and FGF Receptors in the Developing Axial Skeleton (Vertebrae and Skull)

Several FGF ligands, including *Fgf1, Fgf2, Fgf6, Fgf7, Fgf18, and Fgf23*, are expressed in various ways while the axial skeleton is developing (Ornitz and Marie 2015). It has been shown that *Fgfs 2, 4, 9,* and *18* are expressed throughout the early phases of intramembranous bone formation during cranioskeletal mouse development (Britto et al. 2001; Kim et al. 1998; Liu et al. 2002; Ohbayashi et al. 2002; Quarto et al. 2009; Rice et al. 2000). *Fgf8* is expressed in developing calvarial osteoblasts (Xu et al. 1999) and *Fgf9* is expressed in the calvarial mesenchyme from mid to late stages of development (Kim et al. 1998). During the development of cranial bones, *Fgf18* is expressed in mesenchymal cells and developing osteoblasts (Lazarus et al. 2007; Liu et al. 2007), whereas *FGF23* is found to be mainly formed by developed osteoblasts and osteocytes (Liu et al. 2003, 2006).

During skeleton development, in developing craniofacial bones, *Fgfr1 and Fgfr2* are expressed in pre-osteoblasts and osteoblasts (Rice et al. 2003). Later on, mesenchymal progenitors still express FGFR1 and osteoblasts still express FGFR2 (Britto et al. 2001; Coutu et al. 2011; Jacob et al. 2006; Molténi et al. 1999; Ohbayashi et al. 2002). Immature cultured osteoblasts exhibit higher amounts of *Fgfr1* while mature osteoblasts express higher levels of *Fgfr2* (Rice et al. 2003).

FGFs and FGFRs are also expressed in vertebrae. FGF18 was found in the outer annulus fibrosus at 12 weeks, as well as, to a lesser amount, in the hypertrophic cells of the mesenchymal condensations in the vertebral rudiments. In addition to this, FGF-18 expression was widespread in the inner annulus fibrosus, cartilage end plates and hypertrophic cells encircling the ossification centers at 14 weeks. FGFR3 was also broadly expressed in the spinal development. However, the vertebral rudiments of the spine showed somewhat lower levels of FGF-18 and FGFR-3 expression than the disc because they had lower cell densities (Shu et al. 2013). Moreover, the outermost part of the nucleus pulposus (NP) included FGFR3 expression along with co-localized expression of proliferating markers (M. Xu et al. 2022). FGF2, a specific ligand for FGFR3, induces the proliferation of human NP cells (Pratsinis et al. 2012).

8.6 Mutations in FGFRs in Human Skeletal Disease

A large family of osteochondrodysplasia is associated with dysregulation of the FGF signaling. The abnormal activations of FGF/FGFR can affect the axial, appendicular skeleton, and skull and are responsible for both chondrodysplasia and craniosynostoses.

8.6.1 FGFR-Related Chondrodysplasia

A family of chondrodysplasia is due to FGFR3 gain-of-function mutations. The most common form of skeletal dwarfism is achondroplasia (OMIM 100800) (Baujat et al. 2008; Horton et al. 2007; Murdoch et al. 1970; Ornitz and Legeai-Mallet 2017). The patients are characterized by disproportional limb shortening, also known as rhizomelia, macrocephaly with frontal bossing and vertebral anomalies. Affected individuals are prone to significant comorbidities, including sleep apnea, chronic otitis media with conductive hearing loss, spinal stenosis, obesity, and narrowing of the foramen magnum contributing to an increased risk of sudden infant death syndrome and may require prompt surgical intervention during infancy. The autosomal dominant missense mutation (G380R) localized in the transmembrane domain of FGFR3 is responsible for 90% of cases of achondroplasia (Rousseau et al. 1994; Shiang et al. 1994; Vajo et al. 2000; Wilkin et al. 1998). Most of the cases are sporadic (80%) and increased paternal age has been observed to have an association with de novo ACH (Goriely and Wilkie 2012). Hypochondroplasia (OMIM 146000) is a milder form of dwarfism due to FGFR3 mutations. This skeletal dysplasia is characterized by short stature, stocky build, shortening of the proximal segments (rhizomelia), short hands and feet with brachydactyly, mild joint laxity, scoliosis, learning disabilities, temporal lobe epilepsys (Bober et al. 2020). A large number of cases is due to the mutation (N540K) localized in the tyrosine kinase of the protein, some cases of hypochondroplasia are due to mutations localized in the extracellular domain of FGFR3 (Bellus et al. 1995, 2000; Bonaventure et al. 1996; Heuertz et al. 2006; Tavormina et al. 1995). Mutations in FGFR3 also contribute to other autosomal dominant chondrodysplasia with lethal achondroplasia-like features. These severe phenotypes are called thanatophoric dysplasia types I (TDI) (OMIM 187600) and II (TDII) (OMIM 187601) and characterized respectively by curved femurs or straight femur (Rousseau et al. 1996). The clinical and radiological features are evident prenatally, rhizomelic shortening of the long bones, platyspondyly, macrocephaly, ocular proptosis, foramen magnum narrowing, and small chest cavity result in respiratory insufficiency and early neonatal death. Heterozygous FGFR3 de novo missense mutations are identified either in the extracellular or the intracellular domains of the proteins, and no stop codon variant represents fewer than 10% of TDI (Rousseau et al. 1995). A very rare mutation FGFR3 (K650M) is associated with a severe phenotype called SADDAN syndrome (Severe

Achondroplasia with Developing Delay and Acanthosis Nigricans) (OMIM 616482); patients are characterized by extremely short stature, severe tibial bowing, seizure, foramen magnum stenosis and structural brain malformation, developmental delay, and acanthosis nigricans (Bellus et al. 1999; Montone et al. 2018; Tavormina et al. 1999). Lastly, the lethal platyspondylic Dysplasia (San Diego type) (OMIM 151210) is recognized as the same condition as thanatophoric dysplasia (Brodie et al. 1999).

Chondrocytes behavior is affected by this aberrant activation of FGFR3 thus inducing ERK ½ phosphorylation, which causes premature growth stop, ECM loss, abnormal differentiation, and morphological changes in the cells (Dailey et al. 2003; Krejci et al. 2004, 2005, 2007, 2008; Nowroozi et al. 2005; Raucci et al. 2004; Schibler et al. 2009; Yasoda et al. 2004).

At the molecular level, the growth arrest phenotype is induced by the activation of multiple cell cycle inhibitors (p21, p16, p18, p19) (Foldynova-Trantirkova et al. 2012). On the other hand, the loss of ECM results from the inhibition of essential matrix components (collagen type II and aggrecan) (Krejci et al. 2004). The upregulated expression and activation of multiple matrix metalloproteinase family members (MMP) including MMP3, 9, 10, and 13 leads to active matrix degradation (Krejci et al. 2005).

Interestingly, an elevated cell to matrix ratio has been seen across the growth plate in TDI and ACH animal models, indicating either greater turnover or lower matrix production (Yasoda et al. 2004). Indeed, several genes involved in glycosaminoglycan (GAG) and proteoglycan production and sulfation were modulated and one of them was hyaluronan synthase 3 which was upregulated (Schibler et al. 2009).

In addition, a FGFR3 loss-of-function is also associated with a skeletal phenotype. The patients are characterized by camptodactyly, tall stature, and hearing loss (known as the CATSHL syndrome) (OMIM 610474). The kinase domain is inactivated by a missense mutation, which probably has a dominant-negative effect (Makrythanasis et al. 2014). Other skeletal dysplasia exist and are due to different FGFR mutations. For example, Lacrimoauriculodentodigital syndrome (LADD), which is a genetically heterogeneous disorder, caused by heterozygous FGFR3 (LADD-2; OMIM 620192), FGFR2 (LADD-1; OMIM 149730) or FGF10 (LADD-3; OMIM 620193) mutations. FGFR2 and FGFR3 mutations are located in the tyrosine kinase domain of proteins and appear to increase the activity of receptors. LADD patients are characterized by anomalies affecting lacrimal glands and ducts, salivary glands and ducts, ears, teeth, and distal limb segments (Rohmann et al. 2006). Finally, Bent Bone Dysplasia (OMIM 614592), a dysmorphology of the appendicular skeleton, may be brought on by FGFR2 mutations (Merrill et al. 2012). This disease results from FGFR2 mutations (p.M391R or p.Y381D), which diminish ligand-dependent receptor activation and increase the mutant FGFR2's nucleolar activity, increasing osteoprogenitor cell proliferation and decreasing differentiation (Merrill et al. 2012; Neben et al. 2014, 2017). Concerning the role of ECM in these disorders (CATSHL, LADD, Bent Bone Dysplasia), the putative functions are not obvious yet.

8.6.2 FGFR-Related Craniosynostosis

Craniosynostosis is characterized by the early fusion of one or many cranial sutures, thus inducing craniofacial abnormalities (1/2500 births worldwide). Most of the cases (85%) of craniosynostosis are sporadic while 15% of cases are syndromic craniosynostoses. Most common syndromic craniosynostoses are characterized by a more severe phenotype and are defined by the early fusion of one or more cranial sutures, these syndromic craniosynostoses are due mostly to mutations in FGFR1, FGFR2, FGFR3 genes, and for some cases to TWIST and MSX2 genes (Johnson and Wilkie 2011; Ornitz and Marie 2002; Twigg et al. 2015). Mutations in the MSX or TWIST1 genes are responsible for Boston-type craniosynostosis and Saethre Chotzen respectively (el Ghouzzi et al. 1997; Jabs et al. 1993). Mutations in FGFR1-3 are responsible for several craniosynostoses entities described below. These craniosynostoses are mediated by a variety of molecular processes. These include ligand-dependent or independent FGFR activation, loss of function, or altered FGFR trafficking (Ibrahimi et al. 2001; Neben and Merrill 2015; Yu et al. 2000; Yu and Ornitz 2001).

8.6.2.1 FGFR3-Related Craniosynostoses

Muenke syndrome (OMIM 602849) is the most frequent craniosynostoses is due to a specific heterozygous FGFR3 mutation with an incidence of 1 in 30,000 births. Of all patients with craniosynostosis, 8% have Muenke syndrome (Boulet et al. 2008; Wilkie 2000). Both sporadic and familial cases have been reported. Increasing paternal age is a contributing factor for de novo mutations (Rannan-Eliya et al. 2004). Characteristics include bi- or unicoronal synostosis, midfacial hypoplasia, and macrocephaly. Some affected individuals have additional features that may include sensorineural hearing loss, developmental delay, brachydactyly, and coned epiphyses in the hands and feet. A unique autosomal dominant missense mutation (P250R) localized in the extracellular domain of FGFR3 is responsible for most of cases of Muenke syndrome (Moloney et al. 1997). The penetrance is reduced and some individuals with P250R mutation have no clinical feature of Muenke syndrome (Morriss-Kay and Wilkie 2005). The mouse model for Muenke syndrome was created by using a Cre-LoxP system. The Pro244Arg mutation in mouse exon 7 of *Fgfr3*, which is homologous to the Pro250Arg mutation in humans, was inserted into the gene with two LoxP sites (Laurita et al. 2011; Twigg et al. 2015). This Muenke disease model displayed a low penetrance phenotype and a less severe phenotype than in human. In addition, the gender bias was inverted from what it is in humans, with men being more affected than women. It is suggested that *Fgfr3* plays a later and more subtle function in membranous ossification in mice compared to *Fgfr2* and *Fgfr1* role since the craniofacial phenotype of $Fgfr3^{P244R/+}$mice is less severe than that of previously reported mouse models for *Fgfr1*- and *Fgfr2*-related craniosynostosis (Laurita et al. 2011; Twigg et al. 2015).

Patients with Crouzon syndrome (OMIM 612247) with acanthosis nigricans are due to a very rare FGFR3 heterozygous gain-of-function mutation (A391E) localized in the transmembrane domain of FGFR3 (prevalence of 1/1000000 newborns (Meyers et al. 1995). Patients have multiple sagittal or coronal fusions causing brachycephaly, trigonocephaly, rare reports of cloverleaf skull malformation, hypertelorism, a midface hypoplasia, beaked nose, and protrusion of the eyes. Characteristic of this disease is the presence of hyperpigmentation of the skin (acanthosis nigricans), hyperkeratosis, and other skin findings. The A391E mutation is a gain-of-function mutation increasing both the activity of the receptor, its phosphorylation, and Mapkinase signaling (Cornille et al. 2022).

The importance of FGFR3 in membranous ossification is highlighted by the cranial vault abnormalities caused by mutations in this receptor in zebrafish model. It is known that zebrafish exhibit significant levels of FGFR3 expression throughout skull formation. To investigate the function of FGFR3 in cranial vault development, the first *fgfr3* loss-of-function zebrafish (*fgfr3$^{lof/lof}$*) was generated. Significant alterations to the craniofacial skeleton were seen in the mutant fish, including ectopic bones, aberrant frontal and parietal bones, and a lack of sutures (Dambroise et al. 2020). Another *fgfr3* loss-of-function zebrafish (*fgfr3$^{lof/lof}$*) was developed using CRISPR/Cas9 technology, to investigate the developmental pathways and potential treatment targets of CATSHL syndrome, the mutants show craniofacial bone malformation with delayed closure of cranial sutures (Sun et al. 2020).

8.6.2.2 FGFR2-Related Craniosynostoses

Apert syndrome (OMIM 101200), Beare-Stevenson cutis gyrata syndrome (OMIM 123790), Crouzon syndrome (OMIM 123500), Jackson Weiss syndrome (OMIM 123150), Pfeiffer syndrome (OMIM 101600), all these syndromes are due to either missense mutation, deletion/duplication in *FGFR2* gene (Wenger et al. 2020). Apert syndrome is a severe form of craniosynostoses characterized by turribrachycephaly, midface retrusion, ocular anomalies, cleft palate, dental anomalies, hearing loss and syndactyly and broad distal phalanx of the thumb or broad distal hallus. Intellectual disability is often observed in these patients. It is a very rare disease (1 in 65,000 to 200,000 births) due to FGFR2 gain-of-function mutation (Pro253Arg, Ser252Trp), the incidence of the disease is significantly increased with paternal age and the penetrance is complete (Goriely et al. 2003). FGFR2 mutation disturbs the FGF signaling thus changing the balance proliferation differentiation and apoptosis of the osteoblasts (Marie et al. 2002).

Along with FGFR2 mutations, changes in the composition of ECM may also be responsible for pathological differences in cranial development and the altered osteogenic process. In fact, there is evidence available regarding the distinct expressions of ECM macromolecules and cytokines in fibroblasts and osteoblasts produced from craniosynostoses (Carinci et al. 2005). It has been shown in different studies that production and secretion of ECM macromolecules is altered in craniosynostoses. For instance, Apert fibroblasts produced higher amounts of GAGs and hyaluronic acid (HA) compared to the normal fibroblasts. The Apert media also showed an increase in the absolute values of HS, chondroitin sulphate (CS), and dermatan sulphate (DS) (Bodo et al. 1997). In addition, as opposed to normal osteoblasts, Apert osteoblasts secrete more type I collagen, fibronectin, and GAGs.

Concerning animal models of this disease, the Ser252Trp mutation in the murine *Fgfr2* gene was expressed ubiquitously, resulting in the creation of the Apert syndrome mouse model (Chen et al. 2003). This model had a robust craniofacial phenotype, although not resembling the long bone phenotype seen in Apert syndrome. The presence of craniosynostosis in the coronal sutures while expressing the mutation makes them a useful model for studying this syndrome (Chen et al. 2003). Another model for Apert syndrome was obtained by introducing the Pro253Arg mutation into the mouse *Fgfr2* gene (Yin et al. 2008). When comparing the $Fgfr2^{P253R/+}$mouse model to the human syndrome phenotype, an intriguing phenotype is shown. As is commonly seen in individuals with the *Fgfr2* Pro253Arg mutation, it exhibits a severe craniofacial phenotype in addition to syndactyly and shortening of long bones, demonstrating the critical function that *Fgfr2* plays in both the endochondral and membranous ossification processes (Yin et al. 2008). Beare-Stevenson cutis gyrata syndrome is characterized by the furrowed skin disorder of cutis gyrata, acanthosis nigricans, craniofacial dysmorphism, ear defect, and digital anomalies. This syndrome is extremely rare and due to heterozygous *FGFR2* mutation (Fonseca et al. 2008). By introducing the Tyr394Cys mutation into the murine *Fgfr2* gene widely, a mouse model of Beare-Stevenson syndrome with cutis gyrata syndrome was created in order to gain a better understanding of the cellular and molecular causes of this syndrome. By 2 weeks of age, $Fgfr2^{Y394C/+}$animals showed decreased growth and mortality (Y. Wang et al. 2012). The skin and craniofacial phenotypes seen in people with Beare-Stevenson syndrome are strongly correlated with the phenotype of this animal model (Wang et al. 2012).

Crouzon syndrome is an autosomal dominant disorder characterized by premature fusion of the coronal sutures thus inducing alterations of the facial bones and facial structure. Others clinical features are hypertelorism, exophthalmos, hypoplastic maxilla, and mandibular prognathism. Heterozygous gain-of-function mutations were identified in *FGFR2* gene (Reardon et al. 1994). Paternal age effect was also reported in Crouzon syndrome family (Goriely et al. 2010). ECM was altered in Crouzon syndrome. In fact, Crouzon fibroblasts secreted a comparatively higher amount of CS but less total GAG, especially HA and HS. Crouzon fibroblasts produced less fibronectin than normal fibroblasts, and their increased protease activity in cells and medium indicates that there is excess fibronectin catabolism (Bodo et al. 2002), the content of type I and III collagen were increased in Crouzon fibroblast medium (Bodo et al. 1996).

Using cre recombinase controlled by the PGK promoter, the Cys342Tyr mutation, which is frequently present in human Crouzon and Pfeiffer syndromes, was inserted into the murine *Fgfr2* gene to create the first mouse model for Crouzon syndrome (Eswarakumar et al. 2004). While reproducing all significant aspects of Crouzon syndrome, $Fgfr2^{C342Y/+}$ mice also revealed an early function for Fgfr2IIIc signaling in osteoblast proliferation as well as a distinct function in chondrocyte and osteoblast lineages (Eswarakumar et al. 2004). In addition to this first model, the N-ethyl-N-nitrosourea (ENU) mutagenesis screen produced a second animal model for Crouzon syndrome. To develop WT, *Fgfr2*$^{W290R/+}$, and *Fgfr2*$^{W290R/W290R}$ animals, animals heterozygous for the *Fgfr2*$^{W290R/+}$ mutation were collected and bred (Mai et al. 2010). The effect of the mutation on both isoforms accounts for the severe phenotype seen in homozygous animals, whereas in prior *Fgfr2*$^{C342Y/+}$mice, only one isoform was modified. Interestingly, the homozygous phenotype is comparable to observations in Pfeiffer syndrome, tracheal and visceral abnormalities, ankyloses of vertebral bodies. Conversely, the phenotype seen in heterozygous animals can be linked to phenotypic features reported in patients with Crouzon syndrome. This discovery emphasizes the model's considerable clinical applicability in comprehending the function of *Fgfr2* in bone organogenesis and development (Eswarakumar et al. 2004).

Jackson Weiss syndrome is characterized by specific head malformations, proptosis and prognathism, hearing loss, and foot malformation and normal hand. A FGFR2 mutation localized in the immunoglobulin like domain was identified in patients with Jackson Weiss syndrome (Jabs et al. 1994).

Pfeiffer syndrome has been classed in three types, Pfeiffer type I, II, III (Lajeunie et al. 1995). Pfeiffer type II is the most severe form, the patients exhibit proptosis, cloverleaf skull, and humero-radio-ulnar synostosis and characterized by marked disabling and early death. Pfeiffer type II is due to FGFR2 missense mutations (Lajeunie et al. 2006).

8.6.2.3 FGFR1-Related Craniosynostoses

Pfeiffer type I is a relatively mild Pfeiffer phenotype and is the most frequent form usually associated with a benign course and satisfactory prognosis. Missense FGFR1 mutation has been identified in families (Muenke et al. 1994). Utilizing homologous recombination to introduce the P250R mutation into the murine Fgfr1 gene, a mouse model for Pfeiffer syndrome was created (Zhou et al. 2000). This mutation represents the human Pro252Arg mutation that causes the mild form of Pfeiffer syndrome. It is important to note that strong craniofacial phenotypes that are compatible with human pathology are presented by this model, which makes it a good model for studies (Zhou et al. 2000).

8.7 Therapeutic Strategies for Targeting FGF Signaling

Dysregulated FGF signaling has been linked to a number of skeletal diseases such as rare skeletal disorders (osteochondrodysplasia) and common skeletal disorders (osteoarthritis). By altering or reversing the aberrant FGF signaling, it offers a promising therapeutic approach for these diseases. Current therapeutic approaches are many and varied; enhancing or blocking FGF signaling and gene therapy are the three main categories into which FGF-based therapeutics are currently in preclinical and clinical trials.

1. Enhancing FGF signaling

Recombinant FGF or FGF analogs have been developed as first-generation strategies to increase the beneficial effects of FGFs/FGFRs, as FGFs are involved in many pathophysiological processes. For instance, osteoarthritis is a highly prevalent skeletal condition affecting 250 millions of people affected around the world. Osteoarthritis is characterized by the degradation of the cartilage; accumulating data suggest that articular chondrocytes undergo a phenotypic shift and become hypertrophic in osteoarthritis (Aigner et al. 1995). FGF signaling pathway is one of the signaling pathways involved in the regulation of chondrocyte homeostasis. Among FGF ligands, FGF18 is well known as a high affinity ligand for FGFR3 and its critical roles in the regulation of skeletal development. FGF18 have a great therapeutic potential for osteoarthritis. The companies Merck and Nordic Bioscience have developed a recombinant human fibroblast growth factor 18 (rhFGF18) called sprifermin for the treatment of knee OA; sprifermin has been identified as a molecule which can protect articular cartilage from degeneration. Today, rhFGF18 has been approved to treat knee cartilage damage and osteoarthritis (phase II clinical trial). (Eckstein et al. 2020; Gigout et al. 2017; Lohmander et al. 2014; Reker et al. 2020).

2. Blocking FGFs signaling pathways

In recent years, in order to reduce the overactivation of FGF signaling pathway observed in skeletal diseases, several strategies have been developed, including Tyrosine Kinase Inhibitors (TKIs), neutralizing monoclonal antibodies (mAb), FGF aptamer, and soluble FGFR (Legeai-Mallet and Savarirayan 2020; Xie et al. 2020)

(a) Tyrosine kinase inhibitors

The use of tyrosine kinase inhibitor that impair the activity of RTK has been proposed as treatment for rare skeletal disorders (Komla-Ebri et al. 2016)) and osteoarthritis (Ferrao Blanco et al. 2021). In this regard, repositioning drugs is an attractive alternative. The best example is BGJ398/Infigratinib developed in first intention for cancer and which is today a therapeutic approach in clinical trial for patients with achondroplasia (Savarirayan, Hoernschemeyer, et al., 2023). Preclinical studies using a mouse model mimicking achondroplasia had demonstrated the efficacy of low dose of Infigratinib on skeleton development in particular on foramen magnum reduction, growth plate cartilage abnormalities and craniofacial anomalies (Komla-Ebri et al. 2016). Infigratinib prevents FGFR3 phosphorylation, which attenuates downstream signaling pathways. It is the only TKI in clinical development for ACH; it is also the only orally administered medicine in clinical trial for this indication. Recently, the results of the phase 2 clinical trial with Infigratinib were published. The data showed that in children with ACH a significant increase in annualized height velocity, with a mean change of 3.38 cm/year compared with the initial value with no side effect (Savarirayan, Hoernschemeyer, et al., 2023).

(b) Monoclonal antibody therapy neutralization

Monoclonal antibodies (mAbs) as a treatment are more precise than other treatments, since these antibodies are specific for one type of antigen and are used to recognize and bind to it with high affinity. The therapeutic potential of mAbs lies in their ability to selectively target diseased cells while avoiding healthy ones; this improves the effectiveness and can reduce some side effects. Generation of specific FGFR antibodies was also developed for cancers. Various studies have reported the efficacy of FGFR3 antibodies (B-701) on bladder cancer lines, which are FGFR3-dependent tumors (Necchi et al. 2019). The mechanism of action of FGFR3 antibodies is to bind to the extra-cellular domain of FGFR3 and to block ligand binding and receptor dimerization. Recently, Sanofi/Genzyme has generated a selective antibody (KC18) for FGFR3 but not for FGFR1, FGFR2, and FGFR4. Preclinical study showed that subcutaneous injections of KC18 increase limb length and improve bone parameters in a mouse model mimicking achondroplasia. Clinical trial is under development, Phase I in healthy volunteers was completed with ClinicalTrials.gov identifier: NCT06067425.

(c) Soluble FGFRs and aptamer

Using molecules with the capacity to bind and neutralize multiple FGF ligands is an alternative method of controlling the activity of the FGF/FGFR signaling (Xie et al. 2020).

For instance, soluble FGFR3 (sFGFR3), a recombinant protein, functions as a decoy receptor to stop FGF ligand attachment to FGFR3. Subcutaneous injection of sFGFR3 (Recifercept) was shown to rescue chondrodysplasia phenotypes in ACH mice in a dose-dependent manner by competing with endogenous FGFR3 ligands (Garcia et al. 2013). Results of an interim analysis of soluble FGFR3 decoy data from a phase II study have indicated, however, that the study would not reach its primary endpoint and the clinical development of Recifercept has been discontinued.

Aptamers are small DNA or RNA molecules able to neutralize the protein interaction by binding specific epitopes. The aptamer (RBM007) targeting FGF2 was developed, RBM007 blocks the binding of FGF2 to its receptors (FGFR1-4), thereby inhibiting signaling pathways downstream of FGFR. FGF2 neutralization by RBM007 restore the defective skeletal growth in Ach transgenic mice (Kimura et al. 2021). A clinical trial using RBM007 has been initiated for patients with achondroplasia.

Finally, it was discovered that a soluble FGFR2 mutant with S252W, known as sFGFR2IIIc (S252W), could mitigate to some extent by preventing the premature closure of the coronal suture in Apert transgenic mice and cultured calvarias (Morita et al. 2014).

(d) Meclozine and (-)- epicatechin

Inhibition of the Mapkinase pathway downstream to FGFR3 is also a therapeutic strategy. Two molecules such as Meclozine and (-)- epicatechin having this property have been evaluated in cellular and *Fgfr3* mouse models. Meclozine, an antihistamine and motion sickness drug downregulated the phosphorylation of ERK and improved the trabecular bone quality and bone volume in ACH transgenic mice (Matsushita et al. 2017). A clinical trial was developed showing that Meclozine was well tolerated by ACH children, with no serious adverse effects (Kitoh et al. 2020).

(-)-Epicatechin, a phenolic compound, isolated from Theobroma cacao, effectively inhibited FGFR3's downstream signaling pathways. Ex vivo and in vivo experiments showed that (-)-epicatechin eliminated bone growth impairment during 6 days of ex vivo culture and daily subcutaneous injections of (-)-epicatechin to $Fgfr3^{Y367C/+}$ mice increased bone elongation. In addition, (-)-Epicatechin treatment modified the length of the primary cilia of chondrocytes (L. Martin et al. 2022).

3. Gene therapy

Gene therapy is a strategy that is widely spread nowadays. As an example of this method, using AAV9 (AAV9-FGFR2-shRNA), a siRNA that specifically targeted the *Fgfr2-P253R* allele was screened. This siRNA attenuated the premature closure of the coronal suture and the reduced calvaria bone volume when it was applied to the skulls in the Apert syndrome mouse model (Luo et al. 2018). Another example is the one targeting the Fgfr3-G374R mutation in achondroplasia in mice. In fact, a GFP reporter system was created to assess the specificity and cutting efficiency of single guide RNAs (sgRNAs). Then, a targeted therapy of achondroplasia in mice using the sgRNA chosen based on the GFP reporter system was performed. Using Cas9 protein instead of Cas9 mRNA, the Fgfr3-G374R mutation was corrected with a higher frequency of precision. It has been shown also that oligos containing 100 and 200 nucleotides could be targeted to accurately and equally efficiently fix the mutation. This leads to the conclusion that this approach totally inhibited achondroplasia traits, and whole genome sequencing revealed no off-targets effects (Miao et al. 2019).

4. CNP analogs

The C-natriuretic peptide (CNP) and its receptor, natriuretic peptide receptor B (NPR-B), are recognized as key regulators of longitudinal bone growth. An analog of CNP, BMN111 (vosoritide), with an extended plasma half -life due to its resistance to neutral endopeptidase, was generated. Proof of principle studies were conducted in human ACH cells and a mouse model ($Fgfr3^{Y367C/+}$) and confirmed the beneficial effect of BMN111 on long bone growth and the skull (Lorget et al. 2012). Another peptide, TransCon CNP (navepegritide) which is an investigational prodrug of CNP, was designed to provide sustained release of CNP supporting continuous exposure with a once-weekly dosing regimen (Breinholt et al. 2019). Several clinical trials (phase II and III) were conducted in patients with achondroplasia using vosoritide (Savarirayan et al. 2023) or TransCon CNP (Savarirayan, Hoernschemeyer, et al., 2023). At present, only vosoritide (VOXZOGO™), a daily injectable type C natriuretic peptide analog that blocks downstream signaling pathways, is approved in Europe and the USA, as treatment for achondroplasia patients.

8.8 Conclusions and Perspectives

In this chapter, we summarized the function of 22 FGF ligands, the role of FGFR and two co-receptors (HS and klotho) present in ECM. Our knowledge on FGFR-FGF signaling has advanced considerably over the past decades. FGF signaling controls a large spectrum of cellular function including migration, proliferation, differentiation, and survival and plays an essential role in skeletal development and tissue and ECM homeostasis. The spatiotemporal and tissue specific expression of the FGF ligand modulate the signaling pathways and interaction with other partners. The affinity of the ligand for one or for several receptors and the interaction of the coreceptor activate various signaling pathway but the precise mechanisms of these cascades of events need to be clarified. It is well known that FGFR-FGF signaling pathway participates to the pathophysiology of many disorders in human, in particular skeletal disease, metabolic disease, and degenerative disease. FGFR-FGF have also oncogenic function ; aberrant FGF signaling is characterized in almost all cancer types (bladder cancer, breast cancer, glioblastoma, gastrointestinal cancer). It is involved in major step of tumor progression, in particular cell proliferation, angiogenesis and ECM structure and metastatic dissemination. Today, it remains to be determined the exact role of these signaling in tumorigenesis in a particular context.

The role of FGFR-FGF in skeletal disorders is better defined. The expression of FGFs, FGFRs and co-receptor during the different steps of the skeleton development have been well studied, the dysfunctions of their signaling are responsible for the most common chondrodysplasia (e.g., ACH, HCH) and craniosynostoses (e.g., Muenke syndrome, Crouzon syndrome). FGF signaling activates many downstream signaling pathways tightly regulate by diverse transduction specificity and ECM factors. The identification of gain-of-function and loss-of-function mutations in the genes encoding FGF-FGFR has provided a better understanding of the mechanism of action of these signaling pathways in bone and cartilage. Subsequently, this made it possible to generate and study specific cellular models expressing these FGF and FGFR genes providing further information on the different molecular pathways activated or not. Specific and relevant mouse and zebrafish models also provided a lot of information about the development and the formation of the skeleton. These models were used to study, among other things, ECM network, endplate and articular cartilage, bone formation and elongation, bone repair, the synchondroses formation, and fusion of the sutures. A good understanding of the mechanisms of action of this FGF-FGFR/ co-receptor in specific tissues is a key factor in the development of new therapeutic approaches. The identification of specific antibodies, tyrosine kinase inhibitors, small peptides capable of inhibiting the pathological function of FGF or FGFR will permit to consider treatments for either genetic disease or cancers. Several drugs are on clinical trials for rare skeletal disorder such as tyrosine kinase inhibitors, antibodies, aptamer, repurposing drugs; today Voxzogo is the first medicine that has been approved for treating achondroplasia. In the future, new treatments will emerge for FGFR-related dysplasia. Among these, targeting the different components of ECM (Collagen, fibronectin, HS, PG…) could be a therapeutic option for patients.

Acknowledgements This work was supported by the European Community's Horizon-MSCA-DN-01-01 under grant agreement n. 101072766 (CHANGE). Research fund AXA and the Philanthropy Department of Mutuelles AXA through the Head and Heart Chair.

Conflict of Interest The authors declare that they have no conflict of interest.

References

Aigner T, Dietz U, Stöss H, von der Mark K (1995) Differential expression of collagen types I, II, III, and X in human osteophytes. Lab Invest 73(2):236–243

Aigner T, Soeder S, Haag J (2006) IL-1beta and BMPs—interactive players of cartilage matrix degradation and regeneration. Eur Cell Mater 12:49–56.; discussion 56. https://doi.org/10.22203/ecm.v012a06

Allouche M, Bikfalvi A (1995) The role of fibroblast growth factor-2 (FGF-2) in hematopoiesis. Prog Growth Factor Res 6(1):35–48. https://doi.org/10.1016/0955-2235(95)00041-0

Alpdogan Ö, Hubbard VM, Smith OM, Patel N, Lu S, Goldberg GL, Gray DH, Feinman J, Kochman AA, Eng JM, Suh D, Muriglan SJ, Boyd RL, Van Den Brink MRM (2006) Keratinocyte growth factor (KGF) is required for postnatal thymic regeneration. Blood 107(6):2453–2460. https://doi.org/10.1182/blood-2005-07-2831

Armand A-S, Laziz I, Chanoine C (2006) FGF6 in myogenesis. Biochim Biophys Acta 1763(8):773–778. https://doi.org/10.1016/j.bbamcr.2006.06.005

Baird A (1994) Fibroblast growth factors: activities and significance of non-neurotrophin neurotrophic growth factors. Curr Opin Neurobiol 4(1):78–86. https://doi.org/10.1016/0959-4388(94)90035-3

Barak H, Huh S-H, Chen S, Jeanpierre C, Martinovic J, Parisot M, Bole-Feysot C, Nitschké P, Salomon R, Antignac C, Ornitz DM, Kopan R (2012) FGF9 and FGF20 maintain the stemness of nephron progenitors in mice and man. Dev Cell 22(6):1191–1207. https://doi.org/10.1016/j.devcel.2012.04.018

Basilico C, Moscatelli D (1992) The Fgf family of growth factors and oncogenes. In: Advances in cancer research, vol 59. Elsevier, pp 115–165. https://doi.org/10.1016/S0065-230X(08)60305-X

Baujat G, Legeai-Mallet L, Finidori G, Cormier-Daire V, Le Merrer M (2008) Achondroplasia. Best Pract Res Clin Rheumatol 22(1):3–18. https://doi.org/10.1016/j.berh.2007.12.008

Beenken A, Mohammadi M (2009) The FGF family: biology, pathophysiology and therapy. Nat Rev Drug Discov 8(3):235–253. https://doi.org/10.1038/nrd2792

Bellus GA, Bamshad MJ, Przylepa KA, Dorst J, Lee RR, Hurko O, Jabs EW, Curry CJ, Wilcox WR, Lachman RS, Rimoin DL, Francomano CA (1999) Severe achondroplasia with developmental delay and acanthosis nigricans (SADDAN): Phenotypic analysis of a new skeletal dysplasia caused by a Lys650Met mutation in fibroblast growth factor receptor 3. Am J Med Genet 85(1):53–65

Bellus GA, McIntosh I, Smith EA, Aylsworth AS, Kaitila I, Horton WA, Greenhaw GA, Hecht JT, Francomano CA (1995) A recurrent mutation in the tyrosine kinase domain of fibroblast growth factor receptor 3 causes hypochondroplasia. Nat Genet 10(3):357–359. https://doi.org/10.1038/ng0795-357

Bellus GA, Spector EB, Speiser PW, Weaver CA, Garber AT, Bryke CR, Israel J, Rosengren SS, Webster MK, Donoghue DJ, Francomano CA (2000) Distinct missense mutations of the FGFR3 lys650 codon modulate receptor kinase activation and the severity of the skeletal dysplasia phenotype. Am J Hum Genet 67(6):1411–1421. https://doi.org/10.1086/316892

Bellusci S, Grindley J, Emoto H, Itoh N, Hogan BLM (1997) Fibroblast growth factor 10 (FGF10) and branching morphogenesis in the embryonic mouse lung. Development 124(23):4867–4878. https://doi.org/10.1242/dev.124.23.4867

Belov AA, Mohammadi M (2013) Molecular mechanisms of fibroblast growth factor signaling in physiology and pathology. Cold Spring Harb Perspect Biol 5(6):a015958–a015958. https://doi.org/10.1101/cshperspect.a015958

Bénazet J-D, Zeller R (2009) Vertebrate limb development: moving from classical morphogen gradients to an integrated 4-dimensional patterning system. Cold Spring Harb Perspect Biol 1(4):a001339. https://doi.org/10.1101/cshperspect.a001339

Bikfalvi A, Han ZC (1994) Angiogenic factors are hematopoietic growth factors and vice versa. Leukemia 8(3):523–529

Bober M, Bellus G, Nikkel S, Tiller G (2020) Hypochondroplasia. Gene reviews

Bodo M, Carinci F, Baroni T, Becchetti E, Giammarioli M, Bellucci C, Pezzetti F, Evangelisti R, Carinci P (1996) Effect of interleukins on Crouzon fibroblast phenotype in vitro. Release of cytokines and IL-6 mRNA expression. Cytokine 8(10):772–783. https://doi.org/10.1006/cyto.1996.0103

Bodo M, Carinci F, Baroni T, Giammarioli M, Bellucci C, Bosi G, Pezzetti F, Becchetti E, Evangelisti R, Carinci P (1997) Apert's syndrome: differential *in vitro* production of matrix macromolecules and its regulation by interleukins. Eur J Clin Investig 27(1):36–42. https://doi.org/10.1046/j.1365-2362.1997.660618.x

Bodo M, Lilli C, Aisa MC, Scapoli L, Bellucci C, Rinaldi E, Tosi L, Baroni T, Conte C, Bellocchio S, Carinci F, Stabellini G, Carinci P (2002) Basic fibroblast growth factor: effects on matrix remodeling, receptor expression, and transduction pathway in human periosteal fibroblasts with FGFR2 gene mutation. J Interf Cytokine Res 22(6):621–630. https://doi.org/10.1089/10799900260100105

Bonaventure J, Rousseau F, Legeai-Mallet L, Le Merrer M, Munnich A, Maroteaux P (1996) Common mutations in the fibroblast growth factor receptor 3 (FGFR 3) gene account for achondroplasia, hypochondroplasia, and thanatophoric dwarfism. Am J Med Genet 63(1):148–154. https://doi.org/10.1002/(SICI)1096-8628(19960503)63:1<148::AID-AJMG26>3.0.CO;2-N

Bosetti M, Leigheb M, Brooks RA, Boccafoschi F, Cannas MF (2010) Regulation of osteoblast and osteoclast functions by FGF-6. J Cell Physiol 225(2):466–471. https://doi.org/10.1002/jcp.22225

Boulet SL, Rasmussen SA, Honein MA (2008) A population-based study of craniosynostosis in metropolitan Atlanta, 1989-2003. Am J Med Genet A 146A(8):984–991. https://doi.org/10.1002/ajmg.a.32208

Breinholt VM, Rasmussen CE, Mygind PH, Kjelgaard-Hansen M, Faltinger F, Bernhard A, Zettler J, Hersel U (2019) TransCon CNP, a sustained-release C-type natriuretic peptide prodrug, a potentially safe and efficacious new therapeutic modality for the treatment of comorbidities associated with fibroblast growth factor receptor 3-related skeletal Dysplasias. J Pharmacol Exp Ther 370(3):459–471. https://doi.org/10.1124/jpet.119.258251

Brewer JR, Mazot P, Soriano P (2016) Genetic insights into the mechanisms of Fgf signaling. Genes Dev 30(7):751–771. https://doi.org/10.1101/gad.277137.115

Britto JA, Evans RD, Hayward RD, Jones BM (2001) From genotype to phenotype: the differential expression of FGF, FGFR, and TGFbeta genes characterizes human cranioskeletal development and reflects clinical presentation in FGFR syndromes. Plast Reconstr Surg 108(7):2026–2039.; discussion 2040-2046. https://doi.org/10.1097/00006534-200112000-00030

Brodie SG, Kitoh H, Lachman RS, Nolasco LM, Mekikian PB, Wilcox WR (1999) Platyspondylic lethal skeletal dysplasia, San Diego type, is caused by FGFR3 mutations. Am J Med Genet 84(5):476–480

Camps M, Nichols A, Gillieron C, Antonsson B, Muda M, Chabert C, Boschert U, Arkinstall S (1998) Catalytic activation of the phosphatase MKP-3 by ERK2 mitogen-activated protein kinase. Science 280(5367):1262–1265. https://doi.org/10.1126/science.280.5367.1262

Carinci F, Pezzetti F, Locci P, Becchetti E, Carls F, Avantaggiato A, Becchetti A, Carinci P, Baroni T, Bodo M (2005) Apert and Crouzon syndromes: clinical findings, genes and extracellular matrix. J Craniofac Surg 16(3):361–368. https://doi.org/10.1097/01.SCS.0000157078.53871.11

Carter EP, Fearon AE, Grose RP (2015) Careless talk costs lives: fibroblast growth factor receptor signalling and the consequences of pathway malfunction. Trends Cell Biol 25(4):221–233. https://doi.org/10.1016/j.tcb.2014.11.003

Château M-T, Araiz C, Descamps S, Galas S (2010) Klotho interferes with a novel FGF-signalling pathway and insulin/Igf-like signalling to improve longevity and stress resistance in Caenorhabditis elegans. Aging 2(9):567–581. https://doi.org/10.18632/aging.100195

Chen G, Liu Y, Goetz R, Fu L, Jayaraman S, Hu M-C, Moe OW, Liang G, Li X, Mohammadi M (2018) α-Klotho is a non-enzymatic molecular scaffold for FGF23 hormone signalling. Nature 553(7689):461–466. https://doi.org/10.1038/nature25451

Chen L, Li D, Li C, Engel A, Deng C-X (2003) A Ser252Trp [corrected] substitution in mouse fibroblast growth factor receptor 2 (Fgfr2) results in craniosynostosis. Bone 33(2):169–178. https://doi.org/10.1016/s8756-3282(03)00222-9

Cho YM, Woodard GL, Dunbar M, Gocken T, Jimènez JA, Foley J (2003) Hair-cycle-dependent expression of parathyroid hormone-related protein and its type I receptor: evidence for regulation at the Anagen to Catagen transition. J Invest Dermatol 120(5):715–727. https://doi.org/10.1046/j.1523-1747.2003.12147.x

Cholfin JA, Rubenstein JLR (2007) Patterning of frontal cortex subdivisions by Fgf17. Proc Natl Acad Sci USA 104(18):7652–7657. https://doi.org/10.1073/pnas.0702225104

Clements DA, Wang JK, Dionne CA, Goldfarb M (1993) Activation of fibroblast growth factor (FGF) receptors by recombinant human FGF-5. Oncogene 8(5):1311–1316

Colvin JS, White AC, Pratt SJ, Ornitz DM (2001) Lung hypoplasia and neonatal death in Fgf9-null mice identify this gene as an essential regulator of lung mesenchyme. Development 128(11):2095–2106. https://doi.org/10.1242/dev.128.11.2095

Cornille M, Moriceau S, Khonsari RH, Heuzé Y, Loisay L, Boitez V, Morice A, Arnaud E, Collet C, Bensidhoum M, Kaci N, Boddaert N, Paternoster G, Rauschendorfer T, Werner S, Mansour SL, Di Rocco F, Oury F, Legeai-Mallet L (2022) FGFR3 overactivation in the brain is responsible for memory impairments in Crouzon syndrome mouse model. J Exp Med 219(4):e20201879. https://doi.org/10.1084/jem.20201879

Coutu DL, François M, Galipeau J (2011) Inhibition of cellular senescence by developmentally regulated FGF receptors in mesenchymal stem cells. Blood 117(25):6801–6812. https://doi.org/10.1182/blood-2010-12-321539

Dailey L, Laplantine E, Priore R, Basilico C (2003) A network of transcriptional and signaling events is activated by FGF to induce chondrocyte growth arrest and differentiation. J Cell Biol 161(6):1053–1066. https://doi.org/10.1083/jcb.200302075

Dambroise E, Ktorza I, Brombin A, Abdessalem G, Edouard J, Luka M, Fiedler I, Binder O, Pelle O, Patton EE, Busse B, Menager M, Sohm F, Legeai-Mallet L (2020) FGFR3 is a positive regulator of osteoblast expansion and differentiation during zebrafish skull vault development. J Bone Miner Res 35(9):1782–1797. https://doi.org/10.1002/jbmr.4042

Delezoide A-L, Benoist-Lasselin C, Legeai-Mallet L, Le Merrer M, Munnich A, Vekemans M, Bonaventure J (1998) Spatio-temporal expression of FGFR 1, 2 and 3 genes during human embryo-fetal ossification. Mech Dev 77(1):19–30. https://doi.org/10.1016/S0925-4773(98)00133-6

Deng C, Wynshaw-Boris A, Zhou F, Kuo A, Leder P (1996) Fibroblast growth factor receptor 3 is a negative regulator of bone growth. Cell 84(6):911–921. https://doi.org/10.1016/s0092-8674(00)81069-7

Eckstein F, Kraines JL, Aydemir A, Wirth W, Maschek S, Hochberg MC (2020) Intra-articular sprifermin reduces cartilage loss in addition to increasing cartilage gain independent of location in the femorotibial joint: post-hoc analysis of a randomised, placebo-controlled phase II clinical trial. Ann Rheum Dis 79(4):525–528. https://doi.org/10.1136/annrheumdis-2019-216453

el Ghouzzi V, Le Merrer M, Perrin-Schmitt F, Lajeunie E, Benit P, Renier D, Bourgeois P, Bolcato-Bellemin AL, Munnich A, Bonaventure J (1997) Mutations of the TWIST gene in the Saethre-Chotzen syndrome. Nat Genet 15(1):42–46. https://doi.org/10.1038/ng0197-42

Estienne A, Price CA (2018) The fibroblast growth factor 8 family in the female reproductive tract. Reproduction 155(1):R53–R62. https://doi.org/10.1530/REP-17-0542

Eswarakumar VP, Horowitz MC, Locklin R, Morriss-Kay GM, Lonai P (2004) A gain-of-function mutation of *Fgfr2c* demonstrates the roles of this receptor variant in osteogenesis. Proc Natl Acad Sci 101(34):12555–12560. https://doi.org/10.1073/pnas.0405031101

Eswarakumar VP, Monsonego-Ornan E, Pines M, Antonopoulou I, Morriss-Kay GM, Lonai P (2002) The IIIc alternative of Fgfr2 is a positive regulator of bone formation. Development 129(16):3783–3793. https://doi.org/10.1242/dev.129.16.3783

Fallon JF, López A, Ros MA, Savage MP, Olwin BB, Simandl BK (1994) FGF-2: Apical ectodermal ridge growth signal for chick limb development. Science 264(5155):104–107. https://doi.org/10.1126/science.7908145

Farrell B, Breeze AL (2018) Structure, activation and dysregulation of fibroblast growth factor receptor kinases: Perspectives for clinical targeting. Biochemical Society Transactions, 46(6):1753–1770. https://doi.org/10.1042/BST20180004

Faul C, Amaral AP, Oskouei B, Hu M-C, Sloan A, Isakova T, Gutiérrez OM, Aguillon-Prada R, Lincoln J, Hare JM, Mundel P, Morales A, Scialla J, Fischer M, Soliman EZ, Chen J, Go AS, Rosas SE, Nessel L et al (2011) FGF23 induces left ventricular hypertrophy. J Clin Invest 121(11):4393–4408. https://doi.org/10.1172/JCI46122

Ferrao Blanco MN, Bastiaansen-Jenniskens YM, Chambers MG, Pitsillides AA, Narcisi R, van Osch GJVM (2021) Effect of inflammatory signaling on human articular chondrocyte hypertrophy: potential involvement of tissue repair macrophages. Cartilage 13(2_suppl):168S–174S. https://doi.org/10.1177/19476035211021907

Finch PW, Rubin JS, Miki T, Ron D, Aaronson SA (1989) Human KGF is FGF-related with properties of a paracrine effector of epithelial cell growth. Science 245(4919):752–755. https://doi.org/10.1126/science.2475908

Fisher FM, Maratos-Flier E (2016) Understanding the physiology of FGF21. Annu Rev Physiol 78:223–241. https://doi.org/10.1146/annurev-physiol-021115-105339

Foldynova-Trantirkova S, Wilcox WR, Krejci P (2012) Sixteen years and counting: the current understanding of fibroblast growth factor receptor 3 (FGFR3) signaling in skeletal dysplasias. Hum Mutat 33(1):29–41. https://doi.org/10.1002/humu.21636

Fonseca R, Costa-Lima MA, Cosentino V, Orioli IM (2008) Second case of Beare-Stevenson syndrome with an FGFR2 Ser372Cys mutation. Am J Med Genet A 146A(5):658–660. https://doi.org/10.1002/ajmg.a.32176

Garcia S, Dirat B, Tognacci T, Rochet N, Mouska X, Bonnafous S, Patouraux S, Tran A, Gual P, Le Marchand-Brustel Y, Gennero I, Gouze E (2013) Postnatal soluble FGFR3 therapy rescues achondroplasia symptoms and restores bone growth in mice. Sci Transl Med 5(203):203ra124. https://doi.org/10.1126/scitranslmed.3006247

Geng L, Lam KSL, Xu A (2020) The therapeutic potential of FGF21 in metabolic diseases: from bench to clinic. Nat Rev Endocrinol 16(11):654–667. https://doi.org/10.1038/s41574-020-0386-0

Gentili C, Cancedda R (2009) Cartilage and bone extracellular matrix. Curr Pharm Des 15(12):1334–1348. https://doi.org/10.2174/138161209787846739

Gigout A, Guehring H, Froemel D, Meurer A, Ladel C, Reker D, Bay-Jensen AC, Karsdal MA, Lindemann S (2017) Sprifermin (rhFGF18) enables proliferation of chondrocytes producing a hyaline cartilage matrix. Osteoarthr Cartil 25(11):1858–1867. https://doi.org/10.1016/j.joca.2017.08.004

Goldfarb M, Schoorlemmer J, Williams A, Diwakar S, Wang Q, Huang X, Giza J, Tchetchik D, Kelley K, Vega A, Matthews G, Rossi P, Ornitz DM, D'Angelo E (2007) Fibroblast growth factor homologous factors control neuronal excitability through modulation of voltage-gated sodium channels. Neuron 55(3):449–463. https://doi.org/10.1016/j.neuron.2007.07.006

Goriely A, Lord H, Lim J, Johnson D, Lester T, Firth HV, Wilkie AOM (2010) Germline and somatic mosaicism for FGFR2 mutation in the mother of a child with Crouzon syndrome: implications for genetic testing in « paternal age-effect » syndromes. Am J Med Genet A 152A(8):2067–2073. https://doi.org/10.1002/ajmg.a.33513

Goriely A, McVean GAT, Röjmyr M, Ingemarsson B, Wilkie AOM (2003) Evidence for selective advantage of pathogenic FGFR2 mutations in the male germ line. Science 301(5633):643–646. https://doi.org/10.1126/science.1085710

Goriely A, Wilkie AOM (2012) Paternal age effect mutations and selfish spermatogonial selection: causes and consequences for human disease. Am J Hum Genet 90(2):175–200. https://doi.org/10.1016/j.ajhg.2011.12.017

Gotoh N (2008) Regulation of growth factor signaling by FRS2 family docking/scaffold adaptor proteins. Cancer Science, 99(7):1319–1325. https://doi.org/10.1111/j.1349-7006.2008.00840.x

Grabner A, Amaral AP, Schramm K, Singh S, Sloan A, Yanucil C, Li J, Shehadeh LA, Hare JM, David V, Martin A, Fornoni A, Di Marco GS, Kentrup D, Reuter S, Mayer AB, Pavenstädt H, Stypmann J, Kuhn C et al (2015) Activation of cardiac fibroblast growth factor receptor 4 causes left ventricular hypertrophy. Cell Metab 22(6):1020–1032. https://doi.org/10.1016/j.cmet.2015.09.002

Grose R, Dickson C (2005) Fibroblast growth factor signaling in tumorigenesis. Cytokine Growth Factor Rev 16(2):179–186. https://doi.org/10.1016/j.cytogfr.2005.01.003

Hamada T, Suda N, Kuroda T (1999) Immunohistochemical localization of fibroblast growth factor receptors in the rat mandibular condylar cartilage and tibial cartilage. J Bone Miner Metab 17(4):274–282. https://doi.org/10.1007/s007740050095

Hanafusa H, Torii S, Yasunaga T, Nishida E (2002) Sprouty1 and Sprouty2 provide a control mechanism for the Ras/MAPK signalling pathway. Nat Cell Biol 4(11):850–858. https://doi.org/10.1038/ncb867

Hayashi T, Ray CA, Bermingham-McDonogh O (2008) Fgf20 is required for sensory epithelial specification in the developing cochlea. J Neurosci Off J Soc Neurosci 28(23):5991–5999. https://doi.org/10.1523/JNEUROSCI.1690-08.2008

Hecht D, Zimmerman N, Bedford M, Avivi A, Yayon A (1995) Identification of fibroblast growth factor 9 (FGF9) as a high affinity, heparin dependent ligand for FGF receptors 3 and 2 but not for FGF receptors 1 and 4. Growth Factors 12(3):223–233. https://doi.org/10.3109/08977199509036882

Heuertz S, Le Merrer M, Zabel B, Wright M, Legeai-Mallet L, Cormier-Daire V, Gibbs L, Bonaventure J (2006) Novel FGFR3 mutations creating cysteine residues in the extracellular domain of the receptor cause achondroplasia or severe forms of hypochondroplasia. Eur J Hum Genet 14(12):1240–1247. https://doi.org/10.1038/sj.ejhg.5201700

Hoffman MP, Kidder BL, Steinberg ZL, Lakhani S, Ho S, Kleinman HK, Larsen M (2002) Gene expression profiles of mouse submandibular gland development: FGFR1 regulates branching morphogenesis in vitro through BMP- and FGF-dependent mechanisms. Development 129(24):5767–5778. https://doi.org/10.1242/dev.00172

Horton WA, Hall JG, Hecht JT (2007) Achondroplasia. Lancet 370(9582):162–172. https://doi.org/10.1016/S0140-6736(07)61090-3

Hotta Y, Sasaki S, Konishi M, Kinoshita H, Kuwahara K, Nakao K, Itoh N (2008) Fgf16 is required for cardiomyocyte proliferation in the mouse embryonic heart. Dev Dyn 237(10):2947–2954. https://doi.org/10.1002/dvdy.21726

Hsu W-CJ, Nilsson CL, Laezza F (2014) Role of the axonal initial segment in psychiatric disorders: function, dysfunction, and intervention. Front Psych 5:109. https://doi.org/10.3389/fpsyt.2014.00109

Huang Z, Marsiglia WM, Basu Roy U, Rahimi N, Ilghari D, Wang H, Chen H, Gai W, Blais S, Neubert TA, Mansukhani A, Traaseth NJ, Li X, Mohammadi M (2016) Two FGF receptor kinase molecules act in concert to recruit and Transphosphorylate phospholipase Cγ. Mol Cell 61(1):98–110. https://doi.org/10.1016/j.molcel.2015.11.010

Hui Q, Jin Z, Li X, Liu C, Wang X (2018) FGF family: from drug development to clinical application. Int J Mol Sci 19(7):1875. https://doi.org/10.3390/ijms19071875

Hynes RO (2009) The extracellular matrix: not just pretty fibrils. Science 326(5957):1216–1219. https://doi.org/10.1126/science.1176009

Ibrahimi OA, Eliseenkova AV, Plotnikov AN, Yu K, Ornitz DM, Mohammadi M (2001) Structural basis for fibroblast growth factor receptor 2 activation in Apert syndrome. Proc Natl Acad Sci USA 98(13):7182–7187. https://doi.org/10.1073/pnas.121183798

Iijima H, Gilmer G, Wang K, Bean AC, He Y, Lin H, Tang W-Y, Lamont D, Tai C, Ito A, Jones JJ, Evans C, Ambrosio F (2023) Age-related matrix stiffening epigenetically regulates α-klotho expression and compromises chondrocyte integrity. Nat Commun 14(1):18. https://doi.org/10.1038/s41467-022-35359-2

Itoh N (2016) FGF10: A multifunctional mesenchymal–epithelial signaling growth factor in development, health, and disease. Cytokine Growth Factor Rev 28:63–69. https://doi.org/10.1016/j.cytogfr.2015.10.001

Itoh N, Ohta H (2013) Roles of FGF20 in dopaminergic neurons and Parkinson's disease. Front Mol Neurosci 6:15. https://doi.org/10.3389/fnmol.2013.00015

Itoh N, Ohta H, Konishi M (2015) Endocrine FGFs: evolution, physiology, pathophysiology, and pharmacotherapy. Front Endocrinol 6:154. https://doi.org/10.3389/fendo.2015.00154

Itoh N, Ornitz DM (2008) Functional evolutionary history of the mouseFgf gene family. Dev Dyn 237(1):18–27. https://doi.org/10.1002/dvdy.21388

Izvolsky KI, Shoykhet D, Yang Y, Yu Q, Nugent MA, Cardoso WV (2003) Heparan sulfate–FGF10 interactions during lung morphogenesis. Dev Biol 258(1):185–200. https://doi.org/10.1016/S0012-1606(03)00114-3

Jabs EW, Li X, Scott AF, Meyers G, Chen W, Eccles M, Mao JI, Charnas LR, Jackson CE, Jaye M (1994) Jackson-Weiss and Crouzon syndromes are allelic with mutations in fibroblast growth factor receptor 2. Nat Genet 8(3):275–279. https://doi.org/10.1038/ng1194-275

Jabs EW, Müller U, Li X, Ma L, Luo W, Haworth IS, Klisak I, Sparkes R, Warman ML, Mulliken JB (1993) A mutation in the homeodomain of the human MSX2 gene in a family affected with autosomal dominant craniosynostosis. Cell 75(3):443–450. https://doi.org/10.1016/0092-8674(93)90379-5

Jacob AL, Smith C, Partanen J, Ornitz DM (2006) Fibroblast growth factor receptor 1 signaling in the osteo-chondrogenic cell lineage regulates sequential steps of osteoblast maturation. Dev Biol 296(2):315–328. https://doi.org/10.1016/j.ydbio.2006.05.031

Johnson D, Wilkie AOM (2011) Craniosynostosis. Eur J Hum Genet 19(4):369–376. https://doi.org/10.1038/ejhg.2010.235

Jump SS, Childs TE, Zwetsloot KA, Booth FW, Lees SJ (2009) Fibroblast growth factor 2-stimulated proliferation is lower in muscle precursor cells from old rats: fibroblast growth factor 2 in ageing. Exp Physiol 94(6):739–748. https://doi.org/10.1113/expphysiol.2008.046136

Jüppner H (2011) Phosphate and FGF-23. Kidney Int Suppl 79(121):S24–S27. https://doi.org/10.1038/ki.2011.27

Karolak MR, Yang X, Elefteriou F (2015) FGFR1 signaling in hypertrophic chondrocytes is attenuated by the Ras-GAP neurofibromin during endochondral bone formation. Hum Mol Genet 24(9):2552–2564. https://doi.org/10.1093/hmg/ddv019

Karuppaiah K, Yu K, Lim J, Chen J, Smith C, Long F, Ornitz DM (2016) FGF signaling in the osteoprogenitor lineage non-autonomously regulates postnatal chondrocyte proliferation and skeletal growth. Development 143(10):1811–1822. https://doi.org/10.1242/dev.131722

Kettunen P, Furmanek T, Chaulagain R, Hals Kvinnsland I, Luukko K (2011) Developmentally regulated expression of intracellular Fgf11-13, hormone-like Fgf15 and canonical Fgf16, −17 and −20 mRNAs in the developing mouse molar tooth. Acta Odontol Scand 69(6):360–366. https://doi.org/10.3109/00016357.2011.568968

Kim HJ, Rice DP, Kettunen PJ, Thesleff I (1998) FGF-, BMP- and Shh-mediated signalling pathways in the regulation of cranial suture morphogenesis and calvarial bone development. Development 125(7):1241–1251. https://doi.org/10.1242/dev.125.7.1241

Kim Y, Kobayashi A, Sekido R, DiNapoli L, Brennan J, Chaboissier M-C, Poulat F, Behringer RR, Lovell-Badge R, Capel B (2006) Fgf9 and Wnt4 act as antagonistic signals to regulate mammalian sex determination. PLoS Biol 4(6):e187. https://doi.org/10.1371/journal.pbio.0040187

Kimura T, Bosakova M, Nonaka Y, Hruba E, Yasuda K, Futakawa S, Kubota T, Fafilek B, Gregor T, Abraham SP, Gomolkova R, Belaskova S, Pesl M, Csukasi F, Duran I, Fujiwara M, Kavkova M, Zikmund T, Kaiser J et al (2021) An RNA aptamer restores defective bone growth in FGFR3-related skeletal dysplasia in mice. Sci Transl Med 13(592):eaba 4226. https://doi.org/10.1126/scitranslmed.aba4226

Kitoh H, Matsushita M, Mishima K, Nagata T, Kamiya Y, Ueda K, Kuwatsuka Y, Morikawa H, Nakai Y, Ishiguro N (2020) Pharmacokinetics and safety after once and twice a day doses of meclizine hydrochloride administered to children with achondroplasia. PloS One 15(4):e0229639. https://doi.org/10.1371/journal.pone.0229639

Knights V, Cook SJ (2010) De-regulated FGF receptors as therapeutic targets in cancer. Pharmacol Ther 125(1):105–117. https://doi.org/10.1016/j.pharmthera.2009.10.001

Kokkinos J, Tang S, Rye K-A, Ong KL (2017) The role of fibroblast growth factor 21 in atherosclerosis. Atherosclerosis 257:259–265. https://doi.org/10.1016/j.atherosclerosis.2016.11.033

Komla-Ebri D, Dambroise E, Kramer I, Benoist-Lasselin C, Kaci N, Le Gall C, Martin L, Busca P, Barbault F, Graus-Porta D, Munnich A, Kneissel M, Di Rocco F, Biosse-Duplan M, Legeai-Mallet L (2016) Tyrosine kinase inhibitor NVP-BGJ398 functionally improves FGFR3-related dwarfism in mouse model. J Clin Invest 126(5):1871–1884. https://doi.org/10.1172/JCI83926

Kouhara H, Hadari YR, Spivak-Kroizman T, Schilling J, Bar-Sagi D, Lax I, Schlessinger J (1997) A lipid-anchored Grb2-binding protein that links FGF-receptor activation to the Ras/MAPK signaling pathway. Cell 89(5):693–702. https://doi.org/10.1016/s0092-8674(00)80252-4

Krajisnik T, Björklund P, Marsell R, Ljunggren Ö, Åkerström G, Jonsson KB, Westin G, Larsson TE (2007) Fibroblast growth factor-23 regulates parathyroid hormone and 1α-hydroxylase expression in cultured bovine parathyroid cells. J Endocrinol 195(1):125–131. https://doi.org/10.1677/JOE-07-0267

Krejci P, Bryja V, Pachernik J, Hampl A, Pogue R, Mekikian P, Wilcox WR (2004) FGF2 inhibits proliferation and alters the cartilage-like phenotype of RCS cells. Exp Cell Res 297(1):152–164. https://doi.org/10.1016/j.yexcr.2004.03.011

Krejci P, Masri B, Fontaine V, Mekikian PB, Weis M, Prats H, Wilcox WR (2005) Interaction of fibroblast growth factor and C-natriuretic peptide signaling in regulation of chondrocyte proliferation and extracellular matrix homeostasis. J Cell Sci 118(21):5089–5100. https://doi.org/10.1242/jcs.02618

Krejci P, Masri B, Salazar L, Farrington-Rock C, Prats H, Thompson LM, Wilcox WR (2007) Bisindolylmaleimide I suppresses fibroblast growth factor-mediated activation of Erk MAP kinase in chondrocytes by preventing Shp2 association with the Frs2 and Gab1 adaptor proteins. J Biol Chem 282(5):2929–2936. https://doi.org/10.1074/jbc.M606144200

Krejci P, Salazar L, Goodridge HS, Kashiwada TA, Schibler MJ, Jelinkova P, Thompson LM, Wilcox WR (2008) STAT1 and STAT3 do not participate in FGF-mediated growth arrest in chondrocytes. J Cell Sci 121(3):272–281. https://doi.org/10.1242/jcs.017160

Kurosu H, Ogawa Y, Miyoshi M, Yamamoto M, Nandi A, Rosenblatt KP, Baum MG, Schiavi S, Hu M-C, Moe OW, Kuro-o M (2006) Regulation of fibroblast growth Factor-23 signaling by klotho. J Biol Chem 281(10):6120–6123. https://doi.org/10.1074/jbc.C500457200

Lajeunie E, Heuertz S, El Ghouzzi V, Martinovic J, Renier D, Le Merrer M, Bonaventure J (2006) Mutation screening in patients with syndromic craniosynostoses indicates that a limited num-

ber of recurrent FGFR2 mutations accounts for severe forms of Pfeiffer syndrome. Eur J Hum Genet 14(3):289–298. https://doi.org/10.1038/sj.ejhg.5201558
Lajeunie E, Ma HW, Bonaventure J, Munnich A, Le Merrer M, Renier D (1995) FGFR2 mutations in Pfeiffer syndrome. Nat Genet 9(2):108. https://doi.org/10.1038/ng0295-108
Lamothe B, Yamada M, Schaeper U, Birchmeier W, Lax I, Schlessinger J (2004) The docking protein Gab1 is an essential component of an indirect mechanism for fibroblast growth factor stimulation of the phosphatidylinositol 3-kinase/Akt antiapoptotic pathway. Mol Cell Biol 24(13):5657–5666. https://doi.org/10.1128/MCB.24.13.5657-5666.2004
Laurita J, Koyama E, Chin B, Taylor JA, Lakin GE, Hankenson KD, Bartlett SP, Nah H (2011) The Muenke syndrome mutation (*FgfR3* ^{P244R}) causes cranial base shortening associated with growth plate dysfunction and premature perichondrial ossification in murine basicranial synchondroses. Dev Dyn 240(11):2584–2596. https://doi.org/10.1002/dvdy.22752
Lazarus JE, Hegde A, Andrade AC, Nilsson O, Baron J (2007) Fibroblast growth factor expression in the postnatal growth plate. Bone 40(3):577–586. https://doi.org/10.1016/j.bone.2006.10.013
Lee CH, Javed D, Althaus AL, Parent JM, Umemori H (2012) Neurogenesis is enhanced and mossy fiber sprouting arises in FGF7-deficient mice during development. Mol Cell Neurosci 51(3–4):61–67. https://doi.org/10.1016/j.mcn.2012.07.010
Legeai-Mallet L, Savarirayan R (2020) Novel therapeutic approaches for the treatment of achondroplasia. Bone 141:115579. https://doi.org/10.1016/j.bone.2020.115579
Lemmon MA, Schlessinger J (2010) Cell signaling by receptor tyrosine kinases. Cell 141(7):1117–1134. https://doi.org/10.1016/j.cell.2010.06.011
Liotta LA, Tryggvason K, Garbisa S, Hart I, Foltz CM, Shafie S (1980) Metastatic potential correlates with enzymatic degradation of basement membrane collagen. Nature 284(5751):67–68. https://doi.org/10.1038/284067a0
Liu S, Guo R, Simpson LG, Xiao Z-S, Burnham CE, Quarles LD (2003) Regulation of fibroblastic growth factor 23 expression but not degradation by PHEX. J Biol Chem 278(39):37419–37426. https://doi.org/10.1074/jbc.M304544200
Liu S, Zhou J, Tang W, Jiang X, Rowe DW, Quarles LD (2006) Pathogenic role of Fgf23 in Hyp mice. Am J Physiol Endocrinol Metab 291(1):E38–E49. https://doi.org/10.1152/ajpendo.00008.2006
Liu Z, Lavine KJ, Hung IH, Ornitz DM (2007) FGF18 is required for early chondrocyte proliferation, hypertrophy and vascular invasion of the growth plate. Dev Biol 302(1):80–91. https://doi.org/10.1016/j.ydbio.2006.08.071
Liu Z, Xu J, Colvin JS, Ornitz DM (2002) Coordination of chondrogenesis and osteogenesis by fibroblast growth factor 18. Genes Dev 16(7):859–869. https://doi.org/10.1101/gad.965602
Logan A, Frautschy SA, Baird A (1991) Basic fibroblast growth factor and central nervous system injury. Ann N Y Acad Sci 638:474–476. https://doi.org/10.1111/j.1749-6632.1991.tb49073.x
Lohmander LS, Hellot S, Dreher D, Krantz EFW, Kruger DS, Guermazi A, Eckstein F (2014) Intraarticular sprifermin (recombinant human fibroblast growth factor 18) in knee osteoarthritis: a randomized, double-blind, placebo-controlled trial. Arthritis Rheumatol 66(7):1820–1831. https://doi.org/10.1002/art.38614
Lorget F, Kaci N, Peng J, Benoist-Lasselin C, Mugniery E, Oppeneer T, Wendt DJ, Bell SM, Bullens S, Bunting S, Tsuruda LS, O'Neill CA, Di Rocco F, Munnich A, Legeai-Mallet L (2012) Evaluation of the therapeutic potential of a CNP analog in a Fgfr3 mouse model recapitulating achondroplasia. Am J Hum Genet 91(6):1108–1114. https://doi.org/10.1016/j.ajhg.2012.10.014
Lou J-Y, Laezza F, Gerber BR, Xiao M, Yamada KA, Hartmann H, Craig AM, Nerbonne JM, Ornitz DM (2005) Fibroblast growth factor 14 is an intracellular modulator of voltage-gated sodium channels. J Physiol 569(Pt 1):179–193. https://doi.org/10.1113/jphysiol.2005.097220
Luo F, Xie Y, Wang Z, Huang J, Tan Q, Sun X, Li F, Li C, Liu M, Zhang D, Xu M, Su N, Ni Z, Jiang W, Chang J, Chen H, Chen S, Xu X, Deng C, Chen L (2018) Adeno-associated virus-mediated RNAi against mutant alleles attenuates abnormal Calvarial phenotypes in an Apert syndrome mouse model. Mol Ther Nucleic Acids 13:291–302. https://doi.org/10.1016/j.omtn.2018.09.012

Mai S, Wei K, Flenniken A, Adamson SL, Rossant J, Aubin JE, Gong S-G (2010) The missense mutation W290R in *Fgfr2* causes developmental defects from aberrant IIIb and IIIc signaling. Dev Dyn 239(6):1888–1900. https://doi.org/10.1002/dvdy.22314

Makarenkova HP, Ito M, Govindarajan V, Faber SC, Sun L, McMahon G, Overbeek PA, Lang RA (2000) FGF10 is an inducer and Pax 6 a competence factor for lacrimal gland development. Development 127(12):2563–2572. https://doi.org/10.1242/dev.127.12.2563

Makrythanasis P, Temtamy S, Aglan MS, Otaify GA, Hamamy H, Antonarakis SE (2014) A novel homozygous mutation in FGFR3 causes tall stature, severe lateral tibial deviation, scoliosis, hearing impairment, camptodactyly, and arachnodactyly. Hum Mutat 35(8):959–963. https://doi.org/10.1002/humu.22597

Marie PJ, Debiais F, Haÿ E (2002) Regulation of human cranial osteoblast phenotype by FGF-2, FGFR-2 and BMP-2 signaling. Histol Histopathol 17(3):877–885. https://doi.org/10.14670/HH-17.877

Martin GR (1998) The roles of FGFs in the early development of vertebrate limbs. Genes Dev 12(11):1571–1586. https://doi.org/10.1101/gad.12.11.1571

Martin L, Kaci N, Benoist-Lasselin C, Mondoloni M, Decaudaveine S, Estibals V, Cornille M, Loisay L, Flipo J, Demuynck B, de la Luz Cádiz-Gurrea M, Barbault F, Fernández-Arroyo S, Schibler L, Segura-Carretero A, Dambroise E, Legeai-Mallet L (2022) Theobroma cacao improves bone growth by modulating defective ciliogenesis in a mouse model of achondroplasia. Bone Res 10(1):8. https://doi.org/10.1038/s41413-021-00177-7

Mason IJ, Fuller-Pace F, Smith R, Dickson C (1994) FGF-7 (keratinocyte growth factor) expression during mouse development suggests roles in myogenesis, forebrain regionalisation and epithelial-mesenchymal interactions. Mech Dev 45(1):15–30. https://doi.org/10.1016/0925-4773(94)90050-7

Matsumoto E, Sasaki S, Kinoshita H, Kito T, Ohta H, Konishi M, Kuwahara K, Nakao K, Itoh N (2013) Angiotensin II-induced cardiac hypertrophy and fibrosis are promoted in mice lacking Fgf16. Genes Cells 18(7):544–553. https://doi.org/10.1111/gtc.12055

Matsushita M, Esaki R, Mishima K, Ishiguro N, Ohno K, Kitoh H (2017) Clinical dosage of meclozine promotes longitudinal bone growth, bone volume, and trabecular bone quality in transgenic mice with achondroplasia. Sci Rep 7(1):7371. https://doi.org/10.1038/s41598-017-07044-8

McAVOY JW, Chamberlain CG, Iongh RUD, Richardson NA, Lovicu FJ (1991) The role of fibroblast growth factor in eye lens development [a]. Ann N Y Acad Sci 638(1):256–274. https://doi.org/10.1111/j.1749-6632.1991.tb49036.x

Merrill AE, Sarukhanov A, Krejci P, Idoni B, Camacho N, Estrada KD, Lyons KM, Deixler H, Robinson H, Chitayat D, Curry CJ, Lachman RS, Wilcox WR, Krakow D (2012) Bent bone dysplasia-FGFR2 type, a distinct skeletal disorder, has deficient canonical FGF signaling. Am J Hum Genet 90(3):550–557. https://doi.org/10.1016/j.ajhg.2012.02.005

Meyers GA, Orlow SJ, Munro IR, Przylepa KA, Jabs EW (1995) Fibroblast growth factor receptor 3 (FGFR3) transmembrane mutation in Crouzon syndrome with acanthosis nigricans. Nat Genet 11(4):462–464. https://doi.org/10.1038/ng1295-462

Miao K, Zhang X, Su SM, Zeng J, Huang Z, Chan UI, Xu X, Deng C-X (2019) Optimizing CRISPR/Cas9 technology for precise correction of the Fgfr3-G374R mutation in achondroplasia in mice. J Biol Chem 294(4):1142–1151. https://doi.org/10.1074/jbc.RA118.006496

Min H, Danilenko DM, Scully SA, Bolon B, Ring BD, Tarpley JE, DeRose M, Simonet WS (1998) *Fgf-10* is required for both limb and lung development and exhibits striking functional similarity to *drosophila branchless*. Genes Dev 12(20):3156–3161. https://doi.org/10.1101/gad.12.20.3156

Moerlooze LD, Spencer-Dene B, Revest J-M, Hajihosseini M, Rosewell I, Dickson C (2000) An important role for the IIIb isoform of fibroblast growth factor receptor 2 (FGFR2) in mesenchymal-epithelial signalling during mouse organogenesis. Development 127(3):483–492. https://doi.org/10.1242/dev.127.3.483

Mohammadi M, Olsen SK, Goetz R (2005a) A protein canyon in the FGF–FGF receptor dimer selects from an à la carte menu of heparan sulfate motifs. Curr Opin Struct Biol 15(5):506–516. https://doi.org/10.1016/j.sbi.2005.09.002

Mohammadi M, Olsen SK, Ibrahimi OA (2005b) Structural basis for fibroblast growth factor receptor activation. Cytokine Growth Factor Rev 16(2):107–137. https://doi.org/10.1016/j.cytogfr.2005.01.008

Moloney D, Wall S, Ashworth G, Oldridge M, Glass I, Francomano C, Muenke M, Wilkie A (1997) Prevalence of Pro250Arg mutation of fibroblast growth factor receptor 3 in coronal craniosynostosis. Lancet 349(9058):1059–1062. https://doi.org/10.1016/S0140-6736(96)09082-4

Molténi A, Modrowski D, Hott M, Marie PJ (1999) Differential expression of fibroblast growth factor receptor-1, −2, and −3 and syndecan-1, −2, and −4 in neonatal rat mandibular condyle and calvaria during osteogenic differentiation in vitro. Bone 24(4):337–347. https://doi.org/10.1016/s8756-3282(98)00191-4

Montone R, Romanelli MG, Baruzzi A, Ferrarini F, Liboi E, Lievens PM-J (2018) Mutant FGFR3 associated with SADDAN disease causes cytoskeleton disorganization through PLCγ1/Src-mediated paxillin hyperphosphorylation. Int J Biochem Cell Biol 95:17–26. https://doi.org/10.1016/j.biocel.2017.12.008

Morita J, Nakamura M, Kobayashi Y, Deng C-X, Funato N, Moriyama K (2014) Soluble form of FGFR2 with S252W partially prevents craniosynostosis of the apert mouse model. Dev Dyn 243(4):560–567. https://doi.org/10.1002/dvdy.24099

Morriss-Kay GM, Wilkie AOM (2005) Growth of the normal skull vault and its alteration in craniosynostosis: insights from human genetics and experimental studies. J Anat 207(5):637–653. https://doi.org/10.1111/j.1469-7580.2005.00475.x

Muenke M, Schell U, Hehr A, Robin NH, Losken HW, Schinzel A, Pulleyn LJ, Rutland P, Reardon W, Malcolm S (1994) A common mutation in the fibroblast growth factor receptor 1 gene in Pfeiffer syndrome. Nat Genet 8(3):269–274. https://doi.org/10.1038/ng1194-269

Murali SK, Roschger P, Zeitz U, Klaushofer K, Andrukhova O, Erben RG (2016) FGF23 regulates bone Mineralization in a 1,25(OH)$_2$ D$_3$ and klotho-independent manner: FGF23 and bone mineralization. J Bone Miner Res 31(1):129–142. https://doi.org/10.1002/jbmr.2606

Murdoch JL, Walker BA, Hall JG, Abbey H, Smith KK, McKusick VA (1970) Achondroplasia—a genetic and statistical survey. Ann Hum Genet 33(3):227–244. https://doi.org/10.1111/j.1469-1809.1970.tb01648.x

Naski MC, Colvin JS, Coffin JD, Ornitz DM (1998) Repression of hedgehog signaling and BMP4 expression in growth plate cartilage by fibroblast growth factor receptor 3. Development 125(24):4977–4988. https://doi.org/10.1242/dev.125.24.4977

Neben CL, Idoni B, Salva JE, Tuzon CT, Rice JC, Krakow D, Merrill AE (2014) Bent bone dysplasia syndrome reveals nucleolar activity for FGFR2 in ribosomal DNA transcription. Hum Mol Genet 23(21):5659–5671. https://doi.org/10.1093/hmg/ddu282

Neben CL, Merrill AE (2015) Signaling pathways in craniofacial development: insights from rare skeletal disorders. Curr Top Dev Biol 115:493–542. https://doi.org/10.1016/bs.ctdb.2015.09.005

Neben CL, Tuzon CT, Mao X, Lay FD, Merrill AE (2017) FGFR2 mutations in bent bone dysplasia syndrome activate nucleolar stress and perturb cell fate determination. Hum Mol Genet 26(17):3253–3270. https://doi.org/10.1093/hmg/ddx209

Necchi A, Castellano DE, Mellado B, Pang S, Urun Y, Park SH, Vaishampayan UN, Currie G, Abella-Dominicis E, Pal SK (2019) Fierce-21: Phase II study of vofatmab (B-701), a selective inhibitor of FGFR3, as salvage therapy in metastatic urothelial carcinoma (mUC). J Clin Oncol 37(7_suppl):409–409. https://doi.org/10.1200/JCO.2019.37.7_suppl.409

Nowroozi N, Raffioni S, Wang T, Apostol BL, Bradshaw RA, Thompson LM (2005) Sustained ERK1/2 but not STAT1 or 3 activation is required for thanatophoric dysplasia phenotypes in PC12 cells. Hum Mol Genet 14(11):1529–1538. https://doi.org/10.1093/hmg/ddi161

Ohbayashi N, Shibayama M, Kurotaki Y, Imanishi M, Fujimori T, Itoh N, Takada S (2002) FGF18 is required for normal cell proliferation and differentiation during osteogenesis and chondrogenesis. Genes Dev 16(7):870–879. https://doi.org/10.1101/gad.965702

Olsen SK, Garbi M, Zampieri N, Eliseenkova AV, Ornitz DM, Goldfarb M, Mohammadi M (2003) Fibroblast growth factor (FGF) homologous factors share structural but not functional homology with FGFs. J Biol Chem 278(36):34226–34236. https://doi.org/10.1074/jbc.M303183200

Olsen SK, Li JYH, Bromleigh C, Eliseenkova AV, Ibrahimi OA, Lao Z, Zhang F, Linhardt RJ, Joyner AL, Mohammadi M (2006) Structural basis by which alternative splicing modulates the organizer activity of FGF8 in the brain. Genes Dev 20(2):185–198. https://doi.org/10.1101/gad.1365406

Ornitz DM, Itoh N (2015) The fibroblast growth factor signaling pathway. Wiley Interdiscip Rev Dev Biol 4(3):215–266. https://doi.org/10.1002/wdev.176

Ornitz DM, Legeai-Mallet L (2017) Achondroplasia: development, pathogenesis, and therapy. Dev Dyn 246(4):291–309. https://doi.org/10.1002/dvdy.24479

Ornitz DM, Marie PJ (2002) FGF signaling pathways in endochondral and intramembranous bone development and human genetic disease. Genes Dev 16(12):1446–1465. https://doi.org/10.1101/gad.990702

Ornitz DM, Marie PJ (2015) Fibroblast growth factor signaling in skeletal development and disease. Genes Dev 29(14):1463–1486. https://doi.org/10.1101/gad.266551.115

Ornitz DM, Xu J, Colvin JS, McEwen DG, MacArthur CA, Coulier F, Gao G, Goldfarb M (1996) Receptor specificity of the fibroblast growth factor family. J Biol Chem 271(25):15292–15297. https://doi.org/10.1074/jbc.271.25.15292

Orr-Urtreger A, Givol D, Yayon A, Yarden Y, Lonai P (1991) Developmental expression of two murine fibroblast growth factor receptors, *fig* and *bek*. Development 113(4):1419–1434. https://doi.org/10.1242/dev.113.4.1419

Park J-S, Lee J, Lim M-A, Kim E-K, Kim S-M, Ryu J-G, Lee JH, Kwok S-K, Park K-S, Kim H-Y, Park S-H, Cho M-L (2014) JAK2-STAT3 blockade by AG490 suppresses autoimmune arthritis in mice via reciprocal regulation of regulatory T cells and Th17 cells. J Immunol 192(9):4417–4424. https://doi.org/10.4049/jimmunol.1300514

Pena JC (1999) Manipulation of outer root sheath cell survival perturbs the hair-growth cycle. EMBO J 18(13):3596–3603. https://doi.org/10.1093/emboj/18.13.3596

Peters KG, Werner S, Chen G, Williams LT (1992) Two FGF receptor genes are differentially expressed in epithelial and mesenchymal tissues during limb formation and organogenesis in the mouse. Development 114(1):233–243. https://doi.org/10.1242/dev.114.1.233

Peters K, Ornitz D, Werner S, Williams L (1993) Unique expression pattern of the FGF receptor 3 gene during mouse organogenesis. Dev Biol 155(2):423–430. https://doi.org/10.1006/dbio.1993.1040

Pratsinis H, Constantinou V, Pavlakis K, Sapkas G, Kletsas D (2012) Exogenous and autocrine growth factors stimulate human intervertebral disc cell proliferation via the ERK and Akt pathways. J Orthop Res 30(6):958–964. https://doi.org/10.1002/jor.22017

Prud'homme GJ, Kurt M, Wang Q (2022) Pathobiology of the klotho antiaging protein and therapeutic considerations. Front Aging 3:931331. https://doi.org/10.3389/fragi.2022.931331

Purcell P, Joo BW, Hu JK, Tran PV, Calicchio ML, O'Connell DJ, Maas RL, Tabin CJ (2009) Temporomandibular joint formation requires two distinct hedgehog-dependent steps. Proc Natl Acad Sci USA 106(43):18297–18302. https://doi.org/10.1073/pnas.0908836106

Qiao J, Uzzo R, Obara-Ishihara T, Degenstein L, Fuchs E, Herzlinger D (1999) FGF-7 modulates ureteric bud growth and nephron number in the developing kidney. Development 126(3):547–554. https://doi.org/10.1242/dev.126.3.547

Quarto N, Behr B, Li S, Longaker MT (2009) Differential FGF ligands and FGF receptors expression pattern in frontal and parietal calvarial bones. Cells Tissues Organs 190(3):158–169. https://doi.org/10.1159/000202789

Rannan-Eliya SV, Taylor IB, De Heer IM, Van Den Ouweland AMW, Wall SA, Wilkie AOM (2004) Paternal origin of FGFR3 mutations in Muenke-type craniosynostosis. Hum Genet 115(3):200–207. https://doi.org/10.1007/s00439-004-1151-5

Raucci A, Laplantine E, Mansukhani A, Basilico C (2004) Activation of the ERK1/2 and p38 mitogen-activated protein kinase pathways mediates fibroblast growth factor-induced growth arrest of chondrocytes. J Biol Chem 279(3):1747–1756. https://doi.org/10.1074/jbc.M310384200

Reardon W, Winter RM, Rutland P, Pulleyn LJ, Jones BM, Malcolm S (1994) Mutations in the fibroblast growth factor receptor 2 gene cause Crouzon syndrome. Nat Genet 8(1):98–103. https://doi.org/10.1038/ng0994-98

Reker D, Siebuhr AS, Thudium CS, Gantzel T, Ladel C, Michaelis M, Aspberg A, Berchtold MW, Karsdal MA, Gigout A, Bay-Jensen AC (2020) Sprifermin (rhFGF18) versus vehicle induces a biphasic process of extracellular matrix remodeling in human knee OA articular cartilage ex vivo. Sci Rep 10(1):6011. https://doi.org/10.1038/s41598-020-63216-z

Rice DP, Aberg T, Chan Y, Tang Z, Kettunen PJ, Pakarinen L, Maxson RE, Thesleff I (2000) Integration of FGF and TWIST in calvarial bone and suture development. Development 127(9):1845–1855. https://doi.org/10.1242/dev.127.9.1845

Rice DPC, Rice R, Thesleff I (2003) Fgfr mRNA isoforms in craniofacial bone development. Bone 33(1):14–27. https://doi.org/10.1016/s8756-3282(03)00163-7

Riley BB, Savage MP, Simandl BK, Olwin BB, Fallon JF (1993) Retroviral expression of FGF-2 (bFGF) affects patterning in chick limb bud. Development 118(1):95–104. https://doi.org/10.1242/dev.118.1.95

Robinson D, Hasharoni A, Cohen N, Yayon A, Moskowitz RM, Nevo Z (1999) Fibroblast growth factor receptor-3 as a marker for precartilaginous stem cells. Clin Orthop Relat Res 367(Suppl):S163–S175. https://doi.org/10.1097/00003086-199910001-00018

Rohmann E, Brunner HG, Kayserili H, Uyguner O, Nürnberg G, Lew ED, Dobbie A, Eswarakumar VP, Uzumcu A, Ulubil-Emeroglu M, Leroy JG, Li Y, Becker C, Lehnerdt K, Cremers CWRJ, Yüksel-Apak M, Nürnberg P, Kubisch C, Schlessinger J et al (2006) Mutations in different components of FGF signaling in LADD syndrome. Nat Genet 38(4):414–417. https://doi.org/10.1038/ng1757

Rosenquist TA, Martin GR (1996) Fibroblast growth factor signalling in the hair growth cycle: expression of the fibroblast growth factor receptor and ligand genes in the murine hair follicle. Dev Dyn 205(4):379–386. https://doi.org/10.1002/(SICI)1097-0177(199604)205:4<379::AID-AJA2>3.0.CO;2-F

Rousseau F, Bonaventure J, Legeai-Mallet L, Pelet A, Rozet JM, Maroteaux P, Le Merrer M, Munnich A (1994) Mutations in the gene encoding fibroblast growth factor receptor-3 in achondroplasia. Nature 371(6494):252–254. https://doi.org/10.1038/371252a0

Rousseau F, el Ghouzzi V, Delezoide AL, Legeai-Mallet L, Le Merrer M, Munnich A, Bonaventure J (1996) Missense FGFR3 mutations create cysteine residues in thanatophoric dwarfism type I (TD1). Hum Mol Genet 5(4):509–512. https://doi.org/10.1093/hmg/5.4.509

Rousseau F, Saugier P, Le Merrer M, Munnich A, Delezoide AL, Maroteaux P, Bonaventure J, Narcy F, Sanak M (1995) Stop codon FGFR3 mutations in thanatophoric dwarfism type 1. Nat Genet 10(1):11–12. https://doi.org/10.1038/ng0595-11

Savarirayan R, Hoernschemeyer DG, Ljungberg M, Zarate YA, Bacino CA, Bober MB, Legare JM, Högler W, Quattrin T, Abuzzahab MJ, Hofman PL, White KK, Ma NS, Schnabel D, Sousa SB, Mao M, Smith A, Chakraborty M, Giwa A et al (2023a) Once-weekly TransCon CNP (navepegritide) in children with achondroplasia (ACcomplisH): A phase 2, multicentre, randomised, double-blind, placebo-controlled, dose-escalation trial. EClinicalMedicine 65:102258. https://doi.org/10.1016/j.eclinm.2023.102258

Savarirayan R, Wilcox WR, Harmatz P, Phillips J, Polgreen LE, Tofts L, Ozono K, Arundel P, Irving M, Bacino CA, Basel D, Bober MB, Charrow J, Mochizuki H, Kotani Y, Saal HM, Army C, Jeha G, Qi Y, Day J (2023) Vosoritide therapy in children with achondroplasia aged 3-59 months: a multinational, randomised, double-blind, placebo-controlled, phase 2 trial. Lancet Child Adolesc Health S2352-4642(23):00265–00261. https://doi.org/10.1016/S2352-4642(23)00265-1

Schibler L, Gibbs L, Benoist-Lasselin C, Decraene C, Martinovic J, Loget P, Delezoide A-L, Gonzales M, Munnich A, Jais J-P, Legeai-Mallet L (2009) New insight on FGFR3-related

chondrodysplasias molecular physiopathology revealed by human chondrocyte gene expression profiling. PLoS One 4(10):e7633. https://doi.org/10.1371/journal.pone.0007633
Schlessinger J (2000) Cell signaling by receptor tyrosine kinases. Cell 103(2):211–225. https://doi.org/10.1016/s0092-8674(00)00114-8
Schwartz SM, Liaw L (1993) Growth control and morphogenesis in the development and pathology of arteries. J Cardiovasc Pharmacol 21:S31–S49. https://doi.org/10.1097/00005344-199321001-00007
Sekine K, Ohuchi H, Fujiwara M, Yamasaki M, Yoshizawa T, Sato T, Yagishita N, Matsui D, Koga Y, Itoh N, Kato S (1999) Fgf10 is essential for limb and lung formation. Nat Genet 21(1):138–141. https://doi.org/10.1038/5096
Sheeba CJ, Andrade RP, Duprez D, Palmeirim I (2010) Comprehensive analysis of fibroblast growth factor receptor expression patterns during chick forelimb development. Int J Dev Biol 54(10):1515–1524. https://doi.org/10.1387/ijdb.092887cs
Shiang R, Thompson LM, Zhu YZ, Church DM, Fielder TJ, Bocian M, Winokur ST, Wasmuth JJ (1994) Mutations in the transmembrane domain of FGFR3 cause the most common genetic form of dwarfism, achondroplasia. Cell 78(2):335–342. https://doi.org/10.1016/0092-8674(94)90302-6
Shimada T, Kakitani M, Yamazaki Y, Hasegawa H, Takeuchi Y, Fujita T, Fukumoto S, Tomizuka K, Yamashita T (2004) Targeted ablation of Fgf23 demonstrates an essential physiological role of FGF23 in phosphate and vitamin D metabolism. J Clin Invest 113(4):561–568. https://doi.org/10.1172/JCI19081
Shu C, Smith SS, Little CB, Melrose J (2013) Comparative immunolocalisation of perlecan, heparan sulphate, fibroblast growth factor-18, and fibroblast growth factor receptor-3 and their prospective roles in chondrogenic and osteogenic development of the human foetal spine. Eur Spine J 22(8):1774–1784. https://doi.org/10.1007/s00586-013-2680-1
Singh R, Su J, Brooks J, Terauchi A, Umemori H, Fox MA (2012) Fibroblast growth factor 22 contributes to the development of retinal nerve terminals in the dorsal lateral geniculate nucleus. Front Mol Neurosci 4:61. https://doi.org/10.3389/fnmol.2011.00061
Spivak-Kroizman T, Lemmon MA, Dikic I, Ladbury JE, Pinchasi D, Huang J, Jaye M, Crumley G, Schlessinger J, Lax I (1994) Heparin-induced oligomerization of FGF molecules is responsible for FGF receptor dimerization, activation, and cell proliferation. Cell 79(6):1015–1024. https://doi.org/10.1016/0092-8674(94)90032-9
Sun X, Lewandoski M, Meyers EN, Liu YH, Maxson RE, Martin GR (2000) Conditional inactivation of Fgf4 reveals complexity of signalling during limb bud development. Nat Genet 25(1):83–86. https://doi.org/10.1038/75644
Sun X, Zhang R, Chen H, Du X, Chen S, Huang J, Liu M, Xu M, Luo F, Jin M, Su N, Qi H, Yang J, Tan Q, Zhang D, Ni Z, Liang S, Zhang B, Chen D et al (2020) Fgfr3 mutation disrupts chondrogenesis and bone ossification in zebrafish model mimicking CATSHL syndrome partially via enhanced Wnt/β-catenin signaling. Theranostics 10(16):7111–7130. https://doi.org/10.7150/thno.45286
Suzuki M, Uehara Y, Motomura-Matsuzaka K, Oki J, Koyama Y, Kimura M, Asada M, Komi-Kuramochi A, Oka S, Toru T (2008) βKlotho is required for fibroblast growth factor (FGF) 21 signaling through FGF receptor (FGFR) 1c and FGFR3c. Mol Endocrinol 22(4):1006–1014. https://doi.org/10.1210/me.2007-0313
Szebenyi G, Savage MP, Olwin BB, Fallon JF (1995) Changes in the expression of fibroblast growth factor receptors mark distinct stages of chondrogenesis in vitro and during chick limb skeletal patterning. Dev Dyn 204(4):446–456. https://doi.org/10.1002/aja.1002040410
Tabin C, Wolpert L (2007) Rethinking the proximodistal axis of the vertebrate limb in the molecular era. Genes Dev 21(12):1433–1442. https://doi.org/10.1101/gad.1547407
Takase HM, Itoh T, Ino S, Wang T, Koji T, Akira S, Takikawa Y, Miyajima A (2013) FGF7 is a functional niche signal required for stimulation of adult liver progenitor cells that support liver regeneration. Genes Dev 27(2):169–181. https://doi.org/10.1101/gad.204776.112

Tavormina PL, Bellus GA, Webster MK, Bamshad MJ, Fraley AE, McIntosh I, Szabo J, Jiang W, Jabs EW, Wilcox WR, Wasmuth JJ, Donoghue DJ, Thompson LM, Francomano CA (1999) A novel skeletal dysplasia with developmental delay and acanthosis nigricans is caused by a Lys650Met mutation in the fibroblast growth factor receptor 3 gene. Am J Hum Genet 64(3):722–731. https://doi.org/10.1086/302275

Tavormina PL, Shiang R, Thompson LM, Zhu YZ, Wilkin DJ, Lachman RS, Wilcox WR, Rimoin DL, Cohn DH, Wasmuth JJ (1995) Thanatophoric dysplasia (types I and II) caused by distinct mutations in fibroblast growth factor receptor 3. Nat Genet 9(3):321–328. https://doi.org/10.1038/ng0395-321

Tekin M, Hişmi BO, Fitoz S, Ozdağ H, Cengiz FB, Sirmaci A, Aslan I, Inceoğlu B, Yüksel-Konuk EB, Yilmaz ST, Yasun O, Akar N (2007) Homozygous mutations in fibroblast growth factor 3 are associated with a new form of syndromic deafness characterized by inner ear agenesis, microtia, and microdontia. Am J Hum Genet 80(2):338–344. https://doi.org/10.1086/510920

Terauchi A, Johnson-Venkatesh EM, Toth AB, Javed D, Sutton MA, Umemori H (2010) Distinct FGFs promote differentiation of excitatory and inhibitory synapses. Nature 465(7299):783–787. https://doi.org/10.1038/nature09041

Timsah Z, Ahmed Z, Lin C-C, Melo FA, Stagg LJ, Leonard PG, Jeyabal P, Berrout J, O'Neil RG, Bogdanov M, Ladbury JE (2014) Competition between Grb2 and Plcγ1 for FGFR2 regulates basal phospholipase activity and invasion. Nat Struct Mol Biol 21(2):180–188. https://doi.org/10.1038/nsmb.2752

Torii S, Kusakabe M, Yamamoto T, Maekawa M, Nishida E (2004) Sef is a spatial regulator for Ras/MAP kinase signaling. Dev Cell 7(1):33–44. https://doi.org/10.1016/j.devcel.2004.05.019

Trueb B (2011) Biology of FGFRL1, the fifth fibroblast growth factor receptor. Cell Mol Life Sci 68(6):951–964. https://doi.org/10.1007/s00018-010-0576-3

Turner N, Grose R (2010) Fibroblast growth factor signalling: from development to cancer. Nat Rev Cancer 10(2):116–129. https://doi.org/10.1038/nrc2780

Twigg SRF, Forecki J, Goos JAC, Richardson ICA, Hoogeboom AJM, van den Ouweland AMW, Swagemakers SMA, Lequin MH, Van Antwerp D, McGowan SJ, Westbury I, Miller KA, Wall SA, WGS500 Consortium, van der Spek PJ, Mathijssen IMJ, Pauws E, Merzdorf CS, Wilkie AOM (2015) Gain-of-Function Mutations in ZIC1 Are Associated with Coronal Craniosynostosis and Learning Disability. American Journal of Human Genetics *97*(3):378–388. https://doi.org/10.1016/j.ajhg.2015.07.007

Umemori H, Linhoff MW, Ornitz DM, Sanes JR (2004) FGF22 and its close relatives are presynaptic organizing molecules in the mammalian brain. Cell 118(2):257–270. https://doi.org/10.1016/j.cell.2004.06.025

Usui H, Shibayama M, Ohbayashi N, Konishi M, Takada S, Itoh N (2004) Fgf18 is required for embryonic lung alveolar development. Biochem Biophys Res Commun 322(3):887–892. https://doi.org/10.1016/j.bbrc.2004.07.198

Vajo Z, Francomano CA, Wilkin DJ (2000) The molecular and genetic basis of fibroblast growth factor receptor 3 disorders: the achondroplasia family of skeletal dysplasias, Muenke craniosynostosis, and Crouzon syndrome with acanthosis nigricans. Endocr Rev 21(1):23–39. https://doi.org/10.1210/edrv.21.1.0387

Vantaggiato C, Bondioni S, Airoldi G, Bozzato A, Borsani G, Rugarli EI, Bresolin N, Clementi E, Bassi MT (2011) Senataxin modulates neurite growth through fibroblast growth factor 8 signalling. Brain J Neurol 134(Pt 6):1808–1828. https://doi.org/10.1093/brain/awr084

Velleman SG (2000) The role of the extracellular matrix in skeletal development. Poult Sci 79(7):985–989. https://doi.org/10.1093/ps/79.7.985

Wang C, Hennessey JA, Kirkton RD, Wang C, Graham V, Puranam RS, Rosenberg PB, Bursac N, Pitt GS (2011) Fibroblast growth factor homologous factor 13 regulates Na+ channels and conduction velocity in murine hearts. Circ Res 109(7):775–782. https://doi.org/10.1161/CIRCRESAHA.111.247957

Wang Y, Zhou X, Oberoi K, Phelps R, Couwenhoven R, Sun M, Rezza A, Holmes G, Percival CJ, Friedenthal J, Krejci P, Richtsmeier JT, Huso DL, Rendl M, Jabs EW (2012) P38 inhibition ameliorates skin and skull abnormalities in Fgfr2 Beare-Stevenson mice. J Clin Invest 122(6):2153–2164. https://doi.org/10.1172/JCI62644

Webster MK, Donoghue DJ (1997) FGFR activation in skeletal disorders: too much of a good thing. Trends Genet 13(5):178–182. https://doi.org/10.1016/s0168-9525(97)01131-1

Weng T, Yi L, Huang J, Luo F, Wen X, Du X, Chen Q, Deng C, Chen D, Chen L (2012) Genetic inhibition of fibroblast growth factor receptor 1 in knee cartilage attenuates the degeneration of articular cartilage in adult mice. Arthritis Rheum 64(12):3982–3992. https://doi.org/10.1002/art.34645

Wenger, T., Miller, D., & Evans, K. (2020). FGFR Craniosynostosis syndromes overview. Genereviews

Wilkie AO (2000) Epidemiology and genetics of craniosynostosis. Am J Med Genet 90(1):82–84

Wilkie AO, Morriss-Kay GM, Jones EY, Heath JK (1995) Functions of fibroblast growth factors and their receptors. Curr Biol 5(5):500–507. https://doi.org/10.1016/s0960-9822(95)00102-3

Wilkin DJ, Szabo JK, Cameron R, Henderson S, Bellus GA, Mack ML, Kaitila I, Loughlin J, Munnich A, Sykes B, Bonaventure J, Francomano CA (1998) Mutations in fibroblast growth-factor receptor 3 in sporadic cases of achondroplasia occur exclusively on the paternally derived chromosome. Am J Hum Genet 63(3):711–716. https://doi.org/10.1086/302000

Wu X, Ge H, Lemon B, Vonderfecht S, Weiszmann J, Hecht R, Gupte J, Hager T, Wang Z, Lindberg R, Li Y (2010) FGF19-induced hepatocyte proliferation is mediated through FGFR4 activation. J Biol Chem 285(8):5165–5170. https://doi.org/10.1074/jbc.M109.068783

Xiao M, Bosch MK, Nerbonne JM, Ornitz DM (2013) FGF14 localization and organization of the axon initial segment. Mol Cell Neurosci 56:393–403. https://doi.org/10.1016/j.mcn.2013.07.008

Xie Y, Su N, Yang J, Tan Q, Huang S, Jin M, Ni Z, Zhang B, Zhang D, Luo F, Chen H, Sun X, Feng JQ, Qi H, Chen L (2020) FGF/FGFR signaling in health and disease. Signal Transduct Target Ther 5(1):181. https://doi.org/10.1038/s41392-020-00222-7

Xu J, Lawshe A, MacArthur CA, Ornitz DM (1999) Genomic structure, mapping, activity and expression of fibroblast growth factor 17. Mech Dev 83(1–2):165–178. https://doi.org/10.1016/s0925-4773(99)00034-9

Xu M, Huang J, Jin M, Jiang W, Luo F, Tan Q, Zhang R, Luo X, Kuang L, Zhang D, Liang S, Qi H, Chen H, Ni Z, Su N, Yang J, Du X, Chen B, Deng C et al (2022) Expansion of FGFR3-positive nucleus pulposus cells plays important roles in postnatal nucleus pulposus growth and regeneration. Stem Cell Res Ther 13(1):227. https://doi.org/10.1186/s13287-022-02903-2

Yan D, Chen D, Cool SM, van Wijnen AJ, Mikecz K, Murphy G, Im H-J (2011) Fibroblast growth factor receptor 1 is principally responsible for fibroblast growth factor 2-induced catabolic activities in human articular chondrocytes. Arthritis Res Ther 13(4):R130. https://doi.org/10.1186/ar3441

Yao Z-Z, Hu A-X, Liu X-S (2017) DUSP19 regulates IL-1β-induced apoptosis and MMPs expression in rat chondrocytes through JAK2/STAT3 signaling pathway. Biomed Pharmacother 96:1209–1215. https://doi.org/10.1016/j.biopha.2017.11.097

Yasoda A, Komatsu Y, Chusho H, Miyazawa T, Ozasa A, Miura M, Kurihara T, Rogi T, Tanaka S, Suda M, Tamura N, Ogawa Y, Nakao K (2004) Overexpression of CNP in chondrocytes rescues achondroplasia through a MAPK-dependent pathway. Nat Med 10(1):80–86. https://doi.org/10.1038/nm971

Yin L, Du X, Li C, Xu X, Chen Z, Su N, Zhao L, Qi H, Li F, Xue J, Yang J, Jin M, Deng C, Chen L (2008) A Pro253Arg mutation in fibroblast growth factor receptor 2 (Fgfr2) causes skeleton malformation mimicking human Apert syndrome by affecting both chondrogenesis and osteogenesis. Bone 42(4):631–643. https://doi.org/10.1016/j.bone.2007.11.019

Ying L, Wang L, Guo K, Hou Y, Li N, Wang S, Liu X, Zhao Q, Zhou J, Zhao L, Niu J, Chen C, Song L, Hou S, Kong L, Li X, Ren J, Li P, Mohammadi M, Huang Z (2021) Paracrine FGFs target skeletal muscle to exert potent anti-hyperglycemic effects. Nat Commun 12(1):7256. https://doi.org/10.1038/s41467-021-27584-y

Yu K, Herr AB, Waksman G, Ornitz DM (2000) Loss of fibroblast growth factor receptor 2 ligand-binding specificity in Apert syndrome. Proc Natl Acad Sci USA 97(26):14536–14541. https://doi.org/10.1073/pnas.97.26.14536

Yu K, Ornitz DM (2001) Uncoupling fibroblast growth factor receptor 2 ligand binding specificity leads to Apert syndrome-like phenotypes. Proc Natl Acad Sci USA 98(7):3641–3643. https://doi.org/10.1073/pnas.081082498

Yu K, Ornitz DM (2008) FGF signaling regulates mesenchymal differentiation and skeletal patterning along the limb bud proximodistal axis. Development 135(3):483–491. https://doi.org/10.1242/dev.013268

Yu K, Xu J, Liu Z, Sosic D, Shao J, Olson EN, Towler DA, Ornitz DM (2003) Conditional inactivation of FGF receptor 2 reveals an essential role for FGF signaling in the regulation of osteoblast function and bone growth. Development 130(13):3063–3074. https://doi.org/10.1242/dev.00491

Zakrzewska M, Marcinkowska E, Wiedlocha A (2008) FGF-1: from biology through engineering to potential medical applications. Crit Rev Clin Lab Sci 45(1):91–135. https://doi.org/10.1080/10408360701713120

Zhang X, Bao L, Yang L, Wu Q, Li S (2012) Roles of intracellular fibroblast growth factors in neural development and functions. *Science China*. Life Sci 55(12):1038–1044. https://doi.org/10.1007/s11427-012-4412-x

Zhang X, Ibrahimi OA, Olsen SK, Umemori H, Mohammadi M, Ornitz DM (2006) Receptor specificity of the fibroblast growth factor family. The complete mammalian FGF family. J Biol Chem 281(23):15694–15700. https://doi.org/10.1074/jbc.M601252200

Zhou YX, Xu X, Chen L, Li C, Brodie SG, Deng CX (2000) A Pro250Arg substitution in mouse Fgfr1 causes increased expression of Cbfa1 and premature fusion of calvarial sutures. Hum Mol Genet 9(13):2001–2008. https://doi.org/10.1093/hmg/9.13.2001

Zhuang L, Falquet L, Trueb B (2010) Genome-wide comparison of FGFRL1 with structurally related surface receptors. Exp Ther Med 1(1):161–168. https://doi.org/10.3892/etm_00000026

Chapter 9
Impaired Wnt Signaling as a Cause of Skeletal Disorders

Timur A. Yorgan and Thorsten Schinke

Abstract Skeletal development, growth, and remodeling are highly complex processes involving many specific interactions between different cell types, as well as the formation of intact extracellular matrices. Moreover, bone is a highly dynamic tissue undergoing steady remodeling by two fundamentally different cell types, i.e., bone-forming osteoblasts and bone-resorbing osteoclasts. In the last decades, pathogenic variants in a large number of genes have been identified to cause inherited disorders characterized by specific disturbances in the function of distinct skeletal cell types. Whereas several of these skeletal disorders with impaired extracellular matrix production and associated skeletal dysplasia are discussed in other chapters of this book, the focus of the present chapter is related to skeletal disorders caused by pathogenic variants in genes encoding modulators of the Wnt signaling pathway. Importantly, triggered by the genetic analysis of patients with osteosclerotic disorders, i.e., high bone mass due to activation of matrix-producing osteoblasts, these findings have paved the way to develop an efficient osteoanabolic medication. More specifically, Romosozumab, a monoclonal antibody neutralizing the function of Sclerostin was established to increase the activity of the Wnt co-receptor Lrp5, which promotes bone formation by osteoblasts. Besides Sclerostin and Lrp5, there are however other components of the Wnt signaling pathway that play important roles in skeletal biology, as identified by human genetics. This chapter therefore summarizes the current knowledge on impaired Wnt signaling as a cause of skeletal disorders.

T. A. Yorgan · T. Schinke (✉)
Department of Osteology and Biomechanics, University Medical Center Hamburg-Eppendorf, Hamburg, Germany
e-mail: schinke@uke.de

A. Rossi, F. Zaucke (eds.), *The Extracellular Matrix in Genetic Skeletal Disorders*, Biology of Extracellular Matrix 16,
https://doi.org/10.1007/978-3-031-70835-0_9

9.1 Introduction

9.1.1 Skeletal Development, Growth, and Remodeling

Skeletal development, growth, and remodeling are highly complex processes involving many different cell types with unique functions (Berendsen and Olsen 2015). Whereas a small subset of skeletal elements, mostly in the skull, develops by intramembranous ossification, where mesenchymal progenitor cells directly differentiate into bone-forming osteoblasts, the majority of the axial and appendicular skeleton develops by endochondral ossification (Long and Ornitz 2013; Ferguson and Atit 2019). In this latter process, the mesenchymal progenitor cells initially condensate into cell clusters which differentiate into chondrocytes to produce a cartilage intermediate that is subsequently replaced by bone. At the end of puberty, this modeling phase is essentially replaced by bone remodeling, a physiologically relevant process, that is mediated by the coordinated activities of bone-resorbing osteoclasts and osteoblasts (Zaidi 2007). The rate of bone remodeling is controlled, at least in part, by osteocytes, representing terminally differentiated osteoblasts forming a cellular network within the mineralized bone matrix (Niedzwiedzki and Filipowska 2015).

A relative increase in bone resorption is the cause of osteoporosis, one of the most prevalent disorders in the aged population (Johnell and Kanis 2006; Compston et al. 2019). Current treatment options to counteract bone loss and to reduce the risk for osteoporotic fractures are either targeting the osteoclast population (antiresorptives) or aiming at stimulating osteoblast activity (osteoanabolics), which is explained by the fact that osteoblasts and osteoclasts are fundamentally different cell types. More specifically, osteoblasts derive from mesenchymal progenitors and act in large groups of cells secreting an extracellular matrix (osteoid) that gradually mineralizes (Karsenty et al. 2009). A subset of osteoblasts undergoes terminal differentiation into osteocytes, which form a cellular network within the mineralized bone matrix that regulates bone remodeling and mineral homeostasis (Dallas et al. 2013). In contrast, osteoclasts derive by fusion of hematopoietic cells belonging to the monocyte/macrophage lineage and resorb bone by extracellular acidification and secretion of matrix-degrading enzymes (Cappariello et al. 2014).

In the last decades, many physiologically relevant molecules controlling bone development, remodeling, and matrix mineralization were identified by molecular genetics. In fact, as summarized in the latest revision of the nosology of skeletal disorders, there are now 552 listed genes, for which pathogenic variants were found to cause a specific inherited skeletal disorder (Unger et al. 2023). Most importantly, the combined analysis of patients with severe inherited skeletal disorders and the respective mouse deficiency models have led to the discovery of novel drug targets that have paved the way to develop new treatment options for osteoporosis (Denosumab, blocking the pro-osteoclastogenic cytokine RANKL) or hypophosphatemic rickets (Burosumab, blocking the bone-derived phosphaturic hormone FGF23) (Nagy and Penninger 2015; Agoro et al. 2020). One of these targets is the

Wnt signaling inhibitor Sclerostin, which plays a key role as an inhibitor of osteoblast activity. Whereas the blockade of Sclerostin by Romosozumab has now been established as an effective osteoanabolic treatment option (Cosman et al. 2016), it is important to enhance our understanding of Wnt signaling in bone cells, especially since other inherited skeletal disorders are also caused by genetic variants affecting components of this pathway.

9.1.2 The Wnt Signaling Pathway

The Wnt signaling pathway is one of the evolutionary oldest paracrine signal transduction pathways (Loh et al. 2016). Generally, it is facilitated via secreted Wnt ligands and their corresponding receptors of the Frizzled family. Wnt signaling plays an important role in embryonic development and organ homeostasis, including the skeletal system, but also detrimental processes such as cancer development and genetic disorders including musculoskeletal dysplasias (Clevers 2006). On a cellular level, Wnt signals are often involved in modulating differentiation, proliferation, polarity, apoptosis, and cell migration (Saito-Diaz et al. 2013). The complexity of this pathway derives from the large number of different Wnt ligands and receptors as well as the intricate intracellular signal propagation network that allows for modulation and interaction with other cellular signaling pathways before a cellular response is elicited. Broadly, Wnt signaling is categorized into the canonical and non-canonical pathways. Historically, this distinction was based on the ability to induce malignant transformation in a specific cell line, but nowadays it is dependent on the activated intracellular signal transduction pathway (Saito-Diaz et al. 2013). Overall, Wnt signaling is a highly complex and extensive topic that is still not fully elucidated, and new scientific findings regularly continue to expand our knowledge. As such, the following paragraphs will provide a brief overview of the Wnt signaling pathway as it is regarded now. However, for a more detailed description of the pathways and all known molecular components, readers are referred to current excellent reviews on this matter (Saito-Diaz et al. 2013; Rim et al. 2022).

9.1.2.1 Canonical Wnt Signaling

Canonical Wnt signaling is defined by the intracellular signal transduction via β-catenin. In its resting state, cells continuously produce β-catenin that is subsequently phosphorylated by an inactivation complex consisting of several protein subunits (Axin, APC, GSK3β, CK1α) (Clevers 2006). The phosphorylation tag marks β-catenin for proteasomal degradation, preventing the intracellular accumulation of β-catenin (Saito-Diaz et al. 2013). Upon binding of an extracellular Wnt ligand to a receptor of the Frizzled family and recruitment of a co-receptor of the Lrp family (mostly Lrp5 or Lrp6), the destruction complex is bound and inactivated via the coupling protein Dishevelled. This results in an intracellular accumulation and subsequent nuclear translocation of β-catenin, where it acts as a co-activator for transcription factors of the TCF/LEF family (Saito-Diaz et al. 2013). A well-established modulatory mechanism of this pathway is the inhibition of the ligand/receptor interaction. Several inhibitors have been identified, such as Dkk1 and Sclerostin, that are able to bind to Lrp co-receptors and prevent the formation of the activating receptor complex (Balemans et al. 2001; Semënov et al. 2001; Yorgan and Schinke 2014). Furthermore, Lrp4 has been identified as an inhibitory co-receptor that binds Wnt ligands and Fzd receptors without inducing intracellular signal propagation (Ohazama et al. 2008; Li et al. 2010). Likewise, soluble Fzd receptors (e.g., Sfrp1) can scavenge Wnt ligands in the extracellular space before they reach the target cell (Bafico et al. 1999; Rim et al. 2022). Additional signal modulation is possible either in the sending cell via regulation of the proteins responsible for secreting Wnt ligands or providing essential posttranslational modifications or in the receiving cell via modulation and integration of the intracellular signal transduction molecules (Rim et al. 2022).

9.1.2.2 Non-canonical Wnt Signaling

Non-canonical Wnt signaling includes two distinct pathways that markedly differ from the canonical pathway. Similar to the canonical pathway, the Wnt/Ca^{2+} pathway is activated by binding of a Wnt ligand to a Fzd receptor and Lrp co-receptor. However, in the next step, the phospholipase C (PLC) is activated. This leads to an increase in intracellular Ca^{2+} levels causing the activation of further elements of the signal transduction cascade such as PKC and calcineurin (De 2011). The activation of the Wnt/PCP (planar cell polarity) pathway occurs via the interaction of a Wnt ligand with a Fzd receptor and a receptor tyrosine kinase, such as ROR2, PTK7, or NRH1, as a co-receptor instead of a Lrp family member. While this interaction also results in the recruitment of Dishevelled, the further signal transduction is facilitated via the DAAM1 complex and RAX1 resulting in the activation of a kinase cascade including ROCK and JNK that control the cytoskeleton and actin polymerization in addition to regulating target gene transcription (Niehrs 2012; Wang et al. 2014). As such, this pathway is important for determining cell polarity and cell migration. In the context of the skeletal system, these are especially vital aspects for longitudinal bone growth, as the alignment and polarity of the chondrocytes in the growth plate are determined by this pathway (Randall et al. 2012; Hallett et al. 2021). Finally, it needs to be stated that like most biological systems, the Wnt pathways are somewhat malleable, and interactions between canonical and non-canonical signals as well as other pathways such as the BMP signaling pathway or the Yap/Taz pathway have been described (Itasaki and Hoppler 2010; Kim and Jho 2014). Thus, Wnt signals promote essential cellular pathways in many organs, including the musculoskeletal system, but they should not be considered isolated systems.

9.1.2.3 Relevance in Skeletal Dysplasias

The relevance of Wnt signaling for skeletal integrity was first discovered by the analysis of specific genetic disorders, thereby paving the way for numerous investigations that have all greatly contributed to our current state of knowledge. While there are certainly still numerous discoveries to be made on this subject, we have come to a point where the scientific community can appreciate the central role of Wnt signaling in skeletal metabolism. Most importantly, these findings have contributed to the successful development of a highly specific and effective treatment option, i.e., Romosozumab, a neutralizing antibody against the Wnt signaling inhibitor Sclerostin that can be used to treat patients suffering from osteoporosis and potentially other low bone mass disorders (Cosman et al. 2016).

Initially, genetic analysis of specific subtypes of osteosclerosis, namely, sclerosteosis (OMIM: 269500) and van Buchem's disease (OMIM: 239100), revealed variants in the coding region or the upstream region of a gene with previously unknown relevance that would later be termed Sclerostin (encoded by the *SOST* gene) (Balemans et al. 2001; Balemans et al. 2002). Among the most intriguing aspects of this discovery was the expression pattern of this gene which was nearly exclusively found to be expressed in osteocytes (Winkler et al. 2003). While the molecular function of Sclerostin was initially controversial, a substantial body of research ultimately determined that it acts as a Wnt signaling inhibitor, similar to Dkk proteins, which are well-established inhibitory regulators of Wnt signaling binding to Lrp5, for which no pathogenic variants in skeletal diseases have been identified to date (Huang et al. 2018; Tanaka and Matsumoto 2021). Therefore, the importance of Wnt signaling for skeletal homeostasis was further supported. After these seminal discoveries, many other components of the Wnt signaling pathway were found to influence skeletal metabolism or be involved in the pathomechanisms of skeletal dysplasias (Huybrechts et al. 2020). The list is continuously expanding, but even with our current knowledge, it is clear that Wnt signals are involved in many, if not all, aspects of skeletal metabolism, including proliferation and differentiation of osteoblasts and osteoclasts, bone matrix production and resorption, and processing of ECM components, especially type I collagen (Baron and Kneissel 2013; Huybrechts et al. 2020). Thus, understanding Wnt signaling has become an integral part of our attempts to understand and ultimately modulate skeletal metabolism. In the following sections, we will discuss some of the most relevant skeletal dysplasias caused by aberrant Wnt signaling in more detail.

9.2 Wnt Co-receptors with Relevance for the Skeleton

9.2.1 Lrp5, a Wnt Co-receptor Controlling Bone Formation

The initial link between skeletal dysplasias and Wnt signaling was established in the last decade of the twentieth century. Genetic linkage for both, the osteoporosis-pseudoglioma syndrome (OPPG, OMIM: 259770), a low bone mass disorder, and osteosclerosis (OMIM: 607634), a high bone mass disorder, was mapped to the same genetic locus of 11q13, despite the essentially opposite clinical manifestation (Gong et al. 1996; Johnson et al. 1997). The availability of more refined methods allowed for the identification of the affected gene. In both cases, different variants of the *LRP5* gene were determined to be the cause of the apparent skeletal dysplasias (Gong et al. 2001; Boyden et al. 2002; Little et al. 2002). At that point, Lrp5 had been already established as a co-receptor in the Wnt signaling pathway, thus linking Wnt signaling and bone metabolism.

9.2.1.1 Osteoporosis-Pseudoglioma Syndrome

Patients afflicted with osteoporosis-pseudoglioma syndrome (OPPG, OMIM: 259770) suffer from progressive visual loss, usually resulting in complete blindness upon reaching adulthood, as well as early-onset osteoporosis (Sabir and Cole 2019). The osteoporotic phenotype in these patients results in an increased fracture rate, decreased longitudinal bone growth, and skeletal deformities of the limbs, the vertebral column, and the skull (Streeten et al. 2008). Given the extreme rarity of this disorder with less than 100 reported cases, less frequent symptoms cannot be reported with high confidence but may include joint abnormalities, hypotonia, seizures, and potentially mild intellectual disability (Sabir and Cole 2019).

OPPG is inherited in an autosomal recessive pattern, and diagnosis is typically made through genetic testing. Loss-of-function mutations of the LRP5 gene have been identified to be the cause of this syndrome (Gong et al. 2001). As a co-receptor for the Wnt signaling pathway, Lrp5 is required for forming the receptor complex and initiating intracellular signal transduction (Baron and Kneissel 2013). Although Lrp6 can provide similar functionality and has been shown to partially compensate for diminished Lrp5 activity, this redundancy is not sufficient to prevent the manifestation of the syndrome (Kelly et al. 2004). Moreover, while Lrp5 is expressed nearly ubiquitously in most tissues and Wnt signaling has relevant functions in many organs, the symptoms of OPPG mostly occur in the visual and skeletal system (Joiner et al. 2013). While visual impairment in OPPG has been determined to be caused by the persistence of embryonic eye vascularization, experimental evidence indicates that the osteoporotic phenotype is caused by a reduction of osteoblast activity (Kato et al. 2002; Huang et al. 2016). Indeed, detailed analysis of a Lrp5-deficient mouse model revealed that these mice display low bone mass, comparable to OPPG patients, that is fully attributable to decreased bone formation via osteoblasts, whereas osteoclast parameters remain unaffected (Kato et al. 2002). These findings highlight the importance of Lrp5-mediated Wnt signaling for the production of bone matrix. Together with the results from the analysis of osteosclerotic disorders that will be discussed below, the uncovering of the cellular pathomechanism of OPPG greatly contributed to our current appreciation for the relevance of Wnt signaling in bone formation.

While the pathomechanism of OPPG is now revealed and understood on a cellular level, treatment of patients affected by OPPG remains challenging. There is currently no curative treatment for OPPG, and standard therapy includes the administration of bisphosphonates, a class of pharmaceutical compounds that binds to calcium phosphate crystals in the bone matrix and inhibits osteoclastic bone resorption with the aim of increasing bone mass and preventing fractures (Papapoulos 2008; Sabir and Cole 2019). While this approach is met with limited success with regard to the measured bone mass, there is no appreciable improvement in the fracture rate (Streeten et al. 2008). Additional supportive treatment measures include vitamin D and calcium supplementation to help maintain proper bone health and physical therapy to improve strength, balance, and coordination in order to reduce the risk of fractures via falls (Levasseur 2008). If necessary, assistive devices such as walkers, canes, or braces can help improve mobility and reduce the risk of falls as well. Recently, a more targeted therapeutic approach was tested. Lithium is an established and relatively safe drug that has been in use for the treatment of psychiatric disorders for several decades (Schou 2001). In this context, positive effects on bone strength were observed as an unexpected side effect (Zamani et al. 2009; Köhler-Forsberg et al. 2022). On a molecular level, lithium inhibits GSK3β, the enzyme responsible for phosphorylating β-catenin and thus marking it for proteasomal degradation (Hedgepeth et al. 1997). Prevention of this phosphorylation step induces the accumulation of β-catenin, thus artificially activating the canonical Wnt signaling pathway and bypassing the requirement for the formation of a receptor complex involving Lrp5. Treatment of Lrp5-deficient mice with lithium chloride resulted in substantially improved bone mass and a significantly reduced fracture rate (Clement-Lacroix et al. 2005). However, although translation into patient care was attempted, the results remained inconclusive due to the very low number of only two patients completing the full course of the clinical trial (Streeten 2018).

9.2.1.2 Lrp5-Associated High Bone Mass Syndrome

Among the group of osteosclerotic disorders, there is another syndrome that appears phenotypically highly similar to van Buchem disease and sclerosteosis (described in 9.3.1.1) and shares many symptoms and physiological alterations in the affected patients. However, in contrast to the aforementioned skeletal dysplasias, this high bone mass syndrome is caused on a molecular level by variants of the Wnt co-receptor Lrp5 and not mutations of Sclerostin (Boyden et al. 2002; Little et al. 2002). As previously discussed, Lrp5 contributes to Wnt signal transduction by forming a receptor complex with a receptor from the Fzd family and the actual Wnt ligand (Baron and Kneissel 2013). Loss of Lrp5 diminishes this signaling potential and results in decreased osteoblast activity and therefore results in a low bone mass phenotype (Kato et al. 2002). Similarly, the physiological function of the inhibitor Sclerostin is to bind to Lrp5 which prevents the formation of the receptor complex, thus effectively blocking the transduction of Wnt signals (Li et al. 2005; Semënov et al. 2005). Consequently, all Lrp5 variants leading to a high bone mass phenotype need to be considered gain-of-function mutations. Indeed, it was shown that all Lrp5 variants associated with high bone mass syndrome are located in the extracellular region of the co-receptor that is responsible for the interaction with the inhibitor Sclerostin (Ellies et al. 2006; Semënov and He 2006). Thus, the described variants of the Lrp5 co-receptor prevent the binding of Sclerostin and subsequent inhibition of the signaling cascade (Yorgan et al. 2015).

While the symptoms and clinical management of patients suffering from Lrp5-associated high bone mass syndrome (OMIM: 607634) are virtually identical to the measures taken in van Buchem disease and sclerosteosis, the investigation of the underlying pathomechanisms has contributed valuable insights into the role of Wnt signaling in skeletal homeostasis. Indeed, one of the more surprising findings was that mutations of Lrp5 can lead to phenotypes that are essentially opposites of each other. While loss of function causes OPPG, a low bone mass disorder, gain of function increases bone mass (Levasseur et al. 2005). Careful studies, greatly aided by the availability of corresponding mouse lines, have been successful in elucidating the intricate details of the molecular mechanism that was described above, thus further cementing the central role of Wnt signaling and especially of the modulatory function of Sclerostin in bone metabolism (Tanaka and Matsumoto 2021). Nonetheless, despite this great progress, the most central components of the Wnt pathway in the context of skeletal homeostasis remain to be identified. To date, no single Wnt ligand and Fzd receptor combination has been described to exert the prevailing bone-anabolic response in osteoblasts. Furthermore, the cellular signaling cascade downstream of Lrp5 is not fully elucidated. Indeed, several studies have shown that inactivation of β-catenin, the central component of canonical Wnt signaling, specifically in osteoblasts leads to a phenotype distinctly different from Lrp5 inactivation. Indeed, while these mouse models presented with a lower bone mass, this was caused by increased osteoclastic bone resorption and not by decreased bone formation (Glass et al. 2005; Holmen et al. 2005; Kramer et al. 2010). These results suggest that Lrp5- and β-catenin-dependent Wnt signals are not fully overlapping and that more complex signal integration processes may be involved. However, investigations of further skeletal dysplasias associated with impaired Wnt signaling may provide indications toward possible answers to this question. Some of them are discussed in the section below.

9.2.2 *Impact of Variants in Other Lrp Receptors on the Skeleton*

Besides Lrp5, there are also genetically confirmed functions of the other Lrp receptors on the skeleton, although variants in the respective genes, i.e., *LRP4* and *LRP6*, are less frequent. With respect to Lrp6, representing the closest homolog of Lrp5, it was found that heterozygous mutations in the extracellular domain caused high bone mass in two unrelated families (Whyte et al. 2019). On the other hand, heterozygous loss-of-function variants of Lrp6 were identified in individuals with tooth agenesis (OMIM: 616724) or coronary artery disease (OMIM: 616724), and the latter patients also displayed low bone mineral density and/or fractures (Mani et al. 2007; Ockeloen et al. 2016).

While these observations are consistent with the osteoanabolic function also described for Lrp5, the genetic evidence for the role of Lrp4 indicated that this latter protein has an entirely different function in the skeleton. In fact, pathogenic variants of Lrp4 can cause, depending on the type and location of the mutation, sclerosteosis (OMIM: 614305), Cenani-Lenz syndrome (OMIM: 2212780, associated with syndactyly, synostosis, and renal hypoplasia), or congenital myasthenia (OMIM: 616304) (Li et al. 2010; Leupin et al. 2011; Ohkawara et al. 2014). Interestingly, however, the functional analyses, together with the high bone mass phenotype observed in mice with osteoblast-specific deletion of Lrp4, revealed that Lrp4 is required to inhibit Wnt signaling, as it serves as an anchor protein for Sclerostin (Chang et al. 2014; Fijalkowski et al. 2016), a major inhibitor of bone formation, described in more detail in the following section.

9.3 Wnt Signaling Antagonists with Relevance for the Skeleton

9.3.1 *Sclerostin, a Physiologically Relevant Inhibitor of Bone Formation*

9.3.1.1 van Buchem Disease

In contrast to the previously discussed OPPG, patients affected by the van Buchem disease (OMIM: 239100) or the similar and closely related sclerosteosis (OMIM: 269500) present with dramatically increased bone mass (Appelman-Dijkstra et al. 1993). van Buchem disease and sclerosteosis are progressive sclerosing bone dysplasias characterized by generalized osteosclerosis and hyperostosis of the skeleton. In the middle of the twentieth century, this extraordinarily rare family of skeletal disorders was initially described in patients originating from the Dutch region of Urk (Van Buchem et al. 1955). Shortly, thereafter, another accumulation of highly similar cases was identified in South Africa in a population descending from Dutch emigrants (Truswell 1958). Overall, only a few hundred cases have been described so far. While there are several seemingly independent manifestations found all over the world, they usually appear to occur in isolated communities, often on a background of increased consanguinity (van Lierop et al. 2017). The increased bone mass that is characteristic of van Buchem disease and sclerosteosis markedly reduces the probability of fractures in these patients but may lead to other complications (van Lierop et al. 2013). First, symptoms usually occur in early childhood, when bone mass and distinct overgrowth of bones continuously progress until they can stabilize after the third decade. However, depending on the severity of the disease, it may progress throughout the patients' lifetime (van Lierop et al. 2017). These skeletal changes can result in joint stiffness and pain as well as decreased mobility. More importantly, the pathological overgrowth may lead to constriction of nerve exit points in the skull and cause sensory loss (blindness, deafness) and facial palsy. In severe cases, it has been reported that the inward growth of skull bones can lead to increased intracranial pressure and may ultimately lead to the demise of the patients due to damages to the brain stem (Beighton et al. 1976; van Lierop et al. 2017). Syndactyly is another hallmark symptom of sclerosteosis that is interestingly not observed in van Buchem disease and provides, together with the lack of increased stature in van Buchem patients, a distinction between these two disorders (van Lierop et al. 2017).

Both van Buchem disease and sclerosteosis are inherited in an autosomal recessive pattern (Appelman-Dijkstra et al. 1993). Diagnosis can be made via morphological and radiological analysis and nowadays typically genetic testing. Alterations of the gene *SOST* encoding the protein Sclerostin have been identified as the underlying cause for both disorders. While sclerosteosis is caused by loss-of-function mutations of *SOST*, the van Buchem disease is caused by a deletion of a regulatory genomic region approximately 35 kB downstream of *SOST*, resulting in decreased expression levels (Balemans et al. 2001; Balemans et al. 2002). Sclerostin was first discovered in the context of the search for the pathomechanism of the van Buchem disease/sclerosteosis. While its function was initially controversial and inhibitory effects on bone morphogenic protein signaling were postulated, it has become evident that Sclerostin is indeed a potent Wnt signaling inhibitor (Winkler et al. 2003; Li et al. 2005; Semënov et al. 2005). Careful genetic and biochemical studies as well as binding assays have revealed that Sclerostin interacts with Lrp5 and inhibits the formation of the receptor complex, thus preventing the propagation of Wnt signals (Li et al. 2005, Semënov et al. 2005).

Interestingly, the expression of Sclerostin is nearly exclusive to cells of the osteoblast lineage, more specifically osteocytes (Winkler et al. 2003). Therefore, Sclerostin is now considered an essential regulatory component of bone metabolism and one of the major effector proteins secreted by osteocytes (Han et al. 2018). Furthermore, this expression specificity explains the exclusive occurrence of skeletal symptoms in van Buchem disease and sclerosteosis, notwithstanding secondary symptoms caused by excessive bone overgrowth. Interestingly, histopathological analyses revealed that the increased bone mass observed in van Buchem disease and sclerosteosis patients is caused by increased bone-forming activity of osteoblasts, whereas no indication of altered osteoclast activity was observed (van Lierop et al. 2011, 2013, 2017). Indeed, deletion of Sclerostin in mice resulted in increased bone mass caused by increased osteoblastic bone formation, while transgenic overexpression of *Sost* in mouse models led to a strong decrease of bone mass, exclusively mediated via decreased osteoblast activity, thereby further supporting this mechanism (Winkler et al. 2003; Loots et al. 2005; Li et al. 2008; Tu et al. 2012; Yorgan et al. 2015). Importantly, the anti-osteoanabolic influence of transgenic *Sost* overexpression was fully blunted in mice carrying the *Lrp5* high bone mass allele, thereby demonstrating that Sclerostin acts via binding to the extracellular domain of Lrp5 (Fig. 9.1). The excessive bone matrix formed in van Buchem disease and sclerosteosis patients represents structurally healthy bone with proper collagen fiber alignment and mineral integration, thereby supporting the observation that the mechanical

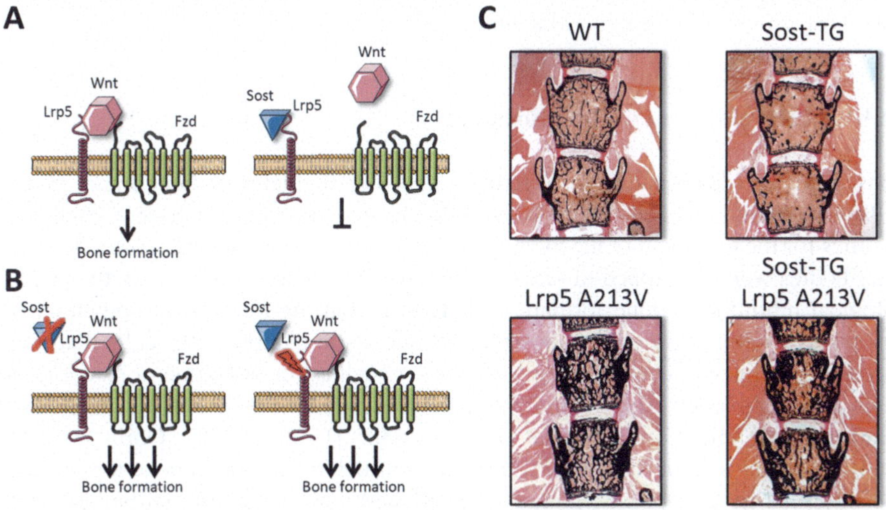

Fig. 9.1 Interaction of Lrp5 and Sclerostin (Sost). (**a**) Physiological regulation of Wnt signaling. After the Wnt ligand binds a Fzd receptor and Lrp5 co-receptor, intracellular signal transduction is induced via the stabilization and nuclear translocation of β-catenin. This results in the increased expression of target genes leading to increased bone formation by osteoblasts. Sost binds Lrp5 and prevents the formation of the receptor complex, thus inhibiting the intracellular Wnt signal transduction. (**b**) Impact of high bone mass causing Sost and Lrp5 variants. Inactivating variants of Sost result in a missing physiological inhibition of the Wnt signal and increased bone formation by osteoblasts. Activating mutations of Lrp5 prevent the binding of Sost and thus the inhibition of the signal, resulting in increased bone formation as well. (**c**) In vivo interaction of Lrp5 and Sost. Shown are von Kossa stained histological sections of the lumbar vertebral bodies L3 and L4 of mice. Mineralized tissue appears in black. Transgenic overexpression of Sost specifically in osteoblasts (Sost-TG) results in a decreased bone mass compared to wildtype (WT) controls. The activating Lrp5 variant A213V results in an increased bone mass compared to WT. The combination of Sost-TG and Lrp5 A213V still results in a high bone mass phenotype, demonstrating the molecular mechanism. Parts of the figure were drawn by using pictures from Servier Medical Art. Servier Medical Art by Servier is licensed under a Creative Commons Attribution 3.0 Unported License (https://creativecommons.org/licenses/by/3.0/)

properties of bones from affected patients are improved and the fracture risk is reduced.

Currently, there is no curative treatment available for van Buchem disease or sclerosteosis patients, and management is focused on minimizing the effect of the occurring symptoms (van Lierop et al. 2017). This approach is usually limited to surgical intervention in order to prevent nerve damage and release intracranial pressure. However, in milder cases, no treatment is necessary as the patients do not experience detrimental effects that would lead to a significant reduction in their quality of life. This aspect indicates that, depending on the Sclerostin dosage, beneficial improvements of the skeletal structure and microarchitecture can be achieved without noteworthy side effects. Therefore, the elucidation of the pathomechanism of van Buchem disease and sclerosteosis has not only further documented the importance of Wnt signaling in skeletal metabolism but has additionally paved the way for the development of one of the most effective osteoanabolic drugs (Romosozumab) (Tanaka and Matsumoto 2021).

9.3.1.2 Development of Romosozumab

Based on the identification of Sclerostin as a key regulator of bone mass, major efforts were undertaken to leverage this knowledge in order to generate novel therapeutics for the treatment of the most common low bone mass disorder, osteoporosis. Antibodies were developed to scavenge Sclerostin and essentially block the physiological inhibitor of bone formation, thereby increasing the inherent potential of osteoblasts without providing supraphysiological stimulation of the cells that may result in detrimental aberrant cellular behavior such as the induction of osteosarcomas (MacNabb et al. 2016). Initial animal studies led to very promising results that warranted the initiation of several clinical studies (Li et al. 2009; Ominsky et al. 2010; Das and Sakthiswary 2013). Here, a remarkable increase in bone mass could be observed in osteoporotic patients treated with anti-Sclerostin antibodies, well beyond the effect of previously available osteoanabolic compounds (McClung et al. 2014; Padhi et al. 2014; Recker et al. 2015). This effect was attributable not only to the expected increase in bone formation, but also to an unexpected but welcomed decrease in bone resorption (McClung et al. 2014; Padhi et al. 2014). To date, Romosozumab is the first and only anti-Sclerostin compound that has successfully completed all clinical trial phases and has been approved for clinical use (Markham 2019). Due to a non-significant, but conspicuous number, of cerebrovascular complications in the treatment group, however, this medication is currently marked with a black box warning and only approved for treatment of postmenopausal osteoporosis in patients with no cardiovascular event in the 12 months prior to treatment initiation (Saag et al. 2017, (FDA) 2019). However, in continued studies and metanalyses of all previously published data, there was no evidence of a significant link between treatment and an increased risk of cardiovascular events (Lv et al. 2020). Nonetheless, an evaluation of the Food and Drug Administration Adverse Event Reporting System database still identified a potential signal for an elevated risk (Vestergaard Kvist et al. 2021). Thus the safety profile of Romosozumab remains to be fully elucidated in future studies.

9.3.2 *Impact of the Wnt Inhibitor Sfrp4 on the Skeleton (Pyle Disease)*

With respect to Wnt signaling antagonists, there is another molecule with physiological relevance for skeletal integrity, i.e., Sfrp4, for which pathogenic variants were identified to cause Pyle disease (OMIM: 265900). Pyle disease or metaphyseal dysplasia is a very rare skeletal disorder with only a few dozen described cases (Chetty et al. 2022). The clinical manifestation of this autosomal recessive disorder is often rather mild, although a marked skeletal phenotype is present. The hallmark characteristics of this disorder are genu valgum and widened metaphyses in the appendicular skeleton, giving the affected bones a morphological appearance reminiscent of a paddle (Kiper et al. 2016). On a microarchitectural level, these dysmorphisms are the result of an expansion of the trabecular bone compartment while cortical bone thickness is reduced (Kiper et al. 2016). These skeletal alterations increase the fragility of the bones and result in an increased fracture risk for patients suffering from Pyle disease. Treatment of Pyle disease is symptomatic and may include management of fractures and corrective orthopedic surgery (Soares et al. 2016). As determined by genetic analysis of affected patients, loss-of-function variants in the gene *SFRP4* have been identified as the cause of Pyle disease (Kiper et al. 2016; Galada et al. 2017; Pawar and Rao 2018). This finding is further supported by the identification of SNPs in the same gene that were linked to bone mineral density (Boudin et al. 2012; Chen et al. 2022).

Sfrp4 is a member of the Sfrp family of Wnt signaling inhibitors. Sfrps are soluble proteins with partial homology to the membrane-bound Fzd receptors (Cruciat and Niehrs 2013). As such, they can act as decoy receptors and scavenge Wnt ligands before they reach the receiving cells, thus attenuating Wnt signal transmission. Varying specificities for different Wnt ligands have been described for the five known Sfrps. However, Sfrp4 is considered to be relatively ligand-agnostic and may interact with many different Wnt ligands (Wawrzak et al. 2007; Agostino et al. 2017). As mice lacking components of the non-canonical Wnt/PCP pathway are partially protected from the effects of Sfrp4 deletion, a predominant preference for Sfrp4 for non-canonical Wnt ligands has been postulated (Chen et al. 2019). While mice that expressed a *Sfrp4* transgene in an osteoblast-specific manner were shown to display decreased trabecular bone mass with no cortical effect (Nakanishi et al. 2008), different mouse lines with a global deletion of Sfrp4 mimicked the phenotype of Pyle disease with increased trabecular bone mass as well as cortical thinning (Brommage et al. 2014; Haraguchi et al. 2016; Kiper et al. 2016). Additionally, a widening of the metaphyseal region of long bones in line with the clinical presentation of Pyle disease was described (Kiper et al. 2016). The contrast between the effect on trabecular and cortical bone is one of the more intriguing aspects of the pathomechanism of Pyle disease, and data from murine studies suggest that Sfrp4 is regulating metaphyseal bone remodeling. More specifically, Sfrp4 deficiency increases trabecular bone formation without affecting resorption in this compartment (Brommage et al. 2014, Haraguchi et al. 2016, Kiper et al. 2016). In contrast, cortical bone formation on the periosteal surface is inhibited by Sfrp4 deficiency, and an additional increase in endosteal bone resorption via osteoclasts was reported (Haraguchi et al. 2016; Kiper et al. 2016; Chen et al. 2019). However, the specific cellular effects of Sfrp4 were likely masked in the global knockout model. Detailed analysis of Sfrp4-deficient mice revealed a locally increased expression of Rankl, a stimulator of osteoclast differentiation, by cortical osteoblast-lineage cells (Kiper et al. 2016). Additionally, it was found that Sfrp4 deletion directly enhances osteoclast differentiation (Chen et al. 2019). Furthermore, specific effects of Sfrp4 have been observed on the periosteal and endocortical surface, suggesting a key role of Sfrp4 in the regulation of cortical remodeling (Chen et al. 2022). Indeed, it has been postulated that Sfrp4 is a messenger between these two bone surfaces (Chen et al. 2022).

9.4 Wnt Ligands with Relevance for the Skeleton

9.4.1 Wnt1, a Physiologically Relevant Osteoanabolic Molecule

9.4.1.1 Early-Onset Osteoporosis

Early-onset osteoporosis (EOOP) is a collective term for osteoporotic disorders occurring before the fifth decade (Mäkitie and Zillikens 2022). As such, vastly different pathological entities that often only share the low bone mass phenotype and age of onset as a commonality are summarized under this hypernym. Of course, this implies an equally expansive pool of causalities that may lead to such disorders. Contributing factors may be lifestyle, other diseases, nutrition, environmental factors, exposure to specific substances, etc. However, most relevant to the current topic are the genetics that can lead to EOOP. Genomic variants that cause an imbalance in skeletal homeostasis leading to a marked reduction of bone mass relatively early in life are always indicative of a relevant function of the affected genes in the skeletal metabolic system. Therefore, the analysis and careful evaluation of cases of EOOP have an enormous potential to lead to the discovery of novel key factors in bone metabolism and the regulation of the involved cell types, i.e., osteoblasts, osteoclasts, and osteocytes.

In this section, we will focus on the relatively recent identification of the Wnt ligand Wnt1 as a major regulator of bone mass. While EOOP is relatively rare, variants of Wnt1 are surprisingly common among larger cohorts of EOOP patients (OMIM: 615221), often found in specialized treatment centers (Keupp et al. 2013;

Hu et al. 2023; Peris et al. 2023). Patients present with the characteristic low bone mass and increased fracture rate as well as often a shortened stature and more symptoms specific to certain mutations, such as the bone marrow fibrosis observed in EOOP caused by the C218G mutation of Wnt1 (Mäkitie et al. 2018). Interestingly, affected patients can be heterozygous for the mutated *WNT1* allele (e.g., in case of the R235W variant), indicating a dominant inheritance pattern (Keupp et al. 2013). Although Wnt ligands do not typically form multimers, it could be speculated that this effect is caused by an occupation of the receptor site by the mutant Wnt1 ligand without the induction of the downstream signaling cascade. As for most cases of EOOP, treatment of Wnt1-associated EOOP is based on regular osteoporosis treatment recommendations, including normalization of bone metabolic parameters such as vitamin D and calcium by substitution, and application of antiresorptive drugs such as bisphosphonates and denosumab as well as osteoanabolic options (e.g., teriparatide), although efficacy of these treatments is variable and may be limited (Välimäki et al. 2017; Yorgan et al. 2020; Qaseem et al. 2023). If necessary, surgical intervention is utilized to stabilize fractures.

In fact, by generating and analyzing a corresponding mouse model, we were not only able to provide evidence that the Wnt1-R235W variant is disease-causing and not only a risk factor, but we were also able to show that osteoanabolic treatment is not impaired despite the presence of the Wnt1 mutation (Yorgan et al. 2020). This is especially relevant as decreased osteoblastic bone formation is a major contributor to the observed phenotype. Indeed, in the corresponding mouse model, it was observed that bone formation is significantly impaired in younger, homozygous animals, while increased osteoclast parameters appear to have an augmented contribution to the phenotype in older animals (Yorgan et al. 2020).

The importance of Wnt1 for bone metabolism began to emerge well before the detailed analysis of the Wnt1-R235W variant. Based on the initial finding of Wnt1 variants in EOOP patients, the analysis of mouse lines lacking Wnt1 specifically in cells of the osteoblast lineage revealed a severe reduction of bone mass and an increased rate of spontaneous fractures (Joeng et al. 2017; Luther et al. 2018; Wang et al. 2019). This effect was predominantly caused by a reduction of bone formation via osteoblasts, although additional effects on osteoclast differentiation were reported (Joeng et al. 2017, Luther et al. 2018, Wang et al. 2019). Conversely, the inducible transgenic expression of Wnt1 in Col1a1-expressing cells (i.e., cells of the osteoblast lineage) led to a rapid and dramatic increase of bone formation and subsequently bone mass in mice (Luther et al. 2018; Vollersen et al. 2021). Indeed, the observed change in bone mass can be considered one of the strongest osteoanabolic responses recorded to date. This observation clearly demonstrated the tremendous osteoanabolic potential of the Wnt1 ligand, making it and its potential receptors the focus of ongoing research efforts with the ultimate aim of modulating bone metabolic processes to treat degenerative skeletal disorders such as osteoporosis. While the Wnt signaling pathway in general has been leveraged to establish a novel osteoanabolic treatment option based on scavenging Sclerostin and thus preventing its interaction with Lrp5, as was discussed in more detail earlier in this chapter, the relevance of Wnt1 and its signaling cascade as an alternative target for osteoanabolic therapy is highlighted by evidence that suggests that Wnt1 acts independent of Lrp5 (Cosman et al. 2016; Saag et al. 2017; Luther et al. 2018). Additionally, this finding further reinforces Wnt signaling as a central regulatory system of skeletal metabolism, as two related but seemingly independent signaling networks from the Wnt family have a major impact on bone mass. However, this raises the question of the underlying cellular signal transduction pathway, if Lrp5 is not involved. Considering some functional redundancy, it is conceivable that Wnt1 is dependent on Lrp6 for signaling, although this hypothesis remains to be tested.

One of the most intriguing aspects of Wnt1 is that this ligand was initially identified as an indispensable factor in neuronal development. Mice with a global deficiency of Wnt1 are not viable due to a severe maldevelopment of the cerebellum (McMahon and Bradley 1990; Thomas and Capecchi 1990). Moreover, the swaying mice, a line harboring a spontaneous mutation of Wnt1, were originally described for their neuronal defects leading to the characteristic swaying gait as well as decreased viability (Thomas et al. 1991). Only later it was discovered that these mice also had a severely osteoporotic skeletal phenotype, in line with the aforementioned skeletal changes observed in mice with osteoblast-specific deficiency of Wnt1 or carrying mutations associated with Wnt1-related skeletal dysplasias (Joeng et al. 2014; Joeng et al. 2017; Luther et al. 2018; Wang et al. 2019; Yorgan et al. 2020; Vollersen et al. 2021). Surprisingly, while neurological symptoms such as hydrocephalus have been described in some patients suffering from Wnt1-associated skeletal dysplasias, this is certainly not the case for all of them (Vollersen et al. 2021). Additionally, the Wnt-R235W mouse model of EOOP and the Wnt1-G177C mouse model of osteogenesis imperfecta, described in more detail in the next section, do not display any discernible neuronal abnormalities (Yorgan et al. 2020, Vollersen et al. 2021). This observation indicates an increased resilience of the nervous system toward alterations of Wnt1-mediated signaling compared to the skeleton, promoting a potential paradigm shift where the skeleton will be considered the primary organ of Wnt1 activity.

9.4.1.2 Osteogenesis Imperfecta Type XV

Osteogenesis imperfecta (OI) is an inherited skeletal disorder characterized by severely increased fracture risk, skeletal deformities, and often shorter stature among many other possible symptoms (Marini et al. 2017). In contrast to the most common forms of OI (type I-IV) that are caused by pathogenic variants of the genes encoding type I collagen, as discussed in previous chapters, or several other rare types that are directly linked to mutations of collagen processing enzymes, OI type XV (OMIM: 615220) is caused by specific variants of *WNT1*(Fahiminiya et al. 2013; Keupp et al. 2013; Laine et al. 2013; Pyott et al. 2013). As mutations in the same gene can also cause EOOP, there is some overlap between the symptoms of both disorders (Keupp et al. 2013; Laine et al. 2013). However, OI type XV is usually caused by mutations that potentially result in a higher degree of functional loss and are either present homozygously or as compound heterozygotes (Keupp et al.

2013). The notion that OI type XV is not simply a more severe form of EOOP, but a separate pathological entity, is not only based on the clinical manifestation observed in patients that shares marked similarities to cases of classical OI but is also supported by findings in a corresponding mouse model (Vollersen et al. 2021; Cavdartepe and Ipek 2023).

In a mouse line harboring the Wnt1-G177C variant, hallmark symptoms of the human disease were recapitulated including low bone mass and the occurrence of spontaneous fractures in the absence of trauma (Vollersen et al. 2021). Furthermore, detailed analysis revealed that in addition to the impact of this Wnt1 mutation on bone density, there were also alterations in the inherent mechanical properties of the bone matrix, as is typical for OI (Vollersen et al. 2021). At this point, it has not been fully elucidated how the diminished activity of a signaling molecule such as Wnt1 can translate into altered mechanical properties of the bone matrix, but data from genome-wide expression analyses indicates that osteomodulin could represent a relevant downstream target of Wnt1 (Vollersen et al. 2021). In vitro studies have shown that osteomodulin is a regulator of collagen fiber thickness and branching, thus suggesting a plausible hypothesis for the observed skeletal phenotype in OI type XV (Tashima et al. 2015, 2018). On the other hand, the analysis of a mouse model with an osteocyte-specific deletion of Wnt1 and in vitro experiments utilizing the mesenchymal ST2 cell line determined that the effect of Wnt1 on the skeleton is at least partially attributable to modulation of the mTORC1 pathway in osteoblasts, a target of non-canonical Wnt signaling (Joeng et al. 2017). Therefore, the truly underlying pathomechanism still remains to be explored, and further research is required. With regard to the treatment of patients suffering from OI type XV, the same strategies as for conventional OI are applied.

9.4.2 The Role of Wnt16 for Skeletal Integrity

Besides Wnt1, there is another Wnt ligand, i.e., Wnt16, which has been extensively studied to provide evidence for an important role, not only for bone mass acquisition, but also for articular cartilage integrity, i.e., protection against osteoarthritis (OA) development. OA represents a chronic joint disease characterized by articular cartilage degeneration and pathogenic subchondral bone hyperplasia and sclerosis, formation of osteophytes, and inflammation of adjacent tissue (Malemud 2015; Abramoff and Caldera 2020). These morphological and metabolic changes result in joint pain and loss of mobility. Given the very high prevalence of OA in the general population, the specific manifestations, underlying pathomechanisms, and potential treatment options of the different subtypes can vary greatly. A detailed description is beyond the scope of this chapter, and interested readers are referred to current reviews on this topic (Malemud 2015; Chen et al. 2017; Abramoff and Caldera 2020).

With regard to Wnt signaling, specific single nucleotide polymorphisms (SNPs) of the ligand *WNT16* have been associated with a predisposition for the development of certain types of OA, while other SNPs of *WNT16* appear to have a protective effect (Garcia-Ibarbia et al. 2017; Huang et al. 2019). Indeed, it has been shown that injury or OA induces the expression of Wnt16 in degenerated articular cartilage (Dell'accio et al. 2008; Nalesso et al. 2017; Tong et al. 2019). Further studies provided evidence that the expression of Wnt16 is protective, as the ligand inhibits cartilage catabolism and stimulates anabolism (Nalesso et al. 2017; Tong et al. 2019; Wu et al. 2021; Ye and Liu 2022). An impressive demonstration of this effect is observed in mice with global Wnt16 deficiency that have an increased susceptibility toward experimentally induced OA (Nalesso et al. 2017).

In addition to its role in OA, genome-wide association studies have shown that SNPs in *WNT16* are also associated with bone mineral density (Garcia-Ibarbia et al. 2013; Hendrickx et al. 2014; Kemp et al. 2014; Zhang et al. 2014; Niu et al. 2016; Su 2020; Ye and Liu 2022). Although no specific skeletal disorder has been linked to Wnt16 mutations to date, experimental studies involving mouse models have revealed a discernible function of Wnt16 in skeletal development and homeostasis (Zheng et al. 2012; Jiang et al. 2014; Moverare-Skrtic et al. 2014; Moverare-Skrtic et al. 2015; Wergedal et al. 2015; Alam et al. 2016, 2017; Zeng et al. 2018; Tong et al. 2019; Ye and Liu 2022). In fact, osteoblast-specific Wnt16 deficiency in mice leads to an unusually specific phenotype impacting exclusively the cortical bone compartment (Moverare-Skrtic et al. 2014). Here, a decrease in cortical thickness, smaller cross-sectional area, and increased fracture risk were observed. These changes were predominantly caused by decreased osteoblastic bone formation (Moverare-Skrtic et al. 2014). Surprisingly, while transgenic overexpression of Wnt16 resulted in increased cortical bone mass, also trabecular bone was positively affected (Moverare-Skrtic et al. 2015; Alam et al. 2016, 2017). In some studies, the trabecular bone mass increased even predominantly with only minor effects on the cortical bone compartment (Moverare-Skrtic et al. 2015; Alam et al. 2017). While these findings raise further questions, the overall consensus is that Wnt16 is a positive regulator of bone mass (Ye and Liu 2022). Wnt16 has been shown to induce the proliferation and activity of bone-forming osteoblasts as well as to inhibit bone resorption via osteoclasts, although the majority of the available data points toward osteoblasts as the major effector cell type (Sebastian et al. 2018; Jin et al. 2020; McGowan et al. 2021). Nonetheless, this dichotomy of Wnt16 function in skeletal homeostasis is further increased in its complexity by data indicating that Wnt16 can function as both a canonical and non-canonical ligand (Hendrickx et al. 2020; Liu et al. 2021). Indeed, it appears as if the receiving cell is a major determinant of the activated pathway. While in osteoblasts and chondrocytes, both pathways can be activated by Wnt16, it exclusively activates the non-canonical PCP pathway in osteoclasts (Moverare-Skrtic et al. 2014; Nalesso et al. 2017; Liu et al. 2021). Finally, the intricacy of the underlying signaling pathway is further elevated by data suggesting extensive crosstalk between different pathways, including canonical and non-canonical Wnt pathways, mTOR signaling, and FoxO signaling among others (Sebastian et al. 2018; Qu et al. 2021).

9.5 Impact of Additional Molecules Involved in Wnt Signaling on the Skeleton

In addition to the previously discussed skeletal dysplasias associated with impaired Wnt signaling, there are numerous skeletal disorders, where a link to components of the Wnt signaling pathway could be demonstrated. However, since these have not been investigated in as much detail or their impact on the field is currently inconclusive, these will be briefly mentioned in a condensed table (Table 9.1). For further details, the reader is referred to an excellent review, where these dysplasias are discussed in more detail (Huybrechts et al. 2020). Importantly, however, it is important to discuss another group of Wnt signaling regulators, for which pathogenic variants have been associated with a specific type of skeletal dysplasia, i.e., Robinow syndrome.

Table 9.1 Additional Wnt signaling components and their effect on the skeleton

Gene/ protein	Function in Wnt signaling	Associated disorder	References
WNT3	Ligand	Tetra-amelia syndrome type 1 (OMIM: 273395)	(Niemann et al. 2004)
WNT6	Ligand	Acro-pectoro-vertebral dysplasia (OMIM: 102510)	(Thiele et al. 2004)
WNT7A	Ligand	Al-Awadi–Raas–Rothschild syndrome (OMIM: 276820), Fuhrmann syndrome (OMIM: 601570)	(Woods et al. 2006, Kantaputra et al. 2010, Eyaid et al. 2011, Garavelli et al. 2011, Mutlu et al. 2016, Lan et al. 2019, Sun et al. 2019)
WNT10B	Ligand	Split-hand/foot malformation type 6 (OMIM: 225300)	(Ugur and Tolun 2008, Blattner et al. 2010)
FZD2	Receptor	Autosomal dominant omodysplasia (OMIM: 164745)	(Saal et al. 2015, Türkmen et al. 2017)
FZD4	Receptor	Bone acquisition	(Kushwaha et al. 2020)
GPC4	Regulator	Keipert syndrome (OMIM: 301026)	(Amor et al. 2019)
GPC6	Regulator	Omodysplasia type 1 (OMIM: 258315)	(Campos-Xavier et al. 2009)
RSPO2	Activator	Tetra-amelia type 2 (OMIM: 618021), humerofemoral hypoplasia (OMIM: 610575)	(Szenker-Ravi et al. 2018)
APC2	Regulator	Sotos syndrome (OMIM: 612034)	(Almuriekhi et al. 2015)
AMER1	Inhibitor	Osteopathia striata with cranial sclerosis (OMIM: 300373)	(Jenkins et al. 2009)
PORCN	Processor	Goltz syndrome (OMIM: 305600)	(Grzeschik et al. 2007, Stevenson et al. 2014)

9.5.1 Robinow Syndrome (Wnt5a, ROR2, Dvl1, Dvl3, NXN)

The Robinow syndrome (RS) is a rare genetic disorder characterized by a specific phenotype of craniofacial dysmorphism, shortened limbs, and genital hypoplasia (Robinow et al. 1969). With regard to the skeletal phenotype, the craniofacial morphology of the affected patients has been described as fetal facies, an apt description of the hypertelorism, midfacial hypoplasia, short, upturned nose, prominent forehead, and flat nasal bridge that result in the overall impression of the facial structures of a fetus (Patton and Afzal 2002). The limbs, especially the radius/ulna and tibia/fibula, of the patients are typically shortened and often accompanied by small feet and hands with occasional observations of ectrodactyly (Patton and Afzal 2002). Furthermore, vertebral segmentation abnormalities are a typical symptom of RS. These can manifest as individual butterfly vertebrae, hemivertebrae, vertebral fusion, or other defects (DeChiara et al. 2000; Patton and Afzal 2002). These symptoms are present from birth and clearly indicate deficiencies during embryonic development. RS can be further subclassified into autosomal dominant RS and autosomal recessive RS (Patton and Afzal 2002). Autosomal dominant RS is the more common form and generally leads to a milder manifestation of the associated symptoms (Mazzeu 2011). Indeed, the recessive form has an increased mortality risk, while no impact on life expectancy has been described for the dominant form (Patton and Afzal 2002).

With the advent of modern molecular genetic methods, researchers were able to identify the underlying mutations causing RS. Interestingly, while mutations in several different genes can cause RS, all of them are part of the non-canonical Wnt pathway, more specifically the Wnt/PCP pathway (Mazzeu and Brunner 2020). Most cases of autosomal recessive RS have been linked to pathogenic variants in ROR2 (OMIM: 268310), a receptor of the Wnt/PCP pathway (Bacino 1993). Conversely, gain-of-function mutations of ROR2 have been linked to another type of skeletal dysplasia termed brachydactyly type B (OMIM: 113000) (Oldridge et al. 2000; Schwabe et al. 2000). Additionally autosomal recessive RS-causing mutations have been described in NXN (OMIM: 618529), an inhibitory regulator of intracellular Wnt signal transduction that acts by binding with the scaffolding proteins of the Dishevelled family (White et al. 2018). Autosomal dominant RS on the other hand is caused by variants in the non-canonical ligand Wnt5a (OMIM: 180700), the receptor Fzd-2 (OMIM: 600667), and the Dishevelled members Dvl1 and Dvl3 (OMIM: 616331, 616894) that are components of the Wnt pathway immediately downstream of the receptor (Roifman et al. 1993; Person et al. 2010; White et al. 2015, 2016, 2018). Considering the clinical manifestation and genetic cause of RS, it is readily apparent that non-canonical Wnt signaling plays an important role during the embryonic development of the skeleton. However, since no persistent or late-onset metabolic symptoms are described, the molecular functions of the affected non-canonical Wnt signaling components for skeletal homeostasis are less transparent.

9.6 Conclusions and Future Perspectives

In conclusion, research throughout the last decades has produced a vast body of evidence linking Wnt signaling to skeletal development and homeostasis. Indeed, it could be shown that many individual components of the Wnt signaling pathways can have a direct influence on multiple aspects of the skeleton. Consistent with these findings, pathological variants of components of the Wnt signaling pathway cause several distinct skeletal dysplasias and disorders. Especially noteworthy are the co-receptor Lrp5, where variants can lead to either low or high bone mass, the inhibitor Sclerostin, where inactivating variants lead to increased bone formation and Wnt1, a potent osteoanabolic ligand whose disruption results in low bone mass pathologies. The intensive research in this field has provided tangible results for patient care with the development of the anti-Sclerostin antibody Romosozumab, an osteoanabolic therapeutic approved for the treatment of postmenopausal osteoporosis. Furthermore, numerous aspects of Wnt signaling in skeletal biology still remain to be explored, and this field has great potential to make further seminal and highly relevant discoveries. One key question in this regard is the identification of specific receptors and downstream mediators of Wnt ligands regulating skeletal biology, such as Wnt1. Assisted by additional genetic findings and molecular biology experiments, this may eventually lead to further treatment options for skeletal disorders.

Compliance with Ethical Standards Both authors have no conflicts of interest to disclose.

References

Abramoff B, Caldera FE (2020) Osteoarthritis: pathology, diagnosis, and treatment options. Med Clin North Am 104(2):293–311. https://doi.org/10.1016/j.mcna.2019.10.007

Agoro R, Ni P, Noonan ML, White KE (2020) Osteocytic FGF23 and its kidney function. Front Endocrinol (Lausanne) 11:592. https://doi.org/10.3389/fendo.2020.00592

Agostino M, Pohl SO, Dharmarajan A (2017) Structure-based prediction of Wnt binding affinities for frizzled-type cysteine-rich domains. J Biol Chem 292(27):11218–11229. https://doi.org/10.1074/jbc.M117.786269

Alam I, Alkhouli M, Gerard-O'Riley RL, Wright WB, Acton D, Gray AK, Patel B, Reilly AM, Lim KE, Robling AG, Econs MJ (2016) Osteoblast-specific overexpression of human WNT16 increases both cortical and trabecular Bone mass and structure in mice. Endocrinology 157(2):722–736. https://doi.org/10.1210/en.2015-1281

Alam I, Reilly AM, Alkhouli M, Gerard-O'Riley RL, Kasipathi C, Oakes DK, Wright WB, Acton D, McQueen AK, Patel B, Lim K-E, Robling AG, Econs MJ (2017) Bone mass and strength are significantly improved in mice overexpressing human WNT16 in osteocytes. Calcif Tissue Int 100(4):361–373. https://doi.org/10.1007/s00223-016-0225-4

Almuriekhi M, Shintani T, Fahiminiya S, Fujikawa A, Kuboyama K, Takeuchi Y, Nawaz Z, Nadaf J, Kamel H, Kitam AK, Samiha Z, Mahmoud L, Ben-Omran T, Majewski J, Noda M (2015) Loss-of-function mutation in APC2 causes Sotos Syndrome features. Cell Rep 10(9):1585–1598. https://doi.org/10.1016/j.celrep.2015.02.011

Amor DJ, Stephenson SEM, Mustapha M, Mensah MA, Ockeloen CW, Lee WS, Tankard RM, Phelan DG, Shinawi M, de Brouwer APM, Pfundt R, Dowling C, Toler TL, Sutton VR,

Agolini E, Rinelli M, Capolino R, Martinelli D, Zampino G, Dumic M, Reardon W, Shaw-Smith C, Leventer RJ, Delatycki MB, Kleefstra T, Mundlos S, Mortier G, Bahlo M, Allen NJ, Lockhart PJ (2019) Pathogenic variants in GPC4 cause Keipert Syndrome. Am J Hum Genet 104(5):914–924. https://doi.org/10.1016/j.ajhg.2019.02.026
Appelman-Dijkstra N, Van Lierop A, Papapoulos S (1993) SOST-related sclerosing bone dysplasias. Gene reviews((R)). MP Adam, GM Mirzaa, RA Pagon et al. Seattle (WA)
Bacino CA (1993) ROR2-related Robinow Syndrome. Gene reviews((R)). MP Adam, GM Mirzaa, RA Pagon et al. Seattle (WA)
Bafico A, Gazit A, Pramila T, Finch PW, Yaniv A, Aaronson SA (1999) Interaction of frizzled related protein (FRP) with Wnt ligands and the frizzled receptor suggests alternative mechanisms for FRP inhibition of Wnt signaling. J Biol Chem 274(23):16180–16187. https://doi.org/10.1074/jbc.274.23.16180
Balemans W, Ebeling M, Patel N, Van Hul E, Olson P, Dioszegi M, Lacza C, Wuyts W, Van Den Ende J, Willems P, Paes-Alves AF, Hill S, Bueno M, Ramos FJ, Tacconi P, Dikkers FG, Stratakis C, Lindpaintner K, Vickery B, Foernzler D, Van Hul W (2001) Increased bone density in sclerosteosis is due to the deficiency of a novel secreted protein (SOST). Hum Mol Genet 10(5):537–543. https://doi.org/10.1093/hmg/10.5.537
Balemans W, Patel N, Ebeling M, Van Hul E, Wuyts W, Lacza C, Dioszegi M, Dikkers FG, Hildering P, Willems PJ, Verheij JB, Lindpaintner K, Vickery B, Foernzler D, Van Hul W (2002) Identification of a 52 kb deletion downstream of the SOST gene in patients with van Buchem disease. J Med Genet 39(2):91–97. https://doi.org/10.1136/jmg.39.2.91
Baron R, Kneissel M (2013) WNT signaling in bone homeostasis and disease: from human mutations to treatments. Nat Med 19(2):179–192. https://doi.org/10.1038/nm.3074
Beighton P, Durr L, Hamersma H (1976) The clinical features of sclerosteosis. A review of the manifestations in twenty-five affected individuals. Ann Intern Med 84(4):393–397. https://doi.org/10.7326/0003-4819-84-4-393
Berendsen AD, Olsen BR (2015) Bone development. Bone 80:14–18. https://doi.org/10.1016/j.bone.2015.04.035
Blattner A, Huber AR, Rothlisberger B (2010) Homozygous nonsense mutation in WNT10B and sporadic split-hand/foot malformation (SHFM) with autosomal recessive inheritance. Am J Med Genet A 152A(8):2053–2056. https://doi.org/10.1002/ajmg.a.33504
Boudin E, Piters E, Nielsen TL, Andersen M, Roef G, Taes Y, Brixen K, Van Hul W (2012) Single nucleotide polymorphisms in sFRP4 are associated with bone and body composition related parameters in Danish but not in Belgian men. Mol Genet Metab 106(3):366–374. https://doi.org/10.1016/j.ymgme.2012.04.021
Boyden LM, Mao J, Belsky J, Mitzner L, Farhi A, Mitnick MA, Wu D, Insogna K, Lifton RP (2002) High bone density due to a mutation in LDL-receptor-related protein 5. N Engl J Med 346(20):1513–1521. https://doi.org/10.1056/NEJMoa013444
Brommage R, Liu J, Hansen GM, Kirkpatrick LL, Potter DG, Sands AT, Zambrowicz B, Powell DR, Vogel P (2014) High-throughput screening of mouse gene knockouts identifies established and novel skeletal phenotypes. Bone Res 2:14034. https://doi.org/10.1038/boneres.2014.34
Campos-Xavier AB, Martinet D, Bateman J, Belluoccio D, Rowley L, Tan TY, Baxova A, Gustavson KH, Borochowitz ZU, Innes AM, Unger S, Beckmann JS, Mittaz L, Ballhausen D, Superti-Furga A, Savarirayan R, Bonafe L (2009) Mutations in the heparan-sulfate proteoglycan glypican 6 (GPC6) impair endochondral ossification and cause recessive omodysplasia. Am J Hum Genet 84(6):760–770. https://doi.org/10.1016/j.ajhg.2009.05.002
Cappariello A, Maurizi A, Veeriah V, Teti A (2014) The great beauty of the osteoclast. Arch Biochem Biophys 558:70–78. https://doi.org/10.1016/j.abb.2014.06.017
Cavdartepe BE, Ipek R (2023) Case report of two siblings diagnosed with osteogenesis imperfecta type XV with a new mutation in the WNT1 gene and Review of the literature. Mol Syndromol 14(2):164–170. https://doi.org/10.1159/000528201
Chang MK, Kramer I, Huber T, Kinzel B, Guth-Gundel S, Leupin O, Kneissel M (2014) Disruption of Lrp 4 function by genetic deletion or pharmacological blockade increases bone

mass and serum sclerostin levels. Proc Natl Acad Sci USA 111(48):E5187–E5195. https://doi.org/10.1073/pnas.1413828111

Chen D, Shen J, Zhao W, Wang T, Han L, Hamilton JL, Im HJ (2017) Osteoarthritis: toward a comprehensive understanding of pathological mechanism. Bone Res 5:16044. https://doi.org/10.1038/boneres.2016.44

Chen K, Ng PY, Chen R, Hu D, Berry S, Baron R, Gori F (2019) Sfrp4 repression of the Ror2/Jnk cascade in osteoclasts protects cortical bone from excessive endosteal resorption. Proc Natl Acad Sci USA 116(28):14138–14143. https://doi.org/10.1073/pnas.1900881116

Chen R, Baron R, Gori F (2022) Sfrp4 and the biology of cortical Bone. Curr Osteoporos Rep 20(2):153–161. https://doi.org/10.1007/s11914-022-00727-w

Chetty M, Roomaney I, Oosterwyk C, Manyisa N, Bope CD, Agenbag G, Wonkam A (2022) Craniofacial, dental, and molecular features of Pyle disease in a south African child. BDJ Open 8(1):28. https://doi.org/10.1038/s41405-022-00120-w

Clement-Lacroix P, Ai M, Morvan F, Roman-Roman S, Vayssiere B, Belleville C, Estrera K, Warman ML, Baron R, Rawadi G (2005) Lrp5-independent activation of Wnt signaling by lithium chloride increases bone formation and bone mass in mice. Proc Natl Acad Sci USA 102(48):17406–17411. https://doi.org/10.1073/pnas.0505259102

Clevers H (2006) Wnt/beta-catenin signaling in development and disease. Cell 127(3):469–480. https://doi.org/10.1016/j.cell.2006.10.018

Compston JE, McClung MR, Leslie WD (2019) Osteoporosis Lancet 393(10169):364–376. https://doi.org/10.1016/S0140-6736(18)32112-3

Cosman F, Crittenden DB, Adachi JD, Binkley N, Czerwinski E, Ferrari S, Hofbauer LC, Lau E, Lewiecki EM, Miyauchi A, Zerbini CA, Milmont CE, Chen L, Maddox J, Meisner PD, Libanati C, Grauer A (2016) Romosozumab treatment in postmenopausal women with Osteoporosis. N Engl J Med 375(16):1532–1543. https://doi.org/10.1056/NEJMoa1607948

Cruciat CM, Niehrs C (2013) Secreted and transmembrane wnt inhibitors and activators. Cold Spring Harb Perspect Biol 5(3):a015081. https://doi.org/10.1101/cshperspect.a015081

Dallas SL, Prideaux M, Bonewald LF (2013) The osteocyte: an endocrine cell … and more. Endocr Rev 34(5):658–690. https://doi.org/10.1210/er.2012-1026

Das S, Sakthiswary R (2013) Bone metabolism and histomorphometric changes in murine models treated with sclerostin antibody: a systematic review. Curr Drug Targets 14(14):1667–1674. https://doi.org/10.2174/1389450114666131217001756

De A (2011) Wnt/Ca2+ signaling pathway: a brief overview. Acta Biochim Biophys Sin Shanghai 43(10):745–756. https://doi.org/10.1093/abbs/gmr079

DeChiara TM, Kimble RB, Poueymirou WT, Rojas J, Masiakowski P, Valenzuela DM, Yancopoulos GD (2000) Ror2, encoding a receptor-like tyrosine kinase, is required for cartilage and growth plate development. Nat Genet 24(3):271–274. https://doi.org/10.1038/73488

Dell'accio F, De Bari C, Eltawil NM, Vanhummelen P, Pitzalis C (2008) Identification of the molecular response of articular cartilage to injury, by microarray screening: Wnt-16 expression and signaling after injury and in osteoarthritis. Arthritis Rheum 58(5):1410–1421. https://doi.org/10.1002/art.23444

Ellies DL, Viviano B, McCarthy J, Rey JP, Itasaki N, Saunders S, Krumlauf R (2006) Bone density ligand, Sclerostin, directly interacts with LRP5 but not LRP5G171V to modulate Wnt activity. J Bone Miner Res 21(11):1738–1749. https://doi.org/10.1359/jbmr.060810

Eyaid W, Al-Qattan MM, Al Abdulkareem I, Fetaini N, Al Balwi M (2011) A novel homozygous missense mutation (c.610G>a, p.Gly204Ser) in the WNT7A gene causes tetra-amelia in two Saudi families. Am J Med Genet A 155A(3):599–604. https://doi.org/10.1002/ajmg.a.33717

Fahiminiya S, Majewski J, Mort J, Moffatt P, Glorieux FH, Rauch F (2013) Mutations in WNT1 are a cause of osteogenesis imperfecta. J Med Genet 50(5):345–348. https://doi.org/10.1136/jmedgenet-2013-101567

Ferguson JW, Atit RP (2019) A tale of two cities: the genetic mechanisms governing calvarial bone development. Genesis 57(1):e23248. https://doi.org/10.1002/dvg.23248

Fijalkowski I, Geets E, Steenackers E, Van Hoof V, Ramos FJ, Mortier G, Fortuna AM, Van Hul W, Boudin E (2016) A novel domain-specific mutation in a Sclerosteosis patient suggests a role of LRP4 as an anchor for Sclerostin in human Bone. J Bone Miner Res 31(4):874–881. https://doi.org/10.1002/jbmr.2782

Galada C, Shah H, Shukla A, Girisha KM (2017) A novel sequence variant in SFRP4 causing Pyle disease. J Hum Genet 62(5):575–576. https://doi.org/10.1038/jhg.2016.166

Garavelli L, Wischmeijer A, Rosato S, Gelmini C, Reverberi S, Sassi S, Ferrari A, Mari F, Zabel B, Lausch E, Unger S, Superti-Furga A (2011) Al-Awadi-Raas-Rothschild (limb/pelvis/uterus-hypoplasia/aplasia) syndrome and WNT7A mutations: genetic homogeneity and nosological delineation. Am J Med Genet A 155A(2):332–336. https://doi.org/10.1002/ajmg.a.33793

Garcia-Ibarbia C, Neila S, Garces C, Alonso MA, Zarrabeitia MT, Valero C, Ortiz F, Riancho JA (2017) Non-synonymous WNT16 polymorphisms alleles are associated with different osteoarthritis phenotypes. Rheumatol Int 37(10):1667–1672. https://doi.org/10.1007/s00296-017-3783-5

Garcia-Ibarbia C, Perez-Nunez MI, Olmos JM, Valero C, Perez-Aguilar MD, Hernandez JL, Zarrabeitia MT, Gonzalez-Macias J, Riancho JA (2013) Missense polymorphisms of the WNT16 gene are associated with bone mass, hip geometry and fractures. Osteoporos Int 24(9):2449–2454. https://doi.org/10.1007/s00198-013-2302-0

Glass DA, Bialek P, Ahn JD, Starbuck M, Patel MS, Clevers H, Taketo MM, Long F, McMahon AP, Lang RA, Karsenty G (2005) Canonical Wnt signaling in differentiated osteoblasts controls osteoclast differentiation. Dev Cell 8(5):751–764. https://doi.org/10.1016/j.devcel.2005.02.017

Gong Y, Slee RB, Fukai N, Rawadi G, Roman-Roman S, Reginato AM, Wang H, Cundy T, Glorieux FH, Lev D, Zacharin M, Oexle K, Marcelino J, Suwairi W, Heeger S, Sabatakos G, Apte S, Adkins WN, Allgrove J, Arslan-Kirchner M, Batch JA, Beighton P, Black GC, Boles RG, Boon LM, Borrone C, Brunner HG, Carle GF, Dallapiccola B, De Paepe A, Floege B, Halfhide ML, Hall B, Hennekam RC, Hirose T, Jans A, Juppner H, Kim CA, Keppler-Noreuil K, Kohlschuetter A, LaCombe D, Lambert M, Lemyre E, Letteboer T, Peltonen L, Ramesar RS, Romanengo M, Somer H, Steichen-Gersdorf E, Steinmann B, Sullivan B, Superti-Furga A, Swoboda W, van den Boogaard MJ, Van Hul W, Vikkula M, Votruba M, Zabel B, Garcia T, Baron R, Olsen BR, Warman ML, Osteoporosis-Pseudoglioma Syndrome Collaborative G (2001) LDL receptor-related protein 5 (LRP5) affects bone accrual and eye development. Cell 107(4):513–523. https://doi.org/10.1016/s0092-8674(01)00571-2

Gong Y, Vikkula M, Boon L, Liu J, Beighton P, Ramesar R, Peltonen L, Somer H, Hirose T, Dallapiccola B, De Paepe A, Swoboda W, Zabel B, Superti-Furga A, Steinmann B, Brunner HG, Jans A, Boles RG, Adkins W, van den Boogaard MJ, Olsen BR, Warman ML (1996) Osteoporosis-pseudoglioma syndrome, a disorder affecting skeletal strength and vision, is assigned to chromosome region 11q12-13. Am J Hum Genet 59(1):146–151

Grzeschik KH, Bornholdt D, Oeffner F, Konig A, del Carmen BM, Enders H, Fritz B, Hertl M, Grasshoff U, Hofling K, Oji V, Paradisi M, Schuchardt C, Szalai Z, Tadini G, Traupe H, Happle R (2007) Deficiency of PORCN, a regulator of Wnt signaling, is associated with focal dermal hypoplasia. Nat Genet 39(7):833–835. https://doi.org/10.1038/ng2052

Hallett SA, Matsushita Y, Ono W, Sakagami N, Mizuhashi K, Tokavanich N, Nagata M, Zhou A, Hirai T, Kronenberg HM, Ono N (2021) Chondrocytes in the resting zone of the growth plate are maintained in a Wnt-inhibitory environment. elife 10. https://doi.org/10.7554/eLife.64513

Han Y, You X, Xing W, Zhang Z, Zou W (2018) Paracrine and endocrine actions of bone-the functions of secretory proteins from osteoblasts, osteocytes, and osteoclasts. Bone Res 6:16. https://doi.org/10.1038/s41413-018-0019-6

Haraguchi R, Kitazawa R, Mori K, Tachibana R, Kiyonari H, Imai Y, Abe T, Kitazawa S (2016) sFRP4-dependent Wnt signal modulation is critical for bone remodeling during postnatal development and age-related bone loss. Sci Rep 6:25198. https://doi.org/10.1038/srep25198

Hedgepeth CM, Conrad LJ, Zhang J, Huang HC, Lee VM, Klein PS (1997) Activation of the Wnt signaling pathway: a molecular mechanism for lithium action. Dev Biol 185(1):82–91. https://doi.org/10.1006/dbio.1997.8552

Hendrickx G, Boudin E, Fijalkowski I, Nielsen TL, Andersen M, Brixen K, Van Hul W (2014) Variation in the Kozak sequence of WNT16 results in an increased translation and is associated with osteoporosis related parameters. Bone 59:57–65. https://doi.org/10.1016/j.bone.2013.10.022

Hendrickx G, Boudin E, Verbeek M, Fransen E, Mortier G, Van Hul W (2020) WNT16 requires Galpha subunits as intracellular Partners for both its Canonical and non-Canonical WNT Signalling activity in osteoblasts. Calcif Tissue Int 106(3):294–302. https://doi.org/10.1007/s00223-019-00633-x

Holmen SL, Zylstra CR, Mukherjee A, Sigler RE, Faugere MC, Bouxsein ML, Deng L, Clemens TL, Williams BO (2005) Essential role of beta-catenin in postnatal bone acquisition. J Biol Chem 280(22):21162–21168. https://doi.org/10.1074/jbc.M501900200

Hu J, Lin X, Gao P, Zhang Q, Zhou B, Wang O, Jiang Y, Xia W, Xing X, Li M (2023) Genotypic and phenotypic Spectrum and pathogenesis of WNT1 variants in a large cohort of patients with OI/Osteoporosis. J Clin Endocrinol Metab 108(7):1776–1786. https://doi.org/10.1210/clinem/dgac752

Huang W, Li Q, Amiry-Moghaddam M, Hokama M, Sardi SH, Nagao M, Warman ML, Olsen BR (2016) Critical endothelial regulation by LRP5 during retinal vascular development. PLoS One 11(3):e0152833. https://doi.org/10.1371/journal.pone.0152833

Huang Y, Jiang L, Yang H, Wu L, Xu N, Zhou X, Li J (2019) Variations of Wnt/beta-catenin pathway-related genes in susceptibility to knee osteoarthritis: a three-Centre case-control study. J Cell Mol Med 23(12):8246–8257. https://doi.org/10.1111/jcmm.14696

Huang Y, Liu L, Liu A (2018) Dickkopf-1: current knowledge and related diseases. Life Sci 209:249–254. https://doi.org/10.1016/j.lfs.2018.08.019

Huybrechts Y, Mortier G, Boudin E, Van Hul W (2020) WNT signaling and Bone: lessons from skeletal Dysplasias and disorders. Front Endocrinol (Lausanne) 11:165. https://doi.org/10.3389/fendo.2020.00165

Itasaki N, Hoppler S (2010) Crosstalk between Wnt and bone morphogenic protein signaling: a turbulent relationship. Dev Dyn 239(1):16–33. https://doi.org/10.1002/dvdy.22009

Jenkins ZA, van Kogelenberg M, Morgan T, Jeffs A, Fukuzawa R, Pearl E, Thaller C, Hing AV, Porteous ME, Garcia-Minaur S, Bohring A, Lacombe D, Stewart F, Fiskerstrand T, Bindoff L, Berland S, Ades LC, Tchan M, David A, Wilson LC, Hennekam RC, Donnai D, Mansour S, Cormier-Daire V, Robertson SP (2009) Germline mutations in WTX cause a sclerosing skeletal dysplasia but do not predispose to tumorigenesis. Nat Genet 41(1):95–100. https://doi.org/10.1038/ng.270

Jiang Z, Von den Hoff JW, Torensma R, Meng L, Bian Z (2014) Wnt16 is involved in intramembranous ossification and suppresses osteoblast differentiation through the Wnt/beta-catenin pathway. J Cell Physiol 229(3):384–392. https://doi.org/10.1002/jcp.24460

Jin Y, Sun X, Pei F, Zhao Z, Mao J (2020) Wnt16 signaling promotes osteoblast differentiation of periosteal derived cells in vitro and in vivo. PeerJ 8:e10374. https://doi.org/10.7717/peerj.10374

Joeng KS, Lee YC, Jiang MM, Bertin TK, Chen Y, Abraham AM, Ding H, Bi X, Ambrose CG, Lee BH (2014) The swaying mouse as a model of osteogenesis imperfecta caused by WNT1 mutations. Hum Mol Genet 23(15):4035–4042. https://doi.org/10.1093/hmg/ddu117

Joeng KS, Lee YC, Lim J, Chen Y, Jiang MM, Munivez E, Ambrose C, Lee BH (2017) Osteocyte-specific WNT1 regulates osteoblast function during bone homeostasis. J Clin Invest 127(7):2678–2688. https://doi.org/10.1172/JCI92617

Johnell O, Kanis JA (2006) An estimate of the worldwide prevalence and disability associated with osteoporotic fractures. Osteoporos Int 17(12):1726–1733. https://doi.org/10.1007/s00198-006-0172-4

Johnson ML, Gong G, Kimberling W, Recker SM, Kimmel DB, Recker RB (1997) Linkage of a gene causing high bone mass to human chromosome 11 (11q12-13). Am J Hum Genet 60(6):1326–1332. https://doi.org/10.1086/515470

Joiner DM, Ke J, Zhong Z, Xu HE, Williams BO (2013) LRP5 and LRP6 in development and disease. Trends Endocrinol Metab 24(1):31–39. https://doi.org/10.1016/j.tem.2012.10.003
Kantaputra PN, Mundlos S, Sripathomsawat W (2010) A novel homozygous Arg222Trp missense mutation in WNT7A in two sisters with severe Al-Awadi/Raas-Rothschild/Schinzel phocomelia syndrome. Am J Med Genet A 152A(11):2832–2837. https://doi.org/10.1002/ajmg.a.33673
Karsenty G, Kronenberg HM, Settembre C (2009) Genetic control of bone formation. Annu Rev Cell Dev Biol 25:629–648. https://doi.org/10.1146/annurev.cellbio.042308.113308
Kato M, Patel MS, Levasseur R, Lobov I, Chang BH, Glass DA, Hartmann C, Li L, Hwang TH, Brayton CF, Lang RA, Karsenty G, Chan L (2002) Cbfa1-independent decrease in osteoblast proliferation, osteopenia, and persistent embryonic eye vascularization in mice deficient in Lrp5, a Wnt coreceptor. J Cell Biol 157(2):303–314. https://doi.org/10.1083/jcb.200201089
Kelly OG, Pinson KI, Skarnes WC (2004) The Wnt co-receptors Lrp5 and Lrp6 are essential for gastrulation in mice. Development 131(12):2803–2815. https://doi.org/10.1242/dev.01137
Kemp JP, Medina-Gomez C, Estrada K, St Pourcain B, Heppe DH, Warrington NM, Oei L, Ring SM, Kruithof CJ, Timpson NJ, Wolber LE, Reppe S, Gautvik K, Grundberg E, Ge B, van der Eerden B, van de Peppel J, Hibbs MA, Ackert-Bicknell CL, Choi K, Koller DL, Econs MJ, Williams FM, Foroud T, Zillikens MC, Ohlsson C, Hofman A, Uitterlinden AG, Davey Smith G, Jaddoe VW, Tobias JH, Rivadeneira F, Evans DM (2014) Phenotypic dissection of bone mineral density reveals skeletal site specificity and facilitates the identification of novel loci in the genetic regulation of bone mass attainment. PLoS Genet 10(6):e1004423. https://doi.org/10.1371/journal.pgen.1004423
Keupp K, Beleggia F, Kayserili H, Barnes AM, Steiner M, Semler O, Fischer B, Yigit G, Janda CY, Becker J, Breer S, Altunoglu U, Grunhagen J, Krawitz P, Hecht J, Schinke T, Makareeva E, Lausch E, Cankaya T, Caparros-Martin JA, Lapunzina P, Temtamy S, Aglan M, Zabel B, Eysel P, Koerber F, Leikin S, Garcia KC, Netzer C, Schonau E, Ruiz-Perez VL, Mundlos S, Amling M, Kornak U, Marini J, Wollnik B (2013) Mutations in WNT1 cause different forms of bone fragility. Am J Hum Genet 92(4):565–574. https://doi.org/10.1016/j.ajhg.2013.02.010
Kim M, Jho EH (2014) Cross-talk between Wnt/beta-catenin and Hippo signaling pathways: a brief review. BMB Rep 47(10):540–545. https://doi.org/10.5483/bmbrep.2014.47.10.177
Kiper POS, Saito H, Gori F, Unger S, Hesse E, Yamana K, Kiviranta R, Solban N, Liu J, Brommage R, Boduroglu K, Bonafe L, Campos-Xavier B, Dikoglu E, Eastell R, Gossiel F, Harshman K, Nishimura G, Girisha KM, Stevenson BJ, Takita H, Rivolta C, Superti-Furga A, Baron R (2016) Cortical-Bone Fragility–Insights from sFRP4 Deficiency in Pyle's Disease. N Engl J Med 374(26):2553–2562. https://doi.org/10.1056/NEJMoa1509342
Köhler-Forsberg O, Rohde C, Nierenberg AA, Ostergaard SD (2022) Association of Lithium Treatment with the risk of Osteoporosis in patients with bipolar disorder. JAMA Psychiatry 79(5):454–463. https://doi.org/10.1001/jamapsychiatry.2022.0337
Kramer I, Halleux C, Keller H, Pegurri M, Gooi JH, Weber PB, Feng JQ, Bonewald LF, Kneissel M (2010) Osteocyte Wnt/beta-catenin signaling is required for normal bone homeostasis. Mol Cell Biol 30(12):3071–3085. https://doi.org/10.1128/MCB.01428-09
Kushwaha P, Kim S, Foxa GE, Michalski MN, Williams BO, Tomlinson RE, Riddle RC (2020) Frizzled-4 is required for normal bone acquisition despite compensation by Frizzled-8. J Cell Physiol 235(10):6673–6683. https://doi.org/10.1002/jcp.29563
Laine CM, Joeng KS, Campeau PM, Kiviranta R, Tarkkonen K, Grover M, Lu JT, Pekkinen M, Wessman M, Heino TJ, Nieminen-Pihala V, Aronen M, Laine T, Kroger H, Cole WG, Lehesjoki AE, Nevarez L, Krakow D, Curry CJ, Cohn DH, Gibbs RA, Lee BH, Makitie O (2013) WNT1 mutations in early-onset osteoporosis and osteogenesis imperfecta. N Engl J Med 368(19):1809–1816. https://doi.org/10.1056/NEJMoa1215458
Lan L, Wang W, Huang Y, Bu X, Zhao C (2019) Roles of Wnt7a in embryo development, tissue homeostasis, and human diseases. J Cell Biochem 120(11):18588–18598. https://doi.org/10.1002/jcb.29217
Leupin O, Piters E, Halleux C, Hu S, Kramer I, Morvan F, Bouwmeester T, Schirle M, Bueno-Lozano M, Fuentes FJ, Itin PH, Boudin E, de Freitas F, Jennes K, Brannetti B, Charara N,

Ebersbach H, Geisse S, Lu CX, Bauer A, Van Hul W, Kneissel M (2011) Bone overgrowth-associated mutations in the LRP4 gene impair sclerostin facilitator function. J Biol Chem 286(22):19489–19500. https://doi.org/10.1074/jbc.M110.190330

Levasseur R (2008) Treatment and management of osteoporosis-pseudoglioma syndrome. Expert Rev Endocrinol Metab 3(3):337–348. https://doi.org/10.1586/17446651.3.3.337

Levasseur R, Lacombe D, de Vernejoul MC (2005) LRP5 mutations in osteoporosis-pseudoglioma syndrome and high-bone-mass disorders. Joint Bone Spine 72(3):207–214. https://doi.org/10.1016/j.jbspin.2004.10.008

Li X, Ominsky MS, Niu QT, Sun N, Daugherty B, D'Agostin D, Kurahara C, Gao Y, Cao J, Gong J, Asuncion F, Barrero M, Warmington K, Dwyer D, Stolina M, Morony S, Sarosi I, Kostenuik PJ, Lacey DL, Simonet WS, Ke HZ, Paszty C (2008) Targeted deletion of the sclerostin gene in mice results in increased bone formation and bone strength. J Bone Miner Res 23(6):860–869. https://doi.org/10.1359/jbmr.080216

Li X, Ominsky MS, Warmington KS, Morony S, Gong J, Cao J, Gao Y, Shalhoub V, Tipton B, Haldankar R, Chen Q, Winters A, Boone T, Geng Z, Niu QT, Ke HZ, Kostenuik PJ, Simonet WS, Lacey DL, Paszty C (2009) Sclerostin antibody treatment increases bone formation, bone mass, and bone strength in a rat model of postmenopausal osteoporosis. J Bone Miner Res 24(4):578–588. https://doi.org/10.1359/jbmr.081206

Li X, Zhang Y, Kang H, Liu W, Liu P, Zhang J, Harris SE, Wu D (2005) Sclerostin binds to LRP5/6 and antagonizes canonical Wnt signaling. J Biol Chem 280(20):19883–19887. https://doi.org/10.1074/jbc.M413274200

Li Y, Pawlik B, Elcioglu N, Aglan M, Kayserili H, Yigit G, Percin F, Goodman F, Nurnberg G, Cenani A, Urquhart J, Chung BD, Ismail S, Amr K, Aslanger AD, Becker C, Netzer C, Scambler P, Eyaid W, Hamamy H, Clayton-Smith J, Hennekam R, Nurnberg P, Herz J, Temtamy SA, Wollnik B (2010) LRP4 mutations alter Wnt/beta-catenin signaling and cause limb and kidney malformations in Cenani-Lenz syndrome. Am J Hum Genet 86(5):696–706. https://doi.org/10.1016/j.ajhg.2010.03.004

Little RD, Carulli JP, Del Mastro RG, Dupuis J, Osborne M, Folz C, Manning SP, Swain PM, Zhao SC, Eustace B, Lappe MM, Spitzer L, Zweier S, Braunschweiger K, Benchekroun Y, Hu X, Adair R, Chee L, FitzGerald MG, Tulig C, Caruso A, Tzellas N, Bawa A, Franklin B, McGuire S, Nogues X, Gong G, Allen KM, Anisowicz A, Morales AJ, Lomedico PT, Recker SM, Van Eerdewegh P, Recker RR, Johnson ML (2002) A mutation in the LDL receptor-related protein 5 gene results in the autosomal dominant high-bone-mass trait. Am J Hum Genet 70(1):11–19. https://doi.org/10.1086/338450

Liu X, Li X, Hua B, Yang X, Zheng J, Liu S (2021) WNT16 is upregulated early in mouse TMJ osteoarthritis and protects fibrochondrocytes against IL-1beta induced inflammatory response by regulation of RUNX2/MMP13 cascade. Bone 143:115793. https://doi.org/10.1016/j.bone.2020.115793

Loh KM, van Amerongen R, Nusse R (2016) Generating cellular diversity and spatial form: Wnt signaling and the evolution of multicellular animals. Dev Cell 38(6):643–655. https://doi.org/10.1016/j.devcel.2016.08.011

Long F, Ornitz DM (2013) Development of the endochondral skeleton. Cold Spring Harb Perspect Biol 5(1):a008334. https://doi.org/10.1101/cshperspect.a008334

Loots GG, Kneissel M, Keller H, Baptist M, Chang J, Collette NM, Ovcharenko D, Plajzer-Frick I, Rubin EM (2005) Genomic deletion of a long-range bone enhancer misregulates sclerostin in Van Buchem disease. Genome Res 15(7):928–935. https://doi.org/10.1101/gr.3437105

Luther J, Yorgan TA, Rolvien T, Ulsamer L, Koehne T, Liao N, Keller D, Vollersen N, Teufel S, Neven M, Peters S, Schweizer M, Trumpp A, Rosigkeit S, Bockamp E, Mundlos S, Kornak U, Oheim R, Amling M, Schinke T, David JP (2018) Wnt1 is an Lrp5-independent bone-anabolic Wnt ligand. Sci Transl Med 10(466). https://doi.org/10.1126/scitranslmed.aau7137

Lv F, Cai X, Yang W, Gao L, Chen L, Wu J, Ji L (2020) Denosumab or romosozumab therapy and risk of cardiovascular events in patients with primary osteoporosis: systematic review and meta- analysis. Bone 130:115121. https://doi.org/10.1016/j.bone.2019.115121

MacNabb C, Patton D, Hayes JS (2016) Sclerostin antibody therapy for the treatment of Osteoporosis: clinical prospects and challenges. J Osteoporos 2016:6217286. https://doi.org/10.1155/2016/6217286

Mäkitie O, Zillikens MC (2022) Early-onset Osteoporosis. Calcif Tissue Int 110(5):546–561. https://doi.org/10.1007/s00223-021-00885-6

Mäkitie RE, Niinimaki R, Kakko S, Honkanen T, Kovanen PE, Makitie O (2018) Defective WNT signaling associates with bone marrow fibrosis-a cross-sectional cohort study in a family with WNT1 osteoporosis. Osteoporos Int 29(2):479–487. https://doi.org/10.1007/s00198-017-4309-4

Malemud CJ (2015) Biologic basis of osteoarthritis: state of the evidence. Curr Opin Rheumatol 27(3):289–294. https://doi.org/10.1097/BOR.0000000000000162

Mani A, Radhakrishnan J, Wang H, Mani A, Mani MA, Nelson-Williams C, Carew KS, Mane S, Najmabadi H, Wu D, Lifton RP (2007) LRP6 mutation in a family with early coronary disease and metabolic risk factors. Science 315(5816):1278–1282. https://doi.org/10.1126/science.1136370

Marini JC, Forlino A, Bachinger HP, Bishop NJ, Byers PH, Paepe A, Fassier F, Fratzl-Zelman N, Kozloff KM, Krakow D, Montpetit K, Semler O (2017) Osteogenesis imperfecta. Nat Rev Dis Primers 3:17052. https://doi.org/10.1038/nrdp.2017.52

Markham A (2019) Romosozumab: First Global Approval. Drugs 79(4):471–476. https://doi.org/10.1007/s40265-019-01072-6

Mazzeu J (2011) Robinow syndrome. Orphanet encyclopedia

Mazzeu JF, Brunner HG (2020) 50 years of Robinow syndrome. Am J Med Genet A 182(9):2005–2007. https://doi.org/10.1002/ajmg.a.61756

McClung MR, Grauer A, Boonen S, Bolognese MA, Brown JP, Diez-Perez A, Langdahl BL, Reginster JY, Zanchetta JR, Wasserman SM, Katz L, Maddox J, Yang YC, Libanati C, Bone HG (2014) Romosozumab in postmenopausal women with low bone mineral density. N Engl J Med 370(5):412–420. https://doi.org/10.1056/NEJMoa1305224

McGowan LM, Kague E, Vorster A, Newham E, Cross S, Hammond CL (2021) Wnt16 elicits a protective effect against fractures and supports Bone repair in zebrafish. JBMR Plus 5(3):e10461. https://doi.org/10.1002/jbm4.10461

McMahon AP, Bradley A (1990) The Wnt-1 (int-1) proto-oncogene is required for development of a large region of the mouse brain. Cell 62(6):1073–1085. https://doi.org/10.1016/0092-8674(90)90385-r

Moverare-Skrtic S, Henning P, Liu X, Nagano K, Saito H, Borjesson AE, Sjogren K, Windahl SH, Farman H, Kindlund B, Engdahl C, Koskela A, Zhang FP, Eriksson EE, Zaman F, Hammarstedt A, Isaksson H, Bally M, Kassem A, Lindholm C, Sandberg O, Aspenberg P, Savendahl L, Feng JQ, Tuckermann J, Tuukkanen J, Poutanen M, Baron R, Lerner UH, Gori F, Ohlsson C (2014) Osteoblast-derived WNT16 represses osteoclastogenesis and prevents cortical bone fragility fractures. Nat Med 20(11):1279–1288. https://doi.org/10.1038/nm.3654

Moverare-Skrtic S, Wu J, Henning P, Gustafsson KL, Sjogren K, Windahl SH, Koskela A, Tuukkanen J, Borjesson AE, Lagerquist MK, Lerner UH, Zhang FP, Gustafsson JA, Poutanen M, Ohlsson C (2015) The bone-sparing effects of estrogen and WNT16 are independent of each other. Proc Natl Acad Sci USA 112(48):14972–14977. https://doi.org/10.1073/pnas.1520408112

Mutlu MB, Cetinkaya A, Koc N, Ceylaner G, Erguner B, Aydin H, Karaman S, Demirci O, Goksu K, Karaman A (2016) A novel missense mutation, p.(R102W) in WNT7A causes Al-Awadi Raas-Rothschild syndrome in a fetus. Eur J Med Genet 59(11):604–606. https://doi.org/10.1016/j.ejmg.2016.09.009

Nagy V, Penninger JM (2015) The RANKL-RANK story. Gerontology 61(6):534–542. https://doi.org/10.1159/000371845

Nakanishi R, Akiyama H, Kimura H, Otsuki B, Shimizu M, Tsuboyama T, Nakamura T (2008) Osteoblast-targeted expression of Sfrp4 in mice results in low bone mass. J Bone Miner Res 23(2):271–277. https://doi.org/10.1359/jbmr.071007

Nalesso G, Thomas BL, Sherwood JC, Yu J, Addimanda O, Eldridge SE, Thorup AS, Dale L, Schett G, Zwerina J, Eltawil N, Pitzalis C, Dell'Accio F (2017) WNT16 antagonises excessive canonical WNT activation and protects cartilage in osteoarthritis. Ann Rheum Dis 76(1):218–226. https://doi.org/10.1136/annrheumdis-2015-208577

Niedzwiedzki T, Filipowska J (2015) Bone remodeling in the context of cellular and systemic regulation: the role of osteocytes and the nervous system. J Mol Endocrinol 55(2):R23–R36. https://doi.org/10.1530/JME-15-0067

Niehrs C (2012) The complex world of WNT receptor signalling. Nat Rev Mol Cell Biol 13(12):767–779. https://doi.org/10.1038/nrm3470

Niemann S, Zhao C, Pascu F, Stahl U, Aulepp U, Niswander L, Weber JL, Muller U (2004) Homozygous WNT3 mutation causes tetra-amelia in a large consanguineous family. Am J Hum Genet 74(3):558–563. https://doi.org/10.1086/382196

Niu T, Liu N, Yu X, Zhao M, Choi HJ, Leo PJ, Brown MA, Zhang L, Pei YF, Shen H, He H, Fu X, Lu S, Chen XD, Tan LJ, Yang TL, Guo Y, Cho NH, Shen J, Guo YF, Nicholson GC, Prince RL, Eisman JA, Jones G, Sambrook PN, Tian Q, Zhu XZ, Papasian CJ, Duncan EL, Uitterlinden AG, Shin CS, Xiang S, Deng HW (2016) Identification of IDUA and WNT16 phosphorylation-related non-synonymous polymorphisms for Bone mineral density in meta-analyses of genome-wide association studies. J Bone Miner Res 31(2):358–368. https://doi.org/10.1002/jbmr.2687

Ockeloen CW, Khandelwal KD, Dreesen K, Ludwig KU, Sullivan R, van Rooij I, Thonissen M, Swinnen S, Phan M, Conte F, Ishorst N, Gilissen C, RoaFuentes L, van de Vorst M, Henkes A, Steehouwer M, van Beusekom E, Bloemen M, Vankeirsbilck B, Berge S, Hens G, Schoenaers J, Poorten VV, Roosenboom J, Verdonck A, Devriendt K, Roeleveldt N, Jhangiani SN, Vissers L, Lupski JR, de Ligt J, Von den Hoff JW, Pfundt R, Brunner HG, Zhou H, Dixon J, Mangold E, van Bokhoven H, Dixon MJ, Kleefstra T, Hoischen A, Carels CEL (2016) Novel mutations in LRP6 highlight the role of WNT signaling in tooth agenesis. Genet Med 18(11):1158–1162. https://doi.org/10.1038/gim.2016.10

Ohazama A, Johnson EB, Ota MS, Choi HY, Porntaveetus T, Oommen S, Itoh N, Eto K, Gritli-Linde A, Herz J, Sharpe PT (2008) Lrp4 modulates extracellular integration of cell signaling pathways in development. PLoS One 3(12):e4092. https://doi.org/10.1371/journal.pone.0004092

Ohkawara B, Cabrera-Serrano M, Nakata T, Milone M, Asai N, Ito K, Ito M, Masuda A, Ito Y, Engel AG, Ohno K (2014) LRP4 third beta-propeller domain mutations cause novel congenital myasthenia by compromising agrin-mediated MuSK signaling in a position-specific manner. Hum Mol Genet 23(7):1856–1868. https://doi.org/10.1093/hmg/ddt578

Oldridge M, Fortuna AM, Maringa M, Propping P, Mansour S, Pollitt C, DeChiara TM, Kimble RB, Valenzuela DM, Yancopoulos GD, Wilkie AO (2000) Dominant mutations in ROR2, encoding an orphan receptor tyrosine kinase, cause brachydactyly type B. Nat Genet 24(3):275–278. https://doi.org/10.1038/73495

Ominsky MS, Vlasseros F, Jolette J, Smith SY, Stouch B, Doellgast G, Gong J, Gao Y, Cao J, Graham K, Tipton B, Cai J, Deshpande R, Zhou L, Hale MD, Lightwood DJ, Henry AJ, Popplewell AG, Moore AR, Robinson MK, Lacey DL, Simonet WS, Paszty C (2010) Two doses of sclerostin antibody in cynomolgus monkeys increases bone formation, bone mineral density, and bone strength. J Bone Miner Res 25(5):948–959. https://doi.org/10.1002/jbmr.14

Padhi D, Allison M, Kivitz AJ, Gutierrez MJ, Stouch B, Wang C, Jang G (2014) Multiple doses of sclerostin antibody romosozumab in healthy men and postmenopausal women with low bone mass: a randomized, double-blind, placebo-controlled study. J Clin Pharmacol 54(2):168–178. https://doi.org/10.1002/jcph.239

Papapoulos SE (2008) Bisphosphonates: how do they work? Best Pract Res Clin Endocrinol Metab 22(5):831–847. https://doi.org/10.1016/j.beem.2008.07.001

Patton MA, Afzal AR (2002) Robinow syndrome. J Med Genet 39(5):305–310. https://doi.org/10.1136/jmg.39.5.305

Pawar NM, Rao P (2018) Secreted frizzled related protein 4 (sFRP4) update: a brief review. Cell Signal 45:63–70. https://doi.org/10.1016/j.cellsig.2018.01.019

Peris P, Monegal A, Makitie RE, Guanabens N, Gonzalez-Roca E (2023) Osteoporosis related to WNT1 variants: a not infrequent cause of osteoporosis. Osteoporos Int 34(2):405–411. https://doi.org/10.1007/s00198-022-06609-2

Person AD, Beiraghi S, Sieben CM, Hermanson S, Neumann AN, Robu ME, Schleiffarth JR, Billington CJ, van Bokhoven H, Hoogeboom JM, Mazzeu JF, Petryk A, Schimmenti LA, Brunner HG, Ekker SC, Lohr JL (2010) WNT5A mutations in patients with autosomal dominant Robinow syndrome. Dev Dyn 239(1):327–337. https://doi.org/10.1002/dvdy.22156

Pyott SM, Tran TT, Leistritz DF, Pepin MG, Mendelsohn NJ, Temme RT, Fernandez BA, Elsayed SM, Elsobky E, Verma I, Nair S, Turner EH, Smith JD, Jarvik GP, Byers PH (2013) WNT1 mutations in families affected by moderately severe and progressive recessive osteogenesis imperfecta. Am J Hum Genet 92(4):590–597. https://doi.org/10.1016/j.ajhg.2013.02.009

Qaseem A, Hicks LA, Etxeandia-Ikobaltzeta I, Shamliyan T, Cooney TG, Clinical Guidelines Committee of the American College of P, Cross JT Jr, Fitterman N, Lin JS, Maroto M, Obley AJ, Tice JA, Tufte JE (2023) Pharmacologic treatment of primary Osteoporosis or low Bone mass to prevent fractures in adults: a living clinical guideline from the American College of Physicians. Ann Intern Med 176(2):224–238. https://doi.org/10.7326/M22-1034

Qu X, Liao M, Liu W, Cai Y, Yi Q, Long J, Tan L, Deng Y, Deng H, Chen X (2021) Loss of Wnt16 leads to skeletal deformities and downregulation of Bone developmental pathway in zebrafish. Int J Mol Sci 22(13). https://doi.org/10.3390/ijms22136673

Randall RM, Shao YY, Wang L, Ballock RT (2012) Activation of Wnt planar cell polarity (PCP) signaling promotes growth plate column formation in vitro. J Orthop Res 30(12):1906–1914. https://doi.org/10.1002/jor.22152

Recker RR, Benson CT, Matsumoto T, Bolognese MA, Robins DA, Alam J, Chiang AY, Hu L, Krege JH, Sowa H, Mitlak BH, Myers SL (2015) A randomized, double-blind phase 2 clinical trial of blosozumab, a sclerostin antibody, in postmenopausal women with low bone mineral density. J Bone Miner Res 30(2):216–224. https://doi.org/10.1002/jbmr.2351

Rim EY, Clevers H, Nusse R (2022) The Wnt pathway: from signaling mechanisms to synthetic modulators. Annu Rev Biochem 91:571–598. https://doi.org/10.1146/annurev-biochem-040320-103615

Robinow M, Silverman FN, Smith HD (1969) A newly recognized dwarfing syndrome. Am J Dis Child 117(6):645–651. https://doi.org/10.1001/archpedi.1969.02100030647005

Roifman M, Brunner H, Lohr J, Mazzeu J, Chitayat D (1993) Autosomal dominant Robinow Syndrome. GeneReviews((R)). MP Adam, GM Mirzaa, RA Pagon et al. Seattle (WA)

Saag KG, Petersen J, Brandi ML, Karaplis AC, Lorentzon M, Thomas T, Maddox J, Fan M, Meisner PD, Grauer A (2017) Romosozumab or alendronate for fracture prevention in women with Osteoporosis. N Engl J Med 377(15):1417–1427. https://doi.org/10.1056/NEJMoa1708322

Saal HM, Prows CA, Guerreiro I, Donlin M, Knudson L, Sund KL, Chang CF, Brugmann SA, Stottmann RW (2015) A mutation in FRIZZLED2 impairs Wnt signaling and causes autosomal dominant omodysplasia. Hum Mol Genet 24(12):3399–3409. https://doi.org/10.1093/hmg/ddv088

Sabir AH, Cole T (2019) The evolving therapeutic landscape of genetic skeletal disorders. Orphanet J Rare Dis 14(1):300. https://doi.org/10.1186/s13023-019-1222-2

Saito-Diaz K, Chen TW, Wang X, Thorne CA, Wallace HA, Page-McCaw A, Lee E (2013) The way Wnt works: components and mechanism. Growth Factors 31(1):1–31. https://doi.org/10.3109/08977194.2012.752737

Schou M (2001) Lithium treatment at 52. J Affect Disord 67(1–3):21–32. https://doi.org/10.1016/s0165-0327(01)00380-9

Schwabe GC, Tinschert S, Buschow C, Meinecke P, Wolff G, Gillessen-Kaesbach G, Oldridge M, Wilkie AO, Komec R, Mundlos S (2000) Distinct mutations in the receptor tyrosine kinase gene ROR2 cause brachydactyly type B. Am J Hum Genet 67(4):822–831. https://doi.org/10.1086/303084

Sebastian A, Hum NR, Morfin C, Murugesh DK, Loots GG (2018) Global gene expression analysis identifies Mef2c as a potential player in Wnt16-mediated transcriptional regulation. Gene 675:312–321. https://doi.org/10.1016/j.gene.2018.06.079

Semënov M, Tamai K, He X (2005) SOST is a ligand for LRP5/LRP6 and a Wnt signaling inhibitor. J Biol Chem 280(29):26770–26775. https://doi.org/10.1074/jbc.M504308200

Semënov MV, He X (2006) LRP5 mutations linked to high bone mass diseases cause reduced LRP5 binding and inhibition by SOST. J Biol Chem 281(50):38276–38284. https://doi.org/10.1074/jbc.M609509200

Semënov MV, Tamai K, Brott BK, Kuhl M, Sokol S, He X (2001) Head inducer Dickkopf-1 is a ligand for Wnt coreceptor LRP6. Curr Biol 11(12):951–961. https://doi.org/10.1016/s0960-9822(01)00290-1

Soares DX, Almeida AM, Barreto AR, Alencar ESIJ, de Castro JD, Magalhaes Pinto FJ, Dias DA, Aguiar LB (2016) Pyle disease (metaphyseal dysplasia) presenting in two adult sisters. Radiol Case Rep 11(4):405–410. https://doi.org/10.1016/j.radcr.2016.10.003

Stevenson DA, Chirpich M, Contreras Y, Hanson H, Dent K (2014) Goltz syndrome and PORCN mosaicism. Int J Dermatol 53(12):1481–1484. https://doi.org/10.1111/ijd.12605

Streeten E (2018) Trial of lithium carbonate for treatment of osteoporosis-pseudoglioma syndrome. https://classic.clinicaltrials.gov/show/NCT01108068

Streeten EA, McBride D, Puffenberger E, Hoffman ME, Pollin TI, Donnelly P, Sack P, Morton H (2008) Osteoporosis-pseudoglioma syndrome: description of 9 new cases and beneficial response to bisphosphonates. Bone 43(3):584–590. https://doi.org/10.1016/j.bone.2008.04.020

Su S-L (2020) Association between bone mineral density and genetic polymorphisms of Wnt signaling pathway among older adults in Taiwan. Bone Reports 13:100496. https://doi.org/10.1016/j.bonr.2020.100496

Sun L, Huang Y, Zhao S, Zhong W, Lin M, Guo Y, Yin Y, Wu N, Wu Z, Tian W (2019) Advances in understanding the genetics of syndromes involving congenital upper limb anomalies. Annals of Joint 4

Szenker-Ravi E, Altunoglu U, Leushacke M, Bosso-Lefevre C, Khatoo M, Thi Tran H, Naert T, Noelanders R, Hajamohideen A, Beneteau C, de Sousa SB, Karaman B, Latypova X, Basaran S, Yucel EB, Tan TT, Vlaminck L, Nayak SS, Shukla A, Girisha KM, Le Caignec C, Soshnikova N, Uyguner ZO, Vleminckx K, Barker N, Kayserili H, Reversade B (2018) RSPO2 inhibition of RNF43 and ZNRF3 governs limb development independently of LGR4/5/6. Nature 557(7706):564–569. https://doi.org/10.1038/s41586-018-0118-y

Tanaka S, Matsumoto T (2021) Sclerostin: from bench to bedside. J Bone Miner Metab 39(3):332–340. https://doi.org/10.1007/s00774-020-01176-0

Tashima T, Nagatoishi S, Caaveiro JMM, Nakakido M, Sagara H, Kusano-Arai O, Iwanari H, Mimuro H, Hamakubo T, Ohnuma SI, Tsumoto K (2018) Molecular basis for governing the morphology of type-I collagen fibrils by Osteomodulin. Commun Biol 1:33. https://doi.org/10.1038/s42003-018-0038-2

Tashima T, Nagatoishi S, Sagara H, Ohnuma S, Tsumoto K (2015) Osteomodulin regulates diameter and alters shape of collagen fibrils. Biochem Biophys Res Commun 463(3):292–296. https://doi.org/10.1016/j.bbrc.2015.05.053

Thiele H, McCann C, van Padje S, Schwabe GC, Hennies HC, Camera G, Opitz J, Laxova R, Mundlos S, Nurnberg P (2004) Acropectorovertebral dysgenesis (F syndrome) maps to chromosome 2q36. J Med Genet 41(3):213–218. https://doi.org/10.1136/jmg.2003.014894

Thomas KR, Capecchi MR (1990) Targeted disruption of the murine int-1 proto-oncogene resulting in severe abnormalities in midbrain and cerebellar development. Nature 346(6287):847–850. https://doi.org/10.1038/346847a0

Thomas KR, Musci TS, Neumann PE, Capecchi MR (1991) Swaying is a mutant allele of the proto-oncogene Wnt-1. Cell 67(5):969–976. https://doi.org/10.1016/0092-8674(91)90369-a

Tong W, Zeng Y, Chow DHK, Yeung W, Xu J, Deng Y, Chen S, Zhao H, Zhang X, Ho KK, Qin L, Mak KK (2019) Wnt16 attenuates osteoarthritis progression through a PCP/JNK-mTORC1-PTHrP cascade. Ann Rheum Dis 78(4):551–561. https://doi.org/10.1136/annrheumdis-2018-214200

Truswell AS (1958) Osteopetrosis with syndactyly; a morphological variant of Albers-Schonberg's disease. J Bone Joint Surg Br 40-B(2):209–218

Tu X, Rhee Y, Condon KW, Bivi N, Allen MR, Dwyer D, Stolina M, Turner CH, Robling AG, Plotkin LI, Bellido T (2012) Sost downregulation and local Wnt signaling are required for the osteogenic response to mechanical loading. Bone 50(1):209–217. https://doi.org/10.1016/j.bone.2011.10.025

Türkmen S, Spielmann M, Gunes N, Knaus A, Flottmann R, Mundlos S, Tuysuz B (2017) A novel de novo FZD2 mutation in a patient with autosomal dominant Omodysplasia. Mol Syndromol 8(6):318–324. https://doi.org/10.1159/000479721

Ugur SA, Tolun A (2008) Homozygous WNT10b mutation and complex inheritance in Split-hand/foot malformation. Hum Mol Genet 17(17):2644–2653. https://doi.org/10.1093/hmg/ddn164

Unger S, Ferreira CR, Mortier GR, Ali H, Bertola DR, Calder A, Cohn DH, Cormier-Daire V, Girisha KM, Hall C, Krakow D, Makitie O, Mundlos S, Nishimura G, Robertson SP, Savarirayan R, Sillence D, Simon M, Sutton VR, Warman ML, Superti-Furga A (2023) Nosology of genetic skeletal disorders: 2023 revision. Am J Med Genet A 191(5):1164–1209. https://doi.org/10.1002/ajmg.a.63132

U.S. Food and Drug Administration (FDA) (2019) Evenity (romosozumab-aqqg) injection, for subcutanous use. Prescribing information. Reference ID: 4416809. Rockville, MD, FDA

Välimäki VV, Makitie O, Pereira R, Laine C, Wesseling-Perry K, Maatta J, Kirjavainen M, Viljakainen H, Valimaki MJ (2017) Teriparatide treatment in patients with WNT1 or PLS3 mutation-related early-onset Osteoporosis: a pilot study. J Clin Endocrinol Metab 102(2):535–544. https://doi.org/10.1210/jc.2016-2423

Van Buchem FS, Hadders HN, Ubbens R (1955) An uncommon familial systemic disease of the skeleton: hyperostosis corticalis generalisata familiaris. Acta Radiol 44(2):109–120

van Lierop AH, Appelman-Dijkstra NM, Papapoulos SE (2017) Sclerostin deficiency in humans. Bone 96:51–62. https://doi.org/10.1016/j.bone.2016.10.010

van Lierop AH, Hamdy NA, Hamersma H, van Bezooijen RL, Power J, Loveridge N, Papapoulos SE (2011) Patients with sclerosteosis and disease carriers: human models of the effect of sclerostin on bone turnover. J Bone Miner Res 26(12):2804–2811. https://doi.org/10.1002/jbmr.474

van Lierop AH, Hamdy NA, van Egmond ME, Bakker E, Dikkers FG, Papapoulos SE (2013) Van Buchem disease: clinical, biochemical, and densitometric features of patients and disease carriers. J Bone Miner Res 28(4):848–854. https://doi.org/10.1002/jbmr.1794

Vestergaard Kvist A, Faruque J, Vallejo-Yague E, Weiler S, Winter EM, Burden AM (2021) Cardiovascular safety profile of Romosozumab: a pharmacovigilance analysis of the US Food and Drug Administration adverse event reporting system (FAERS). J Clin Med 10(8). https://doi.org/10.3390/jcm10081660

Vollersen N, Zhao W, Rolvien T, Lange F, Schmidt FN, Sonntag S, Shmerling D, von Kroge S, Stockhausen KE, Sharaf A, Schweizer M, Karsak M, Busse B, Bockamp E, Semler O, Amling M, Oheim R, Schinke T, Yorgan TA (2021) The WNT1(G177C) mutation specifically affects skeletal integrity in a mouse model of osteogenesis imperfecta type XV. Bone Res 9(1):48. https://doi.org/10.1038/s41413-021-00170-0

Wang F, Tarkkonen K, Nieminen-Pihala V, Nagano K, Majidi RA, Puolakkainen T, Rummukainen P, Lehto J, Roivainen A, Zhang FP, Makitie O, Baron R, Kiviranta R (2019) Mesenchymal cell-derived Juxtacrine Wnt1 signaling regulates osteoblast activity and osteoclast differentiation. J Bone Miner Res 34(6):1129–1142. https://doi.org/10.1002/jbmr.3680

Wang Y, Li YP, Paulson C, Shao JZ, Zhang X, Wu M, Chen W (2014) Wnt and the Wnt signaling pathway in bone development and disease. Front Biosci (Landmark Ed) 19(3):379–407. https://doi.org/10.2741/4214

Wawrzak D, Metioui M, Willems E, Hendrickx M, de Genst E, Leyns L (2007) Wnt3a binds to several sFRPs in the nanomolar range. Biochem Biophys Res Commun 357(4):1119–1123. https://doi.org/10.1016/j.bbrc.2007.04.069

Wergedal JE, Kesavan C, Brommage R, Das S, Mohan S (2015) Role of WNT16 in the regulation of periosteal bone formation in female mice. Endocrinology 156(3):1023–1032. https://doi.org/10.1210/en.2014-1702

White J, Mazzeu JF, Hoischen A, Jhangiani SN, Gambin T, Alcino MC, Penney S, Saraiva JM, Hove H, Skovby F, Kayserili H, Estrella E, Vulto-van Silfhout AT, Steehouwer M, Muzny DM, Sutton VR, Gibbs RA, Baylor-Hopkins Center for Mendelian G, Lupski J R, Brunner H G, van Bon B W, Carvalho C M (2015) DVL1 frameshift mutations clustering in the penultimate exon cause autosomal-dominant Robinow syndrome. Am J Hum Genet 96(4):612–622. https://doi.org/10.1016/j.ajhg.2015.02.015

White JJ, Mazzeu JF, Coban-Akdemir Z, Bayram Y, Bahrambeigi V, Hoischen A, van Bon BWM, Gezdirici A, Gulec EY, Ramond F, Touraine R, Thevenon J, Shinawi M, Beaver E, Heeley J, Hoover-Fong J, Durmaz CD, Karabulut HG, Marzioglu-Ozdemir E, Cayir A, Duz MB, Seven M, Price S, Ferreira BM, Vianna-Morgante AM, Ellard S, Parrish A, Stals K, Flores-Daboub J, Jhangiani SN, Gibbs RA, Baylor-Hopkins Center for Mendelian G, Brunner H G, Sutton V R, Lupski J R, Carvalho C M B (2018) WNT signaling perturbations underlie the genetic heterogeneity of Robinow Syndrome. Am J Hum Genet 102(1):27–43. https://doi.org/10.1016/j.ajhg.2017.10.002

White JJ, Mazzeu JF, Hoischen A, Bayram Y, Withers M, Gezdirici A, Kimonis V, Steehouwer M, Jhangiani SN, Muzny DM, Gibbs RA, Baylor-Hopkins Center for Mendelian G, van Bon B W M, Sutton V R, Lupski J R, Brunner H G, Carvalho C M B (2016) DVL3 alleles resulting in a −1 frameshift of the last exon mediate autosomal-dominant Robinow Syndrome. Am J Hum Genet 98(3):553–561. https://doi.org/10.1016/j.ajhg.2016.01.005

Whyte MP, McAlister WH, Zhang F, Bijanki VN, Nenninger A, Gottesman GS, Lin EL, Huskey M, Duan S, Dahir K, Mumm S (2019) New explanation for autosomal dominant high bone mass: mutation of low-density lipoprotein receptor-related protein 6. Bone 127:228–243. https://doi.org/10.1016/j.bone.2019.05.003

Winkler DG, Sutherland MK, Geoghegan JC, Yu C, Hayes T, Skonier JE, Shpektor D, Jonas M, Kovacevich BR, Staehling-Hampton K, Appleby M, Brunkow ME, Latham JA (2003) Osteocyte control of bone formation via sclerostin, a novel BMP antagonist. EMBO J 22(23):6267–6276. https://doi.org/10.1093/emboj/cdg599

Woods CG, Stricker S, Seemann P, Stern R, Cox J, Sherridan E, Roberts E, Springell K, Scott S, Karbani G, Sharif SM, Toomes C, Bond J, Kumar D, Al-Gazali L, Mundlos S (2006) Mutations in WNT7A cause a range of limb malformations, including Fuhrmann syndrome and Al-Awadi/Raas-Rothschild/Schinzel phocomelia syndrome. Am J Hum Genet 79(2):402–408. https://doi.org/10.1086/506332

Wu C, Yu J, Xu G, Bao G, Zhang J, Xue P, Jiang J, Chen J, Chen C, Hong H, Cui Z (2021) Wnt16 protects chondrocytes from lumbar facet joint osteoarthritis through the Wnt/beta-catenin pathway in low back pain patients. Somatosens Mot Res 38(4):339–346. https://doi.org/10.1080/08990220.2021.1977267

Ye X, Liu X (2022) Wnt16 signaling in bone homeostasis and osteoarthristis. Front Endocrinol (Lausanne) 13:1095711. https://doi.org/10.3389/fendo.2022.1095711

Yorgan TA, Peters S, Jeschke A, Benisch P, Jakob F, Amling M, Schinke T (2015) The anti-Osteoanabolic function of Sclerostin is blunted in mice carrying a high Bone mass mutation of Lrp5. J Bone Miner Res 30(7):1175–1183. https://doi.org/10.1002/jbmr.2461

Yorgan TA, Rolvien T, Sturznickel J, Vollersen N, Lange F, Zhao W, Baranowsky A, Rosenthal L, Hermans-Borgmeyer I, Sharaf A, Karsak M, David JP, Oheim R, Amling M, Schinke T (2020) Mice carrying a ubiquitous R235W mutation of Wnt1 display a Bone-specific phenotype. J Bone Miner Res 35(9):1726–1737. https://doi.org/10.1002/jbmr.4043

Yorgan TA, Schinke T (2014) Relevance of Wnt signaling for osteoanabolic therapy. Mol Cell Ther 2:22. https://doi.org/10.1186/2052-8426-2-22

Zaidi M (2007) Skeletal remodeling in health and disease. Nat Med 13(7):791–801. https://doi.org/10.1038/nm1593

Zamani A, Omrani GR, Nasab MM (2009) Lithium's effect on bone mineral density. Bone 44(2):331–334. https://doi.org/10.1016/j.bone.2008.10.001

Zeng Y, Yueng W, Mak KL, Zhao H (2018) Wnt16 regulates chondrocyte differentiation through Wnt/ planar cell polarity (PCP) pathway. FASEB J 32(S1):533.520-533.520. https://doi.org/10.1096/fasebj.2018.32.1_supplement.533.20

Zhang L, Choi HJ, Estrada K, Leo PJ, Li J, Pei YF, Zhang Y, Lin Y, Shen H, Liu YZ, Liu Y, Zhao Y, Zhang JG, Tian Q, Wang YP, Han Y, Ran S, Hai R, Zhu XZ, Wu S, Yan H, Liu X, Yang TL, Guo Y, Zhang F, Guo YF, Chen Y, Chen X, Tan L, Zhang L, Deng FY, Deng H, Rivadeneira F, Duncan EL, Lee JY, Han BG, Cho NH, Nicholson GC, McCloskey E, Eastell R, Prince RL, Eisman JA, Jones G, Reid IR, Sambrook PN, Dennison EM, Danoy P, Yerges-Armstrong LM, Streeten EA, Hu T, Xiang S, Papasian CJ, Brown MA, Shin CS, Uitterlinden AG, Deng HW (2014) Multistage genome-wide association meta-analyses identified two new loci for bone mineral density. Hum Mol Genet 23(7):1923–1933. https://doi.org/10.1093/hmg/ddt575

Zheng HF, Tobias JH, Duncan E, Evans DM, Eriksson J, Paternoster L, Yerges-Armstrong LM, Lehtimaki T, Bergstrom U, Kahonen M, Leo PJ, Raitakari O, Laaksonen M, Nicholson GC, Viikari J, Ladouceur M, Lyytikainen LP, Medina-Gomez C, Rivadeneira F, Prince RL, Sievanen H, Leslie WD, Mellstrom D, Eisman JA, Moverare-Skrtic S, Goltzman D, Hanley DA, Jones G, St Pourcain B, Xiao Y, Timpson NJ, Smith GD, Reid IR, Ring SM, Sambrook PN, Karlsson M, Dennison EM, Kemp JP, Danoy P, Sayers A, Wilson SG, Nethander M, McCloskey E, Vandenput L, Eastell R, Liu J, Spector T, Mitchell BD, Streeten EA, Brommage R, Pettersson-Kymmer U, Brown MA, Ohlsson C, Richards JB, Lorentzon M (2012) WNT16 influences bone mineral density, cortical bone thickness, bone strength, and osteoporotic fracture risk. PLoS Genet 8(7):e1002745. https://doi.org/10.1371/journal.pgen.1002745

Chapter 10
The Skeletal Ciliopathies

Isabella Collins and Angus K. T. Wann

Abstract Within the broad and growing spectrum of human ciliopathies is a range of linked and overlapping disorders that present with skeletal features. These have been coined the skeletal ciliopathies. The syndromes have, to this point, been largely attributed to alterations in cellular Hedgehog signalling during development. However, a huge surge in fundamental discovery and clinical research has unveiled a plethora of roles for cilia in biology. This implicates a range of molecular and cell processes in the pathogenesis of skeletal ciliopathies. Our understanding of bi-directional interactions between cilia and the extracellular matrix remains in its infancy. The identification of genes and causal mutations defines skeletal ciliopathies. Some of these genes, and the proteins they encode, are now being explored further, by means of cell, animal, and other model approaches, seeking to understand the molecular underpinnings of disease. However, given the relatively recent appreciation of links between cilia biology and human disease, there is much work to be done. This chapter will briefly introduce the primary cilium and its associated 'ciliome', before describing the ciliary-associated skeletal disorders and the genes with which they are associated. Where possible, it will expand upon our current mechanistic understanding.

I. Collins
Kennedy Institute of Rheumatology, University of Oxford, Oxford, UK

A. K. T. Wann (✉)
Kennedy Institute of Rheumatology, University of Oxford, Oxford, UK

School of Biological Sciences, University of Southampton, Southampton, UK
e-mail: a.k.t.wann@soton.ac.uk

A. Rossi, F. Zaucke (eds.), *The Extracellular Matrix in Genetic Skeletal Disorders*, Biology of Extracellular Matrix 16,
https://doi.org/10.1007/978-3-031-70835-0_10

10.1 Introduction

The ciliopathies are a growing class of diseases that stem from mutations in genes established to play an important role in motile or primary cilia function, or in both. In 2023, mutations in at least 247 genes have been established to give rise to pathology (Reiter and Leroux 2017; Wallmeier et al. 2020). The pathologies represent an overlapping spectrum of human syndromes that influence the function of most tissues of the human body. Clear phenotypic-genotypic correlation is a challenge, as mutations in single genes often give rise to multiple ciliopathies. Most ciliopathy genes encode proteins localised at any one time, at least in the cell types studied, to the ciliary structure or the basal body, a modified form of the mature centriole. This has led to the creation of a 'ciliome', a collective of genes encoding proteins with ciliary localisation and/or function. In-direct effects on ciliary function, through protein localisation elsewhere in the cell, have led to the creation of a list of second-order ciliopathy genes. It has been estimated that >300 additional disorders may come 'with ciliary contribution' (Lovera and Luders 2021). Skeletal features are common in human ciliopathies (Handa et al. 2020), including classical dysplasia features such as short arms and legs compared with the rest of the body, a small chest cavity, and changes to joint mobility and structure. However, the skeletal features have received less attention in experimental model systems implemented to unveil pathogenic mechanisms. As such, there has been relatively limited mechanistic exploration of skeletal ciliopathies. Nevertheless, the explosion of ciliopathy identification, coupled with a busy couple of decades of basic science concerning ciliary structure and function, has laid a platform for a greater understanding of disease pathology and the hunt for therapeutic approaches for skeletal ciliopathies.

10.2 Primary Cilia

For an excellent recent review of our current knowledge of primary cilia (non-motile) biology, please consider Mill, Christensen, and Pedersen in 2023 (Mill et al. 2023). For the purposes of contextualising this chapter, the primary cilium, in contrast to motile cilia, is a solitary (one per cell) organelle lacking dynein arm-dependent motility. However primary and motile cilia share many structural features. Primary cilia are microtubule-based organelles (see Fig. 10.1). This compartment is assembled by the vast majority of cell types (Wheatley 1995) upon exit from the cell cycle. Upon exit, the 'mother' centriole of the cell associates with specialised vesicles to form a basal body structure that acts as a template for ciliary axoneme growth. As such, the cilium is positioned proximal to the microtubule organising centre of the cell, often close to the nuclear perimeter, but with access to the extracellular space either through the full length of the axoneme or only partially, when the axoneme is partially invaginated within a ciliary pocket (Molla-Herman et al. 2010).

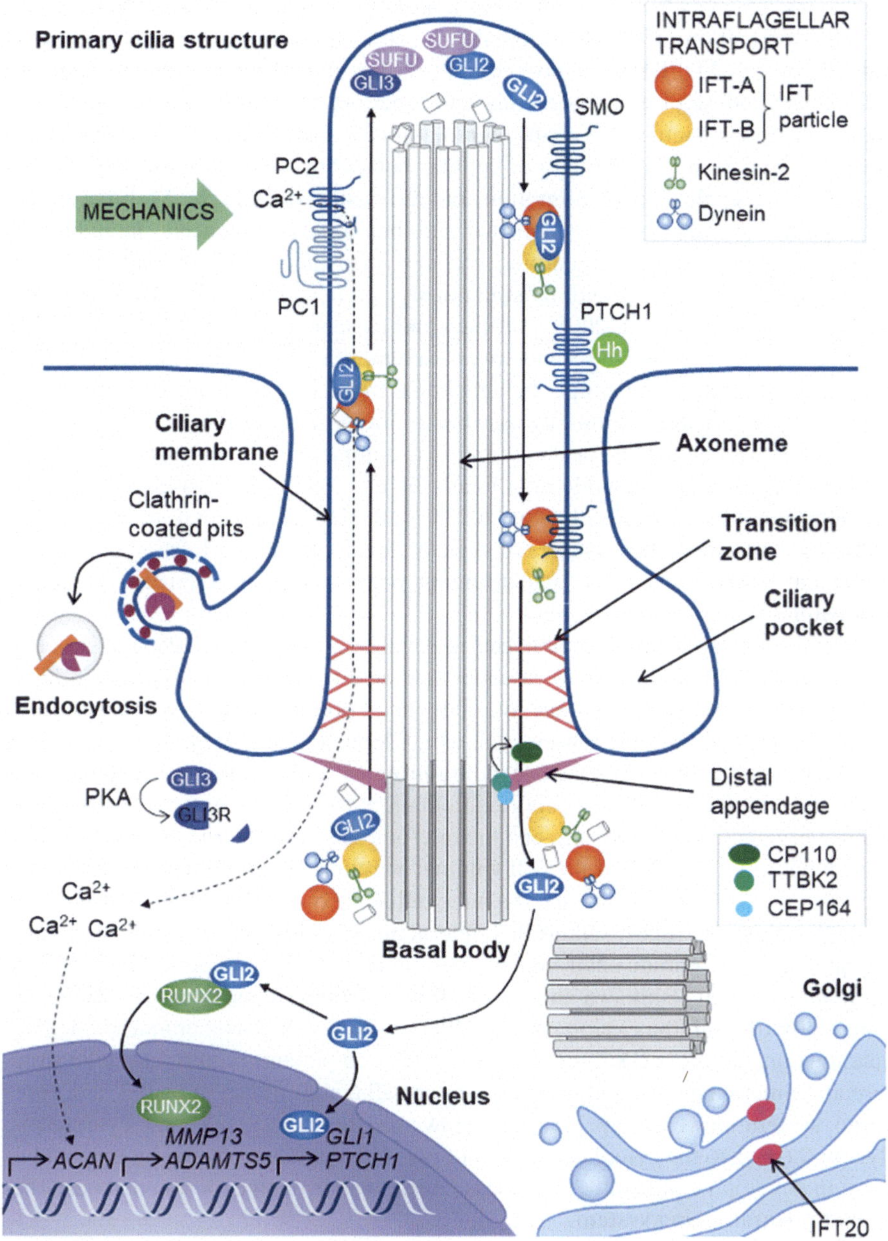

Fig. 10.1 Schematic describing the structure of primary cilia. The ciliome, at some time, inhabits this compartment which is assembled from the basal body, an adapted mother centriole, upon removal of CP110 by TTBK2. The best-described molecular events associated with the cilium include Hedgehog signalling, involving components including the ligand (Hh), receptor (PTCH), transduction effectors such as SMO and SUFU, and the GLIs. It is also associated with calcium signalling, in mechanotransduction, and, through IFT20 and other molecules, the Golgi (Stevenson 2023). From Collins, I. 2021. 'Identification of a Novel Regulator of Extracellular Matrix Degradation in Chondrocytes'. PhD thesis, University of Oxford, an adaptation of (Collins and Wann 2020)

The last 25 years have seen an explosion of cilia research, enthused in large part by the appreciation of human ciliopathies such as autosomal dominant polycystic kidney disease (ADPKD), the most common ciliopathy. This has built a comprehensive understanding of ciliary structure and function. The cilium has been coined a signalling hub, sometimes, perhaps more famously, a signalling antenna, associated with the regulation of a plethora of cellular behaviours, including those as fundamental as cell cycle control, migration, and the control of cell size. Multiple signalling pathways are now established to be ciliary-influenced including G protein-coupled receptor signalling, receptor tyrosine kinase signalling, Notch, Hippo, mTOR, Wnt-PCP, and, most famously and best understood, Hedgehog signalling (Christensen et al. 2008; Pedersen and Rosenbaum 2008; Mill et al. 2023). Despite the common use of the antenna paradigm, the role of cilia in cell signalling is far more complex than simply being a home to the reception of extracellular signals. Another function commonly ascribed to the primary cilium is mechanotransduction (Rita et al. 2019), but how this is accomplished remains unclear. Recent studies in the embryonic node may finally be elucidating mechanisms supporting this long-heralded role (Katoh et al. 2023). Both Hedgehog signalling and mechanobiology are established masters of skeletal development, implicating cilia in this biological context and likely underpinning pathological mechanisms in skeletal ciliopathies.

However, cilia are now established to regulate many critical skeletal development pathways, including those cascades activated by growth factors such as fibroblast growth factors (FGFs), bone morphogenetic protein (BMP), and transforming growth factor (TGF) (Christensen et al. n.d.; Clement et al. 2013).

The ciliome comprises genes encoding proteins found in the centrosome, basal body, the transition zone, a specialised domain at the base of the cilium thought to control organelle entry and exit, and the microtubule-based axoneme. As such, proteins inhabiting 1/10,000 of the cell volume, a compartment often μm in length but only hundreds of nm in width, exert a huge influence over cell biology. This is likely through transient, highly controlled interactions taking place within a unique, receptor-rich, membrane environment (Stilling et al. 2022), supported by a devoted trafficking system (Ishikawa and Marshall n.d.; Ishikawa and Marshall 2017) and potentially through the exploitation of other attributes such as a unique intraciliary calcium concentration (Delling et al. 2013). All components required for cilia structure and function are carried in and out by intraflagellar transport or IFT, a devoted cargo system powered by kinesin-2 (towards the ciliary tip) and cytoplasmic dynein-2 (back to the cytoplasm). Most of our understanding of cilia biology and its roles in development, health, and disease has come through experimental perturbation of this trafficking system.

10.3 Cilia and the Skeleton

Multiple murine models have reflected roles for IFT in skeletal development (Haycraft et al. 2007; McGlashan et al. 2007; Song et al. 2007; Haycraft and Serra 2008; Chang et al. 2012; Chang and Serra 2012; Yuan and Yang 2015), supported by studies characterising the effects of ciliary kinesin disruption (Koyama et al. 2007; Song et al. 2007). Many of these were attributed to the role of cilia in transducing (Bangs and Anderson 2017) and, importantly, acting sometimes as a brake on Hedgehog signalling (Chang and Serra 2012). Thus, most of the mechanistic work exploring skeletal ciliopathies attributes the effects of ciliary mutations to changes (usually a diminishment in response) to the transduction of Hedgehog ligand-induced signalling.

For a review of the literature, concerning the cilium as a regulator for skeletal development, see (Moore and Jacobs 2018). Only a handful of studies have been able to directly tie primary cilia dysfunction to the changes observed during skeletal development in model systems, but, nevertheless, research strongly supports a central role for cilia in the development, health, and disease of the skeleton. Revolutions in microscopy and tissue biology should allow ciliary signalling and ciliary structure to be viewed in concert, allowing the appreciation of how ciliary assembly, position, length, and orientation relate to the molecular events cilia govern and therefore the morphogenetic consequences of ciliary dysfunction.

10.4 Cilia and Matrix

Following an excellent review of how ECM changes reflect in ciliary biology (Seeger-Nukpezah and Golemis 2012), a review in 2020 (Collins and Wann 2020) considered the collective literature at that time concerning evidence and hypotheses for how ciliary structure and function might regulate extracellular matrix. This focussed on musculoskeletal biology and pathology, due to the weight of evidence in this context. Disruption of ciliary signalling through perturbation of ciliary transport (IFT) is associated with changes to matrix composition, for example, the regulation of collagen X in endochondral ossification (McGlashan, Haycraft et al. 2007, Yuan and Yang 2015). The unfolding complex role of the cilium in the cellular and tissue integration of mechanical forces in cartilage, exemplified in the joint by recent studies linking IFT88 to relative mineralisation and structure (Coveney et al. 2021) and tidemark control (Rux et al. 2022) and in the growth plate (Coveney et al. 2022), links ciliary function to matrix regulation through the shaping of skeletal development in response to force. These studies indicate that the skeletal field will have to think beyond simpler paradigms indicated in the kidney (Nauli et al. 2003) and node (Katoh et al. 2023). Cilia are well positioned to integrate multiple different signals, for example, a growth factor signal and compressive forces along with the osmotic differentials loading induced in negatively charged tissues like cartilage. A host of ciliary genes have been linked to both anabolic and catabolic processes in vitro and in vivo (Mangos et al. n.d.; Davis et al. 2007; Malone et al. 2007; Hoey et al. 2012; Wann et al. 2012; Shao et al. 2012; Thompson et al. 2014; Rais et al. 2015; Coveney et al. 2018; Fu et al. 2019; Thompson et al. 2021) in chondrocytes and bone cells. The three-way interplay between cilia, Golgi, and ECM was the focus of a recent review (Stevenson 2023) and is a wealth of up-to-date information concerning cause and effect. There is considerable overlap between ciliopathies and Golgiopathies adding to the complexity of unpicking the mechanisms of such disorders. This chapter will focus on skeletal ciliopathies.

10.5 Skeletal Disorders Linked to Cilia or Ciliary Signalling

For the purposes of focussing on the genetic skeletal disorders associated with mutations in 'ciliary' genes, a recent revision of the nosology lists 61 such disorders organised by the association of phenotype with a single gene. These are linked to the OMIM database by MIM number (Unger et al. 2023). By the grouping used in this nosology, disorders linked with cilia and ciliary signalling constitute the largest group, underlining the powerful influence of ciliary function in skeletal development.

10.6 Bardet-Biedl Syndrome

It is noteworthy that the Bardet-Biedl syndrome (BBS) spectrum (Beales et al. 1997; Forsythe and Beales 2013) is not included in the ciliary grouping within the recent nosology compiled by Unger et al., as the predominant features are non-skeletal. Nevertheless, BBS is associated with polydactyly and skeletal features. Mouse models, for example, *BBS1/2/6* mutants, indicate an influence of the 'BBSome', some 16 genes, in the skeleton. Histological and biochemical differences have been observed in articular cartilage (Kaushik et al. 2009). As is true for many constitutive genetic models, uncoupling systemic and early developmental effects to ascribe tissue-specific and period-specific roles for ciliary genes is challenging. Thus, it is not yet clear what specific roles the BBSome plays in growth plate dynamics. BBS proteins are linked with Hedgehog regulation in limb development. Mis-expression of *BBS1* or *BBS7* results in skeletal pattern changes in zebrafish (Tayeh et al. 2008). Leucine zipper transcription factor-like 1 (*LZTFL1*) or *BBS17* knockout mouse models display early life growth impairment (Jiang et al. 2016), but studies have focussed on retinal phenotypes. It is likely that disruption of ciliary IFT underlies skeletal phenotypes in BBS. Interactions between BBS proteins are localised to the basal body (Ansley et al. 2003) and ciliary cargo (the BBSome is thought to function as an adapter between IFT and protein cargoes (Liu et al. 2023)). Interactions between the protein products of many of the genes associated with the disorders described hereafter may also explain their pathogenesis. The

BBSome is linked with the regulation of several skeletally relevant signalling pathways. These include Wnt (Gerdes et al. 2007) and often in tandem with other ciliary proteins, those associated with skeletal ciliopathies such as Hedgehog signalling (Goetz et al. 2017). Approaches similar to those recently exploited to explore model neurocristopathies (Barrell et al. 2019), using induced pluripotent stem cell (iPSC) differentiation protocols, may be insightful in ciliopathy chondrodysplasia arena, not least given advances in detailed modelling of human limb development in vitro (Richard et al. 2023).

10.7 Short-Rib Polydactyly (SRP)-Related Syndromes

Short-rib thoracic dysplasia (SRTD) may be with or without polydactyly and is characterised by a constricted thoracic cage, shortened ribs and tubular (long) bones, and a 'trident' appearance of the acetabular roof. Many forms of SRTD are lethal in the neonatal period, as a result of limitations to respiration secondary to restrictions of the thoracic cage (Huber and Cormier-Daire 2012). As separated by gene association, 11 disorders fall under the umbrella of SRP syndromes (presenting with polydactyl). The literature describes 5 types: Type I or Saldino-Noonan syndrome, Type II or Majewski syndrome, Type III or Verma-Naumoff syndrome, Type IV or Beemer-Langer syndrome, and Type V. All are heterogenous, of autosomal recessive inheritance, and are characterised by short ribs, polydactyly, a narrow thorax, and shortened long bones. In addition, as is the case with most ciliopathies, they present with a range of extraskeletal phenotypes. There is notable clinical and radiographic overlap between SRP I/III and asphyxiating thoracic dystrophy (ATD), with many genes associated with syndromes with polydactyly and without.

Deficiency in IFT proteins and function underpins a range of the skeletal manifestations of ciliopathies. A heavy chain dynein ATPase/motor subunit DYNC2H1 is among these. Short-rib thoracic dysplasia-3 (SRTD3) is caused by several mutations in the *DYNC2H1* gene on chromosome 11q22 (Dagoneau et al. 2009; Merrill et al. 2009; Thiel et al. 2011; Badiner, Taylor et al. 2017) (MIM 613091). There is evidence that SRTD can be caused by a digenic mutation in *NEK1* (MIM 263520) and *DYNC2H1* (MIM 613091) genes (Thiel et al. 2011), and mutations of other genes including *WDR60* (MIM 615503) have been identified in patients with *NEK1* mutations, the presence of which may underlie spectrums of phenotypes (McInerney-Leo et al. 2015). When first associated with SRP, *WDR60* was a relatively uncharacterised gene (McInerney-Leo et al. 2013). It is expressed in wild-type human chondrocytes at the base of the cilium and linked to Hedgehog function. WDR60 also interacts with ciliary cargo through IFT54 (Hiyamizu et al. 2023a, b). Schmidts et al. (2013a, b, c) described 11 mutations in *WDR34* (MIM 615633), a previously unrecognised link between dynein motors in the cell cytoplasm and IFT dynein-2 motors that regulate ciliary retrograde protein transport (Schmidts, Vodopiutz et al. 2013c). *WDR34* mutations have been identified in dysplasia where *DYNC2H1* and IFT gene mutations have been excluded (Huber et al. 2013). *WDR34* mutations have been linked to NFκB signalling (Huber et al. 2013), a signalling cascade critical to a plethora of cell and tissue biology including that of the skeleton (Novack 2011). NFκB has been linked to cilia in chondrocytes in vitro (Wann et al. 2014), albeit likely through a role independent of the ciliary axoneme (Mc Fie et al. 2020).

Beales et al. (2007) described homozygous mutations in *IFT80* (Beales et al. 2007) (MIM 611263). This has been followed up with murine studies describing the skeletal phenotypes associated with hypomorphic disruption to *IFT80* (Rix et al. 2011). In this model, the thoracic phenotype is similar to that in patients, characterised by ribcage constriction and irregular margins between cartilage and bone. Forelimb bones were correctly patterned, but shortening is seen in distal and proximal long bones where organisation of chondrocytes within the usually distinct layers is disrupted. Cells are described as poorly organised, particularly in the hypertrophic zones where hypertrophic size is qualitatively reduced. Disorganisation of chondrocytic columns in the growth plate is commonly described in ciliopathy models and murine KO experiments (Koyama et al. 2007; Coveney et al. 2022). The phenotype is associated with changes to Hedgehog signalling but not cilia structure (Rix et al. 2011). Further studies have indicated that *IFT80* plays a key role in chondrocyte differentiation (Yuan and Yang 2015). Marked changes in cell shape and zonal organisation were observed in a Col2a1-CreER system and associated with changes to Hedgehog and Wnt signalling (Wang et al. 2013), targeting IFT80 in Xenopus recapitulated ciliopathy phenotypes, including limb deformities (Getwan et al. 2021). Mutations in *IFT81* (MIM 617895) and *IFT122* (MIM 269860) are also associated with SRP (Duran et al. 2016; Pettersson et al. 2018) as are mutations in dynein cytoplasmic 2 light-intermediate chain 1 (*DYNC2LI1*, MIM 617088) (Kessler et al. 2015; Taylor, Dantas et al. 2015; Niceta et al. 2018; Bryson et al. 2021). Destabilisation of the IFT-B core complex due to *IFT81* mutations is associated with a lack of normal progressive enlargement (hypertrophy) within the growth plate and a collagenous band above the proliferative zone, staining potentially also indicative of ectopic mineralisation, which is commonly associated with disrupted Hedgehog signalling of the growth plate (Koyama et al. 2007). Mouse models demonstrate *DYNI1* deletion in limb mesoderm, resulting in severe bone shortening, phenocopying KIF3A/IFT88 models, except for polydactyly and altered Hedgehog signalling (King et al. 2019).

Endocrine-cerebro-osteodysplasia (MIM 612651) is caused by a homozygous mutation in the *ICK* gene encoding intestinal cell kinase. Affected individuals have phenotypic similarities to Majewski syndrome (SRP, type II). Lahiry et al. 2009 (Lahiry et al. 2009) reported facial dysmorphism, limb shortening, and bowing among extensive features. Another family was identified in 2016 (Oud et al. 2016).

10.8 Short-Rib Thoracic Dysplasia or Jeune Asphyxiating Thoracic Dystrophy (JATD)

A genetic module termed CPLANE (ciliogenesis and planar polarity effector) acts to recruit IFT, mutant mice displaying skeletal features, and the gene *INTU* is associated with SRP phenotypes in patients (Toriyama, Lee et al. 2016) (MIM 617925). Mutations in *WDR19*, encoding IFT144 (MIM 614376) and *WDR35* (MIM 614091), have also been identified in patients (Bredrup et al. 2011; Mill et al. 2011). The mutant twinkle-toes (*IFT144*$^{(TWT)}$) phenocopies a number of skeletal ciliopathy features and, like other IFT-A models, is associated with ligand-independent increases in Hedgehog signalling (Ashe et al. 2012).

TRAF3IP1 encodes IFT54, has been localised to the ciliary axoneme, and is associated with both SRP and senior Loken syndrome (MIM 607380), a ciliopathy predominantly associated with nephronophthisis and retinal dystrophy. *TCTEX1D2* (MIM 617405) mutations were identified in SRTD17 by Schmidts et al. in 2015 (Schmidts et al. 2015) who noted classical Jeune syndrome features of the skeleton without renal, hepatic, or ocular features. In 2013, exome sequencing (Schmidts et al. 2013a) identified mutations in dynein subunit *DYNC2H1* as common in Jeune (MIM 613091). Mutations in *IFT40* have been described in 5 JATD families (Schmidts et al. 2013b) (MIM 266920). *TTC21B* (MIM 613819) encodes a protein localised to the ciliary axoneme associated with both nephronophthisis, first described by Davis et al. 2011 (Davis et al. 2011), and JATD (McInerney-Leo et al. 2015) exhibiting narrow thorax, brachydactyly, short long bones, and scoliosis. *TTC21B* is also known as *IFT139* and *THM1* and, through interactions with paralog *Thm2*, alters Hedgehog signalling in postnatal skeletal morphogenesis. *Thm* mutants have altered growth plate organisation, an expanded proliferative zone, reduced hypertrophic, and increased Hedgehog signalling in chondrocytes. Mouse models of loss of *Thm2*, with loss of one allele of *Thm1* or *Gli2* or both, are useful models of osteochondrodysplasia (Allard et al. 2021). Mutations in *IFT43* (MIM 617866, (Duran et al. 2017), *IFT172* (MIM 615630, (Halbritter et al. 2013)), and *IFT52* (MIM 617102, (Girisha et al. 2016) are all also associated with JATD syndromes. IFT52 has been linked to osteogenic differentiation in vitro potentially through regulation of Hedgehog signalling (Guleria et al. 2022). Wheway et al. (2015) cross-compared validated ciliogenesis genes with exome data from patients clinically diagnosed with JATD and identified mutations in the *C21ORF2* gene (Wheway et al. 2015), an example of an unbiased approach to exploring ciliogenesis disease links. The alias for *C21ORF2*, cilia, and flagellar-associated protein 410 (*CFAP410*) is associated with axial spondylometaphyseal dysplasia (MIM602271) and has recently been identified to bind and potentially functionally regulate NEK1 (Gregorczyk et al. 2023), another example of complex interactions underpinning ciliopathy molecular mechanisms. *NEK1* (MIM 252100) is associated with both axial spondylometaphyseal dysplasia but also orofaciodigital syndrome type 2 or Mohr syndrome which typically presents with skeletal abnormalities including short limbs. Numerous NEK1 targets mean that molecular mechanisms are, as yet, unclear, and *NEK1* mutations and variants are associated with a number of skeletal and extraskeletal disorders. *CEP120* encodes a centrosomal protein linked to a number of ciliopathy phenotypes (Shaheen et al. 2015; Roosing et al. 2016) including features of SRTD with polydactyly (MIM 616300). SRTD14 is associated with a nonsense mutation in the *KIAA0586* gene (MIM 616546), also associated with Joubert (Alby et al. 2015). *KIAA0586*, alias *TALPID3*, encodes another centrosomal protein linked to ciliogenesis and Hedgehog signalling. *KIAA0753* is linked with SRTD21 (MIM 619479) (Hammarsjo et al. 2017; Faudi et al. 2020; Sabir et al. 2021). The G protein-coupled receptor kinase (GRK2) is linked to both Jeune and ATD through the identification of loss of function mutations (MIM 109635). *GRK2* null cells assemble cilia but have impaired Hedgehog and Wnt signalling (Bosakova, Abraham et al. 2020).

10.9 Ellis-van Creveld Syndrome (EVC)

Ellis-van Creveld syndrome (EVC) is an autosomal recessive skeletal disease with an occurrence of 1 in 60,000 (MIM 225500). It is characterised by clinical features including short limbs, short ribs, postaxial polydactyly, and dysplastic nails and teeth. Congenital cardiac defects occur in 60% of individuals (Ruiz-Perez et al. 2000). Marked growth retardation has been described (Verbeek et al. 2011). Two genes, *EVC1* and *EVC2*, have been associated with chondroectodermal dysplasia (Tompson et al. 2007). EVC localises at the base of the cilium (Ruiz-Perez et al. 2007) as a key regulator of Indian Hedgehog (Ihh) signal transduction in the skeleton. A relatively large amount of research in cell and murine models is unveiling the functions of the EVC complex in ciliary and Hedgehog biology (Yang et al. 2012). Regulatory links now extend to fibroblast growth factor signalling due to increased FGF18 expression upon inactivation of perichondrial EVC2 (Zhang et al. 2016). Unsurprisingly, characterisation of murine mutant models has demonstrated disruption to the Hedgehog-PTHrP feedback loop and has demonstrated molecular rescue of mouse phenotypes. Recent clinical and molecular analysis (Aubert-Mucca et al. 2023), in the largest cohorts to date, indicates high frequencies of deletions in *EVC1* and *EVC2*, identifying them as the major genes responsible. However, EVC is again highly linked to other ciliary and Hedgehog signalling genes such as *WDR35*, *DYNC2LI1* (MIM 617088) *GLI1*, a GLI family zinc finger transcription factor activated by the Hedgehog signalling cascade and classically used as a pathway readout (MIM 165220) and *SMO* (a single case with compound heterozygosity missense variants found), encoding smoothened a critical signal transducer, smoothened, localised to the cilium and associated with the EVC complex (Corbit et al. 2005; Dorn et al. 2012). Disorganisation of growth plate cell columns is observed in EVC KO mice (Lamuedra et al. 2020). Mechanistic insight from studies focussed on osteoarthritis (Lamuedra et al. 2022) may give insight into pathogenic mechanisms given the shared biology between disease and normal endochondral processes (Javaheri et al. 2018).

10.10 Craniectodermal Dysplasia

In addition to *EVC*, *SMO*- and *GLI*-associated disorders, the genes *TCTN3* or tectonic-3 required for ciliogenesis and regulating ciliary membrane composition (Gong et al. 2018) (MIM 258860), the ciliary trafficking protein *IFT22* (MIM 218330, (Walczak-Sztulpa et al. 2010, Yang et al. 2021)), *WDR35* (MIM 613610 (Gilissen et al. 2010, Bacino et al. 2012)), the IFT-A gene *WDR19* (MIM 614378 (Bredrup et al. 2011)), *IFT40* (MIM 614620), and *IFT43* (MIM 614009) are linked to Levin-Sensenbrenner syndrome. Mouse models are likely again to be the first line of inquiry to explore how these identified genes are disease-causing (Miller et al. 2013). In common with most skeletal ciliopathy genes, the causative mechanisms are likely to be Hedgehog-linked, but increasingly, other signalling mechanisms are implicated directly and indirectly.

10.11 Joubert, Mainzer-Saldino, and Meckel Syndromes

Joubert syndrome is an autosomal disorder characterised by the unique brainstem and cerebellar malformations recognised as the 'molar tooth sign' on MRI. Among a complex and varied phenotypic array (MIM 213300), skeletal features consistent with Jeune syndrome such as short ribs and bell-shaped chest link the gene *CSPP1* (MIM 615636) to Joubert syndrome 21. The morpholino knockdown of the *CSPP1a* gene in zebrafish produced skeletal features consistent with ciliopathies (Tuz et al. 2014), reduced expression of ciliary ARL13B. Analysis of patient-derived fibroblasts revealed ciliary defects, including reduced numbers and cilia shortening.

Mainzer-Saldino syndrome is associated with short-rib thoracic dysplasia (SRTD). SRTD9 is linked specifically with homozygous or compound heterozygous mutations in the ciliary gene *IFT140* (MIM 266920 (Schmidts et al. 2013b)). Lineage-tracing and gene KO approaches indicate roles in bone formation (Tao et al. 2019; Chen et al. 2021). The IFT-B component, IFT172, is linked to Mainzer-Saldino in multiple cases (Halbritter et al. 2013) with broad roles already identified in a plethora of cell biology (Pruski et al. 2019).

Meckel syndrome, also known as Meckel-Gruber, is a severe pleiotropic autosomal recessive developmental disorder linked to primary cilia (MIM 249000). MKS1 is a subunit of the ciliary transition zone, localised to the basal body and required for ciliogenesis with the alias BBS (Bardet-Biedl syndrome) 13. MKS1 has been reported to act as a regulatory substrate-adaptor tightly regulating the Wnt protein β-catenin (Szymanska et al. 2022). MKS2 is linked to the gene *TMEM216* (MIM 603194 (Valente et al. 2010)) and MKS3 by a mutation in the *TMEM67* gene (MIM 607361 (Morgan et al. 2002, Smith et al. 2006). Both are part of the ciliary transition zone (Garcia-Gonzalo et al. 2011; Garcia-Gonzalo and Reiter 2012). MKS4 is linked to homozygous or compound heterozygous mutations in the *CEP290* gene (MIM 611134 (Baala et al. 2007, Frank et al. 2008)). *CEP290*, also named *NPHP6*, encodes a centrosomal protein and through interactions with CEP110 regulates cilia assembly and RAB8A localisation (Tsang et al. 2008). MKS5 is associated with mutations in the ciliary gene *RPGRIP1L* (Delous et al. 2007) (MIM 611561) and MKS6 with the gene *CC2DCA* (Tallila et al. 2008) (MIM 612284).

10.12 Conclusions and Future Perspective

A wealth of human genetics has unveiled the spectrum of cilia-associated skeletal disorders, most of which present with features of chondrodysplasia. The classical approach of mouse KO or mutation models has added some mechanistic insight to the study of such disorders, but there is much work to be done concerning these relatively rare syndromes. Exploring phenotypes reminiscent of ciliopathies, the result of loss of function of ciliary-associated genes (rather than core ciliome components) will be valuable (Hruba et al. 2021).

Key questions remain pertinent to the entire ciliary biology as much as they are to the study of skeletal ciliopathies. To what extent are ciliary structure or protein activities within the cilium directly pertinent to pathology, compared with non-canonical roles for ciliary genes (Boehlke et al. 2015). What Hedgehog-independent mechanisms are relevant to disease? and how can these be uncoupled from the central role the ciliome plays in Hedgehog signal transduction and pathway output tuning? What are the best routes (targets and approaches) to 'ciliotherapeutics'? The implementation of humanised, organoid, or tissue engineering-supported approaches will support the expedition of this work in the skeletal field, as implemented in other systems affected by the broader ciliopathies (Schembs et al. 2022). There is still much to learn, but now, thanks to decades of clinical genetics discovery, we have a detailed genetic guide to follow.

Acknowledgements Authors thank Theana Johnson for proof-reading and content suggestions.

The work of the authors is currently funded by an award from BBSRC (BB/X007049/1), and Collins was funded by Versus Arthritis studentship (21546).

Compliance with Ethical Standards/Ethical Requirements There are no conflicts of interest to declare.

References

Alby C, Piquand K, Huber C, Megarbane A, Ichkou A, Legendre M, Pelluard F, Encha-Ravazi F, Abi-Tayeh G, Bessieres B, El Chehadeh-Djebbar S, Laurent N, Faivre L, Sztriha L, Zombor M, Szabo H, Failler M, Garfa-Traore M, Bole C, Nitschke P, Nizon M, Elkhartoufi N, Clerget-Darpoux F, Munnich A, Lyonnet S, Vekemans M, Saunier S, Cormier-Daire V, Attie-Bitach T, Thomas S (2015) Mutations in KIAA0586 cause lethal ciliopathies ranging from a Hydrolethalus phenotype to short-rib polydactyly syndrome. Am J Hum Genet 97(2):311–318

Allard BA, Wang W, Pottorf TS, Mumtaz H, Jack BM, Wang HH, Silva LM, Jacobs DT, Wang J, Bumann EE, Tran PV (2021) Thm2 interacts with paralog, Thm1, and sensitizes to hedgehog signaling in postnatal skeletogenesis. Cell Mol Life Sci 78(7):3743–3762

Ansley SJ, Badano JL, Blacque OE, Hill J, Hoskins BE, Leitch CC, Kim JC, Ross AJ, Eichers ER, Teslovich TM, Mah AK, Johnsen RC, Cavender JC, Lewis RA, Leroux MR, Beales PL, Katsanis N (2003) Basal body dysfunction is a likely cause of pleiotropic Bardet-Biedl syndrome. Nature 425(6958):628–633

Ashe A, Butterfield NC, Town L, Courtney AD, Cooper AN, Ferguson C, Barry R, Olsson F, Liem KF Jr, Parton RG, Wainwright BJ, Anderson KV, Whitelaw E, Wicking C (2012) Mutations in mouse Ift144 model the craniofacial, limb and rib defects in skeletal ciliopathies. Hum Mol Genet 21(8):1808–1823

Aubert-Mucca M, Huber C, Baujat G, Michot C, Zarhrate M, Bras M, Boutaud L, Malan V, Attie-Bitach T, Clinical C, Cormier-Daire V (2023) Ellis-van Creveld syndrome: Clinical and molecular analysis of 50 individuals. J Med Genet 60(4):337–345

Baala L, Audollent S, Martinovic J, Ozilou C, Babron MC, Sivanandamoorthy S, Saunier S, Salomon R, Gonzales M, Rattenberry E, Esculpavit C, Toutain A, Moraine C, Parent P, Marcorelles P, Dauge MC, Roume J, Le Merrer M, Meiner V, Meir K, Menez F, Beaufrere AM, Francannet C, Tantau J, Sinico M, Dumez Y, MacDonald F, Munnich A, Lyonnet S, Gubler MC, Genin E, Johnson CA, Vekemans M, Encha-Razavi F, Attie-Bitach T (2007) Pleiotropic effects of CEP290 (NPHP6) mutations extend to Meckel syndrome. Am J Hum Genet 81(1):170–179

Bacino CA, Dhar SU, Brunetti-Pierri N, Lee B, Bonnen PE (2012) WDR35 mutation in siblings with Sensenbrenner syndrome: a ciliopathy with variable phenotype. Am J Med Genet A 158A(11):2917–2924

Badiner N, Taylor SP, Forlenza K, Lachman RS, G. University of Washington Center for Mendelian, Bamshad M, Nickerson D, Cohn DH, Krakow D (2017) Mutations in DYNC2H1, the cytoplasmic dynein 2, heavy chain 1 motor protein gene, cause short-rib polydactyly type I, Saldino-Noonan type. Clin Genet 92(2):158–165

Bangs F, Anderson KV (2017) Primary cilia and mammalian hedgehog signaling. Cold Spring Harb Perspect Biol 9(5)

Barrell WB, Griffin JN, Harvey JL, HipSci C, Danovi D, Beales P, Grigoriadis AE, Liu KJ (2019) Induction of neural crest stem cells from Bardet-Biedl syndrome patient derived hiPSCs. Front Mol Neurosci 12:139

Beales PL, Bland E, Tobin JL, Bacchelli C, Tuysuz B, Hill J, Rix S, Pearson CG, Kai M, Hartley J, Johnson C, Irving M, Elcioglu N, Winey M, Tada M, Scambler PJ (2007) IFT80, which encodes a conserved intraflagellar transport protein, is mutated in Jeune asphyxiating thoracic dystrophy. Nat Genet 39(6):727–729

Beales PL, Warner AM, Hitman GA, Thakker R, Flinter FA (1997) Bardet-Biedl syndrome: a molecular and phenotypic study of 18 families. J Med Genet 34(2):92–98

Boehlke C, Janusch H, Hamann C, Powelske C, Mergen M, Herbst H, Kotsis F, Nitschke R, Kuehn EW (2015) A cilia independent role of Ift88/Polaris during cell migration. PLoS One 10(10):e0140378

Bosakova M, Abraham SP, Nita A, Hruba E, Buchtova M, Taylor SP, Duran I, Martin J, Svozilova K, Barta T, Varecha M, Balek L, Kohoutek J, Radaszkiewicz T, Pusapati GV, Bryja V, Rush ET, Thiffault I, Nickerson DA, Bamshad MJ, G. University of Washington Center for Mendelian, Rohatgi R, Cohn DH, Krakow D, Krejci P (2020) Mutations in GRK2 cause Jeune syndrome by impairing hedgehog and canonical Wnt signaling. EMBO Mol Med 12(11):e11739

Bredrup C, Saunier S, Oud MM, Fiskerstrand T, Hoischen A, Brackman D, Leh SM, Midtbo M, Filhol E, Bole-Feysot C, Nitschke P, Gilissen C, Haugen OH, Sanders JS, Stolte-Dijkstra I, Mans DA, Steenbergen EJ, Hamel BC, Matignon M, Pfundt R, Jeanpierre C, Boman H, Rodahl E, Veltman JA, Knappskog PM, Knoers NV, Roepman R, Arts HH (2011) Ciliopathies with skeletal anomalies and renal insufficiency due to mutations in the IFT-A gene WDR19. Am J Hum Genet 89(5):634–643

Bryson LJ, Flynn DM, Sabharwal A, Ahmed SF, Kinning E (2021) A child with congenital short gut associated with DYNC2LI1 ciliopathy. Clin Dysmorphol 30(1):66–68

Chang CF, Ramaswamy G, Serra R (2012) Depletion of primary cilia in articular chondrocytes results in reduced Gli3 repressor to activator ratio, increased hedgehog signaling, and symptoms of early osteoarthritis. Osteoarthr Cartil 20(2):152–161

Chang CF, Serra R (2012) Ift88 regulates hedgehog signaling, Sfrp5 expression, and beta-catenin activity in post-natal growth plate. J Orthop Res 31:350

Chen Y, Fan Q, Zhang H, Tao D, Wang Y, Yue R, Sun Y (2021) Lineage tracing of cells expressing the ciliary gene IFT140 during bone development. Dev Dyn 250(4):574–583

Christensen ST, Clement CA, Satir P, Pedersen LB (n.d.) Primary cilia and coordination of receptor tyrosine kinase (RTK) signalling. J Pathol 226(2):172–184

Christensen ST, Pedersen SF, Satir P, Veland IR, Schneider L (2008) The primary cilium coordinates signaling pathways in cell cycle control and migration during development and tissue repair. Curr Top Dev Biol 85:261–301

Clement CA, Ajbro KD, Koefoed K, Vestergaard ML, Veland IR, Henriques de Jesus MP, Pedersen LB, Benmerah A, Andersen CY, Larsen LA, Christensen ST (2013) TGF-beta signaling is associated with endocytosis at the pocket region of the primary cilium. Cell Rep 3:1806

Collins I, Wann AKT (2020) Regulation of the extracellular matrix by ciliary machinery. Cells 9(2)

Corbit KC, Aanstad P, Singla V, Norman AR, Stainier DY, Reiter JF (2005) Vertebrate smoothened functions at the primary cilium. Nature 437(7061):1018–1021

Coveney CR, Collins I, Mc Fie M, Chanalaris A, Yamamoto K, Wann AKT (2018) Cilia protein IFT88 regulates extracellular protease activity by optimizing LRP-1-mediated endocytosis. FASEB J 32:fj201800334

Coveney CR, Samvelyan HJ, Miotla-Zarebska J, Carnegie J, Chang E, Corrin CJ, Coveney T, Stott B, Parisi I, Duarte C, Vincent TL, Staines KA, Wann AK (2022) Ciliary IFT88 protects coordinated adolescent growth plate ossification from disruptive physiological mechanical forces. J Bone Miner Res

Coveney CR, Zhu L, Miotla-Zarebska J, Stott B, Parisi I, Batchelor V, Duarte C, Chang E, McSorley E, Vincent TL, Wann AK (2021) The ciliary protein IFT88 controls post-natal cartilage thickness and influences development of osteoarthritis. Arthritis Rheumatol

Dagoneau N, Goulet M, Genevieve D, Sznajer Y, Martinovic J, Smithson S, Huber C, Baujat G, Flori E, Tecco L, Cavalcanti D, Delezoide AL, Serre V, Le Merrer M, Munnich A, Cormier-Daire V (2009) DYNC2H1 mutations cause asphyxiating thoracic dystrophy and short rib-polydactyly syndrome, type III. Am J Hum Genet 84(5):706–711

Davis EE, Zhang Q, Liu Q, Diplas BH, Davey LM, Hartley J, Stoetzel C, Szymanska K, Ramaswami G, Logan CV, Muzny DM, Young AC, Wheeler DA, Cruz P, Morgan M, Lewis LR, Cherukuri P, Maskeri B, Hansen NF, Mullikin JC, Blakesley RW, Bouffard GG, Program NCS, Gyapay G, Rieger S, Tonshoff B, Kern I, Soliman NA, Neuhaus TJ, Swoboda KJ, Kayserili H, Gallagher TE, Lewis RA, Bergmann C, Otto EA, Saunier S, Scambler PJ, Beales PL, Gleeson JG, Maher ER, Attie-Bitach T, Dollfus H, Johnson CA, Green ED, Gibbs RA, Hildebrandt F, Pierce EA, Katsanis N (2011) TTC21B contributes both causal and modifying alleles across the ciliopathy spectrum. Nat Genet 43(3):189–196

Davis RE, Swiderski RE, Rahmouni K, Nishimura DY, Mullins RF, Agassandian K, Philp AR, Searby CC, Andrews MP, Thompson S, Berry CJ, Thedens DR, Yang B, Weiss RM, Cassell MD, Stone EM, Sheffield VC (2007) A knockin mouse model of the Bardet-Biedl syndrome 1 M390R mutation has cilia defects, ventriculomegaly, retinopathy, and obesity. Proc Natl Acad Sci USA 104(49):19422–19427

Delling M, DeCaen PG, Doerner JF, Febvay S, Clapham DE (2013) Primary cilia are specialized calcium signalling organelles. Nature 504(7479):311–314

Delous M, Baala L, Salomon R, Laclef C, Vierkotten J, Tory K, Golzio C, Lacoste T, Besse L, Ozilou C, Moutkine I, Hellman NE, Anselme I, Silbermann F, Vesque C, Gerhardt C, Rattenberry E, Wolf MT, Gubler MC, Martinovic J, Encha-Razavi F, Boddaert N, Gonzales M, Macher MA, Nivet H, Champion G, Bertheleme JP, Niaudet P, McDonald F, Hildebrandt

F, Johnson CA, Vekemans M, Antignac C, Ruther U, Schneider-Maunoury S, Attie-Bitach T, Saunier S (2007) The ciliary gene RPGRIP1L is mutated in cerebello-oculo-renal syndrome (Joubert syndrome type B) and Meckel syndrome. Nat Genet 39(7):875–881
Dorn KV, Hughes CE, Rohatgi R (2012) A smoothened-Evc2 complex transduces the hedgehog signal at primary cilia. Dev Cell 23(4):823–835
Duran I, Taylor SP, Zhang W, Martin J, Forlenza KN, Spiro RP, Nickerson DA, Bamshad M, Cohn DH, Krakow D (2016) Destabilization of the IFT-B cilia core complex due to mutations in IFT81 causes a Spectrum of short-rib polydactyly syndrome. Sci Rep 6:34232
Duran I, Taylor SP, Zhang W, Martin J, Qureshi F, Jacques SM, Wallerstein R, Lachman RS, Nickerson DA, Bamshad M, Cohn DH, Krakow D (2017) Mutations in IFT-A satellite core component genes IFT43 and IFT121 produce short rib polydactyly syndrome with distinctive campomelia. Cilia 6:7
Faudi E, Brischoux-Boucher E, Huber C, Dabudyk T, Lenoir M, Baujat G, Michot C, Van Maldergem L, Cormier-Daire V, Piard J (2020) A new case of KIAA0753-related variant of Jeune asphyxiating thoracic dystrophy. Eur J Med Genet 63(4):103823
Forsythe E, Beales PL (2013) Bardet-Biedl syndrome. Eur J Hum Genet 21(1):8–13
Frank V, den Hollander AI, Bruchle NO, Zonneveld MN, Nurnberg G, Becker C, Du Bois G, Kendziorra H, Roosing S, Senderek J, Nurnberg P, Cremers FP, Zerres K, Bergmann C (2008) Mutations of the CEP290 gene encoding a centrosomal protein cause Meckel-Gruber syndrome. Hum Mutat 29(1):45–52
Fu S, Thompson CL, Ali A, Wang W, Chapple JP, Mitchison HM, Beales PL, Wann AKT, Knight MM (2019) Mechanical loading inhibits cartilage inflammatory signalling via an HDAC6 and IFT-dependent mechanism regulating primary cilia elongation. Osteoarthr Cartil 27(7):1064–1074
Garcia-Gonzalo FR, Corbit KC, Sirerol-Piquer MS, Ramaswami G, Otto EA, Noriega TR, Seol AD, Robinson JF, Bennett CL, Josifova DJ, Garcia-Verdugo JM, Katsanis N, Hildebrandt F, Reiter JF (2011) A transition zone complex regulates mammalian ciliogenesis and ciliary membrane composition. Nat Genet 43(8):776–784
Garcia-Gonzalo FR, Reiter JF (2012) Scoring a backstage pass: mechanisms of ciliogenesis and ciliary access. J Cell Biol 197(6):697–709
Gerdes JM, Liu Y, Zaghloul NA, Leitch CC, Lawson SS, Kato M, Beachy PA, Beales PL, DeMartino GN, Fisher S, Badano JL, Katsanis N (2007) Disruption of the basal body compromises proteasomal function and perturbs intracellular Wnt response. Nat Genet 39(11):1350–1360
Getwan M, Hoppmann A, Schlosser P, Grand K, Song W, Diehl R, Schroda S, Heeg F, Deutsch K, Hildebrandt F, Lausch E, Kottgen A, Lienkamp SS (2021) Ttc30a affects tubulin modifications in a model for ciliary chondrodysplasia with polycystic kidney disease. Proc Natl Acad Sci USA 118(39)
Gilissen C, Arts HH, Hoischen A, Spruijt L, Mans DA, Arts P, van Lier B, Steehouwer M, van Reeuwijk J, Kant SG, Roepman R, Knoers NV, Veltman JA, Brunner HG (2010) Exome sequencing identifies WDR35 variants involved in Sensenbrenner syndrome. Am J Hum Genet 87(3):418–423
Girisha KM, Shukla A, Trujillano D, Bhavani GS, Hebbar M, Kadavigere R, Rolfs A (2016) A homozygous nonsense variant in IFT52 is associated with a human skeletal ciliopathy. Clin Genet 90(6):536–539
Goetz SC, Bangs F, Barrington CL, Katsanis N, Anderson KV (2017) The Meckel syndrome-associated protein MKS1 functionally interacts with components of the BBSome and IFT complexes to mediate ciliary trafficking and hedgehog signaling. PLoS One 12(3):e0173399
Gong S, Ji F, Wang B, Zhang Y, Xu X, Sun M (2018) Tectonic proteins are important players in non-motile ciliopathies. Cell Physiol Biochem 50(1):398–409
Gregorczyk M, Pastore G, Munoz I, Carroll T, Streubel J, Munro M, Lis P, Lange S, Lamoliatte F, Macartney T, Toth R, Brown F, Hastie J, Pereira G, Durocher D, Rouse J (2023) Functional characterization of C21ORF2 association with the NEK1 kinase mutated in human in diseases. Life Sci Alliance 6(7):e202201740

Guleria VS, Parit R, Quadri N, Das R, Upadhyai P (2022) The intraflagellar transport protein IFT52 associated with short-rib thoracic dysplasia is essential for ciliary function in osteogenic differentiation in vitro and for sensory perception in drosophila. Exp Cell Res 418(2):113273
Halbritter J, Bizet AA, Schmidts M, Porath JD, Braun DA, Gee HY, McInerney-Leo AM, Krug P, Filhol E, Davis EE, Airik R, Czarnecki PG, Lehman AM, Trnka P, Nitschke P, Bole-Feysot C, Schueler M, Knebelmann B, Burtey S, Szabo AJ, Tory K, Leo PJ, Gardiner B, McKenzie FA, Zankl A, Brown MA, Hartley JL, Maher ER, Li C, Leroux MR, Scambler PJ, Zhan SH, Jones SJ, Kayserili H, Tuysuz B, Moorani KN, Constantinescu A, Krantz ID, Kaplan BS, Shah JV, Consortium UK, Hurd TW, Doherty D, Katsanis N, Duncan EL, Otto EA, Beales PL, Mitchison HM, Saunier S, Hildebrandt F (2013) Defects in the IFT-B component IFT172 cause Jeune and Mainzer-Saldino syndromes in humans. Am J Hum Genet 93(5):915–925
Hammarsjo A, Wang Z, Vaz R, Taylan F, Sedghi M, Girisha KM, Chitayat D, Neethukrishna K, Shannon P, Godoy R, Gowrishankar K, Lindstrand A, Nasiri J, Baktashian M, Newton PT, Guo L, Hofmeister W, Pettersson M, Chagin AS, Nishimura G, Yan L, Matsumoto N, Nordgren A, Miyake N, Grigelioniene G, Ikegawa S (2017) Novel KIAA0753 mutations extend the phenotype of skeletal ciliopathies. Sci Rep 7(1):15585
Handa A, Voss U, Hammarsjo A, Grigelioniene G, Nishimura G (2020) Skeletal ciliopathies: a pattern recognition approach. Jpn J Radiol 38(3):193–206
Haycraft CJ, Serra R (2008) Cilia involvement in patterning and maintenance of the skeleton. Curr Top Dev Biol 85:303–332
Haycraft CJ, Zhang Q, Song B, Jackson WS, Detloff PJ, Serra R, Yoder BK (2007) Intraflagellar transport is essential for endochondral bone formation. Development 134(2):307–316
Hiyamizu S, Qiu H, Tsurumi Y, Hamada Y, Katoh Y, Nakayama K (2023a) Dynein-2-driven intraciliary retrograde trafficking indirectly requires multiple interactions of IFT54 in the IFT-B complex with the dynein-2 complex. Biol Open 12
Hiyamizu S, Qiu H, Vuolo L, Stevenson NL, Shak C, Heesom KJ, Hamada Y, Tsurumi Y, Chiba S, Katoh Y, Stephens DJ, Nakayama K (2023b) Multiple interactions of the dynein-2 complex with the IFT-B complex are required for effective intraflagellar transport. J Cell Sci 136(5)
Hoey DA, Tormey S, Ramcharan S, O'Brien FJ, Jacobs CR (2012) Primary cilia-mediated mechanotransduction in human mesenchymal stem cells. Stem Cells 30(11):2561–2570
Hruba E, Kavkova M, Dalecka L, Macholan M, Zikmund T, Varecha M, Bosakova M, Kaiser J, Krejci P, Hovorakova M, Buchtova M (2021) Loss of Sprouty produces a ciliopathic skeletal phenotype in mice through upregulation of hedgehog signaling. J Bone Miner Res 36(11):2258–2274
Huber C, Cormier-Daire V (2012) Ciliary disorder of the skeleton. Am J Med Genet C Semin Med Genet 160C(3):165–174
Huber C, Wu S, Kim AS, Sigaudy S, Sarukhanov A, Serre V, Baujat G, Le Quan Sang KH, Rimoin DL, Cohn DH, Munnich A, Krakow D, Cormier-Daire V (2013) WDR34 mutations that cause short-rib polydactyly syndrome type III/severe asphyxiating thoracic dysplasia reveal a role for the NF-kappaB pathway in cilia. Am J Hum Genet 93(5):926–931
Ishikawa H, Marshall WF (n.d.) Ciliogenesis: building the cell's antenna. Nat Rev Mol Cell Biol 12(4):222–234
Ishikawa H, Marshall WF (2017) Intraflagellar transport and ciliary dynamics. Cold Spring Harb Perspect Biol 9(3)
Javaheri B, Caetano-Silva SP, Kanakis I, Bou-Gharios G, Pitsillides AA (2018) The Chondro-osseous continuum: is it possible to unlock the potential assigned within? Front Bioeng Biotechnol 6:28
Jiang J, Promchan K, Jiang H, Awasthi P, Marshall H, Harned A, Natarajan V (2016) Depletion of BBS protein LZTFL1 affects growth and causes retinal degeneration in mice. J Genet Genomics 43(6):381–391
Katoh TA, Omori T, Mizuno K, Sai X, Minegishi K, Ikawa Y, Nishimura H, Itabashi T, Kajikawa E, Hiver S, Iwane AH, Ishikawa T, Okada Y, Nishizaka T, Hamada H (2023) Immotile

cilia mechanically sense the direction of fluid flow for left-right determination. Science 379(6627):66–71

Kaushik AP, Martin JA, Zhang Q, Sheffield VC, Morcuende JA (2009) Cartilage abnormalities associated with defects of chondrocytic primary cilia in Bardet-Biedl syndrome mutant mice. J Orthop Res 27(8):1093–1099

Kessler K, Wunderlich I, Uebe S, Falk NS, Giessl A, Brandstatter JH, Popp B, Klinger P, Ekici AB, Sticht H, Dorr HG, Reis A, Roepman R, Seemanova E, Thiel CT (2015) DYNC2LI1 mutations broaden the clinical spectrum of dynein-2 defects. Sci Rep 5:11649

King A, Hoch NC, McGregor NE, Sims NA, Smyth IM, Heierhorst J (2019) Dynll1 is essential for development and promotes endochondral bone formation by regulating intraflagellar dynein function in primary cilia. Hum Mol Genet 28(15):2573–2588

Koyama E, Young B, Nagayama M, Shibukawa Y, Enomoto-Iwamoto M, Iwamoto M, Maeda Y, Lanske B, Song B, Serra R, Pacifici M (2007) Conditional Kif3a ablation causes abnormal hedgehog signaling topography, growth plate dysfunction, and excessive bone and cartilage formation during mouse skeletogenesis. Development 134(11):2159–2169

Lahiry P, Wang J, Robinson JF, Turowec JP, Litchfield DW, Lanktree MB, Gloor GB, Puffenberger EG, Strauss KA, Martens MB, Ramsay DA, Rupar CA, Siu V, Hegele RA (2009) A multiplex human syndrome implicates a key role for intestinal cell kinase in development of central nervous, skeletal, and endocrine systems. Am J Hum Genet 84(2):134–147

Lamuedra A, Gratal P, Calatrava L, Ruiz-Perez VL, Largo R, Herrero-Beaumont G (2020) Disorganization of chondrocyte columns in the growth plate does not aggravate experimental osteoarthritis in mice. Sci Rep 10(1):10745

Lamuedra A, Gratal P, Calatrava L, Ruiz-Perez VL, Palencia-Campos A, Portal-Nunez S, Mediero A, Herrero-Beaumont G, Largo R (2022) Blocking chondrocyte hypertrophy in conditional Evc knockout mice does not modify cartilage damage in osteoarthritis. FASEB J 36(4):e22258

Liu YX, Li WJ, Zhang RK, Sun SN, Fan ZC (2023) Unraveling the intricate cargo-BBSome coupling mechanism at the ciliary tip. Proc Natl Acad Sci USA 120(13):e2218819120

Lovera M, Luders J (2021) The ciliary impact of nonciliary gene mutations. Trends Cell Biol 31(11):876–887

Malone AM, Anderson CT, Tummala P, Kwon RY, Johnston TR, Stearns T, Jacobs CR (2007) Primary cilia mediate mechanosensing in bone cells by a calcium-independent mechanism. Proc Natl Acad Sci USA 104(33):13325–13330

Mangos S, Lam PY, Zhao A, Liu Y, Mudumana S, Vasilyev A, Liu A, Drummond IA (n.d.) The ADPKD genes pkd1a/b and pkd2 regulate extracellular matrix formation. Dis Model Mech 3(5–6):354–365

Mc Fie M, Koneva L, Collins I, Coveney CR, Clube AM, Chanalaris A, Vincent TL, Bezbradica JS, Sansom SN, Wann AKT (2020) Ciliary proteins specify the cell inflammatory response by tuning NFkappaB signalling, independently of primary cilia. J Cell Sci 133(13)

McGlashan SR, Haycraft CJ, Jensen CG, Yoder BK, Poole CA (2007) Articular cartilage and growth plate defects are associated with chondrocyte cytoskeletal abnormalities in Tg737orpk mice lacking the primary cilia protein polaris. Matrix Biol 26(4):234–246

McInerney-Leo AM, Harris JE, Leo PJ, Marshall MS, Gardiner B, Kinning E, Leong HY, McKenzie F, Ong WP, Vodopiutz J, Wicking C, Brown MA, Zankl A, Duncan EL (2015) Whole exome sequencing is an efficient, sensitive and specific method for determining the genetic cause of short-rib thoracic dystrophies. Clin Genet 88(6):550–557

McInerney-Leo AM, Schmidts M, Cortes CR, Leo PJ, Gener B, Courtney AD, Gardiner B, Harris JA, Lu Y, Marshall M, Consortium UK, Scambler PJ, Beales PL, Brown MA, Zankl A, Mitchison HM, Duncan EL, Wicking C (2013) Short-rib polydactyly and Jeune syndromes are caused by mutations in WDR60. Am J Hum Genet 93(3):515–523

Merrill AE, Merriman B, Farrington-Rock C, Camacho N, Sebald ET, Funari VA, Schibler MJ, Firestein MH, Cohn ZA, Priore MA, Thompson AK, Rimoin DL, Nelson SF, Cohn DH, Krakow D (2009) Ciliary abnormalities due to defects in the retrograde transport protein DYNC2H1 in short-rib polydactyly syndrome. Am J Hum Genet 84(4):542–549

Mill P, Christensen ST, Pedersen LB (2023) Primary cilia as dynamic and diverse signalling hubs in development and disease. Nat Rev Genet 24(7):421–441

Mill P, Lockhart PJ, Fitzpatrick E, Mountford HS, Hall EA, Reijns MA, Keighren M, Bahlo M, Bromhead CJ, Budd P, Aftimos S, Delatycki MB, Savarirayan R, Jackson IJ, Amor DJ (2011) Human and mouse mutations in WDR35 cause short-rib polydactyly syndromes due to abnormal ciliogenesis. Am J Hum Genet 88(4):508–515

Miller KA, Ah-Cann CJ, Welfare MF, Tan TY, Pope K, Caruana G, Freckmann ML, Savarirayan R, Bertram JF, Dobbie MS, Bateman JF, Farlie PG (2013) Cauli: a mouse strain with an Ift140 mutation that results in a skeletal ciliopathy modelling Jeune syndrome. PLoS Genet 9(8):e1003746

Molla-Herman A, Ghossoub R, Blisnick T, Meunier A, Serres C, Silbermann F, Emmerson C, Romeo K, Bourdoncle P, Schmitt A, Saunier S, Spassky N, Bastin P, Benmerah A (2010) The ciliary pocket: an endocytic membrane domain at the base of primary and motile cilia. J Cell Sci 123(Pt 10):1785–1795

Moore ER, Jacobs CR (2018) The primary cilium as a signaling nexus for growth plate function and subsequent skeletal development. J Orthop Res 36(2):533–545

Morgan NV, Gissen P, Sharif SM, Baumber L, Sutherland J, Kelly DA, Aminu K, Bennett CP, Woods CG, Mueller RF, Trembath RC, Maher ER, Johnson CA (2002) A novel locus for Meckel-Gruber syndrome, MKS3, maps to chromosome 8q24. Hum Genet 111(4–5):456–461

Nauli SM, Alenghat FJ, Luo Y, Williams E, Vassilev P, Li X, Elia AE, Lu W, Brown EM, Quinn SJ, Ingber DE, Zhou J (2003) Polycystins 1 and 2 mediate mechanosensation in the primary cilium of kidney cells. Nat Genet 33(2):129–137

Niceta M, Margiotti K, Digilio MC, Guida V, Bruselles A, Pizzi S, Ferraris A, Memo L, Laforgia N, Dentici ML, Consoli F, Torrente I, Ruiz-Perez VL, Dallapiccola B, Marino B, De Luca A, Tartaglia M (2018) Biallelic mutations in DYNC2LI1 are a rare cause of Ellis-van Creveld syndrome. Clin Genet 93(3):632–639

Novack DV (2011) Role of NF-kappaB in the skeleton. Cell Res 21(1):169–182

Oud MM, Bonnard C, Mans DA, Altunoglu U, Tohari S, Ng AYJ, Eskin A, Lee H, Rupar CA, de Wagenaar NP, Wu KM, Lahiry P, Pazour GJ, Nelson SF, Hegele RA, Roepman R, Kayserili H, Venkatesh B, Siu VM, Reversade B, Arts HH (2016) A novel ICK mutation causes ciliary disruption and lethal endocrine-cerebro-osteodysplasia syndrome. Cilia 5:8

Pedersen LB, Rosenbaum JL (2008) Intraflagellar transport (IFT) role in ciliary assembly, resorption and signalling. Curr Top Dev Biol 85:23–61

Pettersson M, Vaz R, Hammarsjo A, Eisfeldt J, Carvalho CMB, Hofmeister W, Tham E, Horemuzova E, Voss U, Nishimura G, Klintberg B, Nordgren A, Nilsson D, Grigelioniene G, Lindstrand A (2018) Alu-Alu mediated intragenic duplications in IFT81 and MATN3 are associated with skeletal dysplasias. Hum Mutat 39(10):1456–1467

Pruski M, Hu L, Yang C, Wang Y, Zhang JB, Zhang L, Huang Y, Rajnicek AM, St Clair D, McCaig CD, Lang B, Ding YQ (2019) Roles for IFT172 and primary cilia in cell migration, cell division, and neocortex development. Front Cell Dev Biol 7:287

Rita RF, Fukui H, Chow R, Vilfan A, Vermot J (2019) The cilium as a force sensor-myth versus reality. J Cell Sci 132(14)

Rais Y, Reich A, Simsa-Maziel S, Moshe M, Idelevich A, Kfir T, Miosge N, Monsonego-Ornan E (2015) The growth plate's response to load is partially mediated by mechano-sensing via the chondrocytic primary cilium. Cell Mol Life Sci 72(3):597–615

Reiter JF, Leroux MR (2017) Genes and molecular pathways underpinning ciliopathies. Nat Rev Mol Cell Biol 18(9):533–547

Richard D, Pregizer S, Venkatasubramanian D, Raftery RM, Muthuirulan P, Liu Z, Capellini TD, Craft AM (2023) Lineage-specific differences and regulatory networks governing human chondrocyte development. elife 12

Rix S, Calmont A, Scambler PJ, Beales PL (2011) An Ift80 mouse model of short rib polydactyly syndromes shows defects in hedgehog signalling without loss or malformation of cilia. Hum Mol Genet 20(7):1306–1314

Roosing S, Romani M, Isrie M, Rosti RO, Micalizzi A, Musaev D, Mazza T, Al-Gazali L, Altunoglu U, Boltshauser E, D'Arrigo S, De Keersmaecker B, Kayserili H, Brandenberger S, Kraoua I, Mark PR, McKanna T, Van Keirsbilck J, Moerman P, Poretti A, Puri R, Van Esch H, Gleeson JG, Valente EM (2016) Mutations in CEP120 cause Joubert syndrome as well as complex ciliopathy phenotypes. J Med Genet 53(9):608–615

Ruiz-Perez VL, Blair HJ, Rodriguez-Andres ME, Blanco MJ, Wilson A, Liu YN, Miles C, Peters H, Goodship JA (2007) Evc is a positive mediator of Ihh-regulated bone growth that localises at the base of chondrocyte cilia. Development 134(16):2903–2912

Ruiz-Perez VL, Ide SE, Strom TM, Lorenz B, Wilson D, Woods K, King L, Francomano C, Freisinger P, Spranger S, Marino B, Dallapiccola B, Wright M, Meitinger T, Polymeropoulos MH, Goodship J (2000) Mutations in a new gene in Ellis-van Creveld syndrome and Weyers acrodental dysostosis. Nat Genet 24(3):283–286

Rux D, Helbig K, Han B, Cortese C, Koyama E, Han L, Pacifici M (2022) Primary cilia direct murine articular cartilage tidemark patterning through hedgehog signaling and ambulatory load. J Bone Miner Res

Sabir AH, Sheikh J, Gowda V, Wallis C, Singham S, Durve D, Cocca A, Holder-Espinasse M, Irving M (2021) KIAA0753-related skeletal ciliopathy: a ninth case, extending the phenotype and reporting a novel variant. Clin Dysmorphol 30(3):142–146

Schembs L, Willems A, Hasenpusch-Theil K, Cooper JD, Whiting K, Burr K, Bostrand SMK, Selvaraj BT, Chandran S, Theil T (2022) The ciliary gene INPP5E confers dorsal telencephalic identity to human cortical organoids by negatively regulating sonic hedgehog signaling. Cell Rep 39(7):110811

Schmidts M, Arts HH, Bongers EM, Yap Z, Oud MM, Antony D, Duijkers L, Emes RD, Stalker J, Yntema JB, Plagnol V, Hoischen A, Gilissen C, Forsythe E, Lausch E, Veltman JA, Roeleveld N, Superti-Furga A, Kutkowska-Kazmierczak A, Kamsteeg EJ, Elcioglu N, van Maarle MC, Graul-Neumann LM, Devriendt K, Smithson SF, Wellesley D, Verbeek NE, Hennekam RC, Kayserili H, Scambler PJ, Beales PL, Uk10K, Knoers NV, Roepman R, Mitchison HM (2013a) Exome sequencing identifies DYNC2H1 mutations as a common cause of asphyxiating thoracic dystrophy (Jeune syndrome) without major polydactyly, renal or retinal involvement. J Med Genet 50(5):309–323

Schmidts M, Frank V, Eisenberger T, Al Turki S, Bizet AA, Antony D, Rix S, Decker C, Bachmann N, Bald M, Vinke T, Toenshoff B, Di Donato N, Neuhann T, Hartley JL, Maher ER, Bogdanovic R, Peco-Antic A, Mache C, Hurles ME, Joksic I, Guc-Scekic M, Dobricic J, Brankovic-Magic M, Bolz HJ, Pazour GJ, Beales PL, Scambler PJ, Saunier S, Mitchison HM, Bergmann C (2013b) Combined NGS approaches identify mutations in the intraflagellar transport gene IFT140 in skeletal ciliopathies with early progressive kidney disease. Hum Mutat 34(5):714–724

Schmidts M, Hou Y, Cortes CR, Mans DA, Huber C, Boldt K, Patel M, van Reeuwijk J, Plaza JM, van Beersum SE, Yap ZM, Letteboer SJ, Taylor SP, Herridge W, Johnson CA, Scambler PJ, Ueffing M, Kayserili H, Krakow D, King SM, Uk10K, Beales PL, Al-Gazali L, Wicking C, Cormier-Daire V, Roepman R, Mitchison HM, Witman GB (2015) TCTEX1D2 mutations underlie Jeune asphyxiating thoracic dystrophy with impaired retrograde intraflagellar transport. Nat Commun 6:7074

Schmidts M, Vodopiutz J, Christou-Savina S, Cortes CR, McInerney-Leo AM, Emes RD, Arts HH, Tuysuz B, D'Silva J, Leo PJ, Giles TC, Oud MM, Harris JA, Koopmans M, Marshall M, Elcioglu N, Kuechler A, Bockenhauer D, Moore AT, Wilson LC, Janecke AR, Hurles ME, Emmet W, Gardiner B, Streubel B, Dopita B, Zankl A, Kayserili H, Scambler PJ, Brown MA, Beales PL, Wicking C, Uk10K, Duncan EL, Mitchison HM (2013c) Mutations in the gene encoding IFT dynein complex component WDR34 cause Jeune asphyxiating thoracic dystrophy. Am J Hum Genet 93(5):932–944

Seeger-Nukpezah T, Golemis EA (2012) The extracellular matrix and ciliary signaling. Curr Opin Cell Biol 24(5):652–661

Shaheen R, Schmidts M, Faqeih E, Hashem A, Lausch E, Holder I, Superti-Furga A, Consortium UK, Mitchison HM, Almoisheer A, Alamro R, Alshiddi T, Alzahrani F, Beales PL, Alkuraya FS (2015) A founder CEP120 mutation in Jeune asphyxiating thoracic dystrophy expands the role of centriolar proteins in skeletal ciliopathies. Hum Mol Genet 24(5):1410–1419

Shao YY, Wang L, Welter JF, Ballock RT (2012) Primary cilia modulate Ihh signal transduction in response to hydrostatic loading of growth plate chondrocytes. Bone

Smith UM, Consugar M, Tee LJ, McKee BM, Maina EN, Whelan S, Morgan NV, Goranson E, Gissen P, Lilliquist S, Aligianis IA, Ward CJ, Pasha S, Punyashthiti R, Malik Sharif S, Batman PA, Bennett CP, Woods CG, McKeown C, Bucourt M, Miller CA, Cox P, Algazali L, Trembath RC, Torres VE, Attie-Bitach T, Kelly DA, Maher ER, Gattone VH 2nd, Harris PC, Johnson CA (2006) The transmembrane protein meckelin (MKS3) is mutated in Meckel-Gruber syndrome and the wpk rat. Nat Genet 38(2):191–196

Song B, Haycraft CJ, Seo HS, Yoder BK, Serra R (2007) Development of the post-natal growth plate requires intraflagellar transport proteins. Dev Biol 305(1):202–216

Stevenson NL (2023) The factory, the antenna and the scaffold: the three-way interplay between the Golgi, cilium and extracellular matrix underlying tissue function. Biol Open 12(2)

Stilling S, Kalliakoudas T, Benninghoven-Frey H, Inoue T, Falkenburger BH (2022) PIP(2) determines length and stability of primary cilia by balancing membrane turnovers. Commun Biol 5(1):93

Szymanska K, Boldt K, Logan CV, Adams M, Robinson PA, Ueffing M, Zeqiraj E, Wheway G, Johnson CA (2022) Regulation of canonical Wnt signalling by the ciliopathy protein MKS1 and the E2 ubiquitin-conjugating enzyme UBE2E1. elife 11

Tallila J, Jakkula E, Peltonen L, Salonen R, Kestila M (2008) Identification of CC2D2A as a Meckel syndrome gene adds an important piece to the ciliopathy puzzle. Am J Hum Genet 82(6):1361–1367

Tao D, Xue H, Zhang C, Li G, Sun Y (2019) The role of IFT140 in osteogenesis of adult mice long bone. J Histochem Cytochem 67(8):601–611

Tayeh MK, Yen HJ, Beck JS, Searby CC, Westfall TA, Griesbach H, Sheffield VC, Slusarski DC (2008) Genetic interaction between Bardet-Biedl syndrome genes and implications for limb patterning. Hum Mol Genet 17(13):1956–1967

Taylor SP, Dantas TJ, Duran I, Wu S, Lachman RS, C. University of Washington Center for Mendelian Genomics, Nelson SF, Cohn DH, Vallee RB, Krakow D (2015) Mutations in DYNC2LI1 disrupt cilia function and cause short rib polydactyly syndrome. Nat Commun 6:7092

Thiel C, Kessler K, Giessl A, Dimmler A, Shalev SA, von der Haar S, Zenker M, Zahnleiter D, Stoss H, Beinder E, Abou Jamra R, Ekici AB, Schroder-Kress N, Aigner T, Kirchner T, Reis A, Brandstatter JH, Rauch A (2011) NEK1 mutations cause short-rib polydactyly syndrome type majewski. Am J Hum Genet 88(1):106–114

Thompson CL, Chapple JP, Knight MM (2014) Primary cilia disassembly downregulates mechanosensitive hedgehog signalling: a feedback mechanism controlling ADAMTS-5 expression in chondrocytes. Osteoarthr Cartil 22:490

Thompson CL, McFie M, Chapple JP, Beales P, Knight MM (2021) Polycystin-2 is required for chondrocyte Mechanotransduction and traffics to the primary cilium in response to mechanical stimulation. Int J Mol Sci 22(9)

Tompson SW, Ruiz-Perez VL, Blair HJ, Barton S, Navarro V, Robson JL, Wright MJ, Goodship JA (2007) Sequencing EVC and EVC2 identifies mutations in two-thirds of Ellis-van Creveld syndrome patients. Hum Genet 120(5):663–670

Toriyama M, Lee C, Taylor SP, Duran I, Cohn DH, Bruel AL, Tabler JM, Drew K, Kelly MR, Kim S, Park TJ, Braun DA, Pierquin G, Biver A, Wagner K, Malfroot A, Panigrahi I, Franco B, Al-Lami HA, Yeung Y, Choi YJ, G. University of Washington Center for Mendelian, Duffourd Y, Faivre L, Riviere JB, Chen J, Liu KJ, Marcotte EM, Hildebrandt F, Thauvin-Robinet C, Krakow D, Jackson PK, Wallingford JB (2016) The ciliopathy-associated CPLANE proteins direct basal body recruitment of intraflagellar transport machinery. Nat Genet 48(6):648–656

Tsang WY, Bossard C, Khanna H, Peranen J, Swaroop A, Malhotra V, Dynlacht BD (2008) CP110 suppresses primary cilia formation through its interaction with CEP290, a protein deficient in human ciliary disease. Dev Cell 15(2):187–197

Tuz K, Bachmann-Gagescu R, O'Day DR, Hua K, Isabella CR, Phelps IG, Stolarski AE, O'Roak BJ, Dempsey JC, Lourenco C, Alswaid A, Bonnemann CG, Medne L, Nampoothiri S, Stark Z, Leventer RJ, Topcu M, Cansu A, Jagadeesh S, Done S, Ishak GE, Glass IA, Shendure J, Neuhauss SC, Haldeman-Englert CR, Doherty D, Ferland RJ (2014) Mutations in CSPP1 cause primary cilia abnormalities and Joubert syndrome with or without Jeune asphyxiating thoracic dystrophy. Am J Hum Genet 94(1):62–72

Unger S, Ferreira CR, Mortier GR, Ali H, Bertola DR, Calder A, Cohn DH, Cormier-Daire V, Girisha KM, Hall C, Krakow D, Makitie O, Mundlos S, Nishimura G, Robertson SP, Savarirayan R, Sillence D, Simon M, Sutton VR, Warman ML, Superti-Furga A (2023) Nosology of genetic skeletal disorders: 2023 revision. Am J Med Genet A 191(5):1164–1209

Valente EM, Logan CV, Mougou-Zerelli S, Lee JH, Silhavy JL, Brancati F, Iannicelli M, Travaglini L, Romani S, Illi B, Adams M, Szymanska K, Mazzotta A, Lee JE, Tolentino JC, Swistun D, Salpietro CD, Fede C, Gabriel S, Russ C, Cibulskis K, Sougnez C, Hildebrandt F, Otto EA, Held S, Diplas BH, Davis EE, Mikula M, Strom CM, Ben-Zeev B, Lev D, Sagie TL, Michelson M, Yaron Y, Krause A, Boltshauser E, Elkhartoufi N, Roume J, Shalev S, Munnich A, Saunier S, Inglehearn C, Saad A, Alkindy A, Thomas S, Vekemans M, Dallapiccola B, Katsanis N, Johnson CA, Attie-Bitach T, Gleeson JG (2010) Mutations in TMEM216 perturb ciliogenesis and cause Joubert, Meckel and related syndromes. Nat Genet 42(7):619–625

Verbeek S, Eilers PH, Lawrence K, Hennekam RC, Versteegh FG (2011) Growth charts for children with Ellis-van Creveld syndrome. Eur J Pediatr 170(2):207–211

Walczak-Sztulpa J, Eggenschwiler J, Osborn D, Brown DA, Emma F, Klingenberg C, Hennekam RC, Torre G, Garshasbi M, Tzschach A, Szczepanska M, Krawczynski M, Zachwieja J, Zwolinska D, Beales PL, Ropers HH, Latos-Bielenska A, Kuss AW (2010) Cranioectodermal dysplasia, Sensenbrenner syndrome, is a ciliopathy caused by mutations in the IFT122 gene. Am J Hum Genet 86(6):949–956

Wallmeier J, Nielsen KG, Kuehni CE, Lucas JS, Leigh MW, Zariwala MA, Omran H (2020) Motile ciliopathies. Nat Rev Dis Primers 6(1):77

Wang C, Yuan X, Yang S (2013) IFT80 is essential for chondrocyte differentiation by regulating hedgehog and Wnt signaling pathways. Exp Cell Res 319(5):623–632

Wann AK, Chapple JP, Knight MM (2014) The primary cilium influences interleukin-1beta-induced NFkappaB signalling by regulating IKK activity. Cell Signal 26(8):1735–1742

Wann AK, Zuo N, Haycraft CJ, Jensen CG, Poole CA, McGlashan SR, Knight MM (2012) Primary cilia mediate mechanotransduction through control of ATP-induced Ca2+ signaling in compressed chondrocytes. FASEB J 26(4):1663–1671

Wheatley DN (1995) Primary cilia in normal and pathological tissues. Pathobiology 63(4):222–238

Wheway G, Schmidts M, Mans DA, Szymanska K, Nguyen TT, Racher H, Phelps IG, Toedt G, Kennedy J, Wunderlich KA, Sorusch N, Abdelhamed ZA, Natarajan S, Herridge W, van Reeuwijk J, Horn N, Boldt K, Parry DA, Letteboer SJF, Roosing S, Adams M, Bell SM, Bond J, Higgins J, Morrison EE, Tomlinson DC, Slaats GG, van Dam TJP, Huang L, Kessler K, Giessl A, Logan CV, Boyle EA, Shendure J, Anazi S, Aldahmesh M, Al Hazzaa S, Hegele RA, Ober C, Frosk P, Mhanni AA, Chodirker BN, Chudley AE, Lamont R, Bernier FP, Beaulieu CL, Gordon P, Pon RT, Donahue C, Barkovich AJ, Wolf L, Toomes C, Thiel CT, Boycott KM, McKibbin M, Inglehearn CF, Consortium UK, G. University of Washington Center for Mendelian, Stewart F, Omran H, Huynen MA, Sergouniotis PI, Alkuraya FS, Parboosingh JS, Innes AM, Willoughby CE, Giles RH, Webster AR, Ueffing M, Blacque O, Gleeson JG, Wolfrum U, Beales PL, Gibson T, Doherty D, Mitchison HM, Roepman R, Johnson CA (2015) An siRNA-based functional genomics screen for the identification of regulators of ciliogenesis and ciliopathy genes. Nat Cell Biol 17(8):1074–1087

Yang C, Chen W, Chen Y, Jiang J (2012) Smoothened transduces hedgehog signal by forming a complex with Evc/Evc2. Cell Res 22(11):1593–1604

Yang Q, Zhang Q, Chen F, Yi S, Li M, Yi S, Xu X, Luo J (2021) A novel combination of biallelic IFT122 variants associated with cranioectodermal dysplasia: a case report. Exp Ther Med 21(4):311

Yuan X, Yang S (2015) Deletion of IFT80 impairs epiphyseal and articular cartilage formation due to disruption of chondrocyte differentiation. PLoS One 10(6):e0130618

Zhang H, Kamiya N, Tsuji T, Takeda H, Scott G, Rajderkar S, Ray MK, Mochida Y, Allen B, Lefebvre V, Hung IH, Ornitz DM, Kunieda T, Mishina Y (2016) Elevated fibroblast growth factor signaling is critical for the pathogenesis of the dwarfism in Evc2/Limbin mutant mice. PLoS Genet 12(12):e1006510

Part III
Intracellular Trafficking of ECM Proteins and Related Skeletal Disorders

Chapter 11
Microtubule-Associated Motor Proteins in Skeletal Development and Health

Francesca Manuela Johnson de Sousa Brito, Roufaida Bouchenafa, Rachel Deborah Pearson, and Katarzyna Anna Piróg

Abstract Extracellular changes resulting in perturbed skeletal development can be a direct result of mutations in the structural extracellular matrix (ECM) components and of changes in the receptor activity and signalling pathways, but also of perturbations to intracellular mechanisms that can have secondary extracellular consequences. The developing cartilage and bone are highly secretory and mechanoresponsive tissues, and the intracellular processes governed by the cytoskeletal proteins and by the associated molecular motors aid in protein processing and secretion, recycling of membrane-bound receptors, positioning of the organelles, maintenance of the primary cilia, modulation of signalling pathways, and coordination of the cell cycle. Molecular motors hydrolyse ATP to generate forces that allow them to travel towards either plus or minus ends of the microtubules generating forces that govern the movement of all intracellular cargo as well as the formation of the mitotic spindle and polar ejection forces during cell division. Their complex structure and diversity allow for varied cargo, from membrane-bound vesicles (lysosomes, endosomes, phagosomes, peroxisomes) to organelles (mitochondria, Golgi apparatus, nucleus), lipid droplets, chromosomes, and nucleic acids (DNA and mRNA). In this chapter, we discuss the microtubule-associated molecular motors, dyneins, and kinesins and their emerging roles in skeletal dysplasia.

11.1 Introduction

Extracellular matrix (ECM) changes resulting in skeletal complications can arise directly from intracellular complications. Changes in intracellular signalling, stresses associated with the processing of mutant proteins, cytoskeletal abnormalities, and changes in the secretory and mechanosensing machineries can all lead to

F. M. J. de Sousa Brito · R. Bouchenafa · R. D. Pearson · K. A. Piróg (✉)
Biosciences Institute, Newcastle University, Newcastle upon Tyne, UK
e-mail: katarzyna.pirog@ncl.ac.uk

A. Rossi, F. Zaucke (eds.), *The Extracellular Matrix in Genetic Skeletal Disorders*, Biology of Extracellular Matrix 16,
https://doi.org/10.1007/978-3-031-70835-0_11

skeletal dysplasia and result in perturbed ECM organisation. Intracellular transport along the microtubule cytoskeleton is particularly important for extracellular matrix production and for the function of the highly secretory cells such as the skeletal lineage cells during development. It is also of high importance for the intracellular mechanisms implicated in the aetiology of skeletal dysplasias, such as cell division, ciliogenesis, and organelle positioning and maintenance.

There are three classes of molecular motor proteins that are responsible for molecular motility: myosins (associated with the actin filaments), dyneins, and kinesins (that move along the microtubules) (Rayment 1996). Myosins have originally been discovered in skeletal muscle; however, they are present in almost all eukaryotic cells (Sebé-Pedrós et al. 2014). The actin cytoskeleton and actin-myosin interactions are responsible for the adhesion and movement of non-muscle cells, essential for many developmental processes, including cell shape and cell division (Cooper 2000). Not surprisingly, myosins and the actin crosslinking filamins have been shown to play a role in mechanoresponses (Seppälä et al. 2017), and mutations in *MYH3* result in arthrogryposis and spondylocarpotarsal fusion due to decreased TGFβ signalling (Zieba et al. 2017). Kinesins and dyneins are ATP-powered microtubule-associated motor molecules, responsible for the intracellular transport of organelles and molecules and for organising the microtubule cytoskeleton during cell division. Dyneins travel to the minus end of the microtubule filaments (retrograde transport), localised in the microtubule organising centre near the nucleus of non-dividing cells, whereas most kinesins move towards the plus ends (to the cell periphery, anterograde) (Ferro et al. 2019). The forces generated by the opposite movement of dyneins and kinesins along the microtubules are essential for cell division and transport of intracellular cargo. During cell division, dynein drives the localisation of the spindle to one end of the cell, and kinesins are involved in the regulation of the spindle length, the polar ejection forces, and depolymerizing microtubule minus ends at centrosomes during the anaphase (Drpic et al. 2015).

Most dyneins operate as a dimer, comprising two heavy chains that contain the AAA+ superfamily motor domain and the N-terminal tail that is responsible for oligomerisation and association with cargo and with the adaptor proteins (Fig. 11.1) (Roberts et al. 2013). There are nine main classes of dynein heavy chain, with 16 heavy chain genes identified in the human genome, and five classes of associated subunits (intermediate, light-intermediate, and light chains) which can be further modified by alternative splicing and posttranslational modifications and have been shown to have tissue-specific expression (Pfister et al. 2006). Dynein 1 (DYNC1H1) is the main cytoplasmic dynein required for intracellular movement, whereas dynein 2 (DYNC2H1) is primarily associated with intraflagellar transport in the primary cilium. The remaining seven dyneins are axonemal dyneins involved in the movement of the beating/motile cilia. The cytoplasmic dynein is important in the positioning of the nucleus and in the spindle assembly during cell division, as well as for intracellular transport of the endosomes (Driskell et al. 2007), lysosomes (Jordens et al. 2001), phagosomes (Blocker et al. 1997), melanophores (Gross et al. 2002), peroxisomes (Kural et al. 2005), lipid droplets (Singh et al. 2022), autophagosomes, and vesicles from the endoplasmic reticulum (ER) destined for the Golgi along the

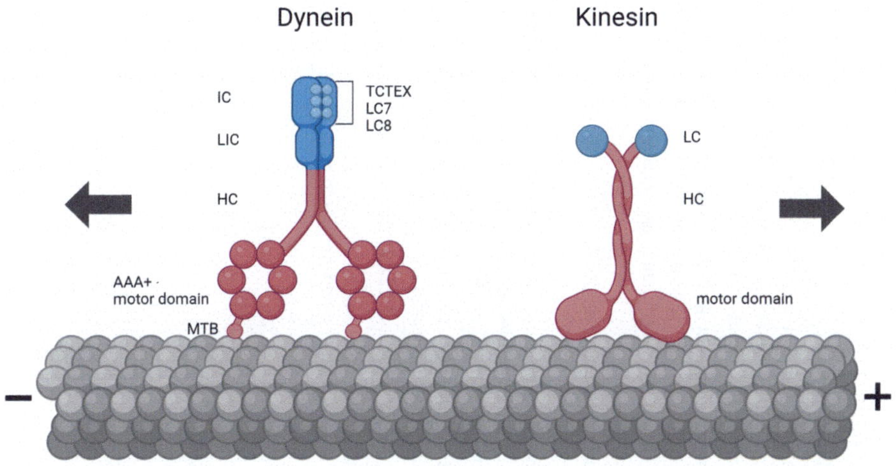

Fig. 11.1 Molecular structure of the dynein and kinesin dimers, showing the heavy chains in red and the intermediate and light chains in blue. The dynein heavy chain comprises a ring composed of six AAA ATPase domains, a microtubule-binding (MTB) stalk, and binding sites for two light-intermediate chains (LIC) and two intermediate chains (IC). The intermediate chains in turn bind the light chains (LC). A kinesin molecule comprises 2 heavy chains (in red) with N-terminal microtubule-binding domain, associating with the light chains binding cargo (created with BioRender.com)

microtubules (Gurel et al. 2014). It has also been shown to position and carry organelles, such as the Golgi apparatus (Yadav et al. 2012) and the mitochondria (Pilling et al. 2006). Dynein can also be involved in the transport of transcription factors (for example, NFκB (Mikenberg et al. 2007)) and mRNA-containing ribonucleoprotein complexes (Reck-Peterson et al. 2018).

To date, there have been over 50 predicted kinesin and kinesin-like molecules identified in the human genome. The general structure of a kinesin comprises two heavy chains composed of a motor domain (where ATP is hydrolysed) responsible for movement along the microtubules and an alpha-helical coiled-coil domain that mediates kinesin dimerisation and non-covalently attached light chains with binding sites for carrying cargo along the microtubules (Fig. 11.1) (Dodding and Way 2011). Interestingly, the kinesin 5 family and KIF25 have also been shown to operate as a bipolar tetramer (Scholey et al. 2014; Decarreau et al. 2017). Kinesin heavy chains have been grouped into 14 distinct groups based on their domain organisation and function (Table 11.1). Kinesins can also be classed into three separate groups based on the localisation of their motor domain, which can be N-terminal (the majority of kinesins are N-kinesins), C-terminal (C-kinesins: KIFC1, KIFC2, and KIFC3), or located in the middle (M-kinesins: KIF2A, KIF2B, and KIF2C). The N- and C-kinesins are responsible for transport to the plus or minus ends of the microtubules, whereas the M-kinesins play a role in microtubule depolymerisation (Wang et al. 2016). The motor domains of kinesins are highly conserved in structure, whereas the variability in the light chains allows for diverse cargos. The light

Table 11.1 Classification of the kinesin family of proteins ((Hirokawa et al. 2009))

Kinesin class	Member kinesins	Expression	Identified roles/processes
Kinesin 1	KIF5A	Neuron-specific (Niclas et al. 1994)	RNA transport (Kanai et al. 2004), mitochondria transport (Yokota et al. 2023), autophagy (Liu et al. 2021), secretory vesicle transport (Kamata et al. 2017)
	KIF5B	Ubiquitous (Niclas et al. 1994)	RNA transport (Kanai et al. 2004), cytokinesis (Gan et al. 2019), vesicle transport (Kamal et al. 2001), peroxisome transport (Kural et al. 2005), nuclear positioning (Metzger et al. 2012)
	KIF5C	Strong expression in the brain, prostate, and testis; moderate in the kidney, small intestine, and ovary (Nagase et al. 1998)	RNA transport (Kanai et al. 2004)
Kinesin 2	KIF3A	Ganglion cells, Muller cells and cell bodies, photoreceptor, ciliary axoneme (Whitehead et al. 1999), strong expression in the mouse brain and kidney (Yamazaki et al. 1995)	Embryonic cilia formation (Morris and Scholey 1997, Cruz et al. 2022), cilia function, anterograde IFT (Lin et al. 2003), hedgehog signalling (Kovacs et al. 2008)
	KIF3B	Ubiquitous (Nagase et al. 1997)	Ciliogenesis (Cruz et al. 2022), hedgehog signalling (Yue et al. 2022), vesicle transport (Alsabban et al. 2020)
	KIF3C	Ubiquitous (Telford et al. 1998)	Ciliogenesis (Zhao et al. 2012), axonal transport (Gumy et al. 2013)
	KIF17	Neuron-specific (Yin et al. 2011)	Vesicle transport, receptor sorting (Setou et al. 2000)

Kinesin 3	KIF1A	Brain, neuron-specific (Rivière et al. 2011)	Vesicle transport (Okada et al. 1995, Yonekawa et al. 1998)
	KIF1B	Ubiquitous, high expression in the nerve cells (Nangaku et al. 1994, Yang et al. 2001)	Mitochondria transport (Nangaku et al. 1994), receptor transport (Xu et al. 2018)
	KIF1C	Ubiquitous, with high expression in the heart and skeletal muscle (Dorner et al. 1998)	Retrograde transport of Golgi vesicles to the ER (Dorner et al. 1998)
	KIF13A	Central nervous system, heart, skeletal muscle, lower expression in other organs and tissues (Jamain et al. 2001)	Vesicle transport, Golgi-to-plasma membrane transport (Nakagawa et al. 2000)
	KIF13B	Ubiquitous (Ishikawa et al. 1998)	VEGFR2 trafficking from the Golgi, angiogenesis (Yamada et al. 2014), secretory vesicle transport (Serra-Marques et al. 2020)
	KIF14	Highly expressed in the brain and kidney (Nakagawa et al. 1997)	Cytokinesis (Carleton et al. 2006)
	KIF16A	Ubiquitous at low levels (Torres et al. 2011)	Cell division and spindle assembly (Torres et al. 2011), pericentriolar material stabilisation (Torres et al. 2011)
	KIF16B	High expression in the brain, also found in the kidney, liver, intestine, placenta, leukocytes, heart, and skeletal muscle (Hoepfner et al. 2005)	Vesicle transport, early-to-late endosomes (Blatner et al. 2007), receptor recycling (Hoepfner et al. 2005)
Kinesin 4	KIF4A	Ubiquitous (Kim et al. 1998)	Chromokinesin (Almeida and Maiato 2018), ciliogenesis (He et al. 2017, Sheng et al. 2018)
	KIF4B	Ubiquitous (Sheng et al. 2018)	Spindle dynamics and cytokinesis (Zhu et al. 2005)
	KIF7	Embryonic stem cells (Katoh and Katoh 2004), lung (Coles et al. 2015), motor neurons (Liem Jr. et al. 2009), cartilage (Hsu et al. 2011)	Primary cilium-associated hedgehog signalling (He et al. 2014)
	KIF21A	Ubiquitous except in the spleen, highly expressed in the nervous system (Marszalek et al. 1999)	Microtubule dynamics (Bianchi et al. 2016)
	KIF21B	Brain, eye, spleen, developing cortex, dendrites (Marszalek et al. 1999, Asselin et al. 2020)	Microtubule dynamics, centrosome polarisation (Hooikaas et al. 2020)
	KIF27	Testis, pancreatic islet, germ cell tumours, and Jurkat T cells (Katoh and Katoh 2004)	Construction of central pair in motile cilia, ciliogenesis, and hedgehog signalling (Wilson et al. 2009)

(continued)

Table 11.1 (continued)

Kinesin class	Member kinesins	Expression	Identified roles/processes
Kinesin 5	KIF11	Neuronal retina, photoreceptor cells, testis (Birtel et al. 2017)	Cytokinesis, spindle formation (Blangy et al. 1995), ciliogenesis (Birtel et al. 2017), translation, association of ribosomes with microtubules (Bartoli et al. 2011)
Kinesin 6	KIF20A	High levels detected in the thymus, bone marrow, and testis, found at lower levels in the heart, placenta, and spleen (Lai et al. 2000)	Cytokinesis (Nguyen et al. 2014), Golgi-derived vesicle transport (Echard et al. 1998)
	KIF20B	Ubiquitous (Lai et al. 2000)	Cell division(Kamimoto et al. 2001)
	KIF23	Ubiquitous (Vikberg et al. 2020)	Cytokinesis (Liljeholm et al. 2013)
Kinesin 7	KIF10	Ubiquitous (Kalantari and Filges 2020)	Cell division, kinetochore function (Mirzaa et al. 2014)
Kinesin 8	KIF18A	Preosteogenic marrow stromal cells (Luboshits and Benayahu 2005)	Mitotic chromosome positioning (Stumpff et al. 2008)
	KIF18B	Ubiquitous, with variable levels (Lee et al. 2010)	Microtubule depolymerisation, cell division (Tanenbaum et al. 2011)
	KIF19A	Strongly expressed in the oviduct and trachea (Niwa et al. 2012)	Cilia length (Niwa et al. 2012)
Kinesin 9	KIF6	Brain, intestine, spinal cord, ovary and testis, heart, and endothelial cells (Hameed et al. 2013, Buchan et al. 2014)	Ciliogenesis, cell division (Iwata et al. 2000, Gualeni et al. 2013, Konjikusic et al. 2018)
	KIF9	Found in the brain, kidney, spleen, lung, and testis, but not in the intestine, liver, or heart (Piddini et al. 2001)	Motile cilia function (Konjikusic et al. 2023), podosome function (Cornfine et al. 2011)
Kinesin 10	KIF22	Bone, cartilage, joint capsule, ligament, skin, liver, small intestine, spleen, and ovary (Min et al. 2011)	Chromokinesin (Almeida and Maiato 2018)
Kinesin 11	KIF26A	Present in the foetal liver and brain and adult kidney and pancreas and at lower levels in the heart, spleen, and ovary (Nagase et al. 1999)	Microtubule stability, RET signalling (Zhou et al. 2009)
	KIF26B	Mesenchymal cells, embryonic limb buds, metanephros, and central nervous system (Uchiyama et al. 2010)	MYH10-mediated cell-cell adhesion (Uchiyama et al. 2010)

Kinesin 12	KIF12	Foetal liver, adult brain, and pancreas (Katoh and Katoh 2005)	Cell polarity (Stalke et al. 2022)
	KIF15	Brain and haematopoietic cells (Sleiman et al. 2017)	Cell division (Brouwers et al. 2017), integrin endocytosis (Eskova et al. 2014)
Kinesin 13	KIF2A	Ubiquitous (Debernardi et al. 1997)	Microtubule depolymerisation, spindle assembly, and chromosome movement (Manning et al. 2007)
	KIF2B	Found in the lung, ovaries, heart, kidney, muscle, and placenta, but not in the brain, liver, or testis (Manning et al. 2007)	Spindle assembly and chromosome movement (Manning et al. 2007)
	KIF2C	Present in the thymus, testis, small intestine, colon, placenta, spleen, and ovary, but not in the prostate, heart, brain, lung, liver, muscle, kidney, or pancreas (Kim et al. 1997)	Microtubule depolymerisation, cell division (Tanenbaum et al. 2011)
	KIF24	Highly expressed in the testis, weaker in the spleen, and little to no expression in the heart, brain, lung, liver, muscle, and kidney (Miki et al. 2001)	Depolymerisation of the centriolar microtubules (Kobayashi et al. 2011)
Kinesin 14a	KIFC1	Ubiquitous, highly expressed in the heart and skeletal muscle (Dorner et al. 1998)	Retrograde transport from Golgi apparatus to the ER (Dorner et al. 1998), retrograde transport of secretory vesicles (Schlager et al. 2010)
Kinesin 14b	KIFC2	Expressed in the thymus and testis; found at low levels in the small intestine, colon, and placenta; and at very low levels in the spleen and ovary (Kim et al. 1997)	Correction of the improper microtubule attachments at kinetochores (Manning et al. 2007)
	KIFC3	Retina, RPE, lung, and kidney (Bost-Usinger et al. 1997)	Maintaining the integrity of the adherens junction (Meng et al. 2008)
	KIF25	Expressed at low levels in several tissues (Decarreau et al. 2017)	Centrosome operation and spindle orientation (Decarreau et al. 2017)

chain genes (*KLC1*, *KLC2*, *KLC3*, and *KLC4*) can undergo alternative splicing to produce isoforms with different C-terminal sequences, able to bind different organelles and structures (Verhey et al. 1998). KLC1 and KLC2 have been shown to associate with kinesin-1 molecules KIF5A and KIF5B (Rahman et al. 1998); however, the role of kinesin light chains in skeletal development is unknown. In addition, kinesins can also use a variety of cargo adaptor molecules that can bind directly to the heavy or light chains, further increasing their diversity (Cross et al. 2021). Kinesins have been implicated in several molecular processes essential for development, such as the movement of organelles, vesicles, and protein complexes, and in the regulation of cell division, movement of chromosomes, and generation and function of primary cilia (Miki et al. 2001). Kinesins bind to various membrane-bound cargos such as intracellular vesicles (Horiguchi et al. 2006; Lo et al. 2011), lysosomes (Matsushita et al. 2004), and mitochondria (Nangaku et al. 1994), but also nucleic acids (messenger RNA and DNA) (Cross et al. 2021) and transmembrane receptor proteins (Nakagawa et al. 2000, Shin et al. 2003, Ueno et al. 2011, Theisen et al. 2012, Xing et al. 2012, Zhou et al. 2013, Yamada et al. 2014, Tanaka et al. 2016), indicating a role in receptor recycling.

Transport along the intracellular microtubules is essential for many cellular processes including protein production, cell motility, cell signalling, and cell division. Microtubules form essential structures inside the cells, often termed "molecular highways" (Balabanian et al. 2018), and are an essential part of the mechanoresponsive organelles such as the primary cilium. In fact, mutations in genes encoding kinesins and dyneins often result in ciliary dyskinesis and spermatogenesis failure. The importance of the motor molecules in developmental processes and organogenesis is demonstrated by the fact that deletion of both non-axonemal dyneins and of several kinesins (KIF1A, KIF1B, KIF2A, KIF3A, KIF3B, KIF5A, KIF5B, KIF7, KIF10, KIF11, KIF18A, KIF20A, KIF20B, KIF21, KIF26B) in mice results in early embryonic or perinatal lethality (Harada et al. 1998; Ocbina et al. 2011; Yao et al. 2022). Microtubule-associated transport and microtubule depolymerisation are essential for cell division and cell cycle progression, and several motor molecules have been studied extensively in the context of cancers. Microtubule-guided axonal transport is also essential for the correct function of neurons, and it is therefore not surprising that several "kinesinopathies" result in neurological complications and microcephaly (Kalantari and Filges 2020). In contrast, the specific roles of microtubule-associated motors in the development of musculoskeletal tissues remain largely unknown. However, motor proteins have been shown to play important roles in cell cycle, cell polarity, primary cilia function, microtubule stability, and intracellular transport, amongst others, and mutations in several dynein- and kinesin-related genes result in skeletal dysplasia (Table 11.2).

Table 11.2 Skeletal dysplasias resulting from mutations in microtubule-associated motor proteins

Gene	Condition	Inheritance	Skeletal features	Extra skeletal complications	References
DYNC2H1	Short-rib thoracic dysplasia 3 with or without polydactyly (SRTD3 OMIM #613091)	AR	Abnormal bone development, including short ribs, small thorax, brachydactyly, and shortened long bones with or without postaxial polydactyly of both hands and feet	Cleft lip/palate; other tissues affected are the brain, eye, heart, kidneys, liver, pancreas, intestines, and genitalia	(Huber and Cormier-Daire 2012, Schmidts et al. 2013)
DYNC2LI1	Short-rib thoracic dysplasia 15 with or without polydactyly (SRTD15 OMIM #617088)	AR	Short ribs, constricted rib cage, variable polydactyly, short tubular bones, narrow thorax	Cardiovascular, cleft lip/palate	(Kessler et al. 2015, Taylor et al. 2015, Niceta et al. 2018)
TCTEX1D2	Short-rib thoracic dysplasia 17 with or without polydactyly (SRTD17, OMIM #617405)	AR	Short ribs, narrow thorax, small ilia with trident acetabulum, handlebar clavicles, brachydactyly, occasional polydactyly	Not reported	(Schmidts et al. 2015)
WDR34	Short-rib thoracic dysplasia 11 with or without polydactyly (SRTD11, OMIM #615633)	AR	Short stature, bell-shaped thorax, brachydactyly, polydactyly, metaphyseal widening, handlebar clavicles	Pulmonary and renal involvement	(Tüysüz et al. 2009, Huber and Cormier-Daire 2012)
WDR60	Short-rib thoracic dysplasia 8 with or without polydactyly (SRTD8, OMIM #615503)	AR	Shortening of long bones, short ribs, bowed femur, postaxial polydactyly, brachydactyly, epiphyseal dysplasia, hypoplastic trabecular bone	Pulmonary, renal, biliary, and cardiac involvement	(McInerney-Leo et al. 2013)
KIF5B	Kyphomelic dysplasia with facial dysmorphism	De novo heterozygous	Sharp angulation of the femora and humeri, distinctive facial features, short stature, postnatal osteoporosis with subsequent fractures, brachycephaly	Neonatal respiratory distress	(Itai et al. 2022)
	Osteogenesis imperfecta	De novo heterozygous	Hypomineralisation of bones, gracile ribs, scoliosis, multiple fractures, and skeletal deformities	Dentinogenesis imperfecta, hearing loss, ptosis, atrial septal defect	(Marom et al. 2023)

(continued)

Table 11.2 (continued)

Gene	Condition	Inheritance	Skeletal features	Extra skeletal complications	References
KIF7	Acrocallosal syndrome (ACLS) or Joubert syndrome 12 (JBTS12), OMIM #200990	AR	Dysmorphic features, postaxial polydactyly of the upper limbs, preaxial polydactyly of the feet, macrocephaly, hallux duplication	Mental retardation	(Schinzel and Schmid 1980, Nelson and Thomson 1982)
	Hydrolethalus syndrome (HLS2, OMIM #614120)	AR	Hydrocephaly or anencephaly, postaxial polydactyly of the upper limbs, and pre- or postaxial polydactyly of the lower limbs, duplication of the hallux	Widened ventricles and midbrain-hindbrain malformation	(Putoux et al. 2011)
	Al-Gazali-Bakalinova syndrome (AGBK, OMIM #607131)	AR	Normal stature, multiple epiphyseal dysplasia, macrocephaly, distinctive facial features including frontal bossing, hypertelorism, flat malar regions, low-set ears, and short neck; pectus excavatum, clinodactyly, prominent joints, and *genu valgum*	Agenesis of the corpus callosum and frontotemporal brain atrophy	(al-Gazali and Bakalinova 1998)
KIF10	Microcephaly 13 (MCPH13, OMIM #616051)	AR	Microcephaly, sloping forehead, prominent nose, mild micrognathia, short stature, small hands and possible metaphyseal sclerosis, osteopenic hands, short metacarpals, no craniosynostosis	Partial agenesis of the corpus callosum and cerebellar hypoplasia, cardiomyopathy	(Mirzaa et al. 2014)

KIF14	Meckel syndrome (MKS12, OMIM #616258)	AR	Intrauterine growth retardation with microcephaly	Renal cystic agenesis, complex brain defects, and genitourinary malformations	(Filges et al. 2014)
	Microcephaly 20 (MCPH20, OMIM #617914)	AR	Microcephaly, short stature, foot deformities	Intellectual disability, speech impairment, renal impairment	(Moawia et al. 2017, Makrythanasis et al. 2018)
KIF15	Braddock-Carey syndrome 2 (BRDCS2, OMIM #619981)	AR	Microcephaly, facial dysmorphisms microretrognathia, cleft palate, downslanting palpebral fissures, bulbous nose, and wide mouth	Microphthalmia, deafness, hypercellular bone marrow	(Sleiman et al. 2017)
KIF22	Spondyloepimetaphyseal dysplasia with joint laxity, type 2 (SEMD-JL, OMIM #603546)	AD	Short stature, distinctive midface retrusion, progressive knee malalignment *(genu valgum* and/or *varum*), early onset osteoarthritis, and mild spinal deformity	Generalised ligamentous laxity	(Hall et al. 2002, Min et al. 2011)

11.2 Microtubule Motors and Cell Cycle Regulation in Skeletogenesis

Chromokinesins are DNA-binding kinesins that are important for chromosome movement during cell division and play roles in chromosome condensation and alignment, stabilisation of the kinetochore, and generation of polar ejection forces necessary for chromosome separation. Several kinesins have been shown to directly bind nucleic acids (mRNA, DNA)—these include KIF4A, KIF4B, KIF10, KIF15, and KIF22. Of these, KIF4A and KIF22 are stronglys expressed in cartilage (Fourriere et al. 2019), and mutations in *KIF10*, *KIF15*, and *KIF22* genes result in skeletal dysplasia. Autosomal recessive mutations in *KIF10* lead to microcephaly, short stature, small hands, metaphyseal sclerosis, short metacarpals, and osteopenia (Mirzaa et al. 2014). KIF10 (also known as centromere protein E, or CENPE) was first identified as a centromere-associated protein (Yen et al. 1991) and is targeted to the kinetochore (connecting the chromosomes to the microtubules of the spindle) through a modification by small ubiquitin-like modifier proteins SUMO2/3 (Wang et al. 2008; Zhang et al. 2008). The importance of KIF10 in development is emphasised by the fact that the deletion of KIF10 leads to embryonic lethality in mice (Putkey et al. 2002). Interestingly, KIF15 can compensate for the lack of KIF10 in the spindle assembly in the cancer cells (Kalantari and Filges 2020), and autosomal recessive changes in *KIF15* also lead to microcephaly, cleft palate, microphthalmia, and facial dysmorphism (Sleiman et al. 2017). KIF22 is a chromokinesin essential for early cell division and important for the shortening of the chromosome mass along the spindle axis during the anaphase (Ohsugi et al. 2008). It likely localises to the mitotic chromosomes through its association with importin-β and importin-α (Tahara et al. 2008). Mutations in *KIF22* result in spondyloepimetaphyseal dysplasia with joint laxity Hall type (SEMDJL2), characterised by short trunk dwarfism, midface retrusion, slender digits, delayed endochondral ossification, and premature osteoarthritis. Deletion and mutations in *KIF22* lead to decreased proliferation and result in multinuclear daughter cells due to disrupted chromosome segregation (Thompson et al. 2022).

Changes to cellular homeostasis can have an indirect effect on extracellular matrix production and organisation. It has been shown in several studies that perturbations to the cell cycle and chondrocyte proliferation result in skeletal dysplasia and downstream ECM effects (hypocellular and disorganised growth plates, decreased type X collagen levels, and disrupted ossification (Iwata et al. 2000, Gualeni et al. 2013, Gan et al. 2019)). It is therefore not surprising that mutations in several motor molecules involved in cell division have skeletal consequences. *KIF5B* mutations result in kyphomelic dysplasia with facial dysmorphism and postnatal osteoporosis with fractures (Itai et al. 2022). Deletion of one copy of *Kif5b* in mice was shown to modulate the organisation of the central spindle in late-stage cytokinesis in chondrocytes and led to growth plate abnormalities, a significant decrease in proliferation and increased apoptosis, and dwarfism reminiscent of human skeletal dysplasia (Gan et al. 2019). The cartilage growth plate of new-born mutant mice was disorganised, with a disrupted columnar structure and decreased type X collagen secretion, suggesting disrupted terminal differentiation. The proliferative chondrocytes were also nearly devoid of primary cilia. KIF5B null cells were often bi-nucleated and multinucleated, indicating a cytokinetic defect, and KIF5B-deficient chondrocytes in vitro demonstrated a marked delay in abscission and failed cytokinesis (Gan et al. 2019). *Kif5a* and *kif5b* compound zebrafish mutants *Kif5b* loss-of-function (Kif5bLoF)), generated through CRISPR/Cas9 mutagenesis, presented with prominent jaw defects and abnormal chondrogenesis, defective perichondral ossification, decreased mineralization, and absent bone collars, suggesting that the mutant cartilage template was unable to support ossification (Santos-Ledo et al. 2017). KIF11, required for establishing a bipolar spindle during mitosis, is used as a marker for bone metastatic prostate cancer, indicating its potential role in skeletal tissues (Wang et al. 2022). Autosomal dominant mutations in *KIF11* lead to microcephaly and mild dysmorphic features (Ostergaard et al. 2012). Autosomal recessive mutations in KIF14 result in microcephaly, short stature, and foot deformities (Filges et al. 2014). KIF14 is localised to the mid-body of dividing cells (Zhernov et al. 2020). This localisation is disrupted in patient fibroblasts leading to disrupted cytokinesis, increased incidence of multinucleated cells, and increased apoptosis. Interestingly, mutations in KIF14 also affect cell migration in a scratch-wound assay, suggesting additional roles for this kinesis in tissue maintenance and homeostasis (Moawia et al. 2017).

Studies in lung tissue have highlighted a role for KIF7 in modulating the cell cycle, showing that KIF7 may be specifically involved in controlling the transition from the G1 to the S phase. Inactivation of KIF7 in mouse lung fibroblasts led to increased cellular proliferation, an increase of cells detected in the S phase of the cell cycle, and affected the timing of G1-to-S and G2-to-M transitions, suggesting that kinesins may regulate the cell cycle via regulating microtubule stability and through their interaction with microtubule-associated proteins (Coles et al. 2015).

11.3 Motor Proteins, Primary Cilia Function, and Microtubule Dynamics

Endochondral ossification is dependent upon chondrocyte proliferation. However, the resting and hypertrophic chondrocytes in cartilage growth plates have slower cell cycle progression, and the resting (G0) phase of the cell cycle is essential in skeletal tissues to produce the extracellular matrix that forms a scaffold for the forming bone (Ağırdil 2020). Primary cilia are mechanosensors and cellular signalling centres present on most mammalian cells including those of skeletal lineage, shown to modulate ECM homeostasis and planar cell polarity, and consisting of nine doublets of microtubules arranged on the periphery of the cilium (Seeger-Nukpezah and Golemis 2012). In contrast to the beating cilia, the primary cilia are non-motile and do not contain the central microtubule doublet (Kim and Dynlacht

2013). They are present on cells in the G1 phase of the cell cycle, and they disassemble in the G2/M phase and reassemble again post division (Plotnikova et al. 2009). Their basal bodies originate from mature mother centrioles which form part of the centrosome during cell division (Kim and Dynlacht 2013). The skeleton is primarily a mechanical organ that adapts to mechanical stimuli, with increased load resulting in increased ECM production and mechanical force adaptations. Osteoblasts, osteocytes, chondrocytes, and mesenchymal stem cells (MSCs) have all been shown to respond to mechanical stimuli (Tummala et al. 2010; Nguyen and Jacobs 2013; Quadri and Upadhyai 2023). Primary cilia present on skeletal cells sense changes in dynamic fluid flow in the tissue fluid caused by the increased load and propagate this response (Huber and Cormier-Daire 2012). It has been shown that the primary cilium acts as a signalling nexus where several developmental signals (IHH, WNT, FGF) converge (Tummala et al. 2010; Quadri and Upadhyai 2023). The IHH signalling is one of the major regulatory pathways that drive chondrocyte differentiation and proliferation at the growth plate, regulating growth plate height (St-Jacques et al. 1999). Dynein and several kinesins are involved in the primary cilia function and in the intraflagellar transport (IFT) and, through modulation of ciliary trafficking, play roles in signalling, mechanosensing, ECM production, and remodelling (Collins and Wann 2020).

Mutations in genes encoding dynein 2 (the heavy chain *DYNC2H1* and the intermediate and light chain genes *WDR60, WDR34*, *DYNC2LI1*, and *TCTEX1D2)*, involved in the intraflagellar transport (IFT) in skeletal cells, have all been linked to skeletal ciliopathies, varying in severity from lethal shortened limbs, cystic kidneys, and retinal degeneration to milder Jeune asphyxiating thoracic dysplasia (Schmidts and Mitchison 2018). Mutations in *DYNC2H1* have been reported in several cases of peri- or neonatally lethal short-rib polydactyly syndrome type II or asphyxiating thoracic dystrophy (El Hokayem et al. 2012; Badiner et al. 2017) and present with short ovoid tibiae or tibial agenesis (El Hokayem et al. 2012). Patient fibroblasts show a defect in the intraflagellar retrograde transport, but the cilia length and number are not affected (Taylor et al. 2015). Deletion of *Dync2h1* in mice resulted in lethality around embryonic day 10.5, due to a block in retrograde ciliary protein transport and the resulting loss of sonic hedgehog (SHH) signalling in the neural tube. Interestingly, in a molecular rescue experiment, lowering the anterograde transport partially restored the phenotype (Ocbina et al. 2011). Mutations in *WDR60* (McInerney-Leo et al. 2013), *WDR34* (Huber et al. 2013; Schmidts et al. 2013), *DYNC2LI1* (Kessler et al. 2015; Taylor et al. 2015; Niceta et al. 2018; Bryson et al. 2021), and *TCTEX1D2* (Schmidts et al. 2015) also result in (milder) forms of short-rib dysplasia with or without polydactyly. WDR60 has been implicated in cilia and retrograde transport, and WDR34 has been found around the centrioles and basal bodies in ATDC5 cells and shown to interact with TAK1, a kinesin essential for the activation of the NFκB pathway (Schmidts et al. 2013). *DYNC2LI1* is expressed in foetal chondrocytes (Kessler et al. 2015) and in the bone, specifically in the perichondrium, periosteum, and primary spongiosa (Taylor et al. 2015), overlapping with the *WDR34* and *DYNC2H1* expression. DYNC2LI1 has been detected in the primary cilia and basal bodies and mutations in *DYNC2LI1* result in the accumulation of several cilia-associated proteins, including a kinesin KIF3A, near the cilia tip in the patient fibroblasts, further highlighting the essential cooperation of dyneins and kinesins in the intraflagellar transport (Taylor et al. 2015). The Jeune asphyxiating thoracic dysplasia mutations in TCTEX1D2 result in impaired retrograde IFT, and deletion of *Tctex1d2* in zebrafish resulted in abnormalities in the craniofacial skeleton development (Schmidts et al. 2015).

Kinesins have also been implicated in the modulation of morphogenetic signals that regulate proliferation and specifically in the regulation of developmental signals that converge on the primary cilium. Interestingly, in the interphase, a pool of KIF11, required for spindle formation during cell division, localises at the basal bodies of the primary cilium, and depletion of KIF11 affects cilium length and organisation (Zalenski et al. 2020). Moreover, KIF7-related skeletal conditions show a phenotyping overlap with ciliopathies. Patients with autosomal recessive *KIF7* mutations present with dysmorphic features (macrocephaly, hydrocephaly, or anencephaly), multiple epiphyseal dysplasia, *genu valgum* (knock knees), polydactyly, clinodactyly, and pectus excavatum (Ali et al. 2012). Acrocallosal syndrome, catalogued under the "Polydactyly, syndactyly and triphalangism" group in the revised nosology (Unger et al. 2023), is an autosomal recessive syndrome, caused by a homozygous mutation in *KIF7*, with both brain abnormalities and skeletal abnormalities, postaxial polydactyly of the hands, and preaxial polydactyly of the feet (Schinzel and Schmid 1980). Acrocallosal syndrome has been classified as a ciliopathy and demonstrates a phenotypic overlap with the Joubert syndrome (Putoux et al. 2011). Al-Gazali-Bakalinova syndrome results from missense mutations in *KIF7* and is characterised by macrocephaly, low-set ears, pectus excavatum, spindle-shaped fingers, joint swelling, clinodactyly, *genu valgum*, multiple epiphyseal dysplasia, and neurological defects (Ali et al. 2012). A microdeletion in the coiled-coil region of KIF7 results in hydrolethalus, characterised by hydrocephaly, upper limb postaxial polydactyly, and hallux duplication (Putoux et al. 2011). KIF7B has been shown to play a role in modulating hedgehog signalling in the cartilage growth plates (Hsu et al. 2011) and to accumulate at the distal tip of the primary cilia upon hedgehog stimulation (He et al. 2014). Mice with *Kif7* deletion showed expanded growth plate hypertrophic zones and decreased chondrocyte proliferation, indicating a dysregulation of the IHH/PTHrP signalling pathway essential for the maintenance of the correct growth plate zones and the growth dynamic and suggesting that KIF7 acts as a positive regulator of IHH signalling (Hsu et al. 2011). Inactivation of KIF7 in mouse cartilage led to a downregulation of HH target genes *Gli1* and *Ptch1*, and an increase in the level of Sufu-Gli complexes suggests that when the HH pathway is activated in chondrocytes, KIF7 plays a key role in excluding them from the primary cilia. Further studies in a chondrocyte-specific Sufu knockout mouse, which presented with decreased ossification and altered growth plate dynamics, showed that KIF7 can also directly inhibit Gli-mediated transcriptional activation in the absence of Sufu (Hsu et al. 2011).

KIF3A, part of the KIF3 complex, is an essential component of the intraflagellar transport (IFT) of primary cilia, important for nodal flow and establishment of the right-left symmetry during embryonic development (Song et al. 2007). KIF3A plays a role in both endochondral and intramembranous ossification. Several KIF3A conditional knockout mouse models presented with shortened skeletal elements, polydactyly, poorly defined or missing distal phalanges, a fusion of the multiple tarsal bones, abnormal distribution of hypertrophic chondrocytes in the growth plates, ectopic cartilage formation, and excessive intramembranous ossification (Kolpakova-Hart et al. 2007; Song et al. 2007; Qiu et al. 2012). Prx1-Cre; *Kif3a* null pups displayed forelimb polydactyly together with complete loss of the anterior-posterior polarity and profound shortening of the limb bones, syndactyly with short and dysmorphic digits, and ill-defined phalanges. Interestingly, SHH pattern nor level was affected; however, the mutants had very low levels of *Ptch1* transcript in the forelimb mesenchyme, and GLI3 processing in the absence of KIF3A was impaired in the KIF3A null limb mesenchyme (Kolpakova-Hart et al. 2007). The chaotic organisation of the mutant chondrocytes in the growth plates suggested that the chondrocytes may be lacking in the ability to determine orientation. Planar cell polarity (PCP) pathway is one of the mechanisms for cell orientation (Devenport 2014). Both PCP and Indian hedgehog (IHH) signalling act via the primary cilia, and it has been suggested that the convergence of these signals in the cilia may allow chondrocytes to control orientation synchronously with response to IHH. Interestingly, deletion of *Kif3a* in bone progenitor cells resulted in decreased osteoblast cilia length and incidence, disrupted hedgehog and WNT signalling, and dysregulated calcium flux and led to osteopenia and disrupted osteoblast differentiation (Qiu et al. 2012).

Kinesins are not only essential for the IFT and cilia-based signalling but are also involved in the maintenance of the primary cilia length and organisation. Depolymerizing kinesins (kinesin-8 and kKinesin-13) influence the rate of the abrupt switch from growth to shortening of the microtubule plus ends (termed "microtubule catastrophe") which is essential for the microtubule function by promoting tubulin dimer removal from filament ends (Moores et al. 2006; Brun et al. 2009; Meadows et al. 2018; Belsham et al. 2022). Kinesin-4, KIF7, has been shown to promote the microtubule catastrophe and regulate primary cilia length by suppressing microtubule polymerisation (He et al. 2014). Mutations in KIF4A and KIF11, which are also involved in ciliogenesis, result in facial dysmorphism, hydrocephalus, and neurological defects. Interestingly, KIF4A is also involved in the translocation of the cytokinesis PRC1 (protein regulating cytokinesis 1) to the microtubule ends in the mitotic spindle, the regulation of PARP1 activity in the developing brain, and the recycling of the L1CAM, further emphasising the pleiotropic roles of the kinesin family (Kurasawa et al. 2004; Zhu and Jiang 2005; Kalantari et al. 2021; Wan et al. 2023).

11.4 Transport of Membrane-Bound Organelles

In the mammalian ciliated cells, the Golgi apparatus is often localised close to the basal body (Collins and Wann 2020). Interestingly, recent studies also show that the Golgi apparatus is interconnected with the primary cilium and that Golgi-driven protein modifications may impact the secretory pathways (Stevenson 2023). Moreover, members of the intraflagellar transport, such as the IFT20 protein, have been shown to be involved in Golgi transport and in the nucleation of the Golgi-associated microtubules (Nishita et al. 2017). Cytoplasmic dynein has been shown to carry membrane-bound organelles and vesicles, including Golgi vesicles and mitochondria, and is involved in the positioning of the Golgi apparatus and the nucleus amongst others (Reck-Peterson et al. 2018). The diversity of the cargo and processes in which the cytoplasmic dynein is involved is modulated by the associated light chains and adaptor proteins. For example, a knockdown of *DYNC1LI1* in HeLa cells resulted in impaired ER export, whereas depletion of DYNC1LI2 affected cytokinesis and endosome distribution (Palmer et al. 2009).

Kinesins have also been implicated in the positioning and intracellular transport of membrane-bound organelles. KIF1C has been shown to be involved in the tethering of the Golgi apparatus and maintaining the integrity of the Golgi membranes (Lee et al. 2015). KIF5B has also been shown to have additional roles, including modulating the intracellular localisation of mitochondria (Wang et al. 2015). The extraembryonic cells null for KIF5B have impaired lysosomal dispersion and mitochondria clustered in the perinuclear region. Kif5bLoF zebrafish mutants had decreased mTOR pathway activation together with a decline in lysosomal function leading to apoptosis and depleted hypertrophic chondrocytes and indicating a role for KIF5B in maintaining not only cell proliferation, but also autophagy (Santos-Ledo et al. 2017). KIF5B has also been implicated in osteogenesis imperfecta, and patients with de novo heterozygous mutations in *KIF5B* presented with defective mitochondria transport and dysregulated autophagy (Marom et al. 2023). It has been postulated that the disease-associated gain of function KIF5B variants affect cilium biogenesis but also disrupt the organisation of several membrane-bound organelles including the lysosomes, autophagosomes, and mitochondria, highlighting the continuity of the membrane-associated cellular machinery and indicating broader roles for KIF5B in endochondral ossification (Flex et al. 2023).

11.5 Kinesins and Dyneins in Membrane Receptor Dynamics

Endosome trafficking of proteins and lipids is an essential step in molecular sorting, important in receptor recycling and degradation (Tu et al. 2020). Dyneins also play a role in the transport of intracellular vesicles and have been shown to transport Rab6-positive vesicles towards the minus ends of the microtubules (Matanis et al. 2002; Wanschers et al. 2008; Schlager et al. 2014). Cytoplasmic dynein (DYNC1H1) plays a role in endosomal transport (Xiang et al. 2015), and several kinesin proteins are also involved in this process. KIF13B, a primary cilium-related kinesin, has been implicated in regulating endocytosis and recruitment of LDL receptor–related protein 1 (LRP1) to the caveolae (Kanai et al. 2014; Mills et al. 2019). Interestingly, IFT88, another intraflagellar transport protein, has also been implicated in the process by supporting the formation of the ciliary pocket (Coveney et al. 2018). During skeletal development, LRP1 is essential for the endocytosis of transmembrane proteins (Theret et al. 2017) and matrix degradation and remodelling through its binding to MMP13 and ADAMTS5 metalloproteinases (Yamamoto et al. 2016; Yamamoto et al. 2022). Deletion of LRP1 has been shown to impact chondrogenesis and development of the articulated joints in mice (Yan et al. 2022), and the recently resolved LRP1 ligandome includes several ECM molecules, as well as members of the WNT, TGFβ, and BMP signalling pathways that are involved in the regulation of skeletal development (Yamamoto et al. 2022).

Furthermore, kinesins have a role in the anterograde transport of the membrane-bound receptors to the cell surface. KIF5B, together with KIF13B, plays a role in the secretory vesicle transport in HeLa cells and in the polarised MDCK and PC12 cells, and KIF13A has been shown to be involved in the trafficking of vesicles containing the mannose-6-phosphate receptor in 3 T3 fibroblasts, NRK epithelial cells, and colorectal L cells (Nakagawa, Setou et al. 2000, Jaulin et al. 2007, Astanina and Jacob 2010, Burgo et al. 2012, Serra-Marques et al. 2020). The concerted action of the two kinesins helps transport the secretory vesicles from the Golgi apparatus to the cell periphery. KIF13B is also involved in trafficking VEGFR2 receptor from the Golgi apparatus to the plasma membrane, which was demonstrated by the knockdown of KIF13B affecting the sprouting of HUVEC cells and leading to persistent VEGFR2 localisation to the Golgi and disrupted translocation of VEGFR2 to the lysosomes (Yamada et al. 2014). It has also been implicated in the anterograde transport of TRPV1, a calcium-permeable cation channel involved in heat sensing, from the Golgi apparatus to the cell periphery in the neurons, especially upon inflammation (Xing et al. 2012). Interestingly, a novel role for the chromokinesin KIF22 in coordinating membrane receptor dynamics has been proposed in epithelial cells and in lung carcinomas. KIF22 has been shown to modulate EGFR internalisation in human adenocarcinoma cell line A549 and to directly bind to the CAR receptor (important for the cell-cell adhesion in various tumours) and to the guanylate-binding protein 2 (GBP2) which plays a role in cancer proliferation and migration. This suggests that chromokinesins, similarly to the other kinesin families, may carry cargo other than the nucleic acids and play different roles outside the meiosis/mitosis (Pike et al. 2018; Ren et al. 2022)). KIF16B, shown to associate with the early endosomes (Blatner et al. 2007), has also been shown to assist in the microtubule-associated traffic and is important for stem cell survival in early development (Ueno et al. 2011). Deletion of KIF16B in mouse embryonic stem cells led to developmental arrest at the blastocyst stage. KIF16B null embryoid bodies (EB) were significantly underdeveloped, with significant cell death and a reduction in the amount of basement membrane and GATA6 expression (Ueno et al. 2011). The surface levels of FGFR2 receptor in the knockout mouse embryonic fibroblasts were reduced, with FGFR2 in the KIF16B-null cells accumulated in perinuclear dots, suggesting that KIF16B is essential for the FGFR2 transport towards the cell periphery. A yeast two-hybrid assay found Rab14 binding directly to the tail domain of KIF16B, indicating that Rab14 modulates this motor-cargo association. Subsequently, KIF16B/Rab14-mediated Golgi-to-endosome transport of FGFR2-carrying vesicles has been shown to critically regulate both the surface presentation of FGFR2 and FGF signalling transduction, which is essential for embryogenesis in early post-implantation development (Ueno et al. 2011). All these studies support a role for kinesins in receptor dynamics and modulation of the developmental signalling pathways.

11.6 Molecular Motors and Extracellular Matrix Homeostasis

Several kinesin molecules have also been implicated in the secretion of extracellular components and specifically in the transport of secretory vesicles containing extracellular matrix components. KIF5B, recently implicated in kyphomelic dysplasia (Itai et al. 2022), was the only kinesin protein increased during TGFβ-stimulated mesothelium to mesenchyme transition in human pleural mesothelium, which coincided with an increase in type I production and secretion, indicative of fibrosis. Using a combination of proximity-ligation assays, siRNA, confocal microscopy, and fluorescent labelling, it has been shown that KIF5B transports type I collagen–containing vesicles along the microtubules in the human pleural mesothelial cells (Kamata et al. 2017). Kinesin-1 and kinesin-3 proteins have also been shown to play a role in protein secretion, specifically in the secretion of basement membrane proteins in *Drosophila melanogaster* epithelium, whereby the kinesin complexes helped move Rab10-positive secretory vesicles along the microtubules of the polarised epithelial cells and control the site of exocytosis of basement membrane proteins (Zajac and Horne-Badovinac 2022). KIF5B, together with KIF13B, has also been shown to carry Rab6-positive secretory vesicles from the Golgi apparatus to the cell periphery in HeLa cells (Serra-Marques et al. 2020). Interestingly, Rab6 vesicles have been implicated in type X collagen exocytosis (Fourriere et al. 2019).

Motor molecules can also modulate matrix degradation and remodelling. KIF5B and KIF3A/KIF3B kinesins, together with dynein, have been shown to be involved in the bi-directional movement of MT1-MMP-containing vesicles along the microtubules in primary macrophages and in human fibrosarcoma cell line (Wiesner et al. 2010; Gifford et al. 2022). The kinesin proteins are essential for the MT1-MMP exposure on the cell surface and important for receptor shedding (of CD44 and syndecan-1) and extracellular matrix degradation. Kinesin proteins, for example, KIF9 and KIF1C, have also been implicated in podosome dynamics (Cornfine et al. 2011; Efimova et al. 2014). Podosomes are actin-rich structures present on the cell periphery of mammalian cells, including macrophages, osteoclasts, and smooth muscle cells, and are involved in matrix metalloproteinase recruitment (Linder 2007). KIF3, KIF9, and KIF13 have been shown to be involved in the MT1-MMP trafficking in the fibrosarcoma HT1080 cells, whereby an siRNA knockdown of KIF9 resulted in increased extracellular collagen degradation, whereas downregulation of KIF3 and KIF13 had the opposite effect (Gifford et al. 2022), demonstrating a role for kinesins in modulation of ECM degradation in musculoskeletal tissues.

11.7 Conclusions

Mutations in several motor proteins result in musculoskeletal complications and impact extracellular matrix production and skeletal cell differentiation by a variety of molecular mechanisms. Specifically, it has been previously shown that mutations in the primary cilia-associated dynein *DYNC2H1*, several dynein light chains, and three kinesin genes (*KIF5B*, *KIF7*, and *KIF22*) lead to chondrodysplasias, affecting limb development and endochondral ossification, whereas mutations in *KIF10*, *KIF11*, *KIF14*, *KIF15*, and *KIF16B* result in multiple tissue complications and dysmorphic cranial development. Microtubule-associated motor proteins have diverse roles in cellular biology and homeostasis, and mutations in dynein and kinesin genes affect skeletal health via a plethora of molecular mechanisms. Several microtubule motors are involved in cytokinesis, and mutations in these lead to a decrease in chondrocyte proliferation. Several kinesins have been implicated in ciliogenesis, which explains the phenotypic overlap of several "kinesinopathies" and ciliopathies. Motor proteins also play essential roles in endocytosis and receptor recycling, playing a role in the modulation of morphogenetic signals during development, and are important for the secretion of extracellular matrix components. Whilst the specific roles of microtubule motors in skeletal development and health remain largely unknown, the complexity of the microtubule motors, compounded with the alternative splicing and posttranslational modification of their adaptor proteins, and differences in the spatial and temporal expression of individual heavy chain genes further emphasise the importance and the diversity of the roles of these molecules in development.

Acknowledgements This work was supported by the European Community's Horizon-MSCA-DN-01-01 under grant agreement n. 101072766 (CHANGE).

Compliance with Ethical Standards/Ethical Requirements We declare no conflict of interest.

References

Ağırdil Y (2020) The growth plate: a physiologic overview. EFORT Open Rev 5(8):498–507

Al-Gazali LI, Bakalinova D (1998) Autosomal recessive syndrome of macrocephaly, multiple epiphyseal dysplasia and distinctive facial appearance. Clin Dysmorphol 7(3):177–184

Ali BR, Silhavy JL, Akawi NA, Gleeson JG, Al-Gazali L (2012) A mutation in KIF7 is responsible for the autosomal recessive syndrome of macrocephaly, multiple epiphyseal dysplasia and distinctive facial appearance. Orphanet J Rare Dis 7:27

Almeida AC, Maiato H (2018) Chromokinesins. Curr Biol 28(19):R1131–r1135

Alsabban AH, Morikawa M, Tanaka Y, Takei Y, Hirokawa N (2020) Kinesin Kif3b mutation reduces NMDAR subunit NR2A trafficking and causes schizophrenia-like phenotypes in mice. EMBO J 39(1):e101090

Asselin L, Rivera Alvarez J, Heide S, Bonnet CS, Tilly P, Vitet H, Weber C, Bacino CA, Baranaño K, Chassevent A, Dameron A, Faivre L, Hanchard NA, Mahida S, McWalter K, Mignot C, Nava C, Rastetter A, Streff H, Thauvin-Robinet C, Weiss MM, Zapata G, Zwijnenburg PJG, Saudou F, Depienne C, Golzio C, Héron D, Godin JD (2020) Mutations in the KIF21B kinesin gene cause neurodevelopmental disorders through imbalanced canonical motor activity. Nat Commun 11(1):2441

Astanina K, Jacob R (2010) KIF5C, a kinesin motor involved in apical trafficking of MDCK cells. Cell Mol Life Sci 67(8):1331–1342

Badiner N, Taylor SP, Forlenza K, Lachman RS, Bamshad M, Nickerson D, Cohn DH, Krakow D (2017) Mutations in DYNC2H1, the cytoplasmic dynein 2, heavy chain 1 motor protein gene, cause short-rib polydactyly type I, Saldino-Noonan type. Clin Genet 92(2):158–165

Balabanian L, Chaudhary AR, Hendricks AG (2018) Traffic control inside the cell: microtubule-based regulation of cargo transport. Biochemist 40(2):14–17

Bartoli KM, Jakovljevic J, Woolford JL Jr, Saunders WS (2011) Kinesin molecular motor Eg5 functions during polypeptide synthesis. Mol Biol Cell 22(18):3420–3430

Belsham HR, Alghamdi HM, Dave N, Rathbone AJ, Wickstead B, Friel CT (2022) A synthetic ancestral kinesin-13 depolymerizes microtubules faster than any natural depolymerizing kinesin. Open Biol 12(8):220133

Bianchi S, van Riel WE, Kraatz SH, Olieric N, Frey D, Katrukha EA, Jaussi R, Missimer J, Grigoriev I, Olieric V, Benoit RM, Steinmetz MO, Akhmanova A, Kammerer RA (2016) Structural basis for misregulation of kinesin KIF21A autoinhibition by CFEOM1 disease mutations. Sci Rep 6:30668

Birtel J, Gliem M, Mangold E, Tebbe L, Spier I, Müller PL, Holz FG, Neuhaus C, Wolfrum U, Bolz HJ, Charbel Issa P (2017) Novel insights into the phenotypical Spectrum of KIF11-associated retinopathy, including a new form of retinal ciliopathy. Invest Ophthalmol Vis Sci 58(10):3950–3959

Blangy A, Lane HA, d'Hérin P, Harper M, Kress M, Nigg EA (1995) Phosphorylation by p34cdc2 regulates spindle association of human Eg5, a kinesin-related motor essential for bipolar spindle formation in vivo. Cell 83(7):1159–1169

Blatner NR, Wilson MI, Lei C, Hong W, Murray D, Williams RL, Cho W (2007) The structural basis of novel endosome anchoring activity of KIF16B kinesin. EMBO J 26(15):3709–3719

Blocker A, Severin FF, Burkhardt JK, Bingham JB, Yu H, Olivo JC, Schroer TA, Hyman AA, Griffiths G (1997) Molecular requirements for bi-directional movement of phagosomes along microtubules. J Cell Biol 137(1):113–129

Bost-Usinger L, Chen RJ, Hillman D, Park H, Burnside B (1997) Multiple kinesin family members expressed in teleost retina and RPE include a novel C-terminal kinesin. Exp Eye Res 64(5):781–794

Brouwers N, Mallol Martinez N, Vernos I (2017) Role of Kif15 and its novel mitotic partner KBP in K-fiber dynamics and chromosome alignment. PLoS One 12(4):e0174819

Brun L, Rupp B, Ward JJ, Nédélec F (2009) A theory of microtubule catastrophes and their regulation. Proc Natl Acad Sci USA 106(50):21173–21178

Bryson LJ, Flynn DM, Sabharwal A, Ahmed SF, Kinning E (2021) A child with congenital short gut associated with DYNC2LI1 ciliopathy. Clin Dysmorphol 30(1):66–68

Buchan JG, Gray RS, Gansner JM, Alvarado DM, Burgert L, Gitlin JD, Gurnett CA, Goldsmith MI (2014) Kinesin family member 6 (kif6) is necessary for spine development in zebrafish. Dev Dyn 243(12):1646–1657

Burgo A, Proux-Gillardeaux V, Sotirakis E, Bun P, Casano A, Verraes A, Liem RK, Formstecher E, Coppey-Moisan M, Galli T (2012) A molecular network for the transport of the TI-VAMP/VAMP7 vesicles from cell center to periphery. Dev Cell 23(1):166–180

Carleton M, Mao M, Biery M, Warrener P, Kim S, Buser C, Marshall CG, Fernandes C, Annis J, Linsley PS (2006) RNA interference-mediated silencing of mitotic kinesin KIF14 disrupts cell cycle progression and induces cytokinesis failure. Mol Cell Biol 26(10):3853–3863

Coles GL, Baglia LA, Ackerman KG (2015) KIF7 controls the proliferation of cells of the respiratory airway through distinct microtubule dependent mechanisms. PLoS Genet 11(10):e1005525

Collins I, Wann AKT (2020) Regulation of the extracellular matrix by ciliary machinery. Cells 9(2)

Cooper GM (2000) Actin, myosin, and cell movement. The cell: a molecular approach. Sinauer Associates, Sunderland (MA)

Cornfine S, Himmel M, Kopp P, El Azzouzi K, Wiesner C, Krüger M, Rudel T, Linder S (2011) The kinesin KIF9 and reggie/flotillin proteins regulate matrix degradation by macrophage podosomes. Mol Biol Cell 22(2):202–215

Coveney CR, Collins I, Mc Fie M, Chanalaris A, Yamamoto K, Wann AKT (2018) Cilia protein IFT88 regulates extracellular protease activity by optimizing LRP-1-mediated endocytosis. FASEB J 32(12):fj201800334

Cross JA, Woolfson DN, Dodding MP (2021) Kinesin-1 captures RNA cargo in its adaptable coils. Genes Dev 35(13–14):937–939

Cruz NM, Reddy R, McFaline-Figueroa JL, Tran C, Fu H, Freedman BS (2022) Modelling ciliopathy phenotypes in human tissues derived from pluripotent stem cells with genetically ablated cilia. Nat Biomed Eng 6(4):463–475

Debernardi S, Fontanella E, De Gregorio L, Pierotti MA, Delia D (1997) Identification of a novel human kinesin-related gene (HK2) by the cDNA differential display technique. Genomics 42(1):67–73

Decarreau J, Wagenbach M, Lynch E, Halpern AR, Vaughan JC, Kollman J, Wordeman L (2017) The tetrameric kinesin Kif25 suppresses pre-mitotic centrosome separation to establish proper spindle orientation. Nat Cell Biol 19(4):384–390

Devenport D (2014) The cell biology of planar cell polarity. J Cell Biol 207(2):171–179

Dodding MP, Way M (2011) Coupling viruses to dynein and kinesin-1. EMBO J 30(17):3527–3539

Dorner C, Ciossek T, Müller S, Møller PH, Ullrich A, Lammers R (1998) Characterization of KIF1C, a new kinesin-like protein involved in vesicle transport from the Golgi apparatus to the endoplasmic reticulum. J Biol Chem 273(32):20267–20275

Driskell OJ, Mironov A, Allan VJ, Woodman PG (2007) Dynein is required for receptor sorting and the morphogenesis of early endosomes. Nat Cell Biol 9(1):113–120

Drpic D, Pereira AJ, Barisic M, Maresca TJ, Maiato H (2015) Polar ejection forces promote the conversion from lateral to end-on kinetochore-microtubule attachments on mono-oriented chromosomes. Cell Rep 13(3):460–468

Echard A, Jollivet F, Martinez O, Lacapère JJ, Rousselet A, Janoueix-Lerosey I, Goud B (1998) Interaction of a Golgi-associated kinesin-like protein with Rab6. Science 279(5350):580–585

Efimova N, Grimaldi A, Bachmann A, Frye K, Zhu X, Feoktistov A, Straube A, Kaverina I (2014) Podosome-regulating kinesin KIF1C translocates to the cell periphery in a CLASP-dependent manner. J Cell Sci 127(Pt 24):5179–5188

El Hokayem J, Huber C, Couvé A, Aziza J, Baujat G, Bouvier R, Cavalcanti DP, Collins FA, Cordier MP, Delezoide AL, Gonzales M, Johnson D, Le Merrer M, Levy-Mozziconacci A, Loget P, Martin-Coignard D, Martinovic J, Mortier GR, Perez MJ, Roume J, Scarano G, Munnich A, Cormier-Daire V (2012) NEK1 and DYNC2H1 are both involved in short rib polydactyly Majewski type but not in Beemer Langer cases. J Med Genet 49(4):227–233

Eskova A, Knapp B, Matelska D, Reusing S, Arjonen A, Lisauskas T, Pepperkok R, Russell R, Eils R, Ivaska J, Kaderali L, Erfle H, Starkuviene V (2014) An RNAi screen identifies KIF15 as a novel regulator of the endocytic trafficking of integrin. J Cell Sci 127(Pt 11):2433–2447

Ferro LS, Can S, Turner MA, ElShenawy MM, Yildiz A (2019) Kinesin and dynein use distinct mechanisms to bypass obstacles. elife 8

Filges I, Nosova E, Bruder E, Tercanli S, Townsend K, Gibson WT, Röthlisberger B, Heinimann K, Hall JG, Gregory-Evans CY, Wasserman WW, Miny P, Friedman JM (2014) Exome sequencing identifies mutations in KIF14 as a novel cause of an autosomal recessive lethal fetal ciliopathy phenotype. Clin Genet 86(3):220–228

Flex E, Albadri S, Radio FC, Cecchetti S, Lauri A, Priolo M, Kissopoulos M, Carpentieri G, Fasano G, Venditti M, Magliocca V, Bellacchio E, Welch CL, Colombo PC, Kochav SM, Chang R, Barrick R, Trivisano M, Micalizzi A, Borghi R, Messina E, Mancini C, Pizzi S, De Santis F, Rosello M, Specchio N, Compagnucci C, McWalter K, Chung WK, Del Bene F, Tartaglia M (2023) Dominantly acting KIF5B variants with pleiotropic cellular consequences cause variable clinical phenotypes. Hum Mol Genet 32(3):473–488

Fourriere L, Kasri A, Gareil N, Bardin S, Bousquet H, Pereira D, Perez F, Goud B, Boncompain G, Miserey-Lenkei S (2019) RAB6 and microtubules restrict protein secretion to focal adhesions. J Cell Biol 218(7):2215–2231

Gan H, Xue W, Gao Y, Zhu G, Chan D, Cheah KSE, Huang J (2019) KIF5B modulates central spindle organization in late-stage cytokinesis in chondrocytes. Cell Biosci 9:85

Gifford V, Woskowicz A, Ito N, Balint S, Lagerholm BC, Dustin ML, Itoh Y (2022) Coordination of two kinesin superfamily motor proteins, KIF3A and KIF13A, is essential for pericellular matrix degradation by membrane-type 1 matrix metalloproteinase (MT1-MMP) in cancer cells. Matrix Biol 107:1–23

Gross SP, Tuma MC, Deacon SW, Serpinskaya AS, Reilein AR, Gelfand VI (2002) Interactions and regulation of molecular motors in Xenopus melanophores. J Cell Biol 156(5):855–865

Gualeni B, Rajpar MH, Kellogg A, Bell PA, Arvan P, Boot-Handford RP, Briggs MD (2013) A novel transgenic mouse model of growth plate dysplasia reveals that decreased chondrocyte proliferation due to chronic ER stress is a key factor in reduced bone growth. Dis Model Mech 6(6):1414–1425

Gumy LF, Chew DJ, Tortosa E, Katrukha EA, Kapitein LC, Tolkovsky AM, Hoogenraad CC, Fawcett JW (2013) The kinesin-2 family member KIF3C regulates microtubule dynamics and is required for axon growth and regeneration. J Neurosci 33(28):11329–11345

Gurel PS, Hatch AL, Higgs HN (2014) Connecting the cytoskeleton to the endoplasmic reticulum and Golgi. Curr Biol 24(14):R660–r672

Hall CM, Elcioglu NH, MacDermot KD, Offiah AC, Winter RM (2002) Spondyloepimetaphyseal dysplasia with multiple dislocations (Hall type): three further cases and evidence of autosomal dominant inheritance. J Med Genet 39(9):666–670

Hameed A, Bennett E, Ciani B, Hoebers LP, Milner R, Lawrie A, Francis SE, Grierson AJ (2013) No evidence for cardiac dysfunction in Kif6 mutant mice. PLoS One 8(1):e54636

Harada A, Takei Y, Kanai Y, Tanaka Y, Nonaka S, Hirokawa N (1998) Golgi vesiculation and lysosome dispersion in cells lacking cytoplasmic dynein. J Cell Biol 141(1):51–59

He M, Agbu S, Anderson KV (2017) Microtubule motors drive hedgehog signaling in primary cilia. Trends Cell Biol 27(2):110–125

He M, Subramanian R, Bangs F, Omelchenko T, Liem KF Jr, Kapoor TM, Anderson KV (2014) The kinesin-4 protein Kif7 regulates mammalian hedgehog signalling by organizing the cilium tip compartment. Nat Cell Biol 16(7):663–672

Hirokawa N, Noda Y, Tanaka Y, Niwa S (2009) Kinesin superfamily motor proteins and intracellular transport. Nat Rev Mol Cell Biol 10(10):682–696

Hoepfner S, Severin F, Cabezas A, Habermann B, Runge A, Gillooly D, Stenmark H, Zerial M (2005) Modulation of receptor recycling and degradation by the endosomal kinesin KIF16B. Cell 121(3):437–450

Hooikaas PJ, Damstra HG, Gros OJ, van Riel WE, Martin M, Smits YT, van Loosdregt J, Kapitein LC, Berger F, Akhmanova A (2020) Kinesin-4 KIF21B limits microtubule growth to allow rapid centrosome polarization in T cells. elife 9

Horiguchi K, Hanada T, Fukui Y, Chishti AH (2006) Transport of PIP3 by GAKIN, a kinesin-3 family protein, regulates neuronal cell polarity. J Cell Biol 174(3):425–436

Hsu SH, Zhang X, Yu C, Li ZJ, Wunder JS, Hui CC, Alman BA (2011) Kif7 promotes hedgehog signaling in growth plate chondrocytes by restricting the inhibitory function of Sufu. Development 138(17):3791–3801

Huber C, Cormier-Daire V (2012) Ciliary disorder of the skeleton. Am J Med Genet C Semin Med Genet 160c(3):165–174

Huber C, Wu S, Kim AS, Sigaudy S, Sarukhanov A, Serre V, Baujat G, Le Quan Sang KH, Rimoin DL, Cohn DH, Munnich A, Krakow D, Cormier-Daire V (2013) WDR34 mutations that cause short-rib polydactyly syndrome type III/severe asphyxiating thoracic dysplasia reveal a role for the NF-κB pathway in cilia. Am J Hum Genet 93(5):926–931

Ishikawa K, Nagase T, Suyama M, Miyajima N, Tanaka A, Kotani H, Nomura N, Ohara O (1998) Prediction of the coding sequences of unidentified human genes. X. The complete sequences of 100 new cDNA clones from brain which can code for large proteins in vitro. DNA Res 5(3):169–176

Itai T, Wang Z, Nishimura G, Ohashi H, Guo L, Wakano Y, Sugiura T, Hayakawa H, Okada M, Saisu T, Kitta A, Doi H, Kurosawa K, Hotta Y, Hosono K, Sato M, Shimizu K, Takikawa K, Watanabe S, Ikeda N, Suzuki M, Fujita A, Uchiyama Y, Tsuchida N, Miyatake S, Miyake N, Matsumoto N, Ikegawa S (2022) De novo heterozygous variants in KIF5B cause kyphomelic dysplasia. Clin Genet 102(1):3–11

Iwata T, Chen L, Li C, Ovchinnikov DA, Behringer RR, Francomano CA, Deng CX (2000) A neonatal lethal mutation in FGFR3 uncouples proliferation and differentiation of growth plate chondrocytes in embryos. Hum Mol Genet 9(11):1603–1613

Jamain S, Quach H, Fellous M, Bourgeron T (2001) Identification of the human KIF13A gene homologous to drosophila kinesin-73 and candidate for schizophrenia. Genomics 74(1):36–44

Jaulin F, Xue X, Rodriguez-Boulan E, Kreitzer G (2007) Polarization-dependent selective transport to the apical membrane by KIF5B in MDCK cells. Dev Cell 13(4):511–522

Jordens I, Fernandez-Borja M, Marsman M, Dusseljee S, Janssen L, Calafat J, Janssen H, Wubbolts R, Neefjes J (2001) The Rab7 effector protein RILP controls lysosomal transport by inducing the recruitment of dynein-dynactin motors. Curr Biol 11(21):1680–1685

Kalantari S, Carlston C, Alsaleh N, Abdel-Salam GMH, Alkuraya F, Kato M, Matsumoto N, Miyatake S, Yamamoto T, Fares-Taie L, Rozet JM, Chassaing N, Vincent-Delorme C, Kang-Bellin A, McWalter K, Bupp C, Palen E, Wagner MD, Niceta M, Cesario C, Milone R, Kaplan J, Wadman E, Dobyns WB, Filges I (2021) Expanding the KIF4A-associated phenotype. Am J Med Genet A 185(12):3728–3739

Kalantari S, Filges I (2020) 'Kinesinopathies': emerging role of the kinesin family member genes in birth defects. J Med Genet 57(12):797–807

Kamal A, Almenar-Queralt A, LeBlanc JF, Roberts EA, Goldstein LS (2001) Kinesin-mediated axonal transport of a membrane compartment containing beta-secretase and presenilin-1 requires APP. Nature 414(6864):643–648

Kamata H, Tsukasaki Y, Sakai T, Ikebe R, Wang J, Jeffers A, Boren J, Owens S, Suzuki T, Higashihara M, Idell S, Tucker TA, Ikebe M (2017) KIF5A transports collagen vesicles of myofibroblasts during pleural fibrosis. Sci Rep 7(1):4556

Kamimoto T, Zama T, Aoki R, Muro Y, Hagiwara M (2001) Identification of a novel kinesin-related protein, KRMP1, as a target for mitotic peptidyl-prolyl isomerase Pin1. J Biol Chem 276(40):37520–37528

Kanai Y, Dohmae N, Hirokawa N (2004) Kinesin transports RNA: isolation and characterization of an RNA-transporting granule. Neuron 43(4):513–525

Kanai Y, Wang D, Hirokawa N (2014) KIF13B enhances the endocytosis of LRP1 by recruiting LRP1 to caveolae. J Cell Biol 204(3):395–408

Katoh M, Katoh M (2005) Characterization of KIF12 gene in silico. Oncol Rep 13(2):367–370

Katoh Y, Katoh M (2004) Characterization of KIF7 gene in silico. Int J Oncol 25(6):1881–1886

Kessler K, Wunderlich I, Uebe S, Falk NS, Gießl A, Brandstätter JH, Popp B, Klinger P, Ekici AB, Sticht H, Dörr HG, Reis A, Roepman R, Seemanová E, Thiel CT (2015) DYNC2LI1 mutations broaden the clinical spectrum of dynein-2 defects. Sci Rep 5:11649

Kim IG, Jun DY, Sohn U, Kim YH (1997) Cloning and expression of human mitotic centromere-associated kinesin gene. Biochim Biophys Acta 1359(3):181–186
Kim S, Dynlacht BD (2013) Assembling a primary cilium. Curr Opin Cell Biol 25(4):506–511
Kim W, Tang Y, Okada Y, Torrey TA, Chattopadhyay SK, Pfleiderer M, Falkner FG, Dorner F, Choi W, Hirokawa N, Morse HC 3rd (1998) Binding of murine leukemia virus gag polyproteins to KIF4, a microtubule-based motor protein. J Virol 72(8):6898–6901
Kobayashi T, Tsang WY, Li J, Lane W, Dynlacht BD (2011) Centriolar kinesin Kif24 interacts with CP110 to remodel microtubules and regulate ciliogenesis. Cell 145(6):914–925
Kolpakova-Hart E, Jinnin M, Hou B, Fukai N, Olsen BR (2007) Kinesin-2 controls development and patterning of the vertebrate skeleton by hedgehog- and Gli3-dependent mechanisms. Dev Biol 309(2):273–284
Konjikusic MJ, Lee C, Yue Y, Shrestha BD, Nguimtsop AM, Horani A, Brody S, Prakash VN, Gray RS, Verhey KJ, Wallingford JB (2023) Kif9 is an active kinesin motor required for ciliary beating and proximodistal patterning of motile axonemes. J Cell Sci 136(5)
Konjikusic MJ, Yeetong P, Boswell CW, Lee C, Roberson EC, Ittiwut R, Suphapeetiporn K, Ciruna B, Gurnett CA, Wallingford JB, Shotelersuk V, Gray RS (2018) Mutations in kinesin family member 6 reveal specific role in ependymal cell ciliogenesis and human neurological development. PLoS Genet 14(11):e1007817
Kovacs JJ, Whalen EJ, Liu R, Xiao K, Kim J, Chen M, Wang J, Chen W, Lefkowitz RJ (2008) Beta-arrestin-mediated localization of smoothened to the primary cilium. Science 320(5884):1777–1781
Kural C, Kim H, Syed S, Goshima G, Gelfand VI, Selvin PR (2005) Kinesin and dynein move a peroxisome in vivo: a tug-of-war or coordinated movement? Science 308(5727):1469–1472
Kurasawa Y, Earnshaw WC, Mochizuki Y, Dohmae N, Todokoro K (2004) Essential roles of KIF4 and its binding partner PRC1 in organized central spindle midzone formation. EMBO J 23(16):3237–3248
Lai F, Fernald AA, Zhao N, Le Beau MM (2000) cDNA cloning, expression pattern, genomic structure and chromosomal location of RAB6KIFL, a human kinesin-like gene. Gene 248(1–2):117–125
Lee PL, Ohlson MB, Pfeffer SR (2015) Rab6 regulation of the kinesin family KIF1C motor domain contributes to Golgi tethering. elife 4
Lee YM, Kim E, Park M, Moon E, Ahn SM, Kim W, Hwang KB, Kim YK, Choi W, Kim W (2010) Cell cycle-regulated expression and subcellular localization of a kinesin-8 member human KIF18B. Gene 466(1–2):16–25
Liem KF Jr, He M, Ocbina PJ, Anderson KV (2009) Mouse Kif7/Costal2 is a cilia-associated protein that regulates sonic hedgehog signaling. Proc Natl Acad Sci USA 106(32):13377–13382
Liljeholm M, Irvine AF, Vikberg AL, Norberg A, Month S, Sandström H, Wahlin A, Mishima M, Golovleva I (2013) Congenital dyserythropoietic anemia type III (CDA III) is caused by a mutation in kinesin family member, KIF23. Blood 121(23):4791–4799
Lin F, Hiesberger T, Cordes K, Sinclair AM, Goldstein LS, Somlo S, Igarashi P (2003) Kidney-specific inactivation of the KIF3A subunit of kinesin-II inhibits renal ciliogenesis and produces polycystic kidney disease. Proc Natl Acad Sci USA 100(9):5286–5291
Linder S (2007) The matrix corroded: podosomes and invadopodia in extracellular matrix degradation. Trends Cell Biol 17(3):107–117
Liu M, Pi H, Xi Y, Wang L, Tian L, Chen M, Xie J, Deng P, Zhang T, Zhou C, Liang Y, Zhang L, He M, Lu Y, Chen C, Yu Z, Zhou Z (2021) KIF5A-dependent axonal transport deficiency disrupts autophagic flux in trimethyltin chloride-induced neurotoxicity. Autophagy 17(4):903–924
Lo KY, Kuzmin A, Unger SM, Petersen JD, Silverman MA (2011) KIF1A is the primary anterograde motor protein required for the axonal transport of dense-core vesicles in cultured hippocampal neurons. Neurosci Lett 491(3):168–173
Luboshits G, Benayahu D (2005) MS-KIF18A, new kinesin; structure and cellular expression. Gene 351:19–28

Makrythanasis P, Maroofian R, Stray-Pedersen A, Musaev D, Zaki MS, Mahmoud IG, Selim L, Elbadawy A, Jhangiani SN, Coban Akdemir ZH, Gambin T, Sorte HS, Heiberg A, McEvoy-Venneri J, James KN, Stanley V, Belandres D, Guipponi M, Santoni FA, Ahangari N, Tara F, Doosti M, Iwaszkiewicz J, Zoete V, Backe PH, Hamamy H, Gleeson JG, Lupski JR, Karimiani EG, Antonarakis SE (2018) Biallelic variants in KIF14 cause intellectual disability with microcephaly. Eur J Hum Genet 26(3):330–339

Manning AL, Ganem NJ, Bakhoum SF, Wagenbach M, Wordeman L, Compton DA (2007) The kinesin-13 proteins Kif2a, Kif2b, and Kif2c/MCAK have distinct roles during mitosis in human cells. Mol Biol Cell 18(8):2970–2979

Marom R, Zhang B, Washington ME, Song IW, Burrage LC, Rossi VC, Berrier AS, Lindsey A, Lesinski J, Nonet ML, Chen J, Baldridge D, Silverman GA, Sutton VR, Rosenfeld JA, Tran AA, Hicks MJ, Murdock DR, Dai H, Weis M, Jhangiani SN, Muzny DM, Gibbs RA, Caswell R, Pottinger C, Cilliers D, Stals K, Eyre D, Krakow D, Schedl T, Pak SC, Lee BH (2023) Dominant negative variants in KIF5B cause osteogenesis imperfecta via downregulation of mTOR signaling. PLoS Genet 19(11):e1011005

Marszalek JR, Weiner JA, Farlow SJ, Chun J, Goldstein LS (1999) Novel dendritic kinesin sorting identified by different process targeting of two related kinesins: KIF21A and KIF21B. J Cell Biol 145(3):469–479

Matanis T, Akhmanova A, Wulf P, Del Nery E, Weide T, Stepanova T, Galjart N, Grosveld F, Goud B, De Zeeuw CI, Barnekow A, Hoogenraad CC (2002) Bicaudal-D regulates COPI-independent Golgi-ER transport by recruiting the dynein-dynactin motor complex. Nat Cell Biol 4(12):986–992

Matsushita M, Tanaka S, Nakamura N, Inoue H, Kanazawa H (2004) A novel kinesin-like protein, KIF1Bbeta3 is involved in the movement of lysosomes to the cell periphery in non-neuronal cells. Traffic 5(3):140–151

McInerney-Leo AM, Schmidts M, Cortés CR, Leo PJ, Gener B, Courtney AD, Gardiner B, Harris JA, Lu Y, Marshall M, Scambler PJ, Beales PL, Brown MA, Zankl A, Mitchison HM, Duncan EL, Wicking C (2013) Short-rib polydactyly and Jeune syndromes are caused by mutations in WDR60. Am J Hum Genet 93(3):515–523

Meadows JC, Messin LJ, Kamnev A, Lancaster TC, Balasubramanian MK, Cross RA, Millar JB (2018) Opposing kinesin complexes queue at plus tips to ensure microtubule catastrophe at cell ends. EMBO Rep 19(11)

Meng W, Mushika Y, Ichii T, Takeichi M (2008) Anchorage of microtubule minus ends to adherens junctions regulates epithelial cell-cell contacts. Cell 135(5):948–959

Metzger T, Gache V, Xu M, Cadot B, Folker ES, Richardson BE, Gomes ER, Baylies MK (2012) MAP and kinesin-dependent nuclear positioning is required for skeletal muscle function. Nature 484(7392):120–124

Mikenberg I, Widera D, Kaus A, Kaltschmidt B, Kaltschmidt C (2007) Transcription factor NF-kappaB is transported to the nucleus via cytoplasmic dynein/dynactin motor complex in hippocampal neurons. PLoS One 2(7):e589

Miki H, Setou M, Kaneshiro K, Hirokawa N (2001) All kinesin superfamily protein, KIF, genes in mouse and human. Proc Natl Acad Sci USA 98(13):7004–7011

Mills J, Hanada T, Hase Y, Liscum L, Chishti AH (2019) LDL receptor related protein 1 requires the I(3) domain of discs-large homolog 1/DLG1 for interaction with the kinesin motor protein KIF13B. Biochim Biophys Acta, Mol Cell Res 1866(12):118552

Min BJ, Kim N, Chung T, Kim OH, Nishimura G, Chung CY, Song HR, Kim HW, Lee HR, Kim J, Kang TH, Seo ME, Yang SD, Kim DH, Lee SB, Kim JI, Seo JS, Choi JY, Kang D, Kim D, Park WY, Cho TJ (2011) Whole-exome sequencing identifies mutations of KIF22 in spondyloepimetaphyseal dysplasia with joint laxity, leptodactylic type. Am J Hum Genet 89(6):760–766

Mirzaa GM, Vitre B, Carpenter G, Abramowicz I, Gleeson JG, Paciorkowski AR, Cleveland DW, Dobyns WB, O'Driscoll M (2014) Mutations in CENPE define a novel kinetochore-centromeric mechanism for microcephalic primordial dwarfism. Hum Genet 133(8):1023–1039

Moawia A, Shaheen R, Rasool S, Waseem SS, Ewida N, Budde B, Kawalia A, Motameny S, Khan K, Fatima A, Jameel M, Ullah F, Akram T, Ali Z, Abdullah U, Irshad S, Höhne W, Noegel AA, Al-Owain M, Hörtnagel K, Stöbe P, Baig SM, Nürnberg P, Alkuraya FS, Hahn A, Hussain MS (2017) Mutations of KIF14 cause primary microcephaly by impairing cytokinesis. Ann Neurol 82(4):562–577

Moores CA, Cooper J, Wagenbach M, Ovechkina Y, Wordeman L, Milligan RA (2006) The role of the kinesin-13 neck in microtubule depolymerization. Cell Cycle 5(16):1812–1815

Morris RL, Scholey JM (1997) Heterotrimeric kinesin-II is required for the assembly of motile 9+2 ciliary axonemes on sea urchin embryos. J Cell Biol 138(5):1009–1022

Nagase T, Ishikawa K, Kikuno R, Hirosawa M, Nomura N, Ohara O (1999) Prediction of the coding sequences of unidentified human genes. XV. The complete sequences of 100 new cDNA clones from brain which code for large proteins in vitro. DNA Res 6(5):337–345

Nagase T, Ishikawa K, Miyajima N, Tanaka A, Kotani H, Nomura N, Ohara O (1998) Prediction of the coding sequences of unidentified human genes. IX. The complete sequences of 100 new cDNA clones from brain which can code for large proteins in vitro. DNA Res 5(1):31–39

Nagase T, Ishikawa K, Nakajima D, Ohira M, Seki N, Miyajima N, Tanaka A, Kotani H, Nomura N, Ohara O (1997) Prediction of the coding sequences of unidentified human genes. VII. The complete sequences of 100 new cDNA clones from brain which can code for large proteins in vitro. DNA Res 4(2):141–150

Nakagawa T, Setou M, Seog D, Ogasawara K, Dohmae N, Takio K, Hirokawa N (2000) A novel motor, KIF13A, transports mannose-6-phosphate receptor to plasma membrane through direct interaction with AP-1 complex. Cell 103(4):569–581

Nakagawa T, Tanaka Y, Matsuoka E, Kondo S, Okada Y, Noda Y, Kanai Y, Hirokawa N (1997) Identification and classification of 16 new kinesin superfamily (KIF) proteins in mouse genome. Proc Natl Acad Sci USA 94(18):9654–9659

Nangaku M, Sato-Yoshitake R, Okada Y, Noda Y, Takemura R, Yamazaki H, Hirokawa N (1994) KIF1B, a novel microtubule plus end-directed monomeric motor protein for transport of mitochondria. Cell 79(7):1209–1220

Nelson MM, Thomson AJ (1982) The acrocallosal syndrome. Am J Med Genet 12(2):195–199

Nguyen AM, Jacobs CR (2013) Emerging role of primary cilia as mechanosensors in osteocytes. Bone 54(2):196–204

Nguyen PA, Groen AC, Loose M, Ishihara K, Wühr M, Field CM, Mitchison TJ (2014) Spatial organization of cytokinesis signaling reconstituted in a cell-free system. Science 346(6206):244–247

Niceta M, Margiotti K, Digilio MC, Guida V, Bruselles A, Pizzi S, Ferraris A, Memo L, Laforgia N, Dentici ML, Consoli F, Torrente I, Ruiz-Perez VL, Dallapiccola B, Marino B, De Luca A, Tartaglia M (2018) Biallelic mutations in DYNC2LI1 are a rare cause of Ellis-van Creveld syndrome. Clin Genet 93(3):632–639

Niclas J, Navone F, Hom-Booher N, Vale RD (1994) Cloning and localization of a conventional kinesin motor expressed exclusively in neurons. Neuron 12(5):1059–1072

Nishita M, Park SY, Nishio T, Kamizaki K, Wang Z, Tamada K, Takumi T, Hashimoto R, Otani H, Pazour GJ, Hsu VW, Minami Y (2017) Ror2 signaling regulates Golgi structure and transport through IFT20 for tumor invasiveness. Sci Rep 7(1):1

Niwa S, Nakajima K, Miki H, Minato Y, Wang D, Hirokawa N (2012) KIF19A is a microtubule-depolymerizing kinesin for ciliary length control. Dev Cell 23(6):1167–1175

Ocbina PJ, Eggenschwiler JT, Moskowitz I, Anderson KV (2011) Complex interactions between genes controlling trafficking in primary cilia. Nat Genet 43(6):547–553

Ohsugi M, Adachi K, Horai R, Kakuta S, Sudo K, Kotaki H, Tokai-Nishizumi N, Sagara H, Iwakura Y, Yamamoto T (2008) Kid-mediated chromosome compaction ensures proper nuclear envelope formation. Cell 132(5):771–782

Okada Y, Yamazaki H, Sekine-Aizawa Y, Hirokawa N (1995) The neuron-specific kinesin superfamily protein KIF1A is a unique monomeric motor for anterograde axonal transport of synaptic vesicle precursors. Cell 81(5):769–780

Ostergaard P, Simpson MA, Mendola A, Vasudevan P, Connell FC, van Impel A, Moore AT, Loeys BL, Ghalamkarpour A, Onoufriadis A, Martinez-Corral I, Devery S, Leroy JG, van Laer L, Singer A, Bialer MG, McEntagart M, Quarrell O, Brice G, Trembath RC, Schulte-Merker S, Makinen T, Vikkula M, Mortimer PS, Mansour S, Jeffery S (2012) Mutations in KIF11 cause autosomal-dominant microcephaly variably associated with congenital lymphedema and chorioretinopathy. Am J Hum Genet 90(2):356–362

Palmer KJ, Hughes H, Stephens DJ (2009) Specificity of cytoplasmic dynein subunits in discrete membrane-trafficking steps. Mol Biol Cell 20(12):2885–2899

Pfister KK, Shah PR, Hummerich H, Russ A, Cotton J, Annuar AA, King SM, Fisher EM (2006) Genetic analysis of the cytoplasmic dynein subunit families. PLoS Genet 2(1):e1

Piddini E, Schmid JA, de Martin R, Dotti CG (2001) The Ras-like GTPase gem is involved in cell shape remodelling and interacts with the novel kinesin-like protein KIF9. EMBO J 20(15):4076–4087

Pike R, Ortiz-Zapater E, Lumicisi B, Santis G, Parsons M (2018) KIF22 coordinates CAR and EGFR dynamics to promote cancer cell proliferation. Sci Signal 11(515)

Pilling AD, Horiuchi D, Lively CM, Saxton WM (2006) Kinesin-1 and dynein are the primary motors for fast transport of mitochondria in drosophila motor axons. Mol Biol Cell 17(4):2057–2068

Plotnikova OV, Pugacheva EN, Golemis EA (2009) Primary cilia and the cell cycle. Methods Cell Biol 94:137–160

Putkey FR, Cramer T, Morphew MK, Silk AD, Johnson RS, McIntosh JR, Cleveland DW (2002) Unstable kinetochore-microtubule capture and chromosomal instability following deletion of CENP-E. Dev Cell 3(3):351–365

Putoux A, Thomas S, Coene KL, Davis EE, Alanay Y, Ogur G, Uz E, Buzas D, Gomes C, Patrier S, Bennett CL, Elkhartoufi N, Frison MH, Rigonnot L, Joyé N, Pruvost S, Utine GE, Boduroglu K, Nitschke P, Fertitta L, Thauvin-Robinet C, Munnich A, Cormier-Daire V, Hennekam R, Colin E, Akarsu NA, Bole-Feysot C, Cagnard N, Schmitt A, Goudin N, Lyonnet S, Encha-Razavi F, Siffroi JP, Winey M, Katsanis N, Gonzales M, Vekemans M, Beales PL, Attié-Bitach T (2011) KIF7 mutations cause fetal hydrolethalus and acrocallosal syndromes. Nat Genet 43(6):601–606

Qiu N, Xiao Z, Cao L, Buechel MM, David V, Roan E, Quarles LD (2012) Disruption of Kif3a in osteoblasts results in defective bone formation and osteopenia. J Cell Sci 125(Pt 8):1945–1957

Quadri N, Upadhyai P (2023) Primary cilia in skeletal development and disease. Exp Cell Res 431(1):113751

Rahman A, Friedman DS, Goldstein LS (1998) Two kinesin light chain genes in mice. Identification and characterization of the encoded proteins. J Biol Chem 273(25):15395–15403

Rayment I (1996) Kinesin and myosin: molecular motors with similar engines. Structure 4(5):501–504

Reck-Peterson SL, Redwine WB, Vale RD, Carter AP (2018) The cytoplasmic dynein transport machinery and its many cargoes. Nat Rev Mol Cell Biol 19(6):382–398

Ren Y, Yang B, Guo G, Zhang J, Sun Y, Liu D, Guo S, Wu Y, Wang X, Wang S, Zhang W, Guo X, Li X, Li R, He J, Zhou Z (2022) GBP2 facilitates the progression of glioma via regulation of KIF22/EGFR signaling. Cell Death Discov 8(1):208

Rivière JB, Ramalingam S, Lavastre V, Shekarabi M, Holbert S, Lafontaine J, Srour M, Merner N, Rochefort D, Hince P, Gaudet R, Mes-Masson AM, Baets J, Houlden H, Brais B, Nicholson GA, Van Esch H, Nafissi S, De Jonghe P, Reilly MM, Timmerman V, Dion PA, Rouleau GA (2011) KIF1A, an axonal transporter of synaptic vesicles, is mutated in hereditary sensory and autonomic neuropathy type 2. Am J Hum Genet 89(2):219–230

Roberts AJ, Kon T, Knight PJ, Sutoh K, Burgess SA (2013) Functions and mechanics of dynein motor proteins. Nat Rev Mol Cell Biol 14(11):713–726

Santos-Ledo A, Garcia-Macia M, Campbell PD, Gronska M, Marlow FL (2017) Kinesin-1 promotes chondrocyte maintenance during skeletal morphogenesis. PLoS Genet 13(7):e1006918

Schinzel A, Schmid W (1980) Hallux duplication, postaxial polydactyly, absence of the corpus callosum, severe mental retardation, and additional anomalies in two unrelated patients: a new syndrome. Am J Med Genet 6(3):241–249

Schlager MA, Kapitein LC, Grigoriev I, Burzynski GM, Wulf PS, Keijzer N, de Graaff E, Fukuda M, Shepherd IT, Akhmanova A, Hoogenraad CC (2010) Pericentrosomal targeting of Rab6 secretory vesicles by Bicaudal-D-related protein 1 (BICDR-1) regulates neuritogenesis. EMBO J 29(10):1637–1651

Schlager MA, Serra-Marques A, Grigoriev I, Gumy LF, Esteves da Silva M, Wulf PS, Akhmanova A, Hoogenraad CC (2014) Bicaudal d family adaptor proteins control the velocity of dynein-based movements. Cell Rep 8(5):1248–1256

Schmidts M, Hou Y, Cortés CR, Mans DA, Huber C, Boldt K, Patel M, van Reeuwijk J, Plaza JM, van Beersum SE, Yap ZM, Letteboer SJ, Taylor SP, Herridge W, Johnson CA, Scambler PJ, Ueffing M, Kayserili H, Krakow D, King SM, Beales PL, Al-Gazali L, Wicking C, Cormier-Daire V, Roepman R, Mitchison HM, Witman GB (2015) TCTEX1D2 mutations underlie Jeune asphyxiating thoracic dystrophy with impaired retrograde intraflagellar transport. Nat Commun 6:7074

Schmidts M, Mitchison HM (2018) 15–Severe skeletal abnormalities caused by defects in retrograde intraflagellar transport dyneins. In: King SM (ed) Dyneins: structure, biology and disease, 2nd edn. Academic Press, pp 356–401

Schmidts M, Vodopiutz J, Christou-Savina S, Cortés CR, McInerney-Leo AM, Emes RD, Arts HH, Tüysüz B, D'Silva J, Leo PJ, Giles TC, Oud MM, Harris JA, Koopmans M, Marshall M, Elçioglu N, Kuechler A, Bockenhauer D, Moore AT, Wilson LC, Janecke AR, Hurles ME, Emmet W, Gardiner B, Streubel B, Dopita B, Zankl A, Kayserili H, Scambler PJ, Brown MA, Beales PL, Wicking C, Duncan EL, Mitchison HM (2013) Mutations in the gene encoding IFT dynein complex component WDR34 cause Jeune asphyxiating thoracic dystrophy. Am J Hum Genet 93(5):932–944

Scholey JE, Nithianantham S, Scholey JM, Al-Bassam J (2014) Structural basis for the assembly of the mitotic motor Kinesin-5 into bipolar tetramers. elife 3:e02217

Sebé-Pedrós A, Grau-Bové X, Richards TA, Ruiz-Trillo I (2014) Evolution and classification of myosins, a paneukaryotic whole-genome approach. Genome Biol Evol 6(2):290–305

Seeger-Nukpezah T, Golemis EA (2012) The extracellular matrix and ciliary signaling. Curr Opin Cell Biol 24(5):652–661

Seppälä J, Bernardi RC, Haataja TJK, Hellman M, Pentikäinen OT, Schulten K, Permi P, Ylänne J, Pentikäinen U (2017) Skeletal dysplasia mutations effect on human Filamins' structure and Mechanosensing. Sci Rep 7(1):4218

Serra-Marques A, Martin M, Katrukha EA, Grigoriev I, Peeters CA, Liu Q, Hooikaas PJ, Yao Y, Solianova V, Smal I, Pedersen LB, Meijering E, Kapitein LC, Akhmanova A (2020) Concerted action of kinesins KIF5B and KIF13B promotes efficient secretory vesicle transport to microtubule plus ends. elife 9

Setou M, Nakagawa T, Seog DH, Hirokawa N (2000) Kinesin superfamily motor protein KIF17 and mLin-10 in NMDA receptor-containing vesicle transport. Science 288(5472):1796–1802

Sheng L, Hao SL, Yang WX, Sun Y (2018) The multiple functions of kinesin-4 family motor protein KIF4 and its clinical potential. Gene 678:90–99

Shin H, Wyszynski M, Huh KH, Valtschanoff JG, Lee JR, Ko J, Streuli M, Weinberg RJ, Sheng M, Kim E (2003) Association of the kinesin motor KIF1A with the multimodular protein liprin-alpha. J Biol Chem 278(13):11393–11401

Singh J, Sanghavi P, Mallik R (2022) Microtubule motor driven interactions of lipid droplets: specificities and opportunities. Front Cell Dev Biol 10:893375

Sleiman PMA, March M, Nguyen K, Tian L, Pellegrino R, Hou C, Dridi W, Sager M, Housawi YH, Hakonarson H (2017) Loss-of-function mutations in KIF15 underlying a Braddock-Carey Genocopy. Hum Mutat 38(5):507–510

Song B, Haycraft CJ, Seo HS, Yoder BK, Serra R (2007) Development of the post-natal growth plate requires intraflagellar transport proteins. Dev Biol 305(1):202–216

St-Jacques B, Hammerschmidt M, McMahon AP (1999) Indian hedgehog signaling regulates proliferation and differentiation of chondrocytes and is essential for bone formation. Genes Dev 13(16):2072–2086

Stalke A, Sgodda M, Cantz T, Skawran B, Lainka E, Hartleben B, Baumann U, Pfister ED (2022) KIF12 variants and disturbed hepatocyte polarity in children with a phenotypic Spectrum of Cholestatic liver disease. J Pediatr 240:284–291.e289

Stevenson NL (2023) The factory, the antenna and the scaffold: the three-way interplay between the Golgi, cilium and extracellular matrix underlying tissue function. Biol Open 12(2)

Stumpff J, von Dassow G, Wagenbach M, Asbury C, Wordeman L (2008) The kinesin-8 motor Kif18A suppresses kinetochore movements to control mitotic chromosome alignment. Dev Cell 14(2):252–262

Tahara K, Takagi M, Ohsugi M, Sone T, Nishiumi F, Maeshima K, Horiuchi Y, Tokai-Nishizumi N, Imamoto F, Yamamoto T, Kose S, Imamoto N (2008) Importin-beta and the small guanosine triphosphatase ran mediate chromosome loading of the human chromokinesin kid. J Cell Biol 180(3):493–506

Tanaka Y, Niwa S, Dong M, Farkhondeh A, Wang L, Zhou R, Hirokawa N (2016) The molecular motor KIF1A transports the TrkA Neurotrophin receptor and is essential for sensory neuron survival and function. Neuron 90(6):1215–1229

Tanenbaum ME, Macurek L, van der Vaart B, Galli M, Akhmanova A, Medema RH (2011) A complex of Kif18b and MCAK promotes microtubule depolymerization and is negatively regulated by Aurora kinases. Curr Biol 21(16):1356–1365

Taylor SP, Dantas TJ, Duran I, Wu S, Lachman RS, Nelson SF, Cohn DH, Vallee RB, Krakow D (2015) Mutations in DYNC2LI1 disrupt cilia function and cause short rib polydactyly syndrome. Nat Commun 6:7092

Telford EA, Wightman P, Leek J, Markham AF, Lench NJ, Bonthron DT (1998) cDNA cloning, genomic organization, and chromosomal localization of a novel human gene that encodes a kinesin-related protein highly similar to mouse Kif3C. Biochem Biophys Res Commun 242(2):407–412

Theisen U, Straube E, Straube A (2012) Directional persistence of migrating cells requires Kif1C-mediated stabilization of trailing adhesions. Dev Cell 23(6):1153–1166

Theret L, Jeanne A, Langlois B, Hachet C, David M, Khrestchatisky M, Devy J, Hervé E, Almagro S, Dedieu S (2017) Identification of LRP-1 as an endocytosis and recycling receptor for β1-integrin in thyroid cancer cells. Oncotarget 8(45):78614–78632

Thompson AF, Blackburn PR, Arons NS, Stevens SN, Babovic-Vuksanovic D, Lian JB, Klee EW, Stumpff J (2022) Pathogenic mutations in the chromokinesin KIF22 disrupt anaphase chromosome segregation. elife 11

Torres JZ, Summers MK, Peterson D, Brauer MJ, Lee J, Senese S, Gholkar AA, Lo YC, Lei X, Jung K, Anderson DC, Davis DP, Belmont L, Jackson PK (2011) The STARD9/Kif16a kinesin associates with mitotic microtubules and regulates spindle pole assembly. Cell 147(6):1309–1323

Tu Y, Zhao L, Billadeau DD, Jia D (2020) Endosome-to-TGN trafficking: organelle-vesicle and organelle-organelle interactions. Front Cell Dev Biol 8:163

Tummala P, Arnsdorf EJ, Jacobs CR (2010) The role of primary cilia in mesenchymal stem cell differentiation: a pivotal switch in guiding lineage commitment. Cell Mol Bioeng 3(3):207–212

Tüysüz B, Bariş S, Aksoy F, Madazli R, Ungür S, Sever L (2009) Clinical variability of asphyxiating thoracic dystrophy (Jeune) syndrome: evaluation and classification of 13 patients. Am J Med Genet A 149a(8):1727–1733

Uchiyama Y, Sakaguchi M, Terabayashi T, Inenaga T, Inoue S, Kobayashi C, Oshima N, Kiyonari H, Nakagata N, Sato Y, Sekiguchi K, Miki H, Araki E, Fujimura S, Tanaka SS, Nishinakamura R (2010) Kif26b, a kinesin family gene, regulates adhesion of the embryonic kidney mesenchyme. Proc Natl Acad Sci USA 107(20):9240–9245

Ueno H, Huang X, Tanaka Y, Hirokawa N (2011) KIF16B/Rab14 molecular motor complex is critical for early embryonic development by transporting FGF receptor. Dev Cell 20(1):60–71

Unger S, Ferreira CR, Mortier GR, Ali H, Bertola DR, Calder A, Cohn DH, Cormier-Daire V, Girisha KM, Hall C, Krakow D, Makitie O, Mundlos S, Nishimura G, Robertson SP, Savarirayan

R, Sillence D, Simon M, Sutton VR, Warman ML, Superti-Furga A (2023) Nosology of genetic skeletal disorders: 2023 revision. Am J Med Genet A 191(5):1164–1209

Verhey KJ, Lizotte DL, Abramson T, Barenboim L, Schnapp BJ, Rapoport TA (1998) Light chain-dependent regulation of Kinesin's interaction with microtubules. J Cell Biol 143(4):1053–1066

Vikberg AL, Malla S, Golovleva I (2020) Differential tissue specific expression of Kif23 alternative transcripts in mice with the human mutation causing congenital dyserythropoietic anemia type III. Blood Cells Mol Dis 85:102483

Wan Y, Morikawa M, Morikawa M, Iwata S, Naseer MI, Ahmed Chaudhary AG, Tanaka Y, Hirokawa N (2023) KIF4 regulates neuronal morphology and seizure susceptibility via the PARP1 signaling pathway. J Cell Biol 222(2)

Wang C, Du W, Su QP, Zhu M, Feng P, Li Y, Zhou Y, Mi N, Zhu Y, Jiang D, Zhang S, Zhang Z, Sun Y, Yu L (2015) Dynamic tubulation of mitochondria drives mitochondrial network formation. Cell Res 25(10):1108–1120

Wang D, Nitta R, Morikawa M, Yajima H, Inoue S, Shigematsu H, Kikkawa M, Hirokawa N (2016) Motility and microtubule depolymerization mechanisms of the Kinesin-8 motor, KIF19A. elife 5

Wang H, Li S, Liu B, Wei S, Wang T, Li T, Lin J, Ni X (2022) KIF11: a potential prognostic biomarker for predicting bone metastasis-free survival of prostate cancer. Oncol Lett 24(3):312

Wang Y, Ladunga I, Miller AR, Horken KM, Plucinak T, Weeks DP, Bailey CP (2008) The small ubiquitin-like modifier (SUMO) and SUMO-conjugating system of Chlamydomonas reinhardtii. Genetics 179(1):177–192

Wanschers B, van de Vorstenbosch R, Wijers M, Wieringa B, King SM, Fransen J (2008) Rab6 family proteins interact with the dynein light chain protein DYNLRB1. Cell Motil Cytoskeleton 65(3):183–196

Whitehead JL, Wang SY, Bost-Usinger L, Hoang E, Frazer KA, Burnside B (1999) Photoreceptor localization of the KIF3A and KIF3B subunits of the heterotrimeric microtubule motor kinesin II in vertebrate retina. Exp Eye Res 69(5):491–503

Wiesner C, Faix J, Himmel M, Bentzien F, Linder S (2010) KIF5B and KIF3A/KIF3B kinesins drive MT1-MMP surface exposure, CD44 shedding, and extracellular matrix degradation in primary macrophages. Blood 116(9):1559–1569

Wilson CW, Nguyen CT, Chen MH, Yang JH, Gacayan R, Huang J, Chen JN, Chuang PT (2009) Fused has evolved divergent roles in vertebrate hedgehog signalling and motile ciliogenesis. Nature 459(7243):98–102

Xiang X, Qiu R, Yao X, Arst HN Jr, Peñalva MA, Zhang J (2015) Cytoplasmic dynein and early endosome transport. Cell Mol Life Sci 72(17):3267–3280

Xing BM, Yang YR, Du JX, Chen HJ, Qi C, Huang ZH, Zhang Y, Wang Y (2012) Cyclin-dependent kinase 5 controls TRPV1 membrane trafficking and the heat sensitivity of nociceptors through KIF13B. J Neurosci 32(42):14709–14721

Xu F, Takahashi H, Tanaka Y, Ichinose S, Niwa S, Wicklund MP, Hirokawa N (2018) KIF1Bβ mutations detected in hereditary neuropathy impair IGF1R transport and axon growth. J Cell Biol 217(10):3480–3496

Yadav S, Puthenveedu MA, Linstedt AD (2012) Golgin160 recruits the dynein motor to position the Golgi apparatus. Dev Cell 23(1):153–165

Yamada KH, Nakajima Y, Geyer M, Wary KK, Ushio-Fukai M, Komarova Y, Malik AB (2014) KIF13B regulates angiogenesis through Golgi to plasma membrane trafficking of VEGFR2. J Cell Sci 127(Pt 20):4518–4530

Yamamoto K, Okano H, Miyagawa W, Visse R, Shitomi Y, Santamaria S, Dudhia J, Troeberg L, Strickland DK, Hirohata S, Nagase H (2016) MMP-13 is constitutively produced in human chondrocytes and co-endocytosed with ADAMTS-5 and TIMP-3 by the endocytic receptor LRP1. Matrix Biol 56:57–73

Yamamoto K, Scavenius C, Meschis MM, Gremida AME, Mogensen EH, Thøgersen IB, Bonelli S, Scilabra SD, Jensen A, Santamaria S, Ahnström J, Bou-Gharios G, Enghild JJ, Nagase H (2022) A top-down approach to uncover the hidden ligandome of low-density lipoprotein receptor-related protein 1 in cartilage. Matrix Biol 112:190–218

Yamazaki H, Nakata T, Okada Y, Hirokawa N (1995) KIF3A/B: a heterodimeric kinesin superfamily protein that works as a microtubule plus end-directed motor for membrane organelle transport. J Cell Biol 130(6):1387–1399

Yan W, Zheng L, Xu X, Hao Z, Zhang Y, Lu J, Sun Z, Dai J, Shi D, Guo B, Jiang Q (2022) Heterozygous LRP1 deficiency causes developmental dysplasia of the hip by impairing triradiate chondrocytes differentiation due to inhibition of autophagy. Proc Natl Acad Sci USA 119(37):e2203557119

Yang HW, Chen YZ, Takita J, Soeda E, Piao HY, Hayashi Y (2001) Genomic structure and mutational analysis of the human KIF1B gene which is homozygously deleted in neuroblastoma at chromosome 1p36.2. Oncogene 20(36):5075–5083

Yao M, Qu H, Han Y, Cheng CY, Xiao X (2022) Kinesins in mammalian spermatogenesis and germ cell transport. Front Cell Dev Biol 10:837542

Yen TJ, Compton DA, Wise D, Zinkowski RP, Brinkley BR, Earnshaw WC, Cleveland DW (1991) CENP-E, a novel human centromere-associated protein required for progression from metaphase to anaphase. EMBO J 10(5):1245–1254

Yin X, Takei Y, Kido MA, Hirokawa N (2011) Molecular motor KIF17 is fundamental for memory and learning via differential support of synaptic NR2A/2B levels. Neuron 70(2):310–325

Yokota S, Shah SH, Huie EL, Wen RR, Luo Z, Goldberg JL (2023) Kif5a regulates mitochondrial transport in developing retinal ganglion cells in vitro. Invest Ophthalmol Vis Sci 64(3):4

Yonekawa Y, Harada A, Okada Y, Funakoshi T, Kanai Y, Takei Y, Terada S, Noda T, Hirokawa N (1998) Defect in synaptic vesicle precursor transport and neuronal cell death in KIF1A motor protein-deficient mice. J Cell Biol 141(2):431–441

Yue Y, Engelke MF, Blasius TL, Verhey KJ (2022) Hedgehog-induced ciliary trafficking of kinesin-4 motor KIF7 requires intraflagellar transport but not KIF7's microtubule binding. Mol Biol Cell 33(1):br1

Zajac AL, Horne-Badovinac S (2022) Kinesin-directed secretion of basement membrane proteins to a subdomain of the basolateral surface in drosophila epithelial cells. Curr Biol 32(4):735–748.e710

Zalenski AA, Majumder S, De K, Venere M (2020) An interphase pool of KIF11 localizes at the basal bodies of primary cilia and a reduction in KIF11 expression alters cilia dynamics. Sci Rep 10(1):13946

Zhang XD, Goeres J, Zhang H, Yen TJ, Porter AC, Matunis MJ (2008) SUMO-2/3 modification and binding regulate the association of CENP-E with kinetochores and progression through mitosis. Mol Cell 29(6):729–741

Zhao C, Omori Y, Brodowska K, Kovach P, Malicki J (2012) Kinesin-2 family in vertebrate ciliogenesis. Proc Natl Acad Sci USA 109(7):2388–2393

Zhernov I, Diez S, Braun M, Lansky Z (2020) Intrinsically disordered domain of Kinesin-3 Kif14 enables unique functional diversity. Curr Biol 30(17):3342–3351.e3345

Zhou R, Niwa S, Guillaud L, Tong Y, Hirokawa N (2013) A molecular motor, KIF13A, controls anxiety by transporting the serotonin type 1A receptor. Cell Rep 3(2):509–519

Zhou R, Niwa S, Homma N, Takei Y, Hirokawa N (2009) KIF26A is an unconventional kinesin and regulates GDNF-ret signaling in enteric neuronal development. Cell 139(4):802–813

Zhu C, Jiang W (2005) Cell cycle-dependent translocation of PRC1 on the spindle by Kif4 is essential for midzone formation and cytokinesis. Proc Natl Acad Sci USA 102(2):343–348

Zhu C, Zhao J, Bibikova M, Leverson JD, Bossy-Wetzel E, Fan JB, Abraham RT, Jiang W (2005) Functional analysis of human microtubule-based motor proteins, the kinesins and dyneins, in mitosis/cytokinesis using RNA interference. Mol Biol Cell 16(7):3187–3199

Zieba J, Zhang W, Chong JX, Forlenza KN, Martin JH, Heard K, Grange DK, Butler MG, Kleefstra T, Lachman RS, Nickerson D, Regnier M, Cohn DH, Bamshad M, Krakow D (2017) A postnatal role for embryonic myosin revealed by MYH3 mutations that alter TGFβ signaling and cause autosomal dominant spondylocarpotarsal synostosis. Sci Rep 7:41803

Chapter 12
COG4 Variant and Saul-Wilson Syndrome

Carlos R. Ferreira

Abstract Saul-Wilson syndrome is a rare disease characterized by profound short stature of prenatal onset with lack of postnatal catch-up growth, characteristic radiographic findings and facial features, ocular manifestations such as lamellar cataracts and rod-cone dystrophy, neutropenia leading to frequent infections (particularly respiratory infections), and elevation of liver transaminases. The disease is caused by a recurrent amino acid substitution in COG4, a subunit of the octameric conserved oligomeric Golgi (COG) complex that participates in vesicular trafficking between the Golgi and endoplasmic reticulum (ER). The variant is known to cause accelerated vesicular trafficking from the Golgi to the ER, with a new steady-state equilibrium leading to decreased Golgi volumetrics. The abnormal glycosylation of proteoglycans is known to be at least partly related to the pathogenesis of the disease. This work summarizes the salient clinical manifestations of the disease, as well as the proposed mechanisms leading to its manifestations.

12.1 History

Dr. Robert A. Saul described the first patient with what is now known as Saul-Wilson syndrome (MIM 618150) in the "Unknown Cases" section of the inaugural issue of the Proceedings of the Greenwood Genetic Center in 1982 (Saul 1982). Years later, Dr. William G. Wilson presented slides to a national meeting that had sessions for patients with unknown conditions. Dr. David Flannery mentioned that he recalled that Dr. Saul had seen a patient with similar findings (Saul and Wilson 2019). Dr. Wilson found a copy of Dr. Saul's report, agreed that the two patients had similar findings, and after communication via mail, they decided to jointly report

C. R. Ferreira (✉)
Skeletal Genomics Unit, Metabolic Medicine Branch, National Human Genome Research Institute, National Institutes of Health, Bethesda, MD, USA
e-mail: carlos.ferreira@nih.gov

A. Rossi, F. Zaucke (eds.), *The Extracellular Matrix in Genetic Skeletal Disorders*, Biology of Extracellular Matrix 16,
https://doi.org/10.1007/978-3-031-70835-0_12

their findings in a paper published in 1990 (Saul and Wilson 1990). A few years later, in 1994, two additional patients were described with the same condition as that described by Drs. Saul and Wilson (Hersh et al. 1994). Then, there was a period of more than 3 decades with no subsequent case reports, and although the condition appeared in the sixth and seventh iterations of the official nosology and classification of genetic skeletal disorders in 2001 and 2006, respectively (Hall 2002; Superti-Furga and Unger 2007), it had been dropped by the eighth version of the Nosology by 2010 (Warman et al. 2011), given doubts about its identity as a unique disorder. A fifth patient was subsequently described (Chinen et al. 2015), although the diagnosis of Saul-Wilson syndrome was questionable on clinical and radiographic grounds, and the patient was subsequently found not to harbor variants in the culprit gene (author, personal communication). In 2018, 14 patients were reported (including two from the original cohort of four patients), with all subjects harboring one of two possible variants in the *COG4* gene (NM_015386.2:c.1546G > A or c.1546G > C); both variants give rise to an identical amino acid substitution, p.G516R (Ferreira et al. 2018). COG4 is a component of the conserved oligomeric Golgi (COG) trafficking octameric complex that controls retrograde and anterograde vesicular protein trafficking and is composed of two lobes, lobe A (COG1–4 subunits) and lobe B (COG5–8 subunits). In all subjects, the variant was heterozygous and inherited de novo. The anesthetic management of an individual from this cohort of 14 (P11.1) was later reported (Matar 2019). Subsequently, additional patients were described (Koruga et al. 2022).

12.2 Clinical Manifestations

12.2.1 Skeletal Features

Patients present with primordial dwarfism, with profound short stature of prenatal onset with no improvement postnatally. There is in fact a sharp decline in standard deviation scores during the first few months of life, with a final height between −4 and −8.5 SD scores (Ferreira et al. 2020b). Growth patterns have been established for stature, weight, and head circumference (as well as weight-for-length and weight velocity for younger children), derived from hundreds of data points obtained by retrospective chart review from the cohort of 14 individuals with molecularly confirmed Saul-Wilson syndrome. These charts were established not only with the intention of evaluating whether Saul-Wilson syndrome seems an appropriate diagnosis and to monitor the growth of affected subjects, but also to reduce morbidity resulting from unnecessarily aggressive nutritional interventions by well-intentioned practitioners trying to promote weight gain (Ferreira et al. 2020a). Statistical comparisons of height-for-age model fits before and after initiation of growth hormone supplementation found that this therapy does not appear to improve height in individuals with Saul-Wilson syndrome (Ferreira et al. 2020a).

Radiographic findings include hypoplasia of the odontoid process (sometimes leading to spinal cord compression), overtubulation of long bones with narrow diaphyses and flared metaphyses (Fig. 12.1a), and large epiphyses. Oval lucencies of the proximal femurs are found during infancy. The vertebral bodies are initially flat but become taller over time, while the vertebral endplates become irregular with age. Hypoplasia of T12 or L1 can be seen in many patients (Fig. 12.1b). Ivory epiphyses of the hands are found in later childhood. Some subjects had fractures with no known or only minimal trauma. Premature degenerative joint disease can be seen in early adulthood, necessitating early joint replacement (Fig. 12.1b) (Ferreira et al. 2020b).

12.2.2 Dysmorphic Features

Progeroid facial features are appreciated during early infancy, although they disappear within the first few months of life. In late infancy or toddlerhood, the scalp veins become particularly prominent (Fig. 12.1c). The columella becomes prominent during the first few years of life. The shape of the forehead is round in early childhood but becomes taller and less round in late childhood (Fig. 12.1c). Shortening of the distal phalanges of the fingers and toes (telebrachydactyly) of variable severity is seen in all subjects, regardless of age (Fig. 12.1d) (Ferreira et al. 2020b).

12.2.3 Ocular Features

Childhood-onset cataracts (particularly lamellar cataracts) and rod-cone dystrophy leading to pigmentary retinopathy are common. Macular cystic changes, delayed dark adaptation, difficulty with night vision, and visual field deficits can also be seen. Based on limited evidence, it appears that the retinal involvement progresses over time, as subjects who were specifically questioned mentioned worsening of their night and peripheral vision with older age (Ferreira et al. 2020b).

12.2.4 Other

All subjects manifest intermittent neutropenia, with no noticeable cyclic periodicity. Neutropenia has not been detected in the neonatal period so far. Granulocyte colony-stimulating factor (G-CSF) has been used successfully in a couple of individuals with infections, with improvement in neutrophil counts.

Asymptomatic elevation of liver transaminases was noted in several subjects, without florid liver disease or other abnormalities in liver function indices. Aspartate

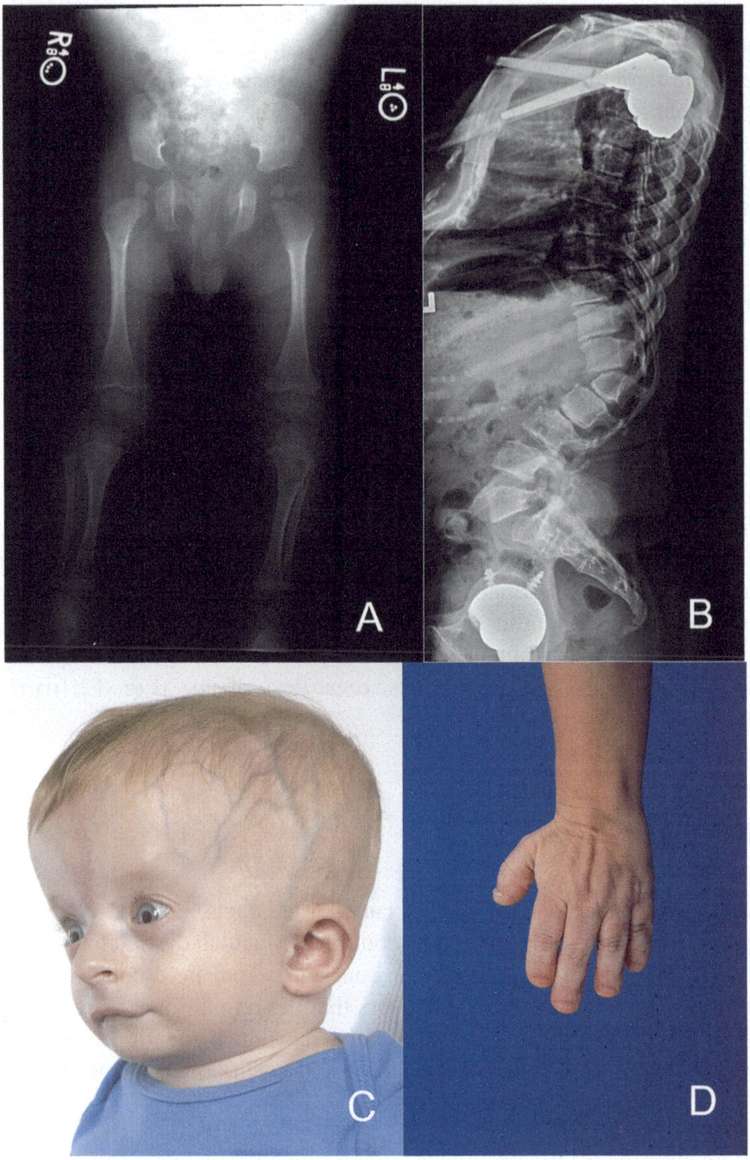

Fig. 12.1 (**a**) Radiograph obtained at 11 months of age showing proximal femoral lucencies and overtubulation of long bones with slender diaphyses and flared metaphyses. (**b**) Radiograph obtained at 29 years old showing hypoplasia of L1 vertebra, and prosthetic replacement of shoulder and hip joints. (**c**) Prominent forehead, prominent scalp veins, and prominent eyes in an infant. (**d**) Shortening of the distal phalanges of the fingers in an adult patient

aminotransferase (AST) was elevated more commonly than alanine aminotransferase (ALT) (Ferreira et al. 2020b).

12.3 Animal Models

Saul-Wilson syndrome is caused by a heterozygous missense pathogenic variant in the *COG4* gene, resulting in a p.Gly516Arg change. Other pathogenic variants, when inherited in a recessive manner, have been associated with a different disorder, a congenital disorder of glycosylation (COG4-CDG) (Ng et al. 2011; Reynders et al. 2009).

A zebrafish was originally established to evaluate the developmental functions of COG4. This model was established using CRISPR-Cas9 gene editing to generate two mutant alleles of the zebrafish ortholog, *cog4*, specifically leading to the one truncating variant in exon 4 (*cog4b1311*, c.352_359del, p.I118Gfs*) and another in exon 12 (*cog4b1312*, c.1499_1511del, p.D500Gfs*6). The Golgi structure was disrupted in mutant cells, and there were defects in the processing of N- and O-linked glycans and decreased glycosphingolipid complexity. This is different from what was found in mutant human fibroblasts (see Pathogenesis below), but Saul-Wilson syndrome is a monoallelic disease, while this zebrafish model better recapitulated the biallelic disease (COG4-CDG). Mutant zebrafish also showed reduced Alcian blue staining of the jaw (indicative of defects in proteoglycan secretion) and reduced *col1a2* expression in the fin. However, expression of collagen type II was normal in the jaw, suggesting that not all secreted proteins are similarly affected by biallelic *cog4* variants (at least not in all tissues where the gene is normally expressed). Chondrocyte stacking was abnormal in mutants. These data supported the fact that COG4 is necessary for normal secretion of proteoglycans, skeletogenesis, and cartilage architecture (Ferreira et al. 2018).

In zebrafish, the semicircular canals develop from epithelial projections emanating from the otic vesicle that grow toward each other and fuse to form pillars. The core of each epithelial projection is filled with extracellular matrix components that are secreted by the epithelial cells lining the projections. In fact, this extracellular matrix drives the growth of the otic vesicle epithelial projections. Alcian blue staining of the epithelial projections at 2 dpf (days post fertilization) was weak in *cog4*$^{-/-}$ zebrafish larvae (suggesting that the extracellular matrix was deficient in proteoglycans), whereas staining was strong in wild-type larvae, indicating secretion of proteoglycans. Collagen type II immunolabeling at 52 hpf (hours post fertilization) revealed abundant secretion into the core of the projections in wild-type larvae, with markedly decreased secretion in *cog4*$^{-/-}$ mutants. However, the level of collagen type II labeling at 5 dpf was similar in the pillars and common crus in wild-type and *cog4*$^{-/-}$ larvae and, in fact, ectopically accumulated in the lumen of the semicircular canals, suggesting that another trafficking pathway may compensate for the lack of *Cog4* function at later developmental stages. The author's interpretation was that defective secretion of extracellular matrix components was responsible for abnormal growth of the epithelial projections of the otic vesicle in *cog4*$^{-/-}$ mutants (Clément et al. 2019).

A different group overexpressed the human mutant allele responsible for Saul-Wilson syndrome in developing zebrafish embryos. Specifically, human *COG4*^{WT} and *COG4*$^{p.G516R}$ mRNAs or DNA constructs were injected into one-cell stage embryos, and the authors confirmed that the *COG4*^{WT} and *COG4*$^{p.G516R}$ injected alleles were expressed at similar levels and that no adverse effects were seen in embryos expressing the *COG4*^{WT} compared to uninjected siblings. An increase in glypicans was seen in embryos expressing *COG4*$^{p.G516R}$ but not *COG4*^{WT} at 3 dpf. The expression of the *COG4*$^{p.G516R}$ mutant allele caused a shortened anterior–posterior body axis at 3 dpf and 6 dpf. Zebrafish at 4 dpf were stained with wheat germ agglutinin (a lectin binding to glycoproteins in the extracellular matrix of cartilage) to visualize chondrocyte morphology. In embryos expressing the mutant *COG4*$^{p.G516R}$ allele, the Meckel's cartilage was deformed, with defects in stacking and more rounded chondrocytes. These defects were observed as late as 7 dpf by Alcian blue staining (Xia et al. 2021).

Still, a different group generated a *C. elegans* model using CRISPR-Cas9 technology, recreating the same glycine-to-arginine substitution in the worm ortholog *cogc-4*. Specifically, animals were homozygous for the orthologous p.G529R variant. The *cogc-4(av107)* worms did not display any obvious differences compared to wild-type worms, despite using a variety of assays including stressing the worms using heat and Paraquat, as well as RNAi against the seven other COG subunits in an attempt to unmask a phenotype (Zafra et al. 2021).

A knock-in mouse model was recently established in our lab using CRISPR-Cas9 technology, with animals heterozygous for the orthologous p.G512R. Further studies are underway to characterize the skeletal phenotype of these mice.

12.4 Pathogenesis

Compared to control cells, fibroblasts from affected patients show normal mRNA expression and protein levels of COG4 and other COG subunits. Fractionation assays by differential centrifugation show normal subcellular distribution of COG4, with the majority in the cytosolic fraction and less than 5% bound to the membrane in both control and patients' cells. The cytosolic versus membrane-bound distribution of other COG subunits is also similar in control vs. mutant cells. Binding of COG4 to other COG subunits both from lobe A (COG3) and lobe B (COG5, COG7) is not altered as assessed via co-immunoprecipitation (Ferreira et al. 2018).

Abnormal Golgi morphology (defined as co-localization of GM130 and TGN46, suggesting the collapse of the cis- and trans-Golgi stacks) was identified in patients' fibroblasts. The Golgi volume (after normalization to the nuclear volume) was significantly decreased, with Golgi volumes 2.8-fold lower in patients' cells compared to that in control cells. Brefeldin A is a fungal metabolite that inhibits vesicular trafficking from the Golgi to the endoplasmic reticulum (ER). Control and patients' fibroblasts were treated with brefeldin A to assess the time it took for giantin staining to localize from the Golgi to the ER. The Golgi-to-ER retrograde transport was significantly faster in patients' cells compared to control cells. On the other hand, when brefeldin A was washed out and the Golgi allowed to reform, the ER-to-Golgi anterograde transport was slower in patients' cells than in control cells (Ferreira et al. 2018).

Despite the marked reduction in volume of the Golgi, an important site for protein glycosylation, glycan analysis in mutant fibroblasts and serum/plasma, and glycosylation status of serum transferrin and apoCIII were not different from that of controls (Ferreira et al. 2018). There is indeed evidence that abnormalities in COG can lead to glycosylation-independent cellular defects (Blackburn et al. 2018). It is possible, however, that glycosylation could be cell-dependent. Since the initiation and polymerization of glycosaminoglycans onto its core proteins (i.e., glycosylation of proteoglycans) also take place mainly in the Golgi apparatus, the glycosylation of decorin was assessed in fibroblasts. A greater proportion of higher molecular weight decorin was seen in patients' cells compared to control cells, indicating either longer glycosaminoglycan chain length, increased sulfation, or both (Ferreira et al. 2018).

Proteoglycans represent a major component of the extracellular matrix, and the aforementioned results encouraged further studies into proteoglycan modification in Saul-Wilson syndrome cells. Glypicans belong to the group of heparan sulfate proteoglycans (HSPGs) and are critical for growth and bone development through multiple signaling pathways, such as for example the Wnt pathway. Glypicans were found to be consistently elevated at the cell surface in patients' cell lines, with an average twofold elevation. Glypicans 1, 4, 5, and 6 were detectable in dermal fibroblasts; however, no significant change was observed in their transcript abundance, suggesting a post-translational increase due to COG4-dependent abnormal trafficking or turnover, instead of transcriptional regulation. Given prior findings in zebrafish (see above), the authors assayed the transcript abundance of *WNT4* and other non-canonical *WNT*s in patients' fibroblasts; interestingly, both *WNT4* transcript and protein levels were increased in patients' fibroblasts as compared to controls. Given similar results between Saul-Wilson syndrome patients' fibroblasts and zebrafish with regards to altered non-canonical Wnt signaling, the authors proposed the involvement of this pathway as one possible pathogenetic mechanism (Xia et al. 2021).

The $COG4^{p.G516R}$ mutant allele was engineered in the chondrocyte-like cell line SW1353 by CRISPR technology. These $COG4^{p.G516R}$ cells show accelerated brefeldin A–induced retrograde vesicular transport, just as seen in patient-derived fibroblasts (see above). The $COG4^{p.G516R}$ cells and COG4-KO cells showed subtle differences in N-glycans; specifically, $COG4^{p.G516R}$ cells showed increased Fuc1HexHexNAc3 and decreased Fuc1Hex5HexNAc4, which are synthesized in medial- and trans-Golgi compartments, respectively. However, these N-glycosylation changes in COG-$COG4^{p.G516R}$ cells were extremely subtle and cell type–dependent. A pellet culture was used to examine the in vitro chondrogenic differentiation; $COG4^{p.G516R}$ cells showed significantly reduced spheroid sizes on day 10 compared to WT and COG4-KO cells. Hematoxylin and eosin (H&E) staining revealed a necrotic core on day 6 in COG4$^{p.G516R}$ cells, with increased apoptosis starting at day 4, probably explaining the reduced aggregate size. Mixed cultures of COG4$^{p.G516R}$ cells with wild-type cells and medium supplement rescued the aforementioned spheroid size and cell death phenotypes. Prompted by this medium complementation experiments, the authors examined the proteins in the conditioned medium by mass spectrometry–based proteomics. The comparisons of COG4$^{p.G516R}$ cells vs. WT cells and COG4$^{p.G516R}$ cells vs. KO cells found a significant over-representation of categories mainly related to the extracellular matrix and some others, with some top candidates being "Naba_Matrisome," "GOCC_Collagen_Containing_Extracellular_Matrix," "Angiogenesis," "Complement," "Coagulation," and "Epithelial_Mesenchymal_Transition." There were 84 proteins with significantly decreased levels and 48 proteins with significantly increased levels belonging to the "Naba Matrisome." These altered matrisome proteins included various proteoglycans (decorin, versican, lumican, hyaluronan, and proteoglycan link protein 1), glycoproteins (IGFBPs, LTBPs, fibulins, laminins, PCOLCE, POSTN), various classes of collagens, extracellular matrix regulators (ADAM metallopeptidases, MMPs, serpin peptidase inhibitor, cathepsins), extracellular matrix–affiliated proteins (glypicans, syndecans, annexins), and secreted factors (BMPs, Wnt proteins, growth factors, chemokines). Many of these deficient matrisome proteins are common regulators of other affected pathways. Prompted by the mass spectrometry data, the authors further examined the mRNA level of differentiation markers MMP13 and COL10A1 of aggregates in pellet culture treated with TGFβ3 and found that COG4$^{p.G516R}$ cells showed profoundly low levels of MMP13 on days 0–4 and COL10A1 on days 2–4 (Xia et al. 2022).

12.5 Conclusions and Future Perspectives

Although the molecular basis of Saul-Wilson syndrome was only discovered 5 years ago, recent literature appears to indicate the involvement of the extracellular matrix in the pathogenesis of this rare disease. Given the known role of the COG complex in protein trafficking and protein glycosylation, either a mistargeting or misglycosylation of proteins is likely to be the culprit for the mechanism of disease. Other disorders of vesicular trafficking are known to result in the involvement of various organ systems affected in patients with Saul-Wilson syndrome. As an example, retinal degeneration can result from aberrant protein trafficking, for instance, of rhodopsin (Bales and Gross 2016). Similarly, abnormal collagen trafficking results in skeletal fragility (Claeys et al. 2021), while aberrant trafficking of the granulocyte colony-stimulating factor receptor (G-CSF) leads to neutropenia (Frey et al. 2021). The precise proteins involved in each of the manifestations of Saul-Wilson syndrome, a multisystemic disorder, remain to be elucidated. The recent discovery of the involved gene opens new avenues of investigation into the pathophysiology of the disease, and future work should uncover further associations between abnormally trafficked proteins and patients' phenotypes.

Acknowledgements This work was supported in part by the Intramural Research Program of the National Human Genome Research Institute (ZIA HG200407).

Conflicts of Interest The author declare no conflict of interest.

References

Bales K, Gross A (2016) Aberrant protein trafficking in retinal degenerations: the initial phase of retinal remodeling. Exp Eye Res 150:71. https://doi.org/10.1016/j.exer.2015.11.007

Blackburn JB, Kudlyk T, Pokrovskaya I, Lupashin VV (2018) More than just sugars: conserved oligomeric Golgi complex deficiency causes glycosylation-independent cellular defects. Traffic (Copenhagen, Denmark) 19(6):463–480. https://doi.org/10.1111/tra.12564

Chinen Y, Kaneshi T, Kamiya T, Hata K, Nishimura G, Kaname T (2015) Progressive hip joint subluxation in Saul-Wilson syndrome. Am J Med Genet A 167A(11):2834–2838. https://doi.org/10.1002/ajmg.a.37278

Claeys L, Storoni S, Eekhoff M, Elting M, Wisse L, Pals G, Bravenboer N, Maugeri A, Micha D (2021) Collagen transport and related pathways in osteogenesis imperfecta. Hum Genet 140(8):1121. https://doi.org/10.1007/s00439-021-02302-2

Clément A, Blanco-Sánchez B, Peirce JL, Westerfield M (2019) Cog4 is required for protrusion and extension of the epithelium in the developing semicircular canals. Mech Dev 155:1–7. https://doi.org/10.1016/j.mod.2018.09.003

Ferreira CR, Niiler T, Duker AL, Jackson AP, Bober MB (2020a) Growth in individuals with Saul-Wilson syndrome. Am J Med Genet A 182(9):2110–2116. https://doi.org/10.1002/ajmg.a.61754

Ferreira CR, Xia Z-J, Clément A, Parry DA, Davids M, Taylan F, Sharma P, Turgeon CT, Blanco-Sánchez B, Ng BG, Logan CV, Wolfe LA, Solomon BD, Cho MT, Douglas G, Carvalho DR, Bratke H, Haug MG, Phillips JB et al (2018) A recurrent De novo heterozygous COG4 substitution leads to Saul-Wilson syndrome, disrupted vesicular trafficking, and altered proteoglycan glycosylation. Am J Hum Genet 103(4):553–567. https://doi.org/10.1016/j.ajhg.2018.09.003

Ferreira CR, Zein WM, Huryn LA, Merker A, Berger SI, Wilson WG, Tiller GE, Wolfe LA, Merideth M, Carvalho DR, Duker AL, Bratke H, Haug MG, Rohena L, Hove HB, Xia Z-J, Ng BG, Freeze HH, Gabriel M et al (2020b) Defining the clinical phenotype of Saul-Wilson syndrome. Genet Med 22(5):857–866. https://doi.org/10.1038/s41436-019-0737-1

Frey L, Ziętara N, Łyszkiewicz M, Marquardt B, Mizoguchi Y, Mi L, Liu Y, Giesert F, Wurst W, Dahlhoff M, Mr S, Wolf E, Somech R, Klein C (2021) Mammalian VPS45 orchestrates trafficking through the endosomal system. Blood 137(14):1932. https://doi.org/10.1182/blood.2020006871

Hall CM (2002) International nosology and classification of constitutional disorders of bone (2001). Am J Med Genet 113(1):65–77. https://doi.org/10.1002/ajmg.10828

Hersh JH, Joyce MR, Spranger J, Goatley EC, Lachman RS, Bhatt S, Rimoin DL (1994) Microcephalic osteodysplastic dysplasia. Am J Med Genet 51(3):194–199. https://doi.org/10.1002/ajmg.1320510304

Koruga N, Pušeljić S, Tomac V, Soldo Koruga A, Marjanac I, Biljan B, Šantić K, Lenz I, Pušeljić N (2022) Severe Cranio-cervical stenosis in a child with Saul-Wilson syndrome: a case report. Children (Basel, Switzerland) 9(4):532. https://doi.org/10.3390/children9040532

Matar MM (2019) Anesthetic considerations in Saul-Wilson syndrome, a rare skeletal dysplasia: a case report. A & A Practice 13(3):93–95. https://doi.org/10.1213/XAA.0000000000001001
Ng BG, Sharma V, Sun L, Loh E, Hong W, Tay SKH, Freeze HH (2011) Identification of the first COG-CDG patient of Indian origin. Mol Genet Metab 102(3):364–367. https://doi.org/10.1016/j.ymgme.2010.11.161
Reynders E, Foulquier F, Leão Teles E, Quelhas D, Morelle W, Rabouille C, Annaert W, Matthijs G (2009) Golgi function and dysfunction in the first COG4-deficient CDG type II patient. Hum Mol Genet 18(17):3244–3256. https://doi.org/10.1093/hmg/ddp262
Saul R (1982) Unknown cases 1:102–105
Saul RA, Wilson WG (1990) A "new" skeletal dysplasia in two unrelated boys. Am J Med Genet 35(3):388–393. https://doi.org/10.1002/ajmg.1320350315
Saul RA, Wilson WG (2019) COMMENTARY-the Saul-Wilson syndrome from its early days until now. Am J Med Genet A 179(2):159–160. https://doi.org/10.1002/ajmg.a.8
Superti-Furga A, Unger S (2007) Nosology and classification of genetic skeletal disorders: 2006 revision. Am J Med Genet A 143A(1):1–18. https://doi.org/10.1002/ajmg.a.31483
Warman ML, Cormier-Daire V, Hall C, Krakow D, Lachman R, LeMerrer M, Mortier G, Mundlos S, Nishimura G, Rimoin DL, Robertson S, Savarirayan R, Sillence D, Spranger J, Unger S, Zabel B, Superti-Furga A (2011) Nosology and classification of genetic skeletal disorders: 2010 revision. Am J Med Genet A 155A(5):943–968. https://doi.org/10.1002/ajmg.a.33909
Xia Z-J, Mahajan S, Paul Daniel EJ, Ng BG, Saraswat M, Campos AR, Murad R, He M, Freeze HH (2022) COG4 mutation in Saul-Wilson syndrome selectively affects secretion of proteins involved in chondrogenesis in chondrocyte-like cells. Front Cell Dev Biol 10:979096. https://doi.org/10.3389/fcell.2022.979096
Xia Z-J, Zeng X-XI, Tambe M, Ng BG, Dong PDS, Freeze HH (2021) A dominant heterozygous mutation in COG4 causes Saul-Wilson syndrome, a primordial dwarfism, and disrupts zebrafish development via Wnt signaling. Front Cell Dev Biol 9:720688. https://doi.org/10.3389/fcell.2021.720688
Zafra I, Nebenfuehr B, Golden A (2021) Saul-Wilson syndrome missense allele does not show obvious Golgi defects in a C. Elegans model. MicroPubl Biol 2021. https://doi.org/10.17912/micropub.biology.000373

Chapter 13
Membrane Trafficking to the Extracellular Matrix and Skeletal Dysplasia

Andrea Maria Guarino, Rossella Venditti, and Cathal Wilson

Abstract The bone extracellular matrix (ECM) is composed mainly of collagen fibrils and many non-collagenous proteins. A large number of skeletal dysplasias are caused by disruption of the ECM architecture due to the mutation of some of these ECM components, particularly collagen, which not only disrupts the bone structure but also alters signaling pathways that emanate from the ECM. Rather than being a static entity, the ECM exhibits dynamic movement and turnover that have important regulatory and structural consequences for bone development and maintenance. The construction and turnover of the ECM relies on the transport of ECM components by membrane trafficking: as a consequence, disruption of the membrane trafficking machinery can also lead to skeletal dysplasia. Here we provide an overview of the mechanisms by which mutations of some of the secretory trafficking machinery lead to human skeletal dysplasia.

13.1 Introduction

Bone mineralized extracellular matrix (ECM) is composed mainly of type I collagen produced by osteoblasts while the ECM of cartilage is composed chiefly of type II collagen produced by chondrocytes. The majority of genetic alterations that cause bone defects affect the collagens themselves, e.g. COL1A1 or COL2A1 (Bateman et al. 2022), enzymes involved in their processing such as P3H1 (Cabral et al. 2007) or P4HB (Rauch et al. 2015), or chaperones such as HSP47 (Syx et al. 2021). Defects in ECM architecture may arise from quantitative defects or an alteration of protein structure, for example by mutations of collagens that affect their folding and thus exit from the endoplasmic reticulum (ER) (Bateman et al. 2022) or processing

A. M. Guarino · R. Venditti · C. Wilson (✉)
Telethon Institute of Genetics and Medicine, TIGEM, Pozzuoli/Naples, Italy

Department of Molecular Medicine and Medical Biotechnology, University of Naples Federico II, Naples, Italy

A. Rossi, F. Zaucke (eds.), *The Extracellular Matrix in Genetic Skeletal Disorders*, Biology of Extracellular Matrix 16,
https://doi.org/10.1007/978-3-031-70835-0_13

enzymes that exhibit normal collagen trafficking but result in disorganization of the ECM, respectively (Rauch et al. 2015; Balasubramanian et al. 2018; Syx et al. 2021). Defects in diverse ECM proteoglycans or glycoproteins cause skeletal dysplasia due to mutations of processing enzymes located along the secretory pathway causing alterations of ECM integrity (Lamandé and Bateman 2020), without affecting protein transport itself. Instead, mutations of components of the protein transport machinery appear to be mainly associated with a block in the transport of procollagen, rather than other ECM components, from the endoplasmic reticulum, especially at ER exit sites (ERES). It is the large size of procollagen molecules that renders them particularly susceptible to changes in the membrane trafficking machinery with consequent skeletal dysplasia. As a result, mutations of several components of the protein transport machinery that result in defects in the ECM and resultant skeletal dysplasia might be considered as representing collagenopathies.

13.2 Membrane Trafficking: ER Exit of Procollagen

Transport of newly synthesized proteins destined for the ECM begins with their exit from the ER. Details of ER exit and ER-to-Golgi transport that are mediated by the COPII complex have been covered comprehensively very recently (Tang and Ginsburg 2023) and will be described only briefly here (Fig. 13.1). The core components of the COPII machinery required for ER exit are SAR1, SEC23/24, SEC13/31 together with the SAR1 activator SEC12. Assembly of COPII on the ER membranes is initiated by guanine nucleotide exchange on the small GTPase SAR1 mediated by SEC12, an integral membrane glycoprotein. The presence of active membrane-bound SAR1-GTP, which triggers positive curvature induction (Paul et al. 2023), leads to successive recruitment of the coat components SEC23-SEC24 (inner layer) and SEC13-SEC31 (outer layer). SEC23 is a GTPase-activating protein (GAP) that promotes the conversion of SAR1-GTP back to SAR1-GDP, which promotes coat disassembly. SEC24, of which there are four paralogs in vertebrates, drives the selective recruitment of cargo proteins into nascent carriers. In addition to cargo binding to COPII, the cooperative binding of COPII components and the influence of accessory factors (see below) all play a role in the modulation of COPII turnover and the formation and dynamics of ER exit sites (D'Arcangelo et al. 2013). This dynamic role of COPII-mediated ER exit is particularly important in the case of large cargos such as long procollagen fibrils or chylomicrons (large lipoproteins produced in enterocytes) (Saito et al. 2017).

COPII subunit mutations in skeletal dysplasia. Although there are multiple COPII subunit paralogs in vertebrates [two SAR1 (SAR1A and B), two SEC23 (SEC23A and B), four SEC24 (SEC24A–D), and two SEC31 (SEC31A and B)], mutations of only two of them are associated with skeletal dysplasia. Mutations of SEC23A cause cranio-lenticulo-sutural dysplasia (CLSD, OMIM #607812), a skeletal defect that manifests as large and late closing fontanels and calvarial hypomineralization, Y-shaped cataracts, skeletal defects, and hypertelorism and other facial

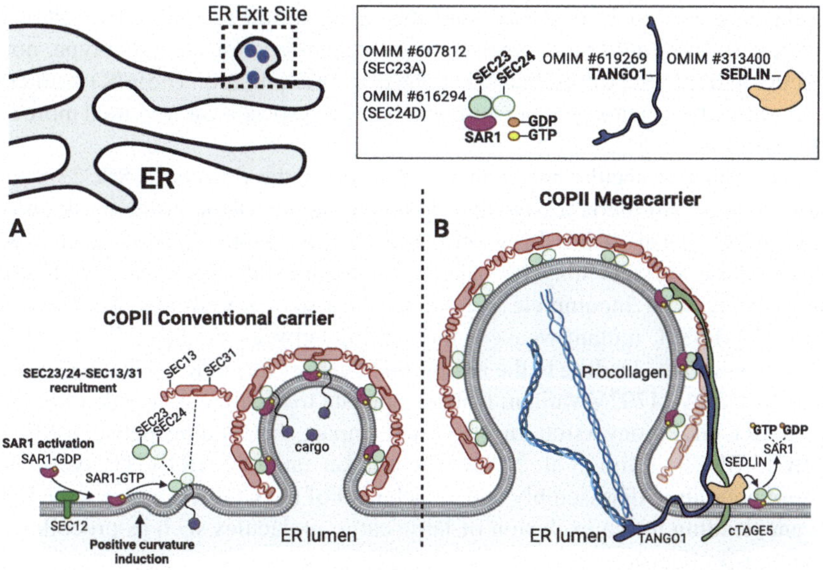

Fig. 13.1 COPII-dependent cargo secretion at ER exit sites. (**a**) SEC12 acts as a GEF for SAR1-GDP, promoting GDP-GTP exchange. Active SAR1-GTP then binds the ER membrane and induces membrane curvature. The inner COPII components SEC23 and SEC24 are recruited, with SEC24 binding cargo proteins within the ER lumen. SEC13/SEC31 are recruited by binding to the inner coat complex and impose further membrane curvature. SAR1-GTP hydrolysis and thus inactivation occurs through the GAP activity of SEC23, which is further enhanced by SEC31, leading to coat disassembly. (**b**) Transport of large cargo molecules in megacarriers may require tight regulation of the SAR1 GTPase cycle. TANGO1 acts as an organizer for procollagen exit. In addition to interacting with procollagen, it also binds SEC23, Sedlin, and recruits SEC16 (not shown), all of which modulate SAR1 activity. When any of these components is defective, the rate the SAR1 cycle is disrupted, with a quicker or slower rate of SAR1-GTP hydrolysis, depending on the type of mutation, so that small cargos are transported but large cargo carriers cannot form. This results in procollagen retention in the ER, disruption of the ECM, and skeletal dysplasia (see associated OMIM numbers in the box). cTAGE5 is reported to associate with TANGO1 at ER exit sites and bind SEC12, promoting SAR1-GTP production and COPII formation. Its association with skeletal dysplasia is unclear

dysmorphisms. A homozygous F382L missense mutation and a heterozygous M702V missense mutation causing CLSD have been described for SEC23A that lead to ER dilation and an impairment of procollagen exit from the ER (Boyadjiev et al. 2006, 2011). An anatomical and morphological correlate was found by knocking down *sec23a* expression in zebrafish (Boyadjiev et al. 2006). In the same issue, Lang et al. (2006) described the zebrafish *crusher* mutant as a loss-of-function mutation in *sec23a* that led to craniofacial phenotypes. *Crusher* mutant chondrocytes accumulated large amounts of ECM proteins in extended endoplasmic reticulum compartments and defective collagen post-translational modifications, leading to extracellular matrix defects, providing further insights into the mechanisms that could underlie the disease. Further, they reported that chondrocytes sense the absence of extracellular matrix by maintaining abnormally high transcription of collagen genes such as *col2a1* and the transcription factor *sox9a*, although it is unclear if this would be compensatory or deleterious for the phenotype, nor if this occurs in humans. A third heterozygous missense mutation (E599K) with craniofacial and ocular phenotypes but without chondrodysplasia was reported more recently (Cisarova et al. 2022).

The exact molecular mechanisms underlying the defects in SEC23A function, also considering the recessive and dominant nature of the different mutations, are not entirely clear. It has been proposed that the phenotypes associated with the monoallelic variants may be caused by digenic inheritance although dominant inheritance with incomplete penetrance is also a possibility. The homozygous SEC23A-F382L mutant weakens the binding between the SEC13–31 and SEC23-SEC24 complex, leading to the inhibition of carrier formation (Fromme et al. 2007). The SEC23A-M702V variant, instead, appears to accelerate the rate of SAR1-GTP to SAR1-GDP conversion and does not impact the binding between SEC23 and SEC13-SEC31 (Kim et al. 2012). The quicker rate of SAR1-GTP hydrolysis may cause premature disassembly and detachment of the COPII coat from the ER membrane, leading to the exclusion of large cargo molecules such as procollagen from the transport carrier.

Lang et al. (2006) also reported that the morpholino-knockdown of *sec23b* in zebrafish leads to craniofacial phenotypes. In humans, however, SEC23B mutation leads to dyserythropoietic anemia type II (CDAII, OMIM #224100) (Schwarz et al. 2009) while in mice it results in a pancreatic phenotype (Tao et al. 2012). The two paralogs SEC23A and SEC23B can functionally compensate for each other when expressed in the appropriate context (Khoriaty et al. 2018; King et al. 2021) or when overexpressed. An active form of the transcription factor BBF2H7 (p60 BBF2H7) that specifically upregulates SEC23A expression (Saito et al. 2009) resulted in a partial correction of the hypo-glycosylation phenotype caused by SEC23B mutation in CDAII (Pellegrin et al. 2019). A mutagenesis screen in zebrafish identified Sec23a mutation as causing not only craniofacial defects but also lymphatic endothelial cell migration in the embryo due to defective Col2a1 transport (Chaudhury et al. 2020), which could broaden the phenotypic spectrum of these diseases. However, these differences in phenotypes in different organisms may be due to differential expression of SEC24A and SEC24B and caution against always extrapolating from animal models to humans.

The second COPII component associated with skeletal dysplasia is SEC24D. Mutations of SEC24D cause Cole-Carpenter Syndrome 2 which is associated with low bone mass or an osteogenesis imperfecta-like syndrome (OMIM #616294). Compound heterozygosity has been described in numerous reports with each of the unaffected parents being heterozygous for one of the mutations (Garbes et al. 2015; Moosa et al. 2016; Zhang et al. 2017; Takeyari et al. 2018; Miao et al. 2022). Analysis of fibroblasts from one individual showed dilated ER tubules and defective procollagen export from the ER (Garbes et al. 2015). A role in procollagen trafficking was supported by a *sec24d/bulldog* zebrafish mutant that had a shorter body length, kinked pectoral fins, and a shorter and misshapen head skeleton that were linked to the accumulation of extracellular matrix proteins in the ER and reduced deposition of type II collagen (Sarmah et al. 2010). Instead, early embryonic lethality of Sec24d knockout was reported in mice (Baines et al. 2013).

The assembly and disassembly of the COPII complex relies on the dynamic cooperation of the individual components in a combinatorial fashion to tailor carrier formation to the cargo to be exported (Stancheva et al. 2020). Indeed, export of large cargo from the ER appears to depend on being able to regulate the rate of the SAR1 GTPase cycle (Saito et al. 2017). SEC31 binds as an extended polypeptide across a composite surface of SEC23 and SAR1-GTP and interacts directly with SEC23 to stimulate its GTPase-activating protein activity, thereby triggering SAR1 GTP hydrolysis (Bi et al. 2007). SEC31 may thus fine-tune the GAP activity of SEC23 to accommodate large cargoes such as collagen. This modulation of the COPII cycle to accommodate large molecules can be observed in the role of COPII accessory proteins that play a role in procollagen transport whose mutation leads to skeletal dysplasia, as discussed in the next section.

13.2.1 Accessory Proteins in COPII-Mediated Procollagen Exit from the ER

Spondyloepiphyseal dysplasia tarda (SEDT, OMIM #313400), an X-linked skeletal disorder caused by mutation of the TRAPPC2/Sedlin gene, is characterized by short stature, disproportionate short trunk, and precocious mild osteoarthritis, which manifests in the early childhood (Whyte et al. 1999). TRAPPC2 is a core component of the highly conserved oligomeric TRAnsport Protein Particle (TRAPP) complex that controls different stages of the membrane trafficking pathway (Riedel et al. 2018). Pathogenic variants in 8 of the 14 TRAPP proteins are known to cause human diseases called TRAPPopathies (Sacher et al. 2019). Even though TRAPPC2 is part of a large complex, it is the only one whose mutation results in skeletal dysplasia. Mutation of other TRAPP components causes early onset neurological or muscular disorders while SEDT is late onset. This may be due in part to an expressed TRAPPC2B pseudogene that could mask some of the effects of TRAPPC2 mutation (Gécz et al. 2000; Ghosh et al. 2001; Vinckenbosch et al. 2006). The analysis of the molecular mechanisms underlying this genetic disease has underlined the role of the dynamics of the GTPase cycle in the export of large cargo molecules (Fig. 13.1). Sedlin and SEC23 coexist in a complex with SAR1, but Sedlin contacts SAR1 in its GTP-bound state thereby favoring the SAR1-GAP activity of SEC23 and accelerating the release of SAR1 from the membranes, thus promoting the efficient cycling of SAR1. In the absence of Sedlin, abnormal carriers are formed that cannot incorporate the large procollagen fibrils (Venditti et al. 2012). Knockout of Sedlin in medaka fish led to phenotypes reminiscent of those observed in humans, with altered deposition of collagen, dilation of the ER, and skeletal phenotypes (platyspondyly/reduced length) (Zappa et al. 2023). Additional phenotypes not observed in humans were defects in the eye and gut, which could be due to a defective ECM, which might not manifest in SEDT patients because of the presence of the expressed pseudogene (Ghosh et al. 2001; Vinckenbosch et al. 2006; Zappa et al. 2023). Analysis of transcriptome changes suggested that impaired collagen secretion inhibits the transcription of other collagens and ECM components. Note that while both Sec23A (Lang et al. 2006) and Sedlin (Zappa et al. 2023) depletion in fish models led to retention of procollagen in the ER, increased transcription of collagen genes such as col2a1 and of the transcription factor sox9a was observed in the former, but lower collagen gene expression and no change in sox9a expression were found in the latter. These animal models may thus be useful in deciphering how specific genetic defects contribute to the associated pathology.

Sedlin is recruited by TANGO1, an important player in the exit of collagen molecules from the ER (Venditti et al. 2012; Raote et al. 2018) (Fig. 13.1). Complete loss of TANGO1 is embryo lethal (Guillemyn et al. 2021). An aborted fetus with almost total bone loss had a homozygous 4-base pair deletion in TANGO1: the heterozygous parents showed decreased TANGO1 mRNA expression and protein levels, but type I collagen secretion and extracellular matrix organization were normal, supporting a threshold model for clinical phenotype development (Guillemyn et al. 2021). A truncated TANGO1 protein was identified as a disease-causing homozygous mutation (OMIM #619269) resulting in severe dentinogenesis imperfecta, short stature, various skeletal abnormalities, insulin-dependent diabetes mellitus, sensorineural hearing loss, and mild intellectual disability (Lekszas et al. 2020). Although there is defective collagen secretion, the other phenotypes suggest that other cargo molecules may be affected, either directly or indirectly (McCaughey et al. 2021). Indeed, TANGO1 may be involved in the secretion not only of different types of collagen (types I, II, III, IV, VII, and IX) in different cell types (chondrocytes, fibroblasts, endothelial cells), but also of other bulky or high volume cargoes (see Raote et al. (2023) for a recent review), while a block of bulky cargo exit can cause ER stress and have secondary effects by inhibiting the transport of other cargo molecules (Ríos-Barrera et al. 2017).

TANGO1 appears to be a key element in regulating the dynamics of COPII carrier formation by creating a scaffold for organizing various elements that mediate the secretion of large cargo. In addition to Sedlin, TANGO1 also recruits SEC16, another peripheral protein involved in COPII coat assembly (Maeda et al. 2017), and it has been proposed that SEC16 counters SEC31-simulated SEC23 GAP activity, stabilizing the coat complex assembled on membranes to facilitate carrier formation (Yorimitsu and Sato 2012). Bi-allelic mutations in SEC16B were identified in a child with osteogenesis imperfecta-like symptoms, with altered collagen trafficking, a general trafficking defect at the level of the ER, and increased ER stress (El-Gazzar et al. 2023).

A member of the TANGO1 family, cTAGE5, associates with TANGO1 at ER exit sites and binds SEC12, thus promoting SAR1-GTP production and COPII formation (Saito et al. 2014). Thus, this assembly of factors acting on the GEF and the GAP of SAR1 can tightly regulate its GTPase cycle to accomplish large cargo secretion (Fig. 13.1). Murine models have shown an involvement of cTAGE5 in proinsulin trafficking in β cells (Fan et al. 2017) and regulating export from the ER of molecules involved in cerebellar development (Wang et al. 2019), showing the broad range of functions of the TANGO1/cTAGE5 proteins that underlie the complex phenotypes observed in TANGO1-associated disease (Lekszas et al. 2020). To distinguish between the roles of TANGO1 and cTAGE5, Clark and Link (2021) used a zebrafish model to study the phenotypic consequences of single and double knockouts of Tango1 and Ctage5. As expected, collagen II export was blocked in *tango1* mutant larvae resulting in distended ER and craniofacial defects. Ctage5 mutants, on the other hand, had normal collagen II trafficking in chondrocytes and normal craniofacial features. Instead, lipoprotein trafficking was affected in *ctage5*. Tango1 seems to have a primary role in collagen trafficking and Ctage5 in lipoprotein trafficking in zebrafish. This differential role of Tango1 and Ctage5 in cargo trafficking supports the observations in mammalian cells, although the differences in the numbers and types of isoforms and their differential expression need to be taken into consideration when comparing different organisms.

Morpholino-mediated knockdown in zebrafish of another COPII component, the outer layer Sec13 (section 2), produces craniofacial skeletal defects similar to those observed in sec23a/crusher mutant embryos (Townley et al. 2008), but thus far no associated human diseases have been associated with SEC13 mutation. However, it underlines that skeletal dysplasia associated with the membrane trafficking pathways is principally due to collagen export from the ER. However, a recent report using zebrafish mutants reported that defects in collagen II trafficking from the Golgi due to loss of Rgp1, together with GTPase Rab8a, could also lead to craniofacial defects (Ritter et al. 2023).

While these studies show the requirement of COPII and accessory proteins for procollagen exit from the ER, the actual visualization of mega carriers that transport large molecules from the ER has been difficult. Recent reports challenge the idea that procollagen exits the ER via distinct transport carriers and suggest that there are direct tubular connections between the ER and the ERGIC/Golgi that transfer folded procollagen, and that cargo is exported from the ER in a tubule/carrier uncoated by COPII which, instead, along with accessory proteins, acts to select and concentrate cargo into carriers (McCaughey et al. 2019; Westrate et al. 2020; Shomron et al. 2021; Weigel et al. 2021; Hirata et al. 2022). ERES may provide a flexible platform that produces vesicles and/or tubules according to the type and quantity of cargo. Regardless of the mechanism, mutations of these proteins block procollagen exit from the ER and arrival at the ECM with consequent skeletal dysplasia.

An additional COPII accessory factor is a SEC23 interacting protein (SEC23ip or p125). p125 was originally identified as an interactor of SEC23, but further studies showed that this protein stably associates with the C-terminal part of SEC31 (Ong et al. 2010). Downregulation of p125 in cells leads to severe inhibition of cargo molecules exiting from the ER, delaying the traffic of trans-membrane and soluble cargoes (Ong et al. 2010). In addition, this specific phenotype is reminiscent of the procollagen secretion defects observed in craniofacial disorders described in humans caused by mutations in SEC23A. A recent study described a truncation variant of the p125 gene in an individual with a distinct phenotype of severe intellectual disability, osseous syndactyly, craniofacial and brain malformations (Reuter et al. 2017). The precise function of p125 remains elusive and additional studies to address its physiological role will be necessary, although it participates actively in COPII activity. p125 can also recognize and bind to phosphatidyl-inositol-4-phosphate at the ERGIC (Klinkenberg et al. 2014), probably to ensure that COPII budding is accompanied by the presence of the ERGIC acceptor membranes.

13.3 Retrograde Trafficking to the ER

Retrograde transport from the Golgi complex back to the ER is required for the proper functioning of the secretory pathway by recycling membranes and proteins from the Golgi to ER. The coatomer complex COPI, which in mammalian cells is composed of seven core subunits (α, β, β′, γ, δ, ε, and ζ) is essential in this process. Deficiency of two of these subunits, β′ and δ, has been associated with skeletal dysplasia (Fig. 13.2a, b). The β′ subunit, also known as COPB2, is associated with osteoporosis and developmental delay (OMIM #619884) (Marom et al. 2021). A complete lack of COPB2 appears to be incompatible with life while homozygous mutations lead to microcephaly (DiStasio et al. 2017) and haploinsufficiency results in osteoporosis and, depending on the subject, variable degrees of developmental delay (Marom et al. 2021). COPI seems to have a particular role in brain development—mutation of COPB1 (β-COP) that resulted in the loss of a motif at an interaction interface with β'-COP was reported to cause severe intellectual disability syndrome with cataracts and variable microcephaly (OMIM # 619255) (Macken et al. 2021). Investigating the underpinnings of the osteoporosis phenotype in COPB2-deficient subjects, Marom et al. (2021) observed delayed collagen secretion with retention of type I collagen in the ER and altered distribution of Golgi markers. This was specific for collagen type I trafficking as other cargo molecules tested were not inhibited. Mutation of another COPI subunit, ARCN1 (δ-COP), was described that led to craniofacial disorder characterized by facial dysmorphisms, severe micrognathia, rhizomelic shortening, microcephalic dwarfism, and mild developmental delay (OMIM #617164) (Izumi et al. 2016). ARCN1 deficiency was found to cause defective type I collagen transport, so collagen secretion could represent the mechanism underlying the skeletal phenotypes in COPI-deficient syndromes. However, the spectrum of COPI-deficient phenotypes makes it difficult to draw firm conclusions about the underlying mechanisms (Izumi et al. 2016; Tidwell et al. 2020; Macken et al. 2021; Marom et al. 2021; Ritter et al. 2022) since ECM defects due to delayed collagen secretion could be exacerbated by other factors such as alterations in proteoglycan or glycoprotein modification/deposition caused by a more general disruption of cellular homeostasis.

HSP47 encodes a collagen-binding protein that functions as a chaperone in the ER and may serve as a guide molecule directing collagens to large COPII transport carriers. Mutation of HSP47 itself can cause severe osteogenesis imperfecta (OMIM # 613848) (Christiansen et al. 2010; Song et al. 2018; Syx et al. 2021) but also mislocalization of the wild-type protein is linked to OI (OMIM #619131). The

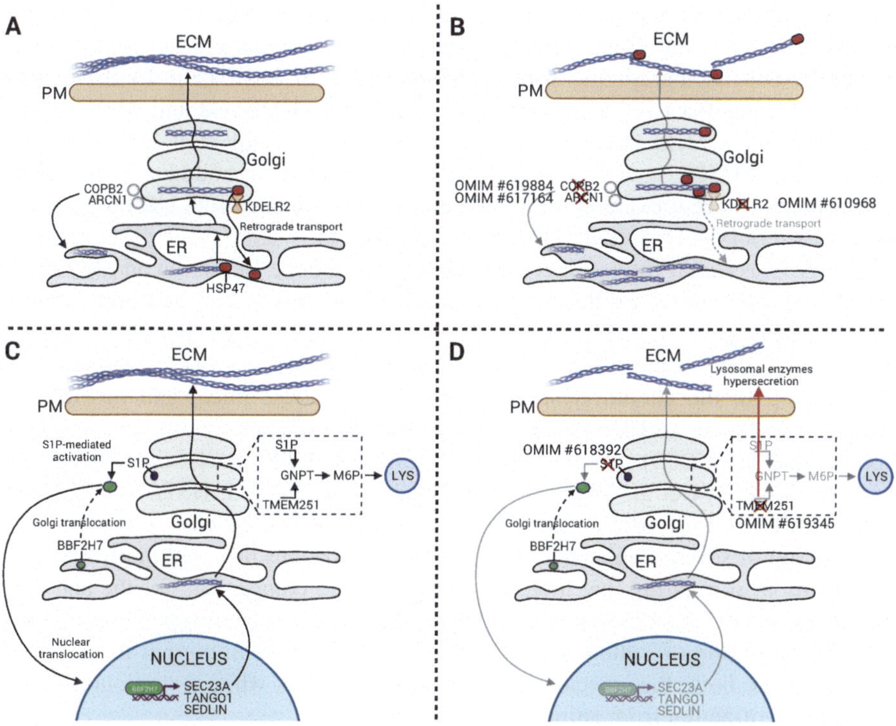

Fig. 13.2 Post-ER membrane trafficking defects that lead to ECM disruption and skeletal dysplasia. Retrograde trafficking from the Golgi to the ER. (**a**) COPB2 or ARCN1 are COPI subunits involved in Golgi-to-ER retrograde trafficking while the KDELR2 receptor recycles HSP47, a collagen-specific molecular chaperone that binds to triple-helical procollagen in the ER, dissociates from collagen in the cis-Golgi, and is recycled to the ER. (**b**) In addition to a number of other phenotypes, mutations of the COPI subunits COPB2 or ARCN1 cause skeletal dysplasia due to collagen retention in the ER, but the underlying mechanisms remain unclear. In the absence of KDELR2, HSP47 remains associated with the collagen molecules, disrupting fibril formation in the ECM. (**c**) The transcription factor BBF2H7 is activated at the Golgi after cleavage by the S1P protease. It translocates to the nucleus inducing the expression of SEC23A, TANGO1, and SEDLIN, which are required for procollagen transport. The S1P protease also forms a complex with TMEM251 to cleave and activate GNPT, which catalyzes the tagging of hydrolases with mannose-6-phosphate (M6P) for their transport to lysosomes after their recognition by receptors in the Golgi. (**d**) Mutation of S1P leads to defective BBF2H7 cleavage and decreased SEC23A, TANGO1, and SEDLIN transcription. S1P mutation also leads to a partial defect in mannose-6-phosphate-dependent Golgi-to-lysosome transport and the extracellular secretion of lysosomal enzymes, probably due to a defect in GNPT activation. Mutation of TMEM251 also leads to the secretion of lysosomal enzymes from the cell due to defective GNPT processing. How the secreted lysosomal enzymes contribute to ECM degradation and skeletal dysplasia has not been fully elucidated. The OMIM number for each skeletal dysplasia is indicated in (**b**) and (**d**)

steady-state localization of HSP47 is maintained by retrograde trafficking from the cis-Golgi to the ER, thanks to the presence of a KDEL-like sequence that facilitates its return to the ER by KDEL receptors (Satoh et al. 1996) (Fig. 13.2a). Bi-allelic variants in the KDEL receptor KDELR2 lead to decreased levels of HSP47, a failure of HSP47 to dissociate from collagen type I, and extracellular HSP47-bound collagen molecules that disrupt fiber formation (van Dijk et al. 2020) (Fig. 13.2b). Levels of FKBP65, another ER chaperone that is retrieved by KDELR2 and whose mutation causes OI (OMIM #610968), were also reduced (van Dijk et al. 2020). In both cases (mutant HSP47 and mutant KDELR2), a disruption of collagen fiber formation appears to occur extracellularly and there is no ER retention, UPR, or associated ER stress (van Dijk et al. 2020; Syx et al. 2021).

Hsp47 is essential for cartilage and endochondral bone formation and KO mice die shortly after birth, exhibiting generalized chondrodysplasia and bone abnormalities (Masago et al. 2012). In these KO mice, collagens and other ECM components accumulate in the ER and induce the UPR, which has been interpreted as contributing to the chondrodysplasia phenotype in these mice, and possibly in humans (Rellmann and Dreier 2018). However, in a disease setting caused by a homozygous p.(R222S) substitution, no ER retention, UPR, or associated ER stress was observed (Syx et al. 2021), as mentioned above. Instead, HSP47-R222S had reduced affinity for type I collagen and was accompanied by elevated expression and binding of several molecular chaperones and enzymes involved in procollagen modification, suggesting that there was an attempt to compensate for the aberrant binding of HSP47-R222S that instead led to overmodification of type I procollagen and the disruption of collagen fiber formation. Thus, it is important to take into consideration both the species and type of mutation when extrapolating the mechanism-of-action from animal disease models.

13.4 The Golgi and Beyond

In general, the post-translational modification and proteolytic processing at the Golgi complex play a vital role in construction of the extracellular matrix, including the generation of proteoglycans via glycosaminoglycan (GAG) synthesis in the Golgi apparatus (Hellicar et al. 2022). Although multiple skeletal phenotypes have been linked to mutations in genes encoding proteins involved in GAG biosynthesis, we do not consider them here as part of the membrane trafficking machinery per se and they have been dealt with in a number of recent reviews (Paganini et al. 2019; Mizumoto and Yamada 2021).

The Golgi-localized site-1 protease (MBTPS1 or S1P) is best known for its function together with site-2 protease (MBTPS2 or S2P) in cleaving and activating the SREBP transcription factors that translocate from the ER to the Golgi when cholesterol levels drop, with the activated SREBPs then translocating to the nucleus to bind cis-acting sterol-responsive elements required for cholesterol biosynthesis. MBTPS1 and MBTPS2 also cleave and activate membrane-bound activating transcription factor 6 (ATF6) when unfolded proteins accumulate in the ER, the activated ATF6 entering the nucleus to drive the transcription of genes encoding ER chaperones (Ye et al. 2000). MBTPS1 mutations can cause skeletal dysplasia

(OMIM #618392) (Kondo et al. 2018; Carvalho et al. 2020; Alotaibi et al. 2022; Yuan et al. 2023) but via a mechanism that is independent of its metabolic role (Kondo et al. 2018). Instead, the skeletal dysplasia appears to be linked to impaired MBTPS1-mediated proteolytic activation of membrane-bound transcription factor BBF2H7 (CREB3L2) (Fig. 13.2c), a UPR transducer that regulates the expression of SEC23A, TANGO1, Sedlin, and HSP47 (Ishikawa et al. 2017) that, as described above, are required for the formation of mega carriers for procollagen transport. These genes were downregulated in MBTPS1-mutated patient fibroblasts and led to impaired collagen secretion (Kondo et al. 2018) (Fig. 13.2d). Additionally, MBTPS1 deficiency led to a partial defect in mannose-6-phosphate-dependent Golgi-to-lysosome transport of lysosomal enzymes (Fig. 13.2c, d), which may contribute to secreted lysosomal enzyme-mediated ECM degradation (Kondo et al. 2018; Carvalho et al. 2020). MBTPS2 mutations can also lead to skeletal dysplasia (OMIM #301014) with a change in the amounts of fatty acids and a reduction in collagen deposition in the extracellular matrix (Kondo et al. 2018; Lindert et al. 2016; Lim et al. 2023) but without an apparent effect on lysosomal enzymes (Kondo et al. 2018). While MBTPS1 and MBTPS2 act together on many common substrates, they may also have independent functions that can lead to disparate phenotypes.

No human skeletal dysplasia is associated with mutation of BBF2H7 itself. BBF2H7 knockout mice display severe chondrodysplasia (Saito et al. 2009), with fibroblasts showing the retention of collagen type II and another matrix molecule, COMP, in a dilated ER, and die shortly after birth. The zebrafish *feelgood* mutation is a hypomorphic mutant of CREB3L2/BBF2H7 that leads to defects in the maturation stage of chondrocyte development with an accumulation of collagens in the ER: a stronger reduction in expression produces a more severe phenotype with increased death (Melville et al. 2011). Melanophore development was also disrupted in *feelgood* mutants but if this can have a wider relevance for other organisms is unknown. The lack of secretion of extracellular matrix proteins promoted by BBF2H7 in chondrocyte differentiation may be lethal in humans.

A role for disrupted mannose-6-phosphate-dependent Golgi-to-lysosome transport of lysosomal enzymes in skeletal dysplasia was also illustrated by the discovery of bi-allelic variants of TMEM251 that cause severe skeletal dysplasia and short stature (OMIM #619345) (Ain et al. 2021). TMEM251 (also known as LYSET) was subsequently found to be required for the cleavage of GlcNAc-1-phopshotransferase (GNPT), the enzyme that catalyzes mannose-6-phosphate modification (Richards et al. 2022; Zhang et al. 2022). GNPT requires cleavage by the site-1-protease (MBTPS1) mentioned above and interacts with both MBTPS1 and TMEM251 at the Golgi (Fig. 13.2c). In the absence of TMEM251, where the symptoms are similar to GNPT deficiency (Mucolipidosis II, OMIM #252500), there is hypersecretion of lysosomal enzymes out of cells (and a resultant shortage of multiple lysosomal enzymes within lysosomes) (Fig. 13.2d). However, it is still unclear how lysosomal enzyme-mediated ECM degradation is involved in generating skeletal manifestations, which are accompanied by other severe symptoms including growth retardation, organomegaly, and cardiorespiratory problems. Lysosomal storage disorders are known to result in skeletal deformity and short stature but the underlying mechanisms, which might include impaired chondrocyte proliferation and hypertrophy, are unclear (Jiang et al. 2020). Knockout of TMEM251 in zebrafish recapitulated skeletal defects observed in human patients with a loss of cartilage and calcified structures (Zhang et al. 2022).

A recurrent theme in discussing genetic disorders is the phenotype–genotype conundrum and how ubiquitously expressed genes result in tissue-specific phenotypes. As mentioned above, one aspect could be whether the relative expression levels of different isoforms are sufficient to compensate for each other in a disease context. However, there is still a vast gap in our knowledge about this subject. An interesting insight was provided by studying mechanisms underlying the skeletal phenotype of achondrogenesis type 1A (ACG1A; OMIM #200600), a severe chondrodystrophy that is invariably lethal before or shortly after birth, which is caused by mutation of the Golgi-associated protein GMAP-210 (TRIP11) (Bird et al. 2018). Conditional GMAP-210 knockout was performed in different cell types (chondrocytes, osteoblasts, osteoclasts, and pancreatic acinar cells) in mice. GMAP-210 inactivation in chondrocytes replicated the ACG1A phenotype, whereas there is no apparent phenotype in the other cell types and the mice were normal. Thus, the defects in osteogenesis appear to be a secondary effect of the chondrocyte dysfunction. Further, using proteomics analysis it was found that the defective cartilage is not the result of a high volume of cargo, since this would also apply in other cell types, but rather that GMAP-210 is essential for trafficking of a select group of cartilage extracellular matrix proteins. These interesting results still leave many aspects to be deciphered such as how GMAP-210 recognizes transport vesicles containing a specific cargo and what are its specific functions in other cell types/tissues.

13.5 Mechanotransduction, Focal Adhesions, and ECM Signaling to the Secretory Pathway

Mechanotransduction (stretching or loading) leads to the upregulation of ECM genes via an integrated network of various signaling pathways, and while some mechanisms and receptors of mechanosensitive signaling pathways in the skeleton have been described (Humphrey et al. 2014; Wang et al. 2022), knowledge about how these are transmitted to changes in gene expression is limited. Mechanical stimulation is important for chondrocyte differentiation and cartilage development (Wang et al. 2022) via these cell receptors and their signal transduction pathways (Prein and Beier 2019; Devos et al. 2023), so changes in chondrocyte differentiation markers and the expression of ECM components (Barbieri et al. 2003; Wai et al. 1998; Komori et al. 2022) might be due to disruption of these mechanosensitive signaling pathways caused by defective ECM deposition.

Since membrane trafficking transports the major components of the ECM, such as collagen, proteoglycans, glycoproteins, and ECM-modifying enzymes, do changes in the ECM that occur by mechanical stimuli, cell–matrix interactions, or defective ECM formation in a disease setting, in turn feedback to modulate membrane trafficking pathways? The degree to which this occurs has probably been underappreciated since cell culture systems do not generally consider the interdependence of altered ECM composition/gene expression and different degrees of substrate stiffness that change under physiological conditions of cartilage or bone remodeling. Some recent studies have addressed this issue by studying how mechanical forces can impact on secretory trafficking pathways.

Evidence for how mechanical forces communicate with the secretory machinery came from the observation that mechanical strain can stimulate COPII-dependent secretory trafficking (Phuyal et al. 2022). Cells grown on micropatterned surfaces or exposed to acute mechanical tension induced the formation of ER exit sites and accelerated ER-to-Golgi trafficking (Fig. 13.3a). Plasma membrane defects were observed when cells with compromised ER exit site function were subjected to mechanical stress. It was found that mechanical stimulation of the secretory pathway required RAC1, a Rho-family GTPase, which on the face of it could have provided a link between the cell surface stimulation and some downstream event since RAC1 is a regulator of the actin cytoskeleton that responds to cell–cell and cell–matrix adhesion (Arthur et al. 2002; Pasapera et al. 2015). However, disrupting actin function did not affect the mechanically stimulated increase in ER exit sites or ER-to-Golgi trafficking. Instead, the authors found that there is a distinct ER-localized pool of RAC1 that interacts directly with SAR1, leading to its activation, possibly via a tripartite complex involving the SAR1 activator SEC12. Thus, as mentioned above, the dynamics of SAR1 activation can be a deciding factor in cargo exit from the ER. The study by Phuyal et al. (2022) used the RUSH (Retention Using Selective Hooks) system (Boncompain et al. 2012) to study the transport of the Golgi-localized protein mannosidase II as a reporter for ER-to Golgi transport, a system that can lead to increased protein retention in the ER and alter ERES size to accommodate a large quantity of proteins (Weigel et al. 2021). Given the importance of mechanical forces in bone remodeling (Wang et al. 2022), it would be of interest to investigate how mechanical force and RAC1 are involved in the transport of endogenous cargo such as ECM components like procollagen in a system relevant for bone or cartilage. Additionally, it will be important to understand the mechanism that relays a signal from mechanical strain to the ER-localized pool of RAC1 and how RAC1 regulation of and interactions with the COPII machinery are controlled.

One possibility is that the transmission of a signal from the ECM to the COPII machinery originates at focal adhesions (FAs). By screening a set of cytoskeleton-associated and related genes by siRNA-mediated knockdown together with siRNA targeting of SEC23A or SEC23B, Jung et al. (Jung et al. 2022) identified a gene signature associated with FAs whose depletion led to a downregulation of SEC23A expression. The KDs led to a change in cell adhesion accompanied by morphological alterations in FAs. Plating cells on substrates such as fibroblast-derived ECM or Matrigel mimicked the conditions of gene KD (which could be rescued by FA kinase inhibition), with a decrease of SEC23A expression and vesicular stomatitis

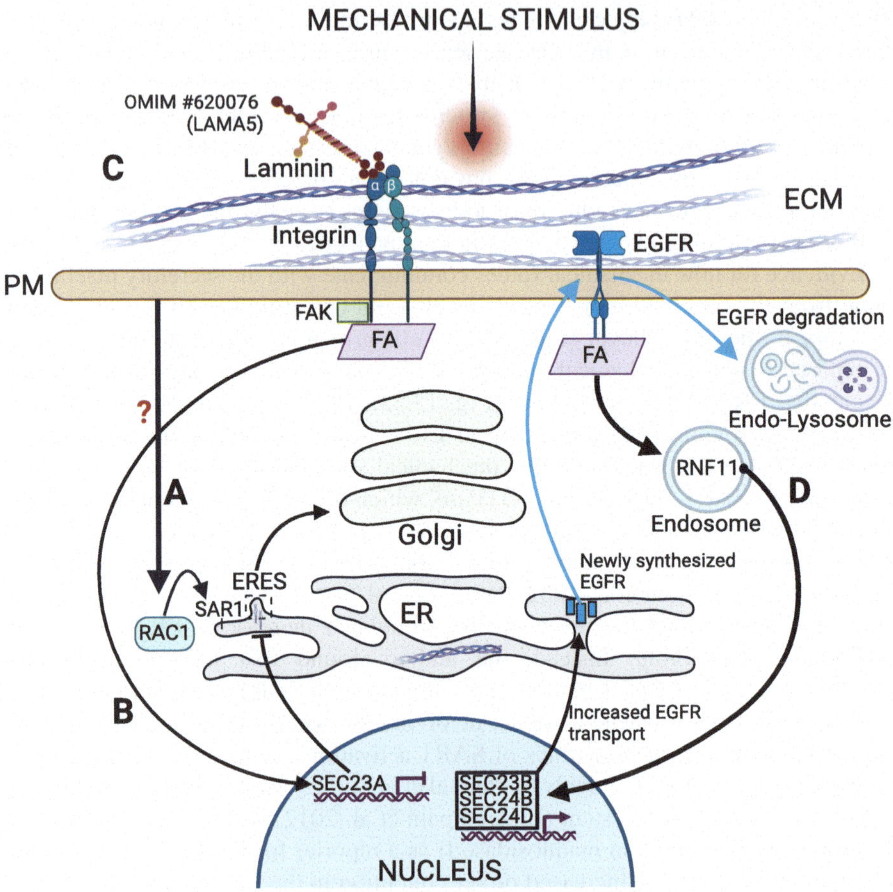

Fig. 13.3 ECM signaling to the secretory pathway. Integrin binding and clustering with FAs serve to communicate mechanical stimuli. (**a**) Mechanical stimulation can lead to SAR1 activation and an increase in ER exit sites via association of SAR1 with an ER-localized pool of the GTPase RAC1, which may stimulate membrane transport. (**b**) The ECM is connected to membrane trafficking by FAK (focal adhesion kinase) signaling to SEC23A expression, which is downregulated when focal adhesions (FA) and FAK signaling are high, for example during cell adhesion and migration, possibly affecting SEC23A-dependent transport. (**c**) LAMA5 (laminin alpha-5, which can mediate ECM–cell interactions) mutation leads to skeletal dysplasia that may operate through disturbed integrin-mediated focal adhesion signaling, which may be modulated in the skeleton through interactions with collagens. (**d**) Under specific ECM stiffness conditions, EGFRs cluster within FAs, and EGF stimulation can lead to EGFR turnover via lysosomal degradation. This can result in RFN11 translocation from endosomes to the nucleus to stimulate SEC23/24 transcription that boosts EGFR transport and replenishment at the PM

virus G protein (VSVG) trafficking (used as a readout for ER to the cell surface transport). This effect on VSVG trafficking was most evident under high cargo-load conditions, in line with the effects of SEC23A defects on collagen trafficking in chondrocytes where, in addition to being a large cargo, collagen is also produced in high amounts. Mechanical forces thus seem to communicate to the COPII secretory machinery from FAs (Fig. 13.3b), but how the signal is transmitted to changes in SEC23A transcription is still unknown. As above, it would be informative to investigate how the trafficking of ECM components is affected under these conditions (especially considering the role of SEC23A in collagen transport).

Therefore, we can add regulation of early steps of the membrane trafficking pathway to the list of functions mediated by FAs in sensing their mechanical environment and transducing a signal across the cell membrane to pathways that regulate extracellular matrix assembly, cell proliferation, and differentiation (Geiger et al. 2001), also in response to different microenvironments in the skeletal system (Choi et al. 2022; Chen et al. 2023). Alterations that influence signaling from FAs can result in skeletal dysplasia. Mutations of a5 laminin (LAMA5; OMIM #620076) that alter laminin interactions with integrin-mediated focal adhesion signaling (Salasznyk et al. 2007) were found to be a cause of bent bone dysplasia (Barad et al. 2020) (Fig. 13.3c). Integrin signaling in the skeleton through interactions with collagens and their association with FAs (Takahashi et al. 2003) is a strong candidate for mediating changes in a highly dynamic environment where FAs change shape, position, and composition according to the rigidity and composition of the substrate, which reciprocally can be dependent on exocytosis along microtubules to the cell surface to areas close to FAs (Fourriere et al. 2019; Nolte et al. 2021; Bjørnestad et al. 2022).

Epidermal growth factor receptor (EGFR), and other growth factor receptors, cluster within FAs under specific ECM stiffness conditions while they are more diffuse on soft substrates (Wang et al. 2007). EGFR signaling has an important role in the transition from cartilage to bone during endochondral ossification and altered signaling may contribute to the degenerative effects in OA (Janssen et al. 2019; Qin and Beier 2019). Egfr$^{-/-}$ mice are growth retarded and exhibit severe bone defects, displaying impaired endochondral and intramembranous ossification (Linder et al. 2018). Turnover of EGFR following stimulation by EGF is partially through degradation by the endo-lysosomal pathway, which requires that the levels of EGFR are replenished by transporting newly synthesized EGFRs to the cell surface. EGF stimulation was found to stimulate the translocation of the transcriptional regulator RNF11 from endosomes to the nucleus which appears to up-regulate expression of the COPII components SEC23B, SEC24B, and SEC24D to increase the transport efficiency of newly synthesized EGFRs from the endoplasmic reticulum to the plasma membrane (Scharaw et al. 2016) (Fig. 13.3d). How these effects may be relevant for ECM composition, signaling, and skeletal homeostasis remain to be tested, but again suggest the possibility of how changes in mechanically sensitive FA signaling might feed back to the membrane trafficking secretory machinery.

13.6 Conclusions and Future Perspectives

Skeletal disorders are associated with over 500 genes (Unger et al. 2023) and show a wide variety of symptoms that to a greater or lesser extent can be confined to skeletal abnormalities or are more pleiotropic, affecting multiple tissues. Given the role of the ECM in chondrocyte differentiation and endochondral bone formation, it is not surprising that mutations of collagens, proteins involved in post-translational modification, or other ECM proteins, cause chondrodysplasia. Genetic mutations affecting proteins that are involved in the membrane trafficking pathway can have a similar effect, by disrupting the delivery of ECM proteins that results in defective matrix formation, with the trafficking of fibrillar collagens being particularly sensitive due to their large size. Nonetheless, there is significant variability in the clinical manifestations of ECM-associated skeletal dysplasias, even in cases of different mutations within the same gene. Understanding this variation is an extremely complex undertaking and we have only scratched the surface in elucidating the underlying mechanisms. In large part this can be attributed to the fact that while these diseases have an altered ECM structure as a primary cause, they also display disturbed intracellular homeostasis that may influence disease progression. These alterations may be the result of compensatory mechanisms that function normally during adaptations to changes in cellular states (e.g. growth, differentiation, mechanical load) that become maladaptive in a mutation state that no longer has the appropriate mechanisms to deal with these changes. In the membrane trafficking pathway, the focus is generally on the primary defect, e.g. collagen transport, but does not consider the broader aspects such as other cargo proteins that may be affected either directly or indirectly that could be important in the dynamic turnover of the ECM. Another linked aspect that is only beginning to be addressed is how changes at the ECM communicate with the membrane trafficking pathway and how this communication is altered by a defective ECM. More systems biology approaches are needed to find commonalities and unravel the cellular effects in different skeletal dysplasias to provide a better understanding of the pathobiological mechanisms underlying these diseases. Integrating high-throughput omics approaches with advanced imaging techniques will further elucidate the complex interactions within the extracellular matrix, paving the way for innovative diagnostic and therapeutic strategies. A greater understanding of each individual's pathobiological profile could lead to the development of targeted therapies to modulate membrane trafficking pathways and new treatments. A final consideration is the application of gene editing technologies, such as CRISPR/Cas9, which hold promise for correcting genetic defects causing these disorders.

Acknowledgments RV acknowledges support from the Italian Association for Cancer Research (grant MFAG 2020_25174) and the Italian Ministry of University and Research (PRIN, 20222MAWZP).

Compliance with Ethical Standards/Ethical Requirements The authors declare no conflict of interest.

References

Ain NU, Muhammad N, Dianatpour M, Baroncelli M, Iqbal M, Fard MAF, Bukhari I, Ahmed S, Hajipour M, Tabatabaie Z, Foroutan H, Nilsson O, Faghihi MA, Makitie O, Naz S (2021) Biallelic TMEM251 variants in patients with severe skeletal dysplasia and extreme short stature. Hum Mutat 42:89–101. https://doi.org/10.1002/humu.24139

Alotaibi M, Aldossari A, Khan I, Alotaibi L (2022) Identification of a new variant of the MBTPS1 gene of the Kondo-Fu type of Spondyloepiphyseal dysplasia (SEDKF) in a Saudi patient. Case Rep Pediatr 2022:5498109. https://doi.org/10.1155/2022/5498109

Arthur WT, Noren NK, Burridge K (2002) Regulation of rho family GTPases by cell-cell and cell-matrix adhesion. Biol Res 35:239–246. https://doi.org/10.4067/s0716-97602002000200016

Baines AC, Adams EJ, Zhang B, Ginsburg D (2013) Disruption of the sec 24d gene results in early embryonic lethality in the mouse. PLoS One 8:e61114. https://doi.org/10.1371/journal.pone.0061114

Balasubramanian M, Padidela R, Pollitt RC, Bishop NJ, Mughal MZ, Offiah AC, Wagner BE, McCaughey J, Stephens DJ (2018) P4HB recurrent missense mutation causing Cole-carpenter syndrome. J Med Genet 55:158–165. https://doi.org/10.1136/jmedgenet-2017-104899

Barad M, Csukasi F, Bosakova M, Martin JH, Zhang W, Paige Taylor S, Lachman RS, Zieba J, Bamshad M, Nickerson D, Chong JX, Cohn DH, Krejci P, Krakow D, Duran I (2020) Biallelic mutations in LAMA5 disrupts a skeletal noncanonical focal adhesion pathway and produces a distinct bent bone dysplasia. EBioMedicine 62:103075. https://doi.org/10.1016/j.ebiom.2020.103075

Barbieri O, Astigiano S, Morini M, Tavella S, Schito A, Corsi A, Di Martino D, Bianco P, Cancedda R, Garofalo S (2003) Depletion of cartilage collagen fibrils in mice carrying a dominant negative Col2a1 transgene affects chondrocyte differentiation. Am J Physiol Cell Physiol 285:C1504–C1512. https://doi.org/10.1152/ajpcell.00579.2002

Bateman JF, Shoulders MD, Lamandé SR (2022) Collagen misfolding mutations: the contribution of the unfolded protein response to the molecular pathology. Connect Tissue Res 63:210–227. https://doi.org/10.1080/03008207.2022.2036735

Bi X, Mancias JD, Goldberg J (2007) Insights into COPII coat nucleation from the structure of Sec23.Sar1 complexed with the active fragment of Sec31. Dev Cell 13:635–645. https://doi.org/10.1016/j.devcel.2007.10.006

Bird IM, Kim SH, Schweppe DK, Caetano-Lopes J, Robling AG, Charles JF, Gygi SP, Warman ML, Smits PJ (2018) The skeletal phenotype of achondrogenesis type 1A is caused exclusively by cartilage defects. Development 145:dev156588. https://doi.org/10.1242/dev.156588

Bjørnestad SA, Guadagno NA, Kjos I, Progida C (2022) Rab33b-exocyst interaction mediates localized secretion for focal adhesion turnover and cell migration. iScience 25:104250. https://doi.org/10.1016/j.isci.2022.104250

Boncompain G, Divoux S, Gareil N, de Forges H, Lescure A, Latreche L, Mercanti V, Jollivet F, Raposo G, Perez F (2012) Synchronization of secretory protein traffic in populations of cells. Nat Methods 9:493–498. https://doi.org/10.1038/nmeth.1928

Boyadjiev SA, Fromme JC, Ben J, Chong SS, Nauta C, Hur DJ, Zhang G, Hamamoto S, Schekman R, Ravazzola M, Orci L, Eyaid W (2006) Cranio-lenticulo-sutural dysplasia is caused by a SEC23A mutation leading to abnormal endoplasmic-reticulum-to-Golgi trafficking. Nat Genet 38:1192–1197. https://doi.org/10.1038/ng1876

Boyadjiev SA, Kim SD, Hata A, Haldeman-Englert C, Zackai EH, Naydenov C, Hamamoto S, Schekman RW, Kim J (2011) Cranio-lenticulo-sutural dysplasia associated with defects in collagen secretion. Clin Genet 80:169–176. https://doi.org/10.1111/j.1399-0004.2010.01550.x

Cabral WA, Chang W, Barnes AM, Weis M, Scott MA, Leikin S, Makareeva E, Kuznetsova NV, Rosenbaum KN, Tifft CJ, Bulas DI, Kozma C, Smith PA, Eyre DR, Marini JC (2007) Prolyl 3-hydroxylase 1 deficiency causes a recessive metabolic bone disorder resembling lethal/severe osteogenesis imperfecta. Nat Genet 39:359–365. https://doi.org/10.1038/ng1968

Carvalho DR, Speck-Martins CE, Brum JM, Ferreira CR, Sobreira NLM (2020) Spondyloepimetaphyseal dysplasia with elevated plasma lysosomal enzymes caused by homozygous variant in MBTPS1. Am J Med Genet A 182:1796–1800. https://doi.org/10.1002/ajmg.a.61614

Chaudhury S, Okuda KS, Koltowska K, Lagendijk AK, Paterson S, Baillie GJ, Simons C, Smith KA, Hogan BM, Bower NI (2020) Localised Collagen2a1 secretion supports lymphatic endothelial cell migration in the zebrafish embryo. Development 147:dev190983. https://doi.org/10.1242/dev.190983

Chen S, He T, Zhong Y, Chen M, Yao Q, Chen D, Shao Z, Xiao G (2023) Roles of focal adhesion proteins in skeleton and diseases. Acta Pharm Sin B 13:998–1013. https://doi.org/10.1016/j.apsb.2022.09.020

Choi JUA, Kijas AW, Lauko J, Rowan AE (2022) The Mechanosensory role of osteocytes and implications for bone health and disease states. Front Cell Dev Biol 9:770143. https://doi.org/10.3389/fcell.2021.770143

Christiansen HE, Schwarze U, Pyott SM, AlSwaid A, Al Balwi M, Alrasheed S, Pepin MG, Weis MA, Eyre DR, Byers PH (2010) Homozygosity for a missense mutation in SERPINH1, which encodes the collagen chaperone protein HSP47, results in severe recessive osteogenesis imperfecta. Am J Hum Genet 86:389–398. https://doi.org/10.1016/j.ajhg.2010.01.034

Cisarova K, Garavelli L, Caraffi SG, Peluso F, Valeri L, Gargano G, Gavioli S, Trimarchi G, Neri A, Campos-Xavier B, Superti-Furga A (2022) A monoallelic SEC23A variant E599K associated with cranio-lenticulo-sutural dysplasia. Am J Med Genet A 188:319–325. https://doi.org/10.1002/ajmg.a.62506

Clark EM, Link BA (2021) Complementary and divergent functions of zebrafish Tango1 and Ctage5 in tissue development and homeostasis. Mol Biol Cell 32:391–401. https://doi.org/10.1091/mbc.E20-11-0745

D'Arcangelo JG, Stahmer KR, Miller EA (2013) Vesicle-mediated export from the ER: COPII coat function and regulation. Biochim Biophys Acta 1833:2464–2472. https://doi.org/10.1016/j.bbamcr.2013.02.003

Devos H, Zoidakis J, Roubelakis MG, Latosinska A, Vlahou A (2023) Reviewing the regulators of COL1A1. Int J Mol Sci 24:10004. https://doi.org/10.3390/ijms241210004

DiStasio A, Driver A, Sund K, Donlin M, Muraleedharan RM, Pooya S, Kline-Fath B, Kaufman KM, Prows CA, Schorry E, Dasgupta B, Stottmann RW (2017) Copb2 is essential for embryogenesis and hypomorphic mutations cause human microcephaly. Hum Mol Genet 26:4836–4848. https://doi.org/10.1093/hmg/ddx362

El-Gazzar A, Voraberger B, Rauch F, Mairhofer M, Schmidt K, Guillemyn B, Mitulović G, Reiterer V, Haun M, Mayr MM, Mayr JA, Kimeswenger S, Drews O, Saraff V, Shaw N, Fratzl-Zelman N, Symoens S, Farhan H, Högler W (2023) Bi-allelic mutation in SEC16B alters collagen trafficking and increases ER stress. EMBO Mol Med 15:e16834. https://doi.org/10.15252/emmm.202216834

Fan J, Wang Y, Liu L, Zhang H, Zhang F, Shi L, Yu M, Gao F, Xu Z (2017) cTAGE5 deletion in pancreatic β cells impairs proinsulin trafficking and insulin biogenesis in mice. J Cell Biol 216:4153–4164. https://doi.org/10.1083/jcb.201705027

Fourriere L, Kasri A, Gareil N, Bardin S, Bousquet H, Pereira D, Perez F, Goud B, Boncompain G, Miserey-Lenkei S (2019) RAB6 and microtubules restrict protein secretion to focal adhesions. J Cell Biol 218:2215–2231. https://doi.org/10.1083/jcb.201805002

Fromme JC, Ravazzola M, Hamamoto S, Al-Balwi M, Eyaid W, Boyadjiev SA, Cosson P, Schekman R, Orci L (2007) The genetic basis of a craniofacial disease provides insight into COPII coat assembly. Dev Cell 13:623–634. https://doi.org/10.1016/j.devcel.2007.10.005

Garbes L, Kim K, Rieß A, Hoyer-Kuhn H, Beleggia F, Bevot A, Kim MJ, Huh YH, Kweon HS, Savarirayan R, Amor D, Kakadia PM, Lindig T, Kagan KO, Becker J, Boyadjiev SA, Wollnik B, Semler O, Bohlander SK, Kim J, Netzer C (2015) Mutations in SEC24D, encoding a component of the COPII machinery, cause a syndromic form of osteogenesis imperfecta. Am J Hum Genet 96:432–439. https://doi.org/10.1016/j.ajhg.2015.01.002

Gécz J, Hillman MA, Gedeon AK, Cox TC, Baker E, Mulley JC (2000) Gene structure and expression study of the SEDL gene for spondyloepiphyseal dysplasia tarda. Genomics 69:242–251. https://doi.org/10.1006/geno.2000.6326

Geiger B, Bershadsky A, Pankov R, Yamada KM (2001) Transmembrane crosstalk between the extracellular matrix–cytoskeleton crosstalk. Nat Rev Mol Cell Biol 2:793–805. https://doi.org/10.1038/35099066

Ghosh AK, Majumder M, Steele R, White RA, Ray RB (2001) A novel 16-kilodalton cellular protein physically interacts with and antagonizes the functional activity of c-myc promoter-binding protein 1. Mol Cell Biol 21:655–662. https://doi.org/10.1128/MCB.21.2.655-662.2001

Guillemyn B, Nampoothiri S, Syx D, Malfait F, Symoens S (2021) Loss of TANGO1 leads to absence of bone mineralization. JBMR Plus 5:e10451. https://doi.org/10.1002/jbm4.10451

Hellicar J, Stevenson NL, Stephens DJ, Lowe M (2022) Supply chain logistics - the role of the Golgi complex in extracellular matrix production and maintenance. J Cell Sci 135:jcs258879. https://doi.org/10.1242/jcs.258879

Hirata Y, Matsui Y, Wada I, Hosokawa N (2022) Endoplasmic reticulum-to-Golgi trafficking of procollagen III via conventional vesicular and tubular carriers. Mol Biol Cell 33:ar21. https://doi.org/10.1091/mbc.E21-07-0372

Humphrey JD, Dufresne ER, Schwartz MA (2014) Mechanotransduction and extracellular matrix homeostasis. Nat Rev Mol Cell Biol 15:802–812. https://doi.org/10.1038/nrm3896

Ishikawa T, Toyama T, Nakamura Y, Tamada K, Shimizu H, Ninagawa S, Okada T, Kamei Y, Ishikawa-Fujiwara T, Todo T, Aoyama E, Takigawa M, Harada A, Mori K (2017) UPR transducer BBF2H7 allows export of type II collagen in a cargo- and developmental stage-specific manner. J Cell Biol 216:1761–1774. https://doi.org/10.1083/jcb.201609100

Izumi K, Brett M, Nishi E, Drunat S, Tan ES, Fujiki K, Lebon S, Cham B, Masuda K, Arakawa M, Jacquinet A, Yamazumi Y, Chen ST, Verloes A, Okada Y, Katou Y, Nakamura T, Akiyama T, Gressens P, Foo R, Passemard S, Tan EC, El Ghouzzi V, Shirahige K (2016) ARCN1 mutations cause a recognizable craniofacial syndrome due to COPI-mediated transport defects. Am J Hum Genet 99:451–459. https://doi.org/10.1016/j.ajhg.2016.06.011

Janssen JN, Batschkus S, Schimmel S, Bode C, Schminke B, Miosge N (2019) The influence of TGF-β3, EGF, and BGN on SOX9 and RUNX2 expression in human Chondrogenic progenitor cells. J Histochem Cytochem 67:117–127. https://doi.org/10.1369/0022155418811645

Jiang Z, Byers S, Casal ML, Smith LJ (2020) Failures of endochondral ossification in the Mucopolysaccharidoses. Curr Osteoporos Rep 18:759–773. https://doi.org/10.1007/s11914-020-00626-y

Jung J, Khan MM, Landry J, Halavatyi A, Machado P, Reiss M, Pepperkok R (2022) Regulation of the COPII secretory machinery via focal adhesions and extracellular matrix signaling. J Cell Biol 221:e202110081. https://doi.org/10.1083/jcb.202110081

Khoriaty R, Hesketh GG, Bernard A, Weyand AC, Mellacheruvu D, Zhu G, Hoenerhoff MJ, McGee B, Everett L, Adams EJ, Zhang B, Saunders TL, Nesvizhskii AI, Klionsky DJ, Shavit JA, Gingras AC, Ginsburg D (2018) Functions of the COPII gene paralogs SEC23A and SEC23B are interchangeable in vivo. Proc Natl Acad Sci USA 115:E7748–E7757. https://doi.org/10.1073/pnas.1805784115

Kim SD, Pahuja KB, Ravazzola M, Yoon J, Boyadjiev SA, Hammamoto S, Schekman R, Orci L, Kim J (2012) The SEC23-SEC31 interface plays critical role for export of procollagen from the endoplasmic reticulum. J Biol Chem 287:10134–10144. https://doi.org/10.1074/jbc.M111.283382

King R, Lin Z, Balbin-Cuesta G, Myers G, Friedman A, Zhu G, McGee B, Saunders TL, Kurita R, Nakamura Y, Engel JD, Reddy P, Khoriaty R (2021) SEC23A rescues SEC23B-deficient congenital dyserythropoietic anemia type II. Sci Adv 7:eabj5293. https://doi.org/10.1126/sciadv.abj5293

Klinkenberg D, Long KR, Shome K, Watkins SC, Aridor M (2014) A cascade of ER exit site assembly that is regulated by p125A and lipid signals. J Cell Sci 127:1765–1778. https://doi.org/10.1242/jcs.138784

Komori T, Ji Y, Pham H, Jani P, Kilts TM, Kram V, Li L, Young MF (2022) Type VI collagen regulates endochondral ossification in the temporomandibular joint. JBMR Plus 6:e10617. https://doi.org/10.1002/jbm4.10617

Kondo Y, Fu J, Wang H, Hoover C, McDaniel JM, Steet R, Patra D, Song J, Pollard L, Cathey S, Yago T, Wiley G, Macwana S, Guthridge J, McGee S, Li S, Griffin C, Furukawa K, James JA, Ruan C, McEver RP, Wierenga KJ, Gaffney PM, Xia L (2018) Site-1 protease deficiency causes human skeletal dysplasia due to defective inter-organelle protein trafficking. JCI Insight 3:e121596. https://doi.org/10.1172/jci.insight.121596

Lamandé SR, Bateman JF (2020) Genetic disorders of the extracellular matrix. Anat Rec (Hoboken) 303:1527–1542. https://doi.org/10.1002/ar.24086

Lang MR, Lapierre LA, Frotscher M, Goldenring JR, Knapik EW (2006) Secretory COPII coat component Sec23a is essential for craniofacial chondrocyte maturation. Nat Genet 38:1198–1203. https://doi.org/10.1038/ng1880

Lekszas C, Foresti O, Raote I, Liedtke D, Konig E-M, Nanda I, Vona B, De Coster P, Cauwels R, Malhotra V, Haaf T (2020) Biallelic TANGO1 mutations cause a novel syndromal disease due to hampered cellular collagen secretion. elife 9:e51319. https://doi.org/10.7554/eLife.51319

Lim PJ, Marcionelli G, Srikanthan P, Ndarugendamwo T, Pinner J, Rohrbach M, Giunta C (2023) Perturbations in fatty acid metabolism and collagen production infer pathogenicity of a novel MBTPS2 variant in osteogenesis imperfecta. Front Endocrinol 14:1195704. https://doi.org/10.3389/fendo.2023.1195704

Linder M, Hecking M, Glitzner E, Zwerina K, Holcmann M, Bakiri L, Ruocco MG, Tuckermann J, Schett G, Wagner EF, Sibilia M (2018) EGFR controls bone development by negatively regulating mTOR-signaling during osteoblast differentiation. Cell Death Differ 25:1094–1106. https://doi.org/10.1038/s41418-017-0054-7

Lindert U, Cabral WA, Ausavarat S, Tongkobpetch S, Ludin K, Barnes AM, Yeetong P, Weis M, Krabichler B, Srichomthong C, Makareeva EN, Janecke AR, Leikin S, Röthlisberger B, Rohrbach M, Kennerknecht I, Eyre DR, Suphapeetiporn K, Giunta C, Marini JC, Shotelersuk V (2016) MBTPS2 mutations cause defective regulated intramembrane proteolysis in X-linked osteogenesis imperfecta. Nat Commun 7:11920. https://doi.org/10.1038/ncomms11920

Macken WL, Godwin A, Wheway G, Stals K, Nazlamova L, Ellard S, Alfares A, Aloraini T, AlSubaie L, Alfadhel M, Alajaji S, Wai HA, Self J, Douglas AGL, Kao AP, Guille M, Baralle D (2021) Biallelic variants in COPB1 cause a novel, severe intellectual disability syndrome with cataracts and variable microcephaly. Genome Med 13:34. https://doi.org/10.1186/s13073-021-00850-w

Maeda M, Katada T, Saito K (2017) TANGO1 recruits Sec16 to coordinately organize ER exit sites for efficient secretion. J Cell Biol 216:1731–1743. https://doi.org/10.1083/jcb.201703084

Marom R, Burrage LC, Venditti R, Clément A, Blanco-Sánchez B, Jain M, Scott DA, Rosenfeld JA, Sutton VR, Shinawi M, Mirzaa G, DeVile C, Roberts R, Calder AD, Allgrove J, Grafe I, Lanza DG, Li X, Joeng KS, Lee YC, Song IW, Sliepka JM, Batkovskyte D, Washington M, Dawson BC, Jin Z, Jiang MM, Chen S, Chen Y, Tran AA, Emrick LT, Murdock DR, Hanchard NA, Zapata GE, Mehta NR, Weis MA, Scott AA, Tremp BA, Phillips JB, Wegner J, Taylor-Miller T, Gibbs RA, Muzny DM, Jhangiani SN, Hicks J, Stottmann RW, Dickinson ME, Seavitt JR, Heaney JD, Eyre DR, Undiagnosed Diseases Network, Westerfield M, De Matteis MA, Lee B (2021) COPB2 loss of function causes a coatopathy with osteoporosis and developmental delay. Am J Hum Genet 108:1710–1724. https://doi.org/10.1016/j.ajhg.2021.08.002

Masago Y, Hosoya A, Kawasaki K, Kawano S, Nasu A, Toguchida J, Fujita K, Nakamura H, Kondoh G, Nagata K (2012) The molecular chaperone Hsp47 is essential for cartilage and endochondral bone formation. J Cell Sci 125:1118–1128. https://doi.org/10.1242/jcs.089748

McCaughey J, Stevenson NL, Cross S, Stephens DJ (2019) ER-to-Golgi trafficking of procollagen in the absence of large carriers. J Cell Biol 218:929–948. https://doi.org/10.1083/jcb.201806035

McCaughey J, Stevenson NL, Mantell JM, Neal CR, Paterson A, Heesom K, Stephens DJ (2021) A general role for TANGO1, encoded by MIA3, in secretory pathway organization and function. J Cell Sci 134:jcs259075. https://doi.org/10.1242/jcs.259075

Melville DB, Montero-Balaguer M, Levic DS, Bradley K, Smith JR, Hatzopoulos AK, Knapik EW (2011) The feelgood mutation in zebrafish dysregulates COPII-dependent secretion of select extracellular matrix proteins in skeletal morphogenesis. Dis Model Mech 4:763–776. https://doi.org/10.1242/dmm.007625

Miao Y, Chen J, Guo X, Wei Y, Wu X, Sang Y, Wu D (2022) Case report: clinical manifestations and genotype analysis of a child with PTPN11 and SEC24D mutations. Front Pediatr 10:973920. https://doi.org/10.3389/fped.2022.973920

Mizumoto S, Yamada S (2021) Congenital disorders of deficiency in glycosaminoglycan biosynthesis. Front Genet 12:717535. https://doi.org/10.3389/fgene.2021.717535

Moosa S, Chung BH-Y, Tung JY-L, Altmuller J, Thiele H, Nurnberg P, Netzer C, Nishimura G, Wollnik B (2016) Mutations in SEC24D cause autosomal recessive osteogenesis imperfecta. Clin Genet 89:517–519. https://doi.org/10.1111/cge.12678

Nolte MA, Nolte-'t Hoen ENM, Margadant C (2021) Integrins control vesicular trafficking; new tricks for old dogs. Trends Biochem Sci 46:124–137. https://doi.org/10.1016/j.tibs.2020.09.001

Ong YS, Tang BL, Loo LS, Hong W (2010) p125A exists as part of the mammalian Sec13/Sec31 COPII subcomplex to facilitate ER-Golgi transport. J Cell Biol 190:331–345. https://doi.org/10.1083/jcb.201003005

Paganini C, Costantini R, Superti-Furga A, Rossi A (2019) Bone and connective tissue disorders caused by defects in glycosaminoglycan biosynthesis: a panoramic view. FEBS J 286:3008–3032. https://doi.org/10.1111/febs.14984

Pasapera AM, Plotnikov SV, Fischer RS, Case LB, Egelhoff TT, Waterman CM (2015) Rac1-dependent phosphorylation and focal adhesion recruitment of myosin IIA regulates migration and mechanosensing. Curr Biol 25:175–186. https://doi.org/10.1016/j.cub.2014.11.043

Paul S, Audhya A, Cui Q (2023) Molecular mechanism of GTP binding- and dimerization-induced enhancement of SAR1-mediated membrane remodeling. Proc Natl Acad Sci USA 120:e2212513120. https://doi.org/10.1073/pnas.2212513120

Pellegrin S, Haydn-Smith KL, Hampton-O'Neil LA, Hawley BR, Heesom KJ, Fermo E, Bianchi P, Toye AM (2019) Transduction with BBF2H7/CREB3L2 upregulates SEC23A protein in erythroblasts and partially corrects the hypo-glycosylation phenotype associated with CDAII. Br J Haematol 184:876–881. https://doi.org/10.1111/bjh.15189

Phuyal S, Djaerff E, Le Roux AL, Baker MJ, Fankhauser D, Mahdizadeh SJ, Reiterer V, Parizadeh A, Felder E, Kahlhofer JC, Teis D, Kazanietz MG, Geley S, Eriksson L, Roca-Cusachs P, Farhan H (2022) Mechanical strain stimulates COPII-dependent secretory trafficking via Rac1. EMBO J 41:e110596. https://doi.org/10.15252/embj.2022110596

Prein C, Beier F (2019) ECM signaling in cartilage development and endochondral ossification. Curr Top Dev Biol 133:25–47. https://doi.org/10.1016/bs.ctdb.2018.11.003

Qin L, Beier F (2019) EGFR signaling: friend or foe for cartilage? JBMR Plus 3:e10177. https://doi.org/10.1002/jbm4.10177

Raote I, Ortega-Bellido M, Santos AJ, Foresti O, Zhang C, Garcia-Parajo MF, Campelo F, Malhotra V (2018) TANGO1 builds a machine for collagen export by recruiting and spatially organizing COPII, tethers and membranes. elife 7:e32723. https://doi.org/10.7554/eLife.32723

Raote I, Saxena S, Malhotra V (2023) Sorting and export of proteins at the endoplasmic reticulum. Cold Spring Harb Perspect Biol 15:a041258. https://doi.org/10.1101/cshperspect.a041258

Rauch F, Fahiminiya S, Majewski J, Carrot-Zhang J, Boudko S, Glorieux F, Mort JS, Bächinger HP, Moffatt P (2015) Cole-carpenter syndrome is caused by a heterozygous missense mutation in P4HB. Am J Hum Genet 96:425–431. https://doi.org/10.1016/j.ajhg.2014.12.027

Rellmann Y, Dreier R (2018) Different forms of ER stress in chondrocytes result in short stature disorders and degenerative cartilage Diseases: new insights by cartilage-specific ERp57 knockout mice. Oxidative Med Cell Longev 2018:8421394. https://doi.org/10.1155/2018/8421394

Reuter MS, Tawamie H, Buchert R, Hosny Gebril O, Froukh T, Thiel C, Uebe S, Ekici AB, Krumbiegel M, Zweier C, Hoyer J, Eberlein K, Bauer J, Scheller U, Strom TM, Hoffjan S, Abdelraouf ER, Meguid NA, Abboud A, Al Khateeb MA, Fakher M, Hamdan S, Ismael A, Muhammad S, Abdallah E, Sticht H, Wieczorek D, Reis A, Abou Jamra R (2017) Diagnostic yield and novel candidate genes by exome sequencing in 152 consanguineous families with neurodevelopmental disorders. JAMA Psychiatry 74:293–299. https://doi.org/10.1001/jamapsychiatry.2016.3798

Richards CM, Jabs S, Qiao W, Varanese LD, Schweizer M, Mosen PR, Riley NM, Klüssendorf M, Zengel JR, Flynn RA, Rustagi A, Widen JC, Peters CE, Ooi YS, Xie X, Shi PY, Bartenschlager R, Puschnik AS, Bogyo M, Bertozzi CR, Blish CA, Winter D, Nagamine CM, Braulke T, Carette JE (2022) The human disease gene LYSET is essential for lysosomal enzyme transport and viral infection. Science 378:eabn5648. https://doi.org/10.1126/science.abn5648

Riedel F, Galindo A, Muschalik N, Munro S (2018) The two TRAPP complexes of metazoans have distinct roles and act on different Rab GTPases. J Cell Biol 217:601–617. https://doi.org/10.1083/jcb.201705068

Ríos-Barrera LD, Sigurbjörnsdóttir S, Baer M, Leptin M (2017) Dual function for Tango1 in secretion of bulky cargo and in ER-Golgi morphology. Proc Natl Acad Sci USA 114:E10389–E10398. https://doi.org/10.1073/pnas.1711408114

Ritter DJ, Choudhary D, Unlu G, Knapik EW (2023) Rgp1 contributes to craniofacial cartilage development and Rab8a-mediated collagen II secretion. Front Endocrinol 14:1120420. https://doi.org/10.3389/fendo.2023.1120420

Ritter AL, Gold J, Hayashi H, Ackermann AM, Hanke S, Skraban C, Cuddapah S, Bhoj E, Li D, Kuroda Y, Wen J, Takeda R, Bibb A, El Chehadeh S, Piton A, Ohl J, Kukolich MK, Nagasaki K, Kato K, Ogi T, Bhatti T, Russo P, Krock B, Murrell JR, Sullivan JA, Shashi V, Stong N, Hakonarson H, Sawano K, Torti E, Willaert R, Si Y, Wilcox WR, Wirgenes KV, Thomassen K, Carlotti K, Erwin A, Lazier J, Marquardt T, He M, Edmondson AC, Izumi K (2022) Expanding the phenotypic spectrum of ARCN1-related syndrome. Genet Med 24:1227–1237. https://doi.org/10.1016/j.gim.2022.02.005

Sacher M, Shahrzad N, Kamel H, Milev MP (2019) TRAPPopathies: an emerging set of disorders linked to variations in the genes encoding transport protein particle (TRAPP)-associated proteins. Traffic 20:5–26. https://doi.org/10.1111/tra.12615

Saito A, Hino S, Murakami T, Kanemoto S, Kondo S, Saitoh M, Nishimura R, Yoneda T, Furuichi T, Ikegawa S, Ikawa M, Okabe M, Imaizumi K (2009) Regulation of endoplasmic reticulum stress response by a BBF2H7-mediated Sec23a pathway is essential for chondrogenesis. Nat Cell Biol 11:1197–1204. https://doi.org/10.1038/ncb1962

Saito K, Maeda M, Katada T (2017) Regulation of the SAR1 GTPase cycle is necessary for large cargo secretion from the endoplasmic reticulum. Front Cell Dev Biol 5:75. https://doi.org/10.3389/fcell.2017.00075

Saito K, Yamashiro K, Shimazu N, Tanabe T, Kontani K, Katada T (2014) Concentration of Sec12 at ER exit sites via interaction with cTAGE5 is required for collagen export. J Cell Biol 206:751–762. https://doi.org/10.1083/jcb.201312062

Salasznyk RM, Klees RF, Boskey A, Plopper GE (2007) Activation of FAK is necessary for the osteogenic differentiation of human mesenchymal stem cells on laminin-5. J Cell Biochem 100:499–514. https://doi.org/10.1002/jcb.21074

Sarmah S, Barrallo-Gimeno A, Melville DB, Topczewski J, Solnica-Krezel L, Knapik EW (2010) Sec24D-dependent transport of extracellular matrix proteins is required for zebrafish skeletal morphogenesis. PLoS One 5:e10367. https://doi.org/10.1371/journal.pone.0010367

Satoh M, Hirayoshi K, Yokota S, Hosokawa N, Nagata K (1996) Intracellular interaction of collagen-specific stress protein HSP47 with newly synthesized procollagen. J Cell Biol 133:469–483. https://doi.org/10.1083/jcb.133.2.469

Scharaw S, Iskar M, Ori A, Boncompain G, Laketa V, Poser I, Lundberg E, Perez F, Beck M, Bork P, Pepperkok R (2016) The endosomal transcriptional regulator RNF11 integrates degradation and transport of EGFR. J Cell Biol 215:543–558. https://doi.org/10.1083/jcb.201601090

Schwarz K, Iolascon A, Verissimo F, Trede NS, Horsley W, Chen W, Paw BH, Hopfner KP, Holzmann K, Russo R, Esposito MR, Spano D, De Falco L, Heinrich K, Joggerst B, Rojewski MT, Perrotta S, Denecke J, Pannicke U, Delaunay J, Pepperkok R, Heimpel H (2009) Mutations affecting the secretory COPII coat component SEC23B cause congenital dyserythropoietic anemia type II. Nat Genet 41:936–940. https://doi.org/10.1038/ng.405

Shomron O, Nevo-Yassaf I, Aviad T, Yaffe Y, Zahavi EE, Dukhovny A, Perlson E, Brodsky I, Yeheskel A, Pasmanik-Chor M, Mironov A, Beznoussenko GV, Mironov AA, Sklan EH, Patterson GH, Yonemura Y, Sannai M, Kaether C, Hirschberg K (2021) COPII collar defines the boundary between ER and ER exit site and does not coat cargo containers. J Cell Biol 220:e201907224. https://doi.org/10.1083/jcb.201907224

Song Y, Zhao D, Xu X, Lv F, Li L, Jiang Y, Wang O, Xia W, Xing X, Li M (2018) Novel compound heterozygous mutations in SERPINH1 cause rare autosomal recessive osteogenesis imperfecta type X. Osteoporos Int 29:1389–1396. https://doi.org/10.1007/s00198-018-4448-2

Stancheva VG, Li XH, Hutchings J, Gomez-Navarro N, Santhanam B, Babu MM, Zanetti G, Miller EA (2020) Combinatorial multivalent interactions drive cooperative assembly of the COPII coat. J Cell Biol 219:e202007135. https://doi.org/10.1083/jcb.202007135

Syx D, Ishikawa Y, Gebauer J, Boudko SP, Guillemyn B, Van Damme T, D'hondt S, Symoens S, Nampoothiri S, Gould DB, Baumann U, Bächinger HP, Malfait F (2021) Aberrant binding of mutant HSP47 affects posttranslational modification of type I collagen and leads to osteogenesis imperfecta. PLoS Genet 17:e1009339. https://doi.org/10.1371/journal.pgen.1009339

Takahashi I, Onodera K, Sasano Y, Mizoguchi I, Bae JW, Mitani H, Kagayama M, Mitani H (2003) Effect of stretching on gene expression of beta1 integrin and focal adhesion kinase and on chondrogenesis through cell-extracellular matrix interactions. Eur J Cell Biol 82:182–192. https://doi.org/10.1078/0171-9335-00307

Takeyari S, Kubota T, Miyata K, Yamamoto K, Nakayama H, Yamamoto K, Ohata Y, Kitaoka T, Yanagi K, Kaname K, Ozono K (2018) Japanese patient with Cole-carpenter syndrome with compound heterozygous variants of SEC24D. Am J Med Genet A 176:2882–2886. https://doi.org/10.1002/ajmg.a.40643

Tang VT, Ginsburg D (2023) Cargo selection in endoplasmic reticulum-to-Golgi transport and relevant diseases. J Clin Invest 133:e163838. https://doi.org/10.1172/JCI163838

Tao J, Zhu M, Wang H, Afelik S, Vasievich MP, Chen XW, Zhu G, Jensen J, Ginsburg D, Zhang B (2012) SEC23B is required for the maintenance of murine professional secretory tissues. Proc Natl Acad Sci USA 109:E2001–E2009. https://doi.org/10.1073/pnas.1209207109

Tidwell T, Deshotel M, Palumbos J, Miller C, Bayrak-Toydemir P, Carey JC (2020) Novel de novo ARCN1 intronic variant causes rhizomelic short stature with microretrognathia and developmental delay. Cold Spring Harb Mol Case Stud 6:a005728. https://doi.org/10.1101/mcs.a005728

Townley AK, Feng Y, Schmidt K, Carter DA, Porter R, Verkade P, Stephens DJ (2008) Efficient coupling of Sec23-Sec24 to Sec13-Sec31 drives COPII-dependent collagen secretion and is essential for normal craniofacial development. J Cell Sci 121:3025–3034. https://doi.org/10.1242/jcs.031070

Unger S, Ferreira CR, Mortier GR, Ali H, Bertola DR, Calder A, Cohn DH, Cormier-Daire V, Girisha KM, Hall C, Krakow D, Makitie O, Mundlos S, Nishimura G, Robertson SP, Savarirayan R, Sillence D, Simon M, Sutton VR, Warman ML, Superti-Furga A (2023) Nosology of genetic skeletal disorders: 2023 revision. Am J Med Genet A 191:1164–1209. https://doi.org/10.1002/ajmg.a.63132

van Dijk FS, Semler O, Etich J, Köhler A, Jimenez-Estrada JA, Bravenboer N, Claeys L, Riesebos E, Gegic S, Piersma SR, Jimenez CR, Waisfisz Q, Flores CL, Nevado J, Harsevoort AJ, Janus GJM, Franken AAM, van der Sar AM, Meijers-Heijboer H, Heath KE, Lapunzina P, Nikkels PGJ, Santen GWE, Nüchel J, Plomann M, Wagener R, Rehberg M, Hoyer-Kuhn H, Eekhoff EMW, Pals G, Mörgelin M, Newstead S, Wilson BT, Ruiz-Perez VL, Maugeri A, Netzer C, Zaucke F, Micha D (2020) Interaction between KDELR2 and HSP47 as a key determinant in osteogenesis imperfecta caused by Bi-allelic variants in KDELR2. Am J Hum Genet 107:989–999. https://doi.org/10.1016/j.ajhg.2020.09.009

Venditti R, Scanu T, Santoro M, Di Tullio G, Spaar A, Gaibisso R, Beznoussenko GV, Mironov AA, Mironov A Jr, Zelante L, Piemontese MR, Notarangelo A, Malhotra V, Vertel BM, Wilson

C, De Matteis MA (2012) Sedlin controls the ER export of procollagen by regulating the Sar1 cycle. Science 337:1668–1672. https://doi.org/10.1126/science.1224947
Vinckenbosch N, Dupanloup I, Kaessmann H (2006) Evolutionary fate of retroposed gene copies in the human genome. Proc Natl Acad Sci USA 103:3220–3225. https://doi.org/10.1073/pnas.0511307103
Wai AW, Ng LJ, Watanabe H, Yamada Y, Tam PP, Cheah KS (1998) Disrupted expression of matrix genes in the growth plate of the mouse cartilage matrix deficiency (cmd) mutant. Dev Genet 22:349–358. https://doi.org/10.1002/(SICI)1520-6408(1998)22:4<349::AID-DVG5>3.0.CO;2-6
Wang XT, Cai XY, Xu FX, Zhou L, Zheng R, Ma KY, Xu ZH, Shen Y (2019) MEA6 deficiency impairs cerebellar development and motor performance by tethering protein trafficking. Front Cell Neurosci 13:250. https://doi.org/10.3389/fncel.2019.00250
Wang JH, Thampatty BP, Lin JS, Im HJ (2007) Mechanoregulation of gene expression in fibroblasts. Gene 391:1–15. https://doi.org/10.1016/j.gene.2007.01.014
Wang L, You X, Zhang L, Zhang C, Zou W (2022) Mechanical regulation of bone remodeling. Bone Res 10:16. https://doi.org/10.1038/s41413-022-00190-4
Weigel AV, Chang CL, Shtengel G, Xu CS, Hoffman DP, Freeman M, Iyer N, Aaron J, Khuon S, Bogovic J, Qiu W, Hess HF, Lippincott-Schwartz J (2021) ER-to-Golgi protein delivery through an interwoven, tubular network extending from ER. Cell 184:2412–2429.e16. https://doi.org/10.1016/j.cell.2021.03.035
Westrate LM, Hoyer MJ, Nash MJ, Voeltz GK (2020) Vesicular and uncoated Rab1-dependent cargo carriers facilitate ER to Golgi transport. J Cell Sci 133:jcs239814. https://doi.org/10.1242/jcs.239814
Whyte MP, Gottesman GS, Eddy MC, McAlister WH (1999) X-linked recessive spondyloepiphyseal dysplasia tarda. Clinical and radiographic evolution in a 6-generation kindred and review of the literature. Medicine (Baltimore) 78:9–25. https://doi.org/10.1097/00005792-199901000-00002
Ye J, Rawson RB, Komuro R, Chen X, Davé UP, Prywes R, Brown MS, Goldstein JL (2000) ER stress induces cleavage of membrane-bound ATF6 by the same proteases that process SREBPs. Mol Cell 6:1355–1364. https://doi.org/10.1016/s1097-2765(00)00133-7
Yorimitsu T, Sato K (2012) Insights into structural and regulatory roles of Sec16 in COPII vesicle formation at ER exit sites. Mol Biol Cell 23:2930–2942. https://doi.org/10.1091/mbc.E12-05-0356
Yuan Y, Zhou Q, Wang C, Zhou W, Gu W, Zheng B (2023) Clinical and molecular characterization of a patient with MBTPS1 related spondyloepiphyseal dysplasia: evidence of pathogenicity for a synonymous variant. Front Pediatr 10:1056141. https://doi.org/10.3389/fped.2022.1056141
Zappa F, Intartaglia D, Guarino AM, De Cegli R, Wilson C, Salierno FG, Polishchuk E, Sorrentino NC, Conte I, De Matteis MA (2023) Role of trafficking protein particle complex 2 in medaka development. Traffic 25:e12924. https://doi.org/10.1111/tra.12924
Zhang W, Yang X, Li Y, Yu L, Zhang B, Zhang J, Cho WJ, Venkatarangan V, Chen L, Burugula BB, Bui S, Wang Y, Duan C, Kitzman JO, Li M (2022) GCAF(TMEM251) regulates lysosome biogenesis by activating the mannose-6-phosphate pathway. Nat Commun 13:5351. https://doi.org/10.1038/s41467-022-33025-1
Zhang H, Yue H, Wang C, Gu J, He J, Fu W, Hu W, Zhang Z (2017) Novel mutations in the SEC24D gene in Chinese families with autosomal recessive osteogenesis imperfecta. Osteoporos Int 28:1473–1480. https://doi.org/10.1007/s00198-016-3866-2